An Introduction to Multivariate Statistical Analysis

A WILEY PUBLICATION IN MATHEMATICAL STATISTICS

Also by T. W. Anderson

The Statistical Analysis of Time Series

John Wiley & Sons, Inc., 1971

Introductory Statistical Analysis

John Wiley & Sons, Inc., 1974

An Introduction to Multivariate Statistical Analysis

T. W. ANDERSON
Professor of Statistics and Economics
Stanford University

John Wiley & Sons, Inc.
New York · London · Sydney

ISBN 0 471 02640 9

20 19 18 17

Library of Congress Catalog Card Number: 58-6068

PRINTED IN THE UNITED STATES OF AMERICA

Preface

THIS BOOK HAS BEEN DESIGNED PRIMARILY AS A TEXT FOR A TWO-semester course in multivariate statistics. It is hoped that the book will also serve as an introduction to many topics in this area to statisticians who are not students and will be used as a reference by other statisticians.

For several years the book in the form of dittoed notes has been used in a two-semester sequence of graduate courses at Columbia University; the first six chapters constituted the text for the first semester, emphasizing correlation theory. It is assumed that the reader is familiar with the usual theory of univariate statistics, particularly methods based on the univariate normal distribution. A knowledge of matrix algebra is also a prerequisite; however, an appendix on this topic has been included.

It is hoped that the more basic and important topics are treated here, though to some extent the coverage is a matter of taste. Some of the more recent and advanced developments are only briefly touched on in the last chapter.

The method of maximum likelihood is used to a large extent. This leads to reasonable procedures; in some cases it can be proved that they are optimal. In many situations, however, the theory of desirable or optimum procedures is lacking.

Over the years this manuscript has been developed, a number of students and colleagues have been of considerable assistance. Allan Birnbaum, Harold Hotelling, Jacob Horowitz, Howard Levene, Ingram Olkin, Gobind Seth, Charles Stein, and Henry Teicher are to be mentioned particularly. Acknowledgments are also due to other members of the Graduate Mathematical Statistics Society at Columbia University for aid in the preparation of the manuscript in dittoed form. The preparation of this manuscript was supported in part by the Office of Naval Research.

T. W. ANDERSON

Center for Advanced Study in the Behavioral Sciences
Stanford, Calif.
December, 1957

Contents

CHAPTER 1

Introduction

1.1. THE MULTIVARIATE NORMAL DISTRIBUTION AS MODEL

In this book we shall concern ourselves with statistical analyses of data that consist of sets of measurements on a number of individuals or objects. For example, the sample data may be heights and weights of some individuals drawn randomly from a population of school children in a given city, or the statistical treatment may be made on a collection of measurements, such as lengths and widths of petals and lengths and widths of sepals of iris plants taken from two species, or one may study the scores on batteries of mental tests administered to a number of students.

The mathematical model on which analysis is based is a multivariate normal distribution or a combination of multivariate normal distributions. The problems we shall treat have to do with inference concerning parameters of such distributions, such problems as testing a hypothesis of equality of means, point estimation of correlation coefficients, and decision between several hypotheses of specified distributions.

Although there are statistical problems of multiple measurement that cannot be based on the model of normal distributions and although there are other statistical methods which are applicable to samples from other types of distributions (for example, the multinomial distribution), we shall consider in this book only statistical analysis based on the normal distribution. The main justification for studying methods relating to the normal distribution so intensively is that this mathematical model is suitable for such a large number of cases where multiple measurements are treated.

As a matter of fact, a review of the development of the theory contained in this book shows that most of it has arisen to meet practical problems. One of the first men to treat such statistical problems was the geneticist, Francis Galton, in the second half of the nineteenth century. Galton's technique of investigation into statistical theory was to study innumerable samples and enunciate the theory of the model as generalizations of observed properties of samples. He particularly studied pairs of measurements, one made on a parent and one on an offspring. It is interesting to note that Galton wrote his friends requesting them to grow sweet peas, make measurements on two generations, and send him the data. The density of the bivariate normal distribution had been studied earlier by Adrian (1808), Laplace (1811), Plana (1813), Gauss (1823), and Bravais (1846). However, none of these writers discovered the correlation coefficient as a measure of association and the characteristics of conditional distributions such as regression lines and homoscedasticity, as did Galton* (1889).

In turn, Karl Pearson and others developed the theory and use of different kinds of correlation coefficients for studying problems in genetics, biology, and other fields. Classification problems in anthropology and botany led to the "coefficient of racial likeness" and the "discriminant function." In another direction, analysis of scores on mental tests led to a theory, including "factor analysis," the sampling theory of which is based on the normal distribution. In these cases, as well as a host of others in agricultural experiments, in engineering problems, in certain economic problems, and in other fields, the multivariate normal distributions have been found to be sufficiently close approximations to the populations so that statistical analyses based on these models are justified.

The univariate normal distribution arises frequently because the effect studied is the sum of many independent random effects. Similarly, the multivariate normal distribution often occurs because the multiple measurements are sums of small independent effects. Just as the central limit theorem leads to the univariate normal distribution for single variables, so does the general central limit theorem for several variables lead to the multivariate normal distribution.

Another reason for confining this book to considerations of normal theory is that multivariate methods based on the normal distribution are extensively developed and can be studied in a rather organized and systematic way. This is due not only to the need of such methods because they are of practical use but also to the fact that normal theory is amenable to exact mathematical treatment. The suitable methods of analysis are mainly based on standard operations of matrix algebra; the distributions of many statistics involved can be obtained exactly or at least characterized by moments; and in some cases optimum properties of procedures can be deduced.

1.2. OUTLINE OF MULTIVARIATE METHODS

We shall find it convenient to classify the statistical methods of multivariate analysis into the following five categories:

* For a detailed study of the development of the ideas of correlation see Walker (1931).

(1) *Correlation.* First of all we need methods of measuring the degree of dependence between two variates in the population and in the sample. The notion of correlation coefficient is extended to a measurement of dependence between one variate and a set of variates by means of the multiple correlation coefficient. The partial correlation coefficient is a measure of dependence when the effects of other correlated variables have been removed. These various correlation coefficients computed from samples are used to estimate corresponding parameters of distributions and to test hypotheses, such as hypotheses of independence.

(2) *Analogues of univariate statistical methods.* A number of problems arising in multivariate populations are straightforward analogues of problems arising in univariate populations; the suitable methods for handling these problems are similarly related. For example, in the univariate case we may wish to test the hypothesis that the mean of a variable is zero; in the multivariate case we may wish to test the hypothesis that the means of several variables are zero. The analogue of the Student t-test for the first hypothesis is the generalized T^2-test. We shall generalize the methods of least squares and analysis of variance.

In most of these problems the coordinate system used is irrelevant; that is, a linear transformation of the set of variables does not affect the methods. More particularly, in many cases of testing hypotheses, a linear transformation does not change the hypothesis or the test procedure.

(3) *Problems of coordinate systems.* These problems are essentially problems of choosing coordinate systems so that the variates have desired statistical properties. One might say that they involve characterizations of inherent properties of normal distributions. These are closely related to the algebraic problems of canonical forms for matrices. An example is finding the normalized linear combination of variables with maximum or minimum variance (finding principal components); this amounts to finding a rotation of axes that carries the covariance matrix to diagonal form. Another example is characterizing the dependence between two sets of variates (finding canonical correlations). These problems involve the characteristic roots and vectors of various matrices.

(4) *More detailed problems.* In many of these problems we divide our set of variables into subsets. One interesting problem is testing hypotheses of independence of these subsets. Other problems concern hypotheses of symmetry between or within subsets. In the category of "more detailed problems" we include factor analysis.

(5) *Dependent observations.* In time series analysis observations are made on variables successively in time. Observations made at one time may be dependent on those made earlier. Such problems lead to the study

of serial correlations and of stochastic difference equations. This is a large subject and, unfortunately, we shall not be able to go into it very far.

REFERENCES

Adrian (1808); Bravais (1846); Galton (1889); Gauss (1823); Laplace (1811); Plana (1813); Walker (1931).

CHAPTER 2

The Multivariate Normal Distribution

2.1. INTRODUCTION

In this chapter we define the multivariate normal distribution and consider some of its properties. In Section 2.2 are considered the fundamental notions of multivariate distributions: the definition by means of multivariate density functions, marginal distributions, conditional distributions, expected values and moments. In the following sections these ideas are studied for normal distributions. One of the interesting characteristics of the multivariate normal distributions is that marginal and conditional distributions are also normal.

2.2. NOTIONS OF MULTIVARIATE DISTRIBUTIONS

2.2.1. Joint Distributions

In this section we shall consider the notions of joint distributions of several variables, derived marginal distributions of subsets of variables, and derived conditional distributions. First consider the case of two (real) random variables* X and Y. Probabilities of events defined in terms of these variables can be obtained by operations involving the *cumulative distribution function* (abbreviated as cdf),

$$F(x, y) = \Pr\{X \leq x,\ Y \leq y\}, \tag{1}$$

defined for every pair of real numbers (x, y). We are interested in cases where $F(x, y)$ is absolutely continuous; this means that the following partial derivative exists almost everywhere:

$$\frac{\partial^2 F(x, y)}{\partial x\, \partial y} = f(x, y), \tag{2}$$

* In Chapter 2 we shall distinguish between random variables and running variables by use of capital and lower-case letters, respectively. In later chapters we shall be unable to hold to this convention because of other complications of notation.

and

$$F(x, y) = \int_{-\infty}^{y} \int_{-\infty}^{x} f(u, v)\, du\, dv. \tag{3}$$

The nonnegative function $f(x, y)$ is called the *density* of X and Y. The pair of random variables (X, Y) defines a random point in a plane. The probability that (X, Y) falls in a rectangle is

$$\begin{aligned} \text{(4)}\quad & \Pr\{x \le X \le x + \Delta x, y \le Y \le y + \Delta y\} \\ &= F(x + \Delta x, y + \Delta y) - F(x + \Delta x, y) - F(x, y + \Delta y) + F(x, y) \\ &= \int_{y}^{y+\Delta y} \int_{x}^{x+\Delta x} f(u, v)\, du\, dv \end{aligned}$$

$(\Delta x > 0, \Delta y > 0)$. The probability of the random point (X, Y) falling in any set E for which the following integral is defined (that is, any measureable set E) is

$$\Pr\{(X, Y) \in E\} = \iint_E f(x, y)\, dx\, dy. \tag{5}$$

This follows from the definition of the integral [as the limit of sums of the sort of (4)]. If $f(x, y)$ is continuous in both variables, the *probability element* $f(x, y)\, \Delta y\, \Delta x$ is approximately the probability that X falls between x and $x + \Delta x$ and Y falls between y and $y + \Delta y$ for

$$\begin{aligned} \text{(6)}\quad \Pr\{x \le X \le x + \Delta x, y \le Y \le y + \Delta y\} &= \int_{y}^{y+\Delta y} \int_{x}^{x+\Delta x} f(u, v)\, du\, dv \\ &= f(x_0, y_0)\, \Delta x\, \Delta y \end{aligned}$$

for some x_0, y_0 $(x \le x_0 \le x + \Delta x, y \le y_0 \le y + \Delta y)$ by the mean value theorem of calculus. Since $f(u, v)$ is continuous, (6) is approximately $f(x, y) \Delta x\, \Delta y$. In fact,

$$\lim_{\substack{\Delta x \to 0 \\ \Delta y \to 0}} \frac{1}{\Delta x\, \Delta y} \left| \Pr\{x \le X \le x + \Delta x, y \le Y \le y + \Delta y\} - f(x, y)\, \Delta x\, \Delta y \right| = 0. \tag{7}$$

Now we consider the case of p random variables $X_1, X_2, \cdots, X_p$. The cdf is

$$F(x_1, \cdots, x_p) = \Pr\{X_1 \le x_1, \cdots, X_p \le x_p\} \tag{8}$$

defined for every set of real numbers $x_1, \cdots, x_p$. The density function, if $F(x_1, \cdots, x_p)$ is absolutely continuous, is

$$\frac{\partial^p F(x_1, \cdots, x_p)}{\partial x_1 \cdots \partial x_p} = f(x_1, \cdots, x_p) \tag{9}$$

(almost everywhere) and

$$(10)\qquad F(x_1,\cdots,x_p)=\int_{-\infty}^{x_p}\cdots\int_{-\infty}^{x_1} f(u_1,\cdots,u_p)\,du_1\cdots du_p.$$

The probability of falling in any (measurable) set R in the p-dimensional Euclidean space is

$$(11)\qquad \Pr\{(X_1,\cdots,X_p)\in R\}=\int\cdots\int_R f(x_1,\cdots,x_p)\,dx_1\cdots dx_p.$$

The probability element $f(x_1,\cdots,x_p)\,\Delta x_1\cdots\Delta x_p$ is approximately the probability $\Pr\{x_1\le X_1\le x_1+\Delta x_1,\cdots,x_p\le X_p\le x_p+\Delta x_p\}$ if $f(x_1,\cdots,x_p)$ is continuous. The joint moments are

$$(12)\qquad \mathscr{E}X_1^{h_1}\cdots X_p^{h_p}=\int_{-\infty}^{\infty}\cdots\int_{-\infty}^{\infty} x_1^{h_1}\cdots x_p^{h_p} f(x_1,\cdots,x_p)\,dx_1\cdots dx_p.$$

2.2.2. Marginal Distributions

Given the cdf of two random variables X, Y as being $F(x,y)$, the marginal cdf of X is

$$(13)\qquad \begin{aligned}\Pr\{X\le x\}&=\Pr\{X\le x,\ Y\le\infty\}\\&=F(x,\infty).\end{aligned}$$

Let this be $F(x)$. Clearly

$$(14)\qquad F(x)=\int_{-\infty}^{x}\int_{-\infty}^{\infty} f(u,v)\,dv\,du.$$

We call

$$(15)\qquad \int_{-\infty}^{\infty} f(u,v)\,dv=f(u),$$

say, the marginal density of X. It is clear from (14) that

$$(16)\qquad F(x)=\int_{-\infty}^{x} f(u)\,du.$$

In a similar fashion we define $G(y)$, the marginal cdf of Y, and $g(y)$, the marginal density of Y.

Now we turn to the general case. Given $F(x_1,\cdots,x_p)$ as the cdf of $X_1,\cdots,X_p$, we wish to find the marginal cdf of some of $X_1,\cdots,X_p$, say, of $X_1,\cdots,X_r$ $(r<p)$. It is

$$(17)\qquad \begin{aligned}&\Pr\{X_1\le x_1,\cdots,X_r\le x_r\}\\&\quad=\Pr\{X_1\le x_1,\cdots,X_r\le x_r,X_{r+1}\le\infty,\cdots,X_p\le\infty\}\\&\quad=F(x_1,\cdots,x_r,\infty,\cdots,\infty).\end{aligned}$$

The marginal density of $X_1, \cdots, X_r$ is

$$\int_{-\infty}^{\infty} \cdots \int_{-\infty}^{\infty} f(u_1, \cdots, u_p)\, du_{r+1} \cdots du_p. \tag{18}$$

The marginal distribution and density of any other subset of $X_1, \cdots, X_p$ are obtained in the obvious similar fashion.

The joint moments of a subset of variates can be computed from the marginal distribution; for example,

$$\begin{aligned} \mathscr{E} X_1^{h_1} \cdots X_r^{h_r} &= \mathscr{E} X_1^{h_1} \cdots X_r^{h_r} X_{r+1}^0 \cdots X_p^0 \\ &= \int_{-\infty}^{\infty} \cdots \int_{-\infty}^{\infty} x_1^{h_1} \cdots x_r^{h_r} f(x_1, \cdots, x_p)\, dx_1 \cdots dx_p \\ &= \int_{-\infty}^{\infty} \cdots \int_{-\infty}^{\infty} x_1^{h_1} \cdots x_r^{h_r} \Bigg[\int_{-\infty}^{\infty} \cdots \\ &\qquad \int_{-\infty}^{\infty} f(x_1, \cdots, x_p) dx_{r+1} \cdots dx_p \Bigg] dx_1 \cdots dx_r. \end{aligned} \tag{19}$$

2.2.3. Statistical Independence

Two random variables X, Y with cdf $F(x, y)$ are said to be *independent* if

$$F(x, y) = F(x)G(y), \tag{20}$$

where $F(x)$ is the marginal cdf of X and $G(y)$ is the marginal cdf of Y. This implies that the density of X, Y is

$$\begin{aligned} f(x, y) &= \frac{\partial^2 F(x, y)}{\partial x\, \partial y} = \frac{\partial^2 F(x) G(y)}{\partial x\, \partial y} \\ &= \frac{dF(x)}{dx} \frac{dG(y)}{dy} \\ &= f(x) g(y). \end{aligned} \tag{21}$$

Conversely, if $f(x, y) = f(x)g(y)$, then

$$\begin{aligned} F(x, y) &= \int_{-\infty}^{y} \int_{-\infty}^{x} f(u, v)\, du\, dv = \int_{-\infty}^{y} \int_{-\infty}^{x} f(u) g(v)\, du\, dv \\ &= \int_{-\infty}^{x} f(u)\, du \int_{-\infty}^{y} g(v)\, dv = F(x)G(y). \end{aligned} \tag{22}$$

Thus an equivalent definition of independence in the case of densities existing is that $f(x, y) = f(x)g(y)$. To see the implications of statistical independence, given any $x_1 < x_2$, $y_1 < y_2$, we consider the probability

$$(23)\quad \Pr\{x_1 \leq X \leq x_2, y_1 \leq Y \leq y_2\}$$
$$= \int_{y_1}^{y_2}\int_{x_1}^{x_2} f(u, v)\, du\, dv = \int_{x_1}^{x_2} f(u)\, du \int_{y_1}^{y_2} g(v)\, dv$$
$$= \Pr\{x_1 \leq X \leq x_2\}\Pr\{y_1 \leq Y \leq y_2\}.$$

The probability of X falling in a given interval and Y falling in a given interval is the product of the probability of X falling in the interval and the probability of Y falling in the other interval.

If the cdf of $X_1, \cdots, X_p$ is $F(x_1, \cdots, x_p)$, the set of random variables is said to be *mutually independent* if

$$(24)\qquad F(x_1, \cdots, x_p) = F_1(x_1) \cdots F_p(x_p),$$

where $F_i(x_i)$ is the marginal cdf of $X_i (i = 1, \cdots, p)$. The set $X_1, \cdots, X_r$ is said to be independent of the set $X_{r+1}, \cdots, X_p$ if

$$(25)\quad F(x_1, \cdots, x_p)$$
$$= F(x_1, \cdots, x_r, \infty, \cdots, \infty) \cdot F(\infty, \cdots, \infty, x_{r+1}, \cdots, x_p).$$

One result of independence is that joint moments factor. For example, if $X_1, \cdots, X_p$ are mutually independent, then

$$(26)\quad \mathscr{E} X_1^{h_1} \cdots X_p^{h_p} = \int_{-\infty}^{\infty} \cdots \int_{-\infty}^{\infty} x_1^{h_1} \cdots x_p^{h_p} f_1(x_1) \cdots f_p(x_p)\, dx_1 \cdots dx_p$$
$$= \prod_{i=1}^{p} \int_{-\infty}^{\infty} x_i^{h_i} f_i(x_i)\, dx_i$$
$$= \prod_{i=1}^{p} \{\mathscr{E} X_i^{h_i}\}.$$

2.2.4. Conditional Distributions

If A and B are two events such that the probability of A and B occurring simultaneously is $P(AB)$ and the probability of B occurring is $P(B)$, then the conditional probability of A occurring given that B has occurred is $P(AB)/P(B)$ [if $P(B) > 0$]. Suppose the event A is X falling in the interval (x_1, x_2) and the event B is Y falling in (y_1, y_2). Then the conditional probability that X fall in (x_1, x_2), given that Y falls in (y_1, y_2), is

$$(27)\quad \Pr\{x_1 \leq X \leq x_2 | y_1 \leq Y \leq y_2\} = \frac{\Pr\{x_1 \leq X \leq x_2, y_1 \leq Y \leq y_2\}}{\Pr\{y_1 \leq Y \leq y_2\}}$$
$$= \frac{\int_{x_1}^{x_2}\int_{y_1}^{y_2} f(u, v)\, dv\, du}{\int_{y_1}^{y_2} g(v)\, dv}.$$

Now let $y_1 = y$, $y_2 = y + \Delta y$. Then for a continuous density

$$\int_y^{y+\Delta y} g(v)\,dv = g(y^*)\,\Delta y, \tag{28}$$

where $y \leq y^* \leq y + \Delta y$. Also

$$\int_y^{y+\Delta y} f(u, v)\,dv = \Delta y\, f[u, y(u)], \tag{29}$$

where $y \leq y(u) \leq y + \Delta y$. Therefore,

$$\Pr\{x_1 \leq X \leq x_2 | y \leq Y \leq y + \Delta y\} = \int_{x_1}^{x_2} \frac{f[u, y(u)]}{g(y^*)}\,du. \tag{30}$$

It will be noticed that for fixed y and Δy (> 0), the integrand of (30) behaves as a univariate density function. Now for y such that $g(y) > 0$, we define $\Pr\{x_1 \leq X \leq x_2 | Y = y\}$, the probability that X lies between x_1 and x_2, given that Y is y, as the limit of (30) as $\Delta y \to 0$. Thus

$$\Pr\{x_1 \leq X \leq x_2 | Y = y\} = \int_{x_1}^{x_2} f(u|y)\,du, \tag{31}$$

where $f(u|y) = f(u, y)/g(y)$. For given y, $f(u|y)$ is a density function and is called the conditional density of X given y. We note that if X and Y are independent, $f(x|y) = f(x)$.

In the general case of $X_1, \cdots, X_p$ with cdf $F(x_1, \cdots, x_p)$, the conditional density of $X_1, \cdots, X_r$, given $X_{r+1} = x_{r+1}, \cdots, X_p = x_p$, is

$$\frac{f(x_1, \cdots, x_p)}{\int_{-\infty}^{\infty} \cdots \int_{-\infty}^{\infty} f(u_1, \cdots, u_r, x_{r+1}, \cdots, x_p)\,du_1 \cdots du_r}. \tag{32}$$

For a more general discussion of conditional probabilities the reader is referred to Kolmogorov (1950).

2.2.5. Transformation of Variables

Let the density of $X_1, \cdots, X_p$ be $f(x_1, \cdots, x_p)$. Consider the p real-valued functions

$$y_i = y_i(x_1, \cdots, x_p), \qquad i = 1, \cdots, p. \tag{33}$$

We assume that the transformation from the x-space to the y-space is one-to-one;* the inverse transformation is

$$x_i = x_i(y_1, \cdots, y_p), \qquad i = 1, \cdots, p. \tag{34}$$

* More precisely, we assume this is true for the part of the x-space for which $f(x_1, \cdots, x_p)$ is positive.

Let the random variables $Y_1, \cdots, Y_p$ be defined by

$$Y_i = y_i(X_1, \cdots, X_p), \qquad i = 1, \cdots, p. \tag{35}$$

Then the density of $Y_1, \cdots, Y_p$ is

$$g(y_1, \cdots, y_p) = f[x_1(y_1, \cdots, y_p), \cdots, x_p(y_1, \cdots, y_p)]J(y_1, \cdots, y_p), \tag{36}$$

where $J(y_1, \cdots, y_p)$ is the Jacobian

$$J(y_1, \cdots, y_p) = \text{mod} \begin{vmatrix} \frac{\partial x_1}{\partial y_1} & \frac{\partial x_1}{\partial y_2} & \cdots & \frac{\partial x_1}{\partial y_p} \\ \frac{\partial x_2}{\partial y_1} & \frac{\partial x_2}{\partial y_2} & \cdots & \frac{\partial x_2}{\partial y_p} \\ \vdots & \vdots & & \vdots \\ \frac{\partial x_p}{\partial y_1} & \frac{\partial x_p}{\partial y_2} & \cdots & \frac{\partial x_p}{\partial y_p} \end{vmatrix}. \tag{37}$$

We assume the derivatives exist, and "mod" means absolute value of the expression following it. The probability that $(X_1, \cdots, X_p)$ falls in a region R is given by (11); the probability that $(Y_1, \cdots, Y_p)$ falls in a region S is

$$\Pr\{(Y_1, \cdots, Y_p) \in S\} = \int \underset{S}{\cdots} \int g(y_1, \cdots, y_p)\, dy_1 \cdots dy_p. \tag{38}$$

If S is the transform of R, that is, if each point of R transforms by (33) into a point of S and if each point of S transforms into R by (34), then (11) is equal to (38) by the usual theory of transformation of multiple integrals. From this follows the assertion that (36) is the density of $Y_1, \cdots, Y_p$.

2.3. THE MULTIVARIATE NORMAL DISTRIBUTION

The univariate normal density function can be written

$$ke^{-\frac{1}{2}\alpha(x-\beta)^2} = ke^{-\frac{1}{2}(x-\beta)\alpha(x-\beta)}, \tag{1}$$

where α is positive and k is chosen so that the integral of (1) over the entire x-axis is unity. The density function of a multivariate normal distribution of $X_1, \cdots, X_p$ has an analogous form. The scalar variable x is replaced by a vector

$$\boldsymbol{x} = \begin{pmatrix} x_1 \\ \cdot \\ \cdot \\ \cdot \\ x_p \end{pmatrix}; \tag{2}$$

the scalar constant β is replaced by a vector

$$(3) \qquad \boldsymbol{b} = \begin{pmatrix} b_1 \\ \cdot \\ \cdot \\ \cdot \\ b_p \end{pmatrix};$$

and the positive constant α is replaced by a positive definite (symmetric) matrix

$$(4) \qquad A = \begin{pmatrix} a_{11} & a_{12} & \cdots & a_{1p} \\ a_{21} & a_{22} & \cdots & a_{2p} \\ \cdot & \cdot & & \cdot \\ \cdot & \cdot & & \cdot \\ \cdot & \cdot & & \cdot \\ a_{p1} & a_{p2} & \cdots & a_{pp} \end{pmatrix}.$$

The square $\alpha(x - \beta)^2 = (x - \beta)\alpha(x - \beta)$ is replaced by the quadratic form

$$(5) \qquad (\boldsymbol{x} - \boldsymbol{b})'A(\boldsymbol{x} - \boldsymbol{b}) = \sum_{i,j=1}^{p} a_{ij}(x_i - b_i)(x_j - b_j).$$

Thus the density function of a p-variate normal distribution is

$$(6) \qquad f(x_1, \cdots, x_p) = Ke^{-\frac{1}{2}(\boldsymbol{x}-\boldsymbol{b})'A(\boldsymbol{x}-\boldsymbol{b})},$$

where $K(> 0)$ is chosen so that the integral over the entire p-dimensional Euclidean space of $x_1, \cdots, x_p$ is unity.

Written in matrix notation, the similarity of the multivariate normal density (6) to the univariate density (1) is clear. Throughout this book we shall use matrix notation and operations. The reader is referred to Appendix 1 for a review of matrix theory and for definitions of our notation for matrix operations.

We observe that $f(x_1, \cdots, x_p)$ is nonnegative. Since A is positive definite,

$$(7) \qquad (\boldsymbol{x} - \boldsymbol{b})'A(\boldsymbol{x} - \boldsymbol{b}) \geq 0,$$

and therefore the density is bounded; that is,

$$(8) \qquad f(x_1, \cdots, x_p) \leq K.$$

Now let us determine K so that the integral of (6) over the p-dimensional space is one. We shall evaluate

$$(9) \qquad K^* = \int_{-\infty}^{\infty} \cdots \int_{-\infty}^{\infty} e^{-\frac{1}{2}(\boldsymbol{x}-\boldsymbol{b})'A(\boldsymbol{x}-\boldsymbol{b})}\, dx_p \cdots dx_1.$$

We use the fact (see Corollary 4 of Appendix 1) that if $\boldsymbol{A}$ is positive definite there exists a nonsingular matrix $\boldsymbol{C}$ such that

$$C'AC = I, \tag{10}$$

where $\boldsymbol{I}$ denotes the identity and $\boldsymbol{C}'$ the transpose of $\boldsymbol{C}$. Let

$$x - b = Cy, \tag{11}$$

where

$$y = \begin{pmatrix} y_1 \\ \cdot \\ \cdot \\ \cdot \\ y_p \end{pmatrix}. \tag{12}$$

Then

$$(x - b)'A(x - b) = y'C'ACy = y'y. \tag{13}$$

The Jacobian of the transformation is

$$J = \text{mod}|C|, \tag{14}$$

where "mod$|\boldsymbol{C}|$" indicates the absolute value of the determinant of $\boldsymbol{C}$. Thus (9) becomes

$$K^* = \text{mod}|C| \int_{-\infty}^{\infty} \cdots \int_{-\infty}^{\infty} e^{-\frac{1}{2}y'y}\, dy_p \cdots dy_1. \tag{15}$$

Since

$$e^{-\frac{1}{2}y'y} = e^{-\frac{1}{2}\sum_{i=1}^{p} y_i^2} = \prod_{i=1}^{p} e^{-\frac{1}{2}y_i^2}, \tag{16}$$

we can write (15) as

$$\begin{aligned} K^* &= \text{mod}|C| \int_{-\infty}^{\infty} \cdots \int_{-\infty}^{\infty} e^{-\frac{1}{2}y_1^2} \cdots e^{-\frac{1}{2}y_p^2}\, dy_p \cdots dy_1 \\ &= \text{mod}|C| \prod_{i=1}^{p} \left\{ \int_{-\infty}^{\infty} e^{-\frac{1}{2}y_i^2}\, dy_i \right\} \\ &= \text{mod}|C| \prod_{i=1}^{p} \{\sqrt{2\pi}\} \\ &= \text{mod}|C|(2\pi)^{\frac{1}{2}p}, \end{aligned} \tag{17}$$

by virtue of

$$\frac{1}{\sqrt{2\pi}} \int_{-\infty}^{\infty} e^{-\frac{1}{2}t^2}\, dt = 1. \tag{18}$$

Corresponding to (10) is the determinantal equation

$$|C'| \cdot |A| \cdot |C| = |I|. \tag{19}$$

Since

$$|C'| = |C|, \tag{20}$$

and since $|I| = 1$, we deduce from (19) that

$$\operatorname{mod}|C| = 1/\sqrt{|A|}. \tag{21}$$

Thus

$$K = 1/K^* = \sqrt{|A|}\,(2\pi)^{-\frac{1}{2}p}. \tag{22}$$

The normal density function is

$$\frac{\sqrt{|A|}}{(2\pi)^{\frac{1}{2}p}}\, e^{-\frac{1}{2}(x-b)'A(x-b)}. \tag{23}$$

We shall now show the significance of b and A by finding the first and second moments of $X_1, \cdots, X_p$. It will be convenient to consider these random variables as constituting a random vector

$$X = \begin{pmatrix} X_1 \\ \cdot \\ \cdot \\ \cdot \\ X_p \end{pmatrix}. \tag{24}$$

We shall define generally a random matrix and the expected value of a random matrix; a random vector is considered as a special case of a random matrix with one column.

DEFINITION 2.3.1. *A random matrix Z is a matrix*

$$Z = (Z_{gh}), \qquad g = 1, \cdots, m; \; h = 1, \cdots, n, \tag{25}$$

of random variables $Z_{11}, \cdots, Z_{mn}$.

If the random variables $Z_{11}, \cdots, Z_{mn}$ can take on only a finite number of values, the random matrix Z can be one of a finite number of matrices, say $Z(1), \cdots, Z(q)$. If the probability of $Z = Z(i)$ is p_i, then we should like to define $\mathscr{E}Z$ as $\sum_{i=1}^{q} Z(i)p_i$. It is clear then that $\mathscr{E}Z = (\mathscr{E}Z_{gh})$. If the random variables $Z_{11}, \cdots, Z_{mn}$ have a joint density, then operating with Riemann sums we can define $\mathscr{E}Z$ as the limit (if the limit exists) of approximating sums of the kind occurring in the discrete case; then again $\mathscr{E}Z = (\mathscr{E}Z_{gh})$, Therefore, in general we shall use the following definition:

DEFINITION 2.3.2. *The expected value of a random matrix Z is*

$$\mathscr{E}Z = (\mathscr{E}Z_{gh}), \qquad g = 1, \cdots, m; \; h = 1, \cdots, n. \tag{26}$$

The operation of taking the expected value of a random matrix (or vector) satisfies certain rules which we can summarize in the following lemma:

LEMMA 2.3.1. *If* $\mathbf{Z}$ *is an* $m \times n$ *random matrix,* $\mathbf{D}$ *is an* $l \times m$ *real matrix,* $\mathbf{E}$ *is an* $n \times q$ *real matrix and* $\mathbf{F}$ *is an* $l \times q$ *real matrix, then*

$$\mathscr{E}(\mathbf{DZE} + \mathbf{F}) = \mathbf{D}(\mathscr{E}\mathbf{Z})\mathbf{E} + \mathbf{F}. \tag{27}$$

Proof. The element in the ith row and jth column of $\mathscr{E}(\mathbf{DZE} + \mathbf{F})$ is

$$\mathscr{E}\Big(\sum_{h,g} d_{ih}Z_{hg}e_{gj} + f_{ij}\Big) = \sum_{h,g} d_{ih}(\mathscr{E}Z_{hg})e_{gj} + f_{ij}, \tag{28}$$

which is the element in the ith row and jth column of $\mathbf{D}(\mathscr{E}\mathbf{Z})\mathbf{E} + \mathbf{F}$. Q.E.D.

Thus, if

$$\boldsymbol{X} = \boldsymbol{C}\boldsymbol{Y} + \boldsymbol{b}, \tag{29}$$

where $\boldsymbol{C}$ and $\boldsymbol{b}$ are as before and $\boldsymbol{Y}$ is a p-dimensional random vector, then

$$\mathscr{E}\boldsymbol{X} = \boldsymbol{C}\mathscr{E}\boldsymbol{Y} + \boldsymbol{b}. \tag{30}$$

By the transformation theory given in Section 2.2, the density of $\boldsymbol{Y}$ is proportional to (16); that is, it is

$$\frac{1}{(2\pi)^{\frac{1}{2}p}} e^{-\frac{1}{2}\boldsymbol{y}'\boldsymbol{y}} = \prod_{j=1}^{p} \left\{\frac{1}{\sqrt{2\pi}} e^{-\frac{1}{2}y_j^2}\right\}. \tag{31}$$

The expected value of the ith component of $\boldsymbol{Y}$ is

$$\begin{aligned} \mathscr{E}Y_i &= \int_{-\infty}^{\infty} \cdots \int_{-\infty}^{\infty} y_i \prod_{j=1}^{p} \left\{\frac{1}{\sqrt{2\pi}} e^{-\frac{1}{2}y_j^2}\right\} dy_1 \cdots dy_p \\ &= \frac{1}{\sqrt{2\pi}} \int_{-\infty}^{\infty} y_i e^{-\frac{1}{2}y_i^2} dy_i \prod_{\substack{j=1 \\ j \neq i}}^{p} \left\{\int_{-\infty}^{\infty} \frac{1}{\sqrt{2\pi}} e^{-\frac{1}{2}y_j^2} dy_j\right\} \\ &= \frac{1}{\sqrt{2\pi}} \int_{-\infty}^{\infty} y_i e^{-\frac{1}{2}y_i^2} dy_i \\ &= 0. \end{aligned} \tag{32}$$

The last equality follows because* $y_i e^{-\frac{1}{2}y_i^2}$ is an odd function of y_i. Thus $\mathscr{E}\boldsymbol{Y} = \mathbf{0}$. Therefore, the mean of $\boldsymbol{X}$, denoted by $\boldsymbol{\mu}$, is

$$\boldsymbol{\mu} = \mathscr{E}\boldsymbol{X} = \boldsymbol{b}. \tag{33}$$

* Alternatively, the last equality follows because the next to last expression is the expected value of a normally distributed variable with mean 0.

The *covariance matrix* of X is defined as

$$\mathscr{C}(X, X') = \mathscr{E}(X - \mu)(X - \mu)' = (\mathscr{E}(X_i - \mu_i)(X_j - \mu_j)). \tag{34}$$

A diagonal element of this matrix, $\mathscr{E}(X_i - \mu_i)^2$, is the variance of X_i; a non-diagonal element, $\mathscr{E}(X_i - \mu_i)(X_j - \mu_j)$, is the covariance of X_i and X_j.

Using (29), we see that

$$\begin{aligned} \mathscr{E}(X - \mu)(X - \mu)' &= \mathscr{E}CYY'C' \\ &= C(\mathscr{E}YY')C'. \end{aligned} \tag{35}$$

The i, jth element of $\mathscr{E}YY'$ is

$$\mathscr{E}Y_iY_j = \int_{-\infty}^{\infty} \cdots \int_{-\infty}^{\infty} y_i y_j \prod_{h=1}^{p} \left\{\frac{1}{\sqrt{2\pi}} e^{-\frac{1}{2}y_h^2}\right\} dy_1 \cdots dy_p \tag{36}$$

because the density of Y is (31). If $i = j$, we have

$$\begin{aligned} \mathscr{E}Y_i^2 &= \frac{1}{\sqrt{2\pi}} \int_{-\infty}^{\infty} y_i^2 e^{-\frac{1}{2}y_i^2} dy_i \prod_{\substack{h=1 \\ h \neq i}}^{p} \left\{\int_{-\infty}^{\infty} \frac{1}{\sqrt{2\pi}} e^{-\frac{1}{2}y_h^2} dy_h\right\} \\ &= \frac{1}{\sqrt{2\pi}} \int_{-\infty}^{\infty} y_i^2 e^{-\frac{1}{2}y_i^2} dy_i \\ &= 1. \end{aligned} \tag{37}$$

The last equality follows because the next to last expression is the expected value of the square of a variable normally distributed with mean 0 and variance 1. If $i \neq j$, (36) becomes

$$\begin{aligned} \mathscr{E}Y_iY_j &= \frac{1}{\sqrt{2\pi}} \int_{-\infty}^{\infty} y_i e^{-\frac{1}{2}y_i^2} dy_i \cdot \frac{1}{\sqrt{2\pi}} \int_{-\infty}^{\infty} y_j e^{-\frac{1}{2}y_j^2} dy_j \\ &\qquad \cdot \prod_{\substack{h=1 \\ h \neq i, j}}^{p} \left\{\frac{1}{\sqrt{2\pi}} \int_{-\infty}^{\infty} e^{-\frac{1}{2}y_h^2} dy_h\right\} \\ &= 0 \qquad (i \neq j) \end{aligned} \tag{38}$$

since the first integration gives 0. We can summarize (37) and (38) as

$$\mathscr{E}YY' = I. \tag{39}$$

Thus

$$\mathscr{E}(X - \mu)(X - \mu)' = CIC' = CC'. \tag{40}$$

From (10) we obtain $A = (C')^{-1}C^{-1}$ by multiplication by $(C')^{-1}$ on the left and by C^{-1} on the right. Taking inverses on both sides of the equality gives us

$$CC' = A^{-1}. \tag{41}$$

Thus, the covariance matrix of X, which we shall denote by $\mathbf{\Sigma}$, is

$$\mathbf{\Sigma} = \mathscr{E}(X - \boldsymbol{\mu})(X - \boldsymbol{\mu})' = A^{-1}. \tag{42}$$

From (41) we see that $\mathbf{\Sigma}$ is positive definite. Let us summarize these results.

THEOREM 2.3.1. *If the density of a p-dimensional random vector X is* (23), *the expected value of X is $\mathbf{b}$ and the covariance matrix is A^{-1}. Conversely, given a vector $\boldsymbol{\mu}$ and a positive definite matrix $\mathbf{\Sigma}$, there is a multivariate normal density*

$$(2\pi)^{-\frac{1}{2}p}|\mathbf{\Sigma}|^{-\frac{1}{2}}e^{-\frac{1}{2}(x-\boldsymbol{\mu})'\mathbf{\Sigma}^{-1}(x-\boldsymbol{\mu})} \tag{43}$$

such that the expected value of the vector with this density is $\boldsymbol{\mu}$ and the covariance matrix is $\mathbf{\Sigma}$.

We shall denote the density (43) as $n(x|\boldsymbol{\mu}, \mathbf{\Sigma})$ and the distribution law as $N(\boldsymbol{\mu}, \mathbf{\Sigma})$.

The ith diagonal element of the covariance matrix, σ_{ii}, is the variance of the ith component of X; we may sometimes denote this by σ_i^2. The *correlation coefficient* between X_i and X_j is defined as

$$\rho_{ij} = \frac{\sigma_{ij}}{\sqrt{\sigma_{ii}}\sqrt{\sigma_{jj}}} = \frac{\sigma_{ij}}{\sigma_i\sigma_j}. \tag{44}$$

Obviously $\rho_{ij} = \rho_{ji}$. Since

$$\begin{pmatrix} \sigma_{ii} & \sigma_{ij} \\ \sigma_{ji} & \sigma_{jj} \end{pmatrix} = \begin{pmatrix} \sigma_i^2 & \sigma_i\sigma_j\rho_{ij} \\ \sigma_i\sigma_j\rho_{ij} & \sigma_j^2 \end{pmatrix} \tag{45}$$

is positive definite (see Appendix 1), the determinant

$$\begin{vmatrix} \sigma_i^2 & \sigma_i\sigma_j\rho_{ij} \\ \sigma_i\sigma_j\rho_{ij} & \sigma_j^2 \end{vmatrix} = \sigma_i^2\sigma_j^2(1 - \rho_{ij}^2) \tag{46}$$

is positive. Therefore, $-1 < \rho_{ij} < 1$ (for nonsingular distributions, see Section 2.4).

As a special case of the preceding theory we consider the bivariate normal distribution. The mean vector is

$$\mathscr{E}\begin{pmatrix} X_1 \\ X_2 \end{pmatrix} = \begin{pmatrix} \mu_1 \\ \mu_2 \end{pmatrix}; \tag{47}$$

the covariance matrix may be written

$$\begin{aligned} \mathbf{\Sigma} &= \mathscr{E}\begin{pmatrix} (X_1 - \mu_1)^2 & (X_1 - \mu_1)(X_2 - \mu_2) \\ (X_2 - \mu_2)(X_1 - \mu_1) & (X_2 - \mu_2)^2 \end{pmatrix} \\ &= \begin{pmatrix} \sigma_{11} & \sigma_{12} \\ \sigma_{21} & \sigma_{22} \end{pmatrix} = \begin{pmatrix} \sigma_1^2 & \sigma_1\sigma_2\rho \\ \sigma_2\sigma_1\rho & \sigma_2^2 \end{pmatrix}, \end{aligned} \tag{48}$$

where σ_1^2 is the variance of X_1, σ_2^2 the variance of X_2, and ρ the correlation between X_1 and X_2. It is easily verified that the inverse of (48) is

$$\boldsymbol{\Sigma}^{-1} = \frac{1}{1-\rho^2}\begin{pmatrix} \dfrac{1}{\sigma_1^2} & -\dfrac{\rho}{\sigma_1\sigma_2} \\ -\dfrac{\rho}{\sigma_2\sigma_1} & \dfrac{1}{\sigma_2^2} \end{pmatrix}. \tag{49}$$

The density function of X_1 and X_2 is

$$\frac{1}{2\pi\sigma_1\sigma_2\sqrt{1-\rho^2}} \exp\left\{-\frac{1}{2(1-\rho^2)}\left[\frac{(x_1-\mu_1)^2}{\sigma_1^2} - 2\rho\frac{(x_1-\mu_1)(x_2-\mu_2)}{\sigma_1\sigma_2} + \frac{(x_2-\mu_2)^2}{\sigma_2^2}\right]\right\}. \tag{50}$$

It will be seen later that if $\rho = 0$, X_1 and X_2 are independent; if $\rho > 0$, X_1 and X_2 tend to be positively related; and if $\rho < 0$, X_1 and X_2 tend to be negatively related.

It will be noticed that the density function (43) is constant on ellipsoids

$$(\boldsymbol{x} - \boldsymbol{\mu})'\boldsymbol{\Sigma}^{-1}(\boldsymbol{x} - \boldsymbol{\mu}) = c \tag{51}$$

for every positive value of c in a p-dimensional Euclidean space. The center of each ellipsoid is at the point $\boldsymbol{\mu}$. The shape and orientation of the ellipsoid are determined by $\boldsymbol{\Sigma}$, and the size (given $\boldsymbol{\Sigma}$) is determined by c.

Let us consider in detail the bivariate case of the density (50). We transform coordinates by $(x_i - \mu_i)/\sigma_i = y_i$ $(i = 1, 2)$ so that the centers of the loci of constant density are at the origin. These loci are defined by

$$\frac{1}{1-\rho^2}(y_1^2 - 2\rho y_1 y_2 + y_2^2) = c. \tag{52}$$

The intercepts on the y_1-axis and y_2-axis are equal. If $\rho > 0$, the major axis of the ellipse is along the 45° line with a length of $2\sqrt{c(1+\rho)}$ and the minor axis has a length of $2\sqrt{c(1-\rho)}$. If $\rho < 0$, the major axis is along the 135° line with a length of $2\sqrt{c(1-\rho)}$ and the minor axis has a length of $2\sqrt{c(1+\rho)}$. In this bivariate case we can think of the density function as a surface above the plane. The contours of equal density are contours of equal altitude on a topographical map; they indicate the shape of the hill (or probability surface). If $\rho > 0$, the hill will tend to run along a line with a positive slope; most of the hill will be in the first and third quadrants. When we transform back to $x_i = \sigma_i y_i + \mu_i$, we expand each contour by a factor of σ_i in the direction of the ith axis and shift the center to (μ_1, μ_2).

The numerical values of the cdf of the univariate normal variable are obtained from tables found in most statistical texts. The numerical values of

$$(53)\qquad F(x_1, x_2) = \Pr\{X_1 \le x_1, X_2 \le x_2\} = \Pr\left\{\frac{X_1 - \mu_1}{\sigma_1} \le y_1, \frac{X_2 - \mu_2}{\sigma_2} \le y_2\right\},$$

where $y_1 = (x_1 - \mu_1)/\sigma_1$ and $y_2 = (x_2 - \mu_2)/\sigma_2$, can be found in Pearson (1930, 1931). Pearson has also shown that

$$(54)\qquad F(x_1, x_2) = \sum_{j=0}^{\infty} \rho_{12}^j \, \tau_j(y_1) \, \tau_j(y_2),$$

where the so-called *tetrachoric functions* $\tau_j(y)$ are tabulated in Pearson (1930, 1931) up to $\tau_{19}(y)$. Kendall (1941) has shown that expression (54) can be generalized for $F(x_1, \cdots, x_n)$.

2.4. THE DISTRIBUTION OF LINEAR COMBINATIONS OF NORMALLY DISTRIBUTED VARIATES; INDEPENDENCE OF VARIATES; MARGINAL DISTRIBUTIONS

One of the reasons the study of normal multivariate distributions is worthwhile is that marginal distributions and conditional distributions derived from multivariate normal distributions are also normal distributions. Moreover, linear combinations of normal variates are again normally distributed. First we shall show that if we make a nonsingular linear transformation of a vector whose components have a joint distribution with a normal density we obtain a vector whose components are jointly distributed with a normal density.

THEOREM 2.4.1. *Let* $\boldsymbol{X}$ *(with p components) be distributed according to* $N(\boldsymbol{\mu}, \boldsymbol{\Sigma})$. *Then*

$$(1)\qquad \boldsymbol{Y} = \boldsymbol{C}\boldsymbol{X}$$

is distributed according to $N(\boldsymbol{C\mu}, \boldsymbol{C\Sigma C'})$ *for* $\boldsymbol{C}$ *nonsingular.*

Proof. The density of $\boldsymbol{Y}$ is obtained from the density of $\boldsymbol{X}$, $n(\boldsymbol{x}|\boldsymbol{\mu}, \boldsymbol{\Sigma})$, by replacing $\boldsymbol{x}$ by

$$(2)\qquad \boldsymbol{x} = \boldsymbol{C}^{-1}\boldsymbol{y},$$

and multiplying by the Jacobian of the transformation (2), which is $\operatorname{mod}|\boldsymbol{C}^{-1}|$. This is

$$(3)\qquad \operatorname{mod}|\boldsymbol{C}^{-1}| = \frac{1}{\operatorname{mod}|\boldsymbol{C}|} = \sqrt{\frac{1}{|\boldsymbol{C}|^2}} = \sqrt{\frac{|\boldsymbol{\Sigma}|}{|\boldsymbol{C}| \cdot |\boldsymbol{\Sigma}| \cdot |\boldsymbol{C}'|}} = \frac{|\boldsymbol{\Sigma}|^{\frac{1}{2}}}{|\boldsymbol{C\Sigma C'}|^{\frac{1}{2}}}.$$

The quadratic form in the exponent of $n(\boldsymbol{x}|\boldsymbol{\mu},\boldsymbol{\Sigma})$ is

$$Q = (\boldsymbol{x} - \boldsymbol{\mu})'\boldsymbol{\Sigma}^{-1}(\boldsymbol{x} - \boldsymbol{\mu}). \tag{4}$$

The transformation (2) carries Q into

$$\begin{aligned} Q &= (C^{-1}\boldsymbol{y} - \boldsymbol{\mu})'\boldsymbol{\Sigma}^{-1}(C^{-1}\boldsymbol{y} - \boldsymbol{\mu}) \\ &= (C^{-1}\boldsymbol{y} - C^{-1}C\boldsymbol{\mu})'\boldsymbol{\Sigma}^{-1}(C^{-1}\boldsymbol{y} - C^{-1}C\boldsymbol{\mu}) \\ &= [C^{-1}(\boldsymbol{y} - C\boldsymbol{\mu})]'\boldsymbol{\Sigma}^{-1}[C^{-1}(\boldsymbol{y} - C\boldsymbol{\mu})] \\ &= (\boldsymbol{y} - C\boldsymbol{\mu})'(C^{-1})'\boldsymbol{\Sigma}^{-1}C^{-1}(\boldsymbol{y} - C\boldsymbol{\mu}) \\ &= (\boldsymbol{y} - C\boldsymbol{\mu})'(C\boldsymbol{\Sigma}C')^{-1}(\boldsymbol{y} - C\boldsymbol{\mu}) \end{aligned} \tag{5}$$

since $(C^{-1})' = (C')^{-1}$ by virtue of transposition of $CC^{-1} = I$. Thus the density of $\boldsymbol{Y}$ is

$$\begin{aligned} & n(C^{-1}\boldsymbol{y}|\boldsymbol{\mu}, \boldsymbol{\Sigma}) \bmod |C|^{-1} \\ &\qquad = (2\pi)^{-\frac{1}{2}p}|C\boldsymbol{\Sigma}C'|^{-\frac{1}{2}} \exp[-\tfrac{1}{2}(\boldsymbol{y} - C\boldsymbol{\mu})'(C\boldsymbol{\Sigma}C')^{-1}(\boldsymbol{y} - C\boldsymbol{\mu})] \\ &\qquad = n(\boldsymbol{y}|C\boldsymbol{\mu}, C\boldsymbol{\Sigma}C'). \end{aligned} \tag{6}$$

This proves the theorem.

Now let us consider two sets of random variables $X_1, \cdots, X_q$ and $X_{q+1}, \cdots, X_p$ forming the vectors

$$\boldsymbol{X}^{(1)} = \begin{pmatrix} X_1 \\ \cdot \\ \cdot \\ \cdot \\ X_q \end{pmatrix}, \qquad \boldsymbol{X}^{(2)} = \begin{pmatrix} X_{q+1} \\ \cdot \\ \cdot \\ \cdot \\ X_p \end{pmatrix}. \tag{7}$$

These variables form the random vector

$$\boldsymbol{X} = \begin{pmatrix} \boldsymbol{X}^{(1)} \\ \boldsymbol{X}^{(2)} \end{pmatrix} = \begin{pmatrix} X_1 \\ \cdot \\ \cdot \\ \cdot \\ X_p \end{pmatrix}. \tag{8}$$

Now let us assume that the p variates have a joint normal distribution with means

$$\mathscr{E}\boldsymbol{X}^{(1)} = \boldsymbol{\mu}^{(1)}, \qquad \mathscr{E}\boldsymbol{X}^{(2)} = \boldsymbol{\mu}^{(2)}, \tag{9}$$

and covariances

$$\mathscr{E}(\boldsymbol{X}^{(1)} - \boldsymbol{\mu}^{(1)})(\boldsymbol{X}^{(1)} - \boldsymbol{\mu}^{(1)})' = \boldsymbol{\Sigma}_{11}, \tag{10}$$

$$\mathscr{E}(\boldsymbol{X}^{(2)} - \boldsymbol{\mu}^{(2)})(\boldsymbol{X}^{(2)} - \boldsymbol{\mu}^{(2)})' = \boldsymbol{\Sigma}_{22}, \tag{11}$$

$$\mathscr{E}(\boldsymbol{X}^{(1)} - \boldsymbol{\mu}^{(1)})(\boldsymbol{X}^{(2)} - \boldsymbol{\mu}^{(2)})' = \boldsymbol{\Sigma}_{12} = \boldsymbol{0} \tag{12}$$

($\mathbf{\Sigma}_{21} = \mathbf{\Sigma}'_{12} = \mathbf{0}$). We say that the random vector $\boldsymbol{X}$ has been partitioned in (8) into sub-vectors, that

$$\boldsymbol{\mu} = \begin{pmatrix} \boldsymbol{\mu}^{(1)} \\ \boldsymbol{\mu}^{(2)} \end{pmatrix} \tag{13}$$

has been partitioned similarly into subvectors, and that

$$\mathbf{\Sigma} = \begin{pmatrix} \mathbf{\Sigma}_{11} & \mathbf{\Sigma}_{12} \\ \mathbf{\Sigma}_{21} & \mathbf{\Sigma}_{22} \end{pmatrix} = \begin{pmatrix} \mathbf{\Sigma}_{11} & \mathbf{0} \\ \mathbf{0} & \mathbf{\Sigma}_{22} \end{pmatrix} \tag{14}$$

has been similarly partitioned into submatrices (see Appendix 1, Section 3).

We shall show that $\boldsymbol{X}^{(1)}$ and $\boldsymbol{X}^{(2)}$ are independently normally distributed. The inverse of $\mathbf{\Sigma}$ is

$$\mathbf{\Sigma}^{-1} = \begin{pmatrix} \mathbf{\Sigma}_{11}^{-1} & \mathbf{0} \\ \mathbf{0} & \mathbf{\Sigma}_{22}^{-1} \end{pmatrix}. \tag{15}$$

Thus the quadratic form in the exponent of $n(\boldsymbol{x}|\boldsymbol{\mu}, \mathbf{\Sigma})$ is

$$\begin{aligned} Q &= (\boldsymbol{x} - \boldsymbol{\mu})'\mathbf{\Sigma}^{-1}(\boldsymbol{x} - \boldsymbol{\mu}) \\ &= [(\boldsymbol{x}^{(1)} - \boldsymbol{\mu}^{(1)})', (\boldsymbol{x}^{(2)} - \boldsymbol{\mu}^{(2)})'] \begin{pmatrix} \mathbf{\Sigma}_{11}^{-1} & \mathbf{0} \\ \mathbf{0} & \mathbf{\Sigma}_{22}^{-1} \end{pmatrix} \begin{pmatrix} \boldsymbol{x}^{(1)} - \boldsymbol{\mu}^{(1)} \\ \boldsymbol{x}^{(2)} - \boldsymbol{\mu}^{(2)} \end{pmatrix} \\ &= [(\boldsymbol{x}^{(1)} - \boldsymbol{\mu}^{(1)})'\mathbf{\Sigma}_{11}^{-1}, (\boldsymbol{x}^{(2)} - \boldsymbol{\mu}^{(2)})'\mathbf{\Sigma}_{22}^{-1}] \begin{pmatrix} \boldsymbol{x}^{(1)} - \boldsymbol{\mu}^{(1)} \\ \boldsymbol{x}^{(2)} - \boldsymbol{\mu}^{(2)} \end{pmatrix} \\ &= (\boldsymbol{x}^{(1)} - \boldsymbol{\mu}^{(1)})'\mathbf{\Sigma}_{11}^{-1}(\boldsymbol{x}^{(1)} - \boldsymbol{\mu}^{(1)}) + (\boldsymbol{x}^{(2)} - \boldsymbol{\mu}^{(2)})'\mathbf{\Sigma}_{22}^{-1}(\boldsymbol{x}^{(2)} - \boldsymbol{\mu}^{(2)}) \\ &= Q_1 + Q_2, \end{aligned} \tag{16}$$

say, where

$$\begin{aligned} Q_1 &= (\boldsymbol{x}^{(1)} - \boldsymbol{\mu}^{(1)})'\mathbf{\Sigma}_{11}^{-1}(\boldsymbol{x}^{(1)} - \boldsymbol{\mu}^{(1)}), \\ Q_2 &= (\boldsymbol{x}^{(2)} - \boldsymbol{\mu}^{(2)})'\mathbf{\Sigma}_{22}^{-1}(\boldsymbol{x}^{(2)} - \boldsymbol{\mu}^{(2)}). \end{aligned} \tag{17}$$

Also we note that $|\mathbf{\Sigma}| = |\mathbf{\Sigma}_{11}| \cdot |\mathbf{\Sigma}_{22}|$. The density of $\boldsymbol{X}$ can be written

$$\begin{aligned} n(\boldsymbol{x}|\boldsymbol{\mu}, \mathbf{\Sigma}) &= \frac{1}{(2\pi)^{\frac{1}{2}p}|\mathbf{\Sigma}|^{\frac{1}{2}}} e^{-\frac{1}{2}Q} \\ &= \frac{1}{(2\pi)^{\frac{1}{2}q}|\mathbf{\Sigma}_{11}|^{\frac{1}{2}}} e^{-\frac{1}{2}Q_1} \cdot \frac{1}{(2\pi)^{\frac{1}{2}(p-q)}|\mathbf{\Sigma}_{22}|^{\frac{1}{2}}} e^{-\frac{1}{2}Q_2} \\ &= n(\boldsymbol{x}^{(1)}|\boldsymbol{\mu}^{(1)}, \mathbf{\Sigma}_{11})n(\boldsymbol{x}^{(2)}|\boldsymbol{\mu}^{(2)}, \mathbf{\Sigma}_{22}). \end{aligned} \tag{18}$$

The marginal density of $X^{(1)}$ is given by the integral

$$(19)\quad \int_{-\infty}^{\infty}\cdots\int_{-\infty}^{\infty} n(x|\mu, \Sigma)\, dx_{q+1}\cdots dx_p$$
$$= n(x^{(1)}|\mu^{(1)}, \Sigma_{11})\int_{-\infty}^{\infty}\cdots\int_{-\infty}^{\infty} n(x^{(2)}|\mu^{(2)}, \Sigma_{22})\, dx_{q+1}\cdots dx_p$$
$$= n(x^{(1)}|\mu^{(1)}, \Sigma_{11}).$$

Thus the marginal distribution of $X^{(1)}$ is $N(\mu^{(1)}, \Sigma_{11})$; similarly the marginal distribution of $X^{(2)}$ is $N(\mu^{(2)}, \Sigma_{22})$. Thus the joint density of $X_1, \cdots, X_p$ is the product of the marginal density of $X_1, \cdots, X_q$ and the marginal density of $X_{q+1}, \cdots, X_p$, and, therefore, the two sets of variates are independent. Since the numbering of variates can always be done so that $X^{(1)}$ consists of any subset of the variates, we have proved the sufficiency in the following theorem:

THEOREM 2.4.2. *If $X_1, \cdots, X_p$ have a joint normal distribution, a necessary and sufficient condition that one subset of the random variables and the subset consisting of the remaining variables be independent is that each covariance of a variable from one set and a variable from the other set be* 0.

The necessity follows from that fact that if X_i is from one set and X_j from the other, then for any density (see Section 2.2.3)

$$(20)\quad \sigma_{ij} = \mathscr{E}(X_i - \mu_i)(X_j - \mu_j)$$
$$= \int_{-\infty}^{\infty}\cdots\int_{-\infty}^{\infty}(x_i - \mu_i)(x_j - \mu_j)f(x_1, \cdots, x_q)\, f(x_{q+1}, \cdots, x_p)dx_1\cdots dx_p$$
$$= \int_{-\infty}^{\infty}\cdots\int_{-\infty}^{\infty}(x_i - \mu_i)f(x_1, \cdots, x_q)\, dx_1\cdots dx_q \int_{-\infty}^{\infty}\cdots\int_{-\infty}^{\infty}(x_j - \mu_j)f(x_{q+1}, \cdots, x_p)\, dx_{q+1}\cdots dx_p$$
$$= 0.$$

Since $\sigma_{ij} = \sigma_i\sigma_j\rho_{ij}$, and $\sigma_i \neq 0$ (we tacitly assume that Σ is nonsingular), the condition $\sigma_{ij} = 0$ is equivalent to $\rho_{ij} = 0$. Thus if one set of variates is uncorrelated with the remaining variates, the two sets are independent. It should be emphasized that the implication of independence by lack of correlation depends on the assumption of normality, but the converse is always true.

Let us consider the special case of the bivariate normal distribution. Then $X^{(1)} = X_1$, $X^{(2)} = X_2$, $\mu^{(1)} = \mu_1$, $\mu^{(2)} = \mu_2$, $\Sigma_{11} = \sigma_{11} = \sigma_1^2$, $\Sigma_{22} = \sigma_{22} = \sigma_2^2$, and $\Sigma_{12} = \Sigma_{21} = \sigma_{12} = \sigma_1\sigma_2\rho_{12}$. Thus if X_1 and X_2 have a bivariate normal distribution they are independent if and only if they are uncorrelated. If they are uncorrelated, the marginal distribution of X_i is normal with mean μ_i and variance σ_i^2. The above discussion also proves the following corollary:

COROLLARY 2.4.1. *If $\mathbf{X}$ is distributed according to $N(\boldsymbol{\mu}, \boldsymbol{\Sigma})$ and if a set of components of $\mathbf{X}$ is uncorrelated with the other components, the marginal distribution of the set is multivariate normal with means, variances, and covariances obtained by taking the proper components of $\boldsymbol{\mu}$ and $\boldsymbol{\Sigma}$ respectively.*

Now let us show that the corollary holds even if the two sets are not independent. We partition $\mathbf{X}$, $\boldsymbol{\mu}$, and $\boldsymbol{\Sigma}$ as before. We shall make a nonsingular linear transformation to subvectors

$$\mathbf{Y}^{(1)} = \mathbf{X}^{(1)} + \mathbf{M}\mathbf{X}^{(2)}, \tag{21}$$

$$\mathbf{Y}^{(2)} = \mathbf{X}^{(2)}, \tag{22}$$

choosing $\mathbf{M}$ so that the components of $\mathbf{Y}^{(1)}$ are uncorrelated with the components of $\mathbf{Y}^{(2)} = \mathbf{X}^{(2)}$. The matrix $\mathbf{M}$ must satisfy the equation

$$\begin{aligned} \mathbf{0} &= \mathscr{E}(\mathbf{Y}^{(1)} - \mathscr{E}\mathbf{Y}^{(1)})(\mathbf{Y}^{(2)} - \mathscr{E}\mathbf{Y}^{(2)})' \\ &= \mathscr{E}(\mathbf{X}^{(1)} + \mathbf{M}\mathbf{X}^{(2)} - \mathscr{E}\mathbf{X}^{(1)} - \mathbf{M}\mathscr{E}\mathbf{X}^{(2)})(\mathbf{X}^{(2)} - \mathscr{E}\mathbf{X}^{(2)})' \\ &= \mathscr{E}[(\mathbf{X}^{(1)} - \mathscr{E}\mathbf{X}^{(1)}) + \mathbf{M}(\mathbf{X}^{(2)} - \mathscr{E}\mathbf{X}^{(2)})](\mathbf{X}^{(2)} - \mathscr{E}\mathbf{X}^{(2)})' \\ &= \boldsymbol{\Sigma}_{12} + \mathbf{M}\boldsymbol{\Sigma}_{22}. \end{aligned} \tag{23}$$

Thus $\mathbf{M} = -\boldsymbol{\Sigma}_{12}\boldsymbol{\Sigma}_{22}^{-1}$ and

$$\mathbf{Y}^{(1)} = \mathbf{X}^{(1)} - \boldsymbol{\Sigma}_{12}\boldsymbol{\Sigma}_{22}^{-1}\mathbf{X}^{(2)}. \tag{24}$$

The vector

$$\begin{pmatrix} \mathbf{Y}^{(1)} \\ \mathbf{Y}^{(2)} \end{pmatrix} = \mathbf{Y} = \begin{pmatrix} \mathbf{I} & -\boldsymbol{\Sigma}_{12}\boldsymbol{\Sigma}_{22}^{-1} \\ \mathbf{0} & \mathbf{I} \end{pmatrix} \mathbf{X} \tag{25}$$

is a nonsingular transform of $\mathbf{X}$, and therefore has a normal distribution with

$$\begin{aligned} \mathscr{E}\begin{pmatrix} \mathbf{Y}^{(1)} \\ \mathbf{Y}^{(2)} \end{pmatrix} &= \mathscr{E}\begin{pmatrix} \mathbf{I} & -\boldsymbol{\Sigma}_{12}\boldsymbol{\Sigma}_{22}^{-1} \\ \mathbf{0} & \mathbf{I} \end{pmatrix} \mathbf{X} \\ &= \begin{pmatrix} \mathbf{I} & -\boldsymbol{\Sigma}_{12}\boldsymbol{\Sigma}_{22}^{-1} \\ \mathbf{0} & \mathbf{I} \end{pmatrix} \begin{pmatrix} \boldsymbol{\mu}^{(1)} \\ \boldsymbol{\mu}^{(2)} \end{pmatrix} \\ &= \begin{pmatrix} \boldsymbol{\mu}^{(1)} - \boldsymbol{\Sigma}_{12}\boldsymbol{\Sigma}_{22}^{-1}\boldsymbol{\mu}^{(2)} \\ \boldsymbol{\mu}^{(2)} \end{pmatrix} = \begin{pmatrix} \boldsymbol{\nu}^{(1)} \\ \boldsymbol{\nu}^{(2)} \end{pmatrix} \\ &= \boldsymbol{\nu}, \end{aligned} \tag{26}$$

say, and

$$(27)\quad \mathscr{C}(Y, Y') = \mathscr{E}(Y - \nu)(Y - \nu)'$$
$$= \begin{pmatrix} \mathscr{E}(Y^{(1)} - \nu^{(1)})(Y^{(1)} - \nu^{(1)})' & \mathscr{E}(Y^{(1)} - \nu^{(1)})(Y^{(2)} - \nu^{(2)})' \\ \mathscr{E}(Y^{(2)} - \nu^{(2)})(Y^{(1)} - \nu^{(1)})' & \mathscr{E}(Y^{(2)} - \nu^{(2)})(Y^{(2)} - \nu^{(2)})' \end{pmatrix}$$
$$= \begin{pmatrix} \Sigma_{11} - \Sigma_{12}\Sigma_{22}^{-1}\Sigma_{21} & 0 \\ 0 & \Sigma_{22} \end{pmatrix}$$

since

$$(28)\quad \mathscr{E}(Y^{(1)} - \nu^{(1)})(Y^{(1)} - \nu^{(1)})'$$
$$= \mathscr{E}[(X^{(1)} - \mu^{(1)}) - \Sigma_{12}\Sigma_{22}^{-1}(X^{(2)} - \mu^{(2)})][(X^{(1)} - \mu^{(1)}) - \Sigma_{12}\Sigma_{22}^{-1}(X^{(2)} - \mu^{(2)})]'$$
$$= \Sigma_{11} - \Sigma_{12}\Sigma_{22}^{-1}\Sigma_{21} - \Sigma_{12}\Sigma_{22}^{-1}\Sigma_{21} + \Sigma_{12}\Sigma_{22}^{-1}\Sigma_{22}\Sigma_{22}^{-1}\Sigma_{21}$$
$$= \Sigma_{11} - \Sigma_{12}\Sigma_{22}^{-1}\Sigma_{21}.$$

Thus $Y^{(1)}$ and $Y^{(2)}$ are independent and by Corollary 2.4.1 $X^{(2)} = Y^{(2)}$ has the marginal distribution $N(\mu^{(2)}, \Sigma_{22})$. Because the numbering of the components of X is arbitrary, we can state the following theorem:

THEOREM 2.4.3. *If X is distributed according to $N(\mu, \Sigma)$, the marginal distribution of any set of components of X is multivariate normal with means, variances, and covariances obtained by taking the proper components of μ and Σ respectively.*

Now consider any transformation

$$(29)\qquad Z = DX,$$

where Z has q components and D is a $q \times p$ real matrix. The expected value of Z is

$$(30)\qquad \mathscr{E}Z = D\mu$$

and the covariance matrix is

$$(31)\qquad \mathscr{E}(Z - D\mu)(Z - D\mu)' = D\Sigma D'.$$

The case $q = p$ and D nonsingular has been treated above. If $q \leq p$ and D is of rank q, we can find a $(p - q) \times p$ matrix E such that

$$(32)\qquad \begin{pmatrix} Z \\ W \end{pmatrix} = \begin{pmatrix} D \\ E \end{pmatrix} X$$

is a nonsingular transformation (see Appendix 1, Section 3). Then Z and W have a joint normal distribution and Z has a marginal normal distribution by Theorem 2.4.3. Thus for D of rank q (and X having a nonsingular distribution, that is, a density) we have proved the following theorem:

THEOREM 2.4.4. *If X is distributed according to $N(\boldsymbol{\mu}, \boldsymbol{\Sigma})$, then $Z = DX$ is distributed according to $N(D\boldsymbol{\mu}, D\boldsymbol{\Sigma}D')$, where D is a $q \times p$ matrix of rank $q \leq p$.*

The remainder of this section is devoted to the *singular* or *degenerate* normal distribution and the extension of Theorem 2.4.4 to the case of any matrix D. A singular distribution is a distribution in p-space which is concentrated on a lower dimensional set; that is, the probability associated with any set not intersecting the given set is 0. In the case of the singular normal distribution the mass is concentrated on a given linear set [that is, the intersection of a number of $(p - 1)$-dimensional hyperplanes]. Let y be a set of coordinates in the linear set (the number of coordinates equaling the dimensionality of the linear set); then the "parametric" definition of the linear set can be given as $x = Ay + \boldsymbol{\lambda}$, where A is a $p \times q$ matrix and $\boldsymbol{\lambda}$ is a p-vector. Suppose that Y is normally distributed in the q-dimensional linear set; then we say that

$$X = AY + \boldsymbol{\lambda} \tag{33}$$

has a singular or degenerate normal distribution in p-space. If $\mathscr{E}Y = \boldsymbol{\nu}$, then $\mathscr{E}X = A\boldsymbol{\nu} + \boldsymbol{\lambda} = \boldsymbol{\mu}$, say. If $\mathscr{E}(Y - \boldsymbol{\nu})(Y - \boldsymbol{\nu})' = T$, then

$$\mathscr{E}(X - \boldsymbol{\mu})(X - \boldsymbol{\mu})' = \mathscr{E}A(Y - \boldsymbol{\nu})(Y - \boldsymbol{\nu})'A' = ATA' = \boldsymbol{\Sigma},$$

say. It should be noticed that if $p > q$ then $\boldsymbol{\Sigma}$ is singular and, therefore, has no inverse and thus we cannot write the normal density for X. In fact, X cannot have a density at all because the fact that the probability of any set not intersecting the q-set is 0 implies that the density is 0 almost everywhere.

Now conversely let us see that if X has mean $\boldsymbol{\mu}$ and covariance matrix $\boldsymbol{\Sigma}$ of rank r it can be written as (33) (except for 0 probabilities), where X has an arbitrary distribution and Y of r $(\leq p)$ components has a suitable distribution. If $\boldsymbol{\Sigma}$ is of rank r, there is a $p \times p$ nonsingular matrix B such that

$$B\boldsymbol{\Sigma}B' = \begin{pmatrix} I & 0 \\ 0 & 0 \end{pmatrix}, \tag{34}$$

where the identity is of order r (see Theorem 6 of Appendix 1). The transformation

$$BX = V = \begin{pmatrix} V^{(1)} \\ V^{(2)} \end{pmatrix} \tag{35}$$

defines a random vector V with covariance matrix (34) and a mean vector

$$\mathscr{E}V = B\boldsymbol{\mu} = \boldsymbol{\nu} = \begin{pmatrix} \boldsymbol{\nu}^{(1)} \\ \boldsymbol{\nu}^{(2)} \end{pmatrix}, \tag{36}$$

say. Since the variances of the elements of $\mathbf{V}^{(2)}$ are zero, $\mathbf{V}^{(2)} = \boldsymbol{\nu}^{(2)}$ with probability 1. Now partition

$$\mathbf{B}^{-1} = (\mathbf{C}\ \mathbf{D}), \tag{37}$$

where $\mathbf{C}$ consists of r columns. Then (35) is

$$\mathbf{X} = \mathbf{B}^{-1}\mathbf{V} = (\mathbf{C}\ \mathbf{D})\begin{pmatrix}\mathbf{V}^{(1)}\\ \mathbf{V}^{(2)}\end{pmatrix} = \mathbf{C}\mathbf{V}^{(1)} + \mathbf{D}\mathbf{V}^{(2)}. \tag{38}$$

Thus with probability 1

$$\mathbf{X} = \mathbf{C}\mathbf{V}^{(1)} + \mathbf{D}\boldsymbol{\nu}^{(2)}, \tag{39}$$

which is of the form of (33) with $\mathbf{C}$ as $\mathbf{A}$, $\mathbf{V}^{(1)}$ as $\mathbf{Y}$, and $\mathbf{D}\boldsymbol{\nu}^{(2)}$ as $\boldsymbol{\lambda}$.

Now we give a formal definition of a normal distribution that includes the singular distribution.

DEFINITION 2.4.1. *A random vector $\mathbf{X}$ of p components with $\mathscr{E}\mathbf{X} = \boldsymbol{\mu}$ and $\mathscr{E}(\mathbf{X} - \boldsymbol{\mu})(\mathbf{X} - \boldsymbol{\mu})' = \boldsymbol{\Sigma}$ is said to be normally distributed [or is said to be distributed according to $N(\boldsymbol{\mu}, \boldsymbol{\Sigma})$] if there is a transformation* (33) *where the number of rows of $\mathbf{A}$ is p and the number of columns is the rank of $\boldsymbol{\Sigma}$, say r, and $\mathbf{Y}$ (of r components) has a nonsingular normal distribution, that is, has a density*

$$ke^{-\frac{1}{2}(\mathbf{y}-\boldsymbol{\nu})'\mathbf{T}^{-1}(\mathbf{y}-\boldsymbol{\nu})}. \tag{40}$$

It is clear that if $\boldsymbol{\Sigma}$ has rank p, then $\mathbf{A}$ can be taken to be $\mathbf{I}$ and $\boldsymbol{\lambda}$ to be $\mathbf{0}$; then $\mathbf{X} = \mathbf{Y}$ and Definition 2.4.1 agrees with Section 2.3.

THEOREM 2.4.5. *If $\mathbf{X}$ is distributed according to $N(\boldsymbol{\mu}, \boldsymbol{\Sigma})$, then $\mathbf{Z} = \mathbf{D}\mathbf{X}$ is distributed according to $N(\mathbf{D}\boldsymbol{\mu}, \mathbf{D}\boldsymbol{\Sigma}\mathbf{D}')$.*

This theorem includes the cases where $\mathbf{X}$ may have a nonsingular or a singular distribution and $\mathbf{D}$ may be nonsingular or of rank less than q. Since $\mathbf{X}$ can be represented by (33) where $\mathbf{Y}$ has a nonsingular distribution, $N(\boldsymbol{\nu}, \mathbf{T})$, we can write

$$\mathbf{Z} = \mathbf{D}\mathbf{A}\mathbf{Y} + \mathbf{D}\boldsymbol{\lambda}, \tag{41}$$

where $\mathbf{D}\mathbf{A}$ is $q \times r$. If the rank of $\mathbf{D}\mathbf{A}$ is r the theorem is proved. If the rank is less than r, say s, the covariance matrix of $\mathbf{Z}$,

$$\mathbf{D}\mathbf{A}\mathbf{T}\mathbf{A}'\mathbf{D}' = \mathbf{E}, \tag{42}$$

say, is of rank s. By Theorem 6 of Appendix 1 there is a nonsingular matrix

$$\mathbf{F} = \begin{pmatrix}\mathbf{F}_1\\ \mathbf{F}_2\end{pmatrix} \tag{43}$$

such that

$$(44)\quad \boldsymbol{FEF}' = \begin{pmatrix} \boldsymbol{F}_1\boldsymbol{EF}_1' & \boldsymbol{F}_1\boldsymbol{EF}_2' \\ \boldsymbol{F}_2\boldsymbol{EF}_1' & \boldsymbol{F}_2\boldsymbol{EF}_2' \end{pmatrix}$$

$$= \begin{pmatrix} (\boldsymbol{F}_1\boldsymbol{DA})\boldsymbol{T}(\boldsymbol{F}_1\boldsymbol{DA})' & (\boldsymbol{F}_1\boldsymbol{DA})\boldsymbol{T}(\boldsymbol{F}_2\boldsymbol{DA})' \\ (\boldsymbol{F}_2\boldsymbol{DA})\boldsymbol{T}(\boldsymbol{F}_1\boldsymbol{DA})' & (\boldsymbol{F}_2\boldsymbol{DA})\boldsymbol{T}(\boldsymbol{F}_2\boldsymbol{DA})' \end{pmatrix} = \begin{pmatrix} \boldsymbol{I} & \boldsymbol{0} \\ \boldsymbol{0} & \boldsymbol{0} \end{pmatrix}.$$

Thus $\boldsymbol{F}_1\boldsymbol{DA}$ is of rank s (by the converse of Theorem 1 of Appendix 1) and $\boldsymbol{F}_2\boldsymbol{DA} = \boldsymbol{0}$ because each diagonal element of $(\boldsymbol{F}_2\boldsymbol{DA})\boldsymbol{T}(\boldsymbol{F}_2\boldsymbol{DA})'$ is a quadratic form in a row of $\boldsymbol{F}_2\boldsymbol{DA}$ with positive definite matrix $\boldsymbol{T}$. Thus the covariance matrix of $\boldsymbol{FZ}$ is (44), and

$$(45)\quad \boldsymbol{FZ} = \begin{pmatrix} \boldsymbol{F}_1 \\ \boldsymbol{F}_2 \end{pmatrix}\boldsymbol{DAY} + \boldsymbol{FD\lambda} = \begin{pmatrix} \boldsymbol{F}_1\boldsymbol{DAY} \\ \boldsymbol{0} \end{pmatrix} + \boldsymbol{FD\lambda} = \begin{pmatrix} \boldsymbol{U}_1 \\ \boldsymbol{0} \end{pmatrix} + \boldsymbol{FD\lambda},$$

say. Clearly $\boldsymbol{U}_1$ has a nonsingular normal distribution. Let $\boldsymbol{F}^{-1} = (\boldsymbol{G}_1\boldsymbol{G}_2)$. Then

$$(46)\quad \boldsymbol{Z} = \boldsymbol{G}_1\boldsymbol{U}_1 + \boldsymbol{D\lambda}$$

which is of the form (33). Q.E.D.

All the developments in this section can be illuminated by considering the geometric interpretation put forward in the previous section. The density of $\boldsymbol{X}$ is constant on the ellipsoids (51) of Section 2.3. Since the transformation (2) is a linear transformation (that is, a change of coordinate axes), the density of $\boldsymbol{Y}$ is constant on ellipsoids

$$(47)\quad (\boldsymbol{y} - \boldsymbol{C\mu})'(\boldsymbol{C\Sigma C}')^{-1}(\boldsymbol{y} - \boldsymbol{C\mu}) = k.$$

The marginal distribution of $\boldsymbol{X}^{(1)}$ is the projection of the mass of the distribution of $\boldsymbol{X}$ onto the q-dimensional space of the first q coordinate axes. The surfaces of constant density are again ellipsoids. The projection of mass on any line is clearly normal.

2.5. CONDITIONAL DISTRIBUTIONS AND MULTIPLE CORRELATION COEFFICIENT

2.5.1. Conditional Distributions

In this section we find that conditional distributions derived from joint normal distributions are normal. The conditional distributions are of a particularly simple nature because the means depend only linearly on the variates held fixed and the variances and covariances do not depend at all on the values of the fixed variates. The theory of partial and multiple correlation discussed in this section was originally developed by Karl Pearson (1896) for three variables and extended by Yule (1897a, 1897b).

Let $\boldsymbol{X}$ be distributed according to $N(\boldsymbol{\mu}, \boldsymbol{\Sigma})$ (with $\boldsymbol{\Sigma}$ nonsingular). Let us partition

$$(1) \qquad \boldsymbol{X} = \begin{pmatrix} \boldsymbol{X}^{(1)} \\ \boldsymbol{X}^{(2)} \end{pmatrix}$$

as before into q- and $(p-q)$-component subvectors respectively. We shall use the algebra developed in Section 2.4 here. The joint density of $\boldsymbol{Y}^{(1)} = \boldsymbol{X}^{(1)} - \boldsymbol{\Sigma}_{12}\boldsymbol{\Sigma}_{22}^{-1}\boldsymbol{X}^{(2)}$ and $\boldsymbol{Y}^{(2)} = \boldsymbol{X}^{(2)}$ is

$$n(\boldsymbol{y}^{(1)}|\boldsymbol{\mu}^{(1)} - \boldsymbol{\Sigma}_{12}\boldsymbol{\Sigma}_{22}^{-1}\boldsymbol{\mu}^{(2)}, \boldsymbol{\Sigma}_{11} - \boldsymbol{\Sigma}_{12}\boldsymbol{\Sigma}_{22}^{-1}\boldsymbol{\Sigma}_{21})n(\boldsymbol{y}^{(2)}|\boldsymbol{\mu}^{(2)}, \boldsymbol{\Sigma}_{22}).$$

The density of $\boldsymbol{X}^{(1)}$ and $\boldsymbol{X}^{(2)}$ then can be obtained from this expression by substituting $\boldsymbol{x}^{(1)} - \boldsymbol{\Sigma}_{12}\boldsymbol{\Sigma}_{22}^{-1}\boldsymbol{x}^{(2)}$ for $\boldsymbol{y}^{(1)}$ and $\boldsymbol{x}^{(2)}$ for $\boldsymbol{y}^{(2)}$ (the Jacobian of this transformation being 1); the resulting density of $\boldsymbol{X}^{(1)}$ and $\boldsymbol{X}^{(2)}$ is

$$(2) \quad f(\boldsymbol{x}^{(1)}, \boldsymbol{x}^{(2)}) = \frac{1}{(2\pi)^{\frac{1}{2}q}\sqrt{|\boldsymbol{\Sigma}_{11\cdot 2}|}} \exp\{-\tfrac{1}{2}[(\boldsymbol{x}^{(1)} - \boldsymbol{\mu}^{(1)})$$
$$- \boldsymbol{\Sigma}_{12}\boldsymbol{\Sigma}_{22}^{-1}(\boldsymbol{x}^{(2)} - \boldsymbol{\mu}^{(2)})]'\boldsymbol{\Sigma}_{11\cdot 2}^{-1}[(\boldsymbol{x}^{(1)} - \boldsymbol{\mu}^{(1)}) - \boldsymbol{\Sigma}_{12}\boldsymbol{\Sigma}_{22}^{-1}(\boldsymbol{x}^{(2)} - \boldsymbol{\mu}^{(2)})]\}$$
$$\cdot \frac{1}{(2\pi)^{\frac{1}{2}(p-q)}\sqrt{|\boldsymbol{\Sigma}_{22}|}} \exp[-\tfrac{1}{2}(\boldsymbol{x}^{(2)} - \boldsymbol{\mu}^{(2)})'\boldsymbol{\Sigma}_{22}^{-1}(\boldsymbol{x}^{(2)} - \boldsymbol{\mu}^{(2)})],$$

where

$$(3) \qquad \boldsymbol{\Sigma}_{11\cdot 2} = \boldsymbol{\Sigma}_{11} - \boldsymbol{\Sigma}_{12}\boldsymbol{\Sigma}_{22}^{-1}\boldsymbol{\Sigma}_{21}.$$

This density must be $n(\boldsymbol{x}|\boldsymbol{\mu}, \boldsymbol{\Sigma})$. The conditional density of $\boldsymbol{X}^{(1)}$ given that $\boldsymbol{X}^{(2)} = \boldsymbol{x}^{(2)}$ is the quotient of (2) and the marginal density of $\boldsymbol{X}^{(2)}$ at the point $\boldsymbol{x}^{(2)}$, which is $n(\boldsymbol{x}^{(2)}|\boldsymbol{\mu}^{(2)}, \boldsymbol{\Sigma}_{22})$, the second factor of (2). The quotient is

$$(4) \quad f(\boldsymbol{x}^{(1)}|\boldsymbol{x}^{(2)}) = \frac{1}{(2\pi)^{\frac{1}{2}q}\sqrt{|\boldsymbol{\Sigma}_{11\cdot 2}|}} \exp\{-\tfrac{1}{2}[(\boldsymbol{x}^{(1)} - \boldsymbol{\mu}^{(1)})$$
$$- \boldsymbol{\Sigma}_{12}\boldsymbol{\Sigma}_{22}^{-1}(\boldsymbol{x}^{(2)} - \boldsymbol{\mu}^{(2)})]'\boldsymbol{\Sigma}_{11\cdot 2}^{-1}[(\boldsymbol{x}^{(1)} - \boldsymbol{\mu}^{(1)}) - \boldsymbol{\Sigma}_{12}\boldsymbol{\Sigma}_{22}^{-1}(\boldsymbol{x}^{(2)} - \boldsymbol{\mu}^{(2)})]\}.$$

It is understood that $\boldsymbol{x}^{(2)}$ consists of $p - q$ numbers. The density $f(\boldsymbol{x}^{(1)}|\boldsymbol{x}^{(2)})$ is clearly a q-variate normal density with mean

$$(5) \qquad \mathscr{E}(\boldsymbol{X}^{(1)}|\boldsymbol{x}^{(2)}) = \boldsymbol{\mu}^{(1)} + \boldsymbol{\Sigma}_{12}\boldsymbol{\Sigma}_{22}^{-1}(\boldsymbol{x}^{(2)} - \boldsymbol{\mu}^{(2)}) = \boldsymbol{\nu}(\boldsymbol{x}^{(2)}),$$

say, and covariance matrix

$$(6) \quad \mathscr{E}\{[\boldsymbol{X}^{(1)} - \boldsymbol{\nu}(\boldsymbol{x}^{(2)})][\boldsymbol{X}^{(1)} - \boldsymbol{\nu}(\boldsymbol{x}^{(2)})]'|\boldsymbol{x}^{(2)}\} = \boldsymbol{\Sigma}_{11\cdot 2} = \boldsymbol{\Sigma}_{11} - \boldsymbol{\Sigma}_{12}\boldsymbol{\Sigma}_{22}^{-1}\boldsymbol{\Sigma}_{21}.$$

It should be noted that the mean of $\boldsymbol{X}^{(1)}$ given $\boldsymbol{x}^{(2)}$ is simply a linear function of $\boldsymbol{x}^{(2)}$, and the covariance matrix of $\boldsymbol{X}^{(1)}$ given $\boldsymbol{x}^{(2)}$ does not depend on $\boldsymbol{x}^{(2)}$ at all.

DEFINITION 2.5.1. *The matrix* $\mathbf{\Sigma}_{12}\mathbf{\Sigma}_{22}^{-1}$ *is the matrix of regression coefficients of* $\mathbf{X}^{(1)}$ *on* $\mathbf{x}^{(2)}$.

The $i, j - q$th element of $\mathbf{\Sigma}_{12}\mathbf{\Sigma}_{22}^{-1}$ is often denoted by

$$\beta_{ij \cdot q+1, \cdots, j-1, j+1, \cdots, p}.$$

The vector $\boldsymbol{\mu}^{(1)} + \mathbf{\Sigma}_{12}\mathbf{\Sigma}_{22}^{-1}(\mathbf{x}^{(2)} - \boldsymbol{\mu}^{(2)})$ is often called the *regression function.*

Let $\sigma_{ij \cdot q+1, \cdots, p}$ be the i, jth element of $\mathbf{\Sigma}_{11 \cdot 2}$. We call these *partial covariances.*

DEFINITION 2.5.2.

$$\rho_{ij \cdot q+1, \cdots, p} = \frac{\sigma_{ij \cdot q+1, \cdots, p}}{\sqrt{\sigma_{ii \cdot q+1, \cdots, p}}\sqrt{\sigma_{jj \cdot q+1, \cdots, p}}} \tag{7}$$

is the partial correlation between X_i *and* X_j *holding* $X_{q+1}, \cdots, X_p$ *fixed.*

The numbering of the components of $\mathbf{X}$ is arbitrary and q is arbitrary. Hence, the above serves to define the conditional distribution of any q components of $\mathbf{X}$ given any other $p - q$ components. Indeed, we can consider the marginal distribution of any r components of $\mathbf{X}$ and define the conditional distribution of any q components given the other $r - q$ components.

THEOREM 2.5.1. *Let the components of* $\mathbf{X}$ *be divided into two groups composing the subvectors* $\mathbf{X}^{(1)}$ *and* $\mathbf{X}^{(2)}$. *Suppose the mean* $\boldsymbol{\mu}$ *is similarly divided into* $\boldsymbol{\mu}^{(1)}$ *and* $\boldsymbol{\mu}^{(2)}$, *and suppose the covariance matrix* $\mathbf{\Sigma}$ *of* $\mathbf{X}$ *is divided into* $\mathbf{\Sigma}_{11}$, $\mathbf{\Sigma}_{12}$, $\mathbf{\Sigma}_{22}$, *the covariance matrices of* $\mathbf{X}^{(1)}$, *of* $\mathbf{X}^{(1)}$ *and* $\mathbf{X}^{(2)}$, *and of* $\mathbf{X}^{(2)}$ *respectively. Then if the distribution of* $\mathbf{X}$ *is normal, the conditional distribution of* $\mathbf{X}^{(1)}$ *given* $\mathbf{X}^{(2)} = \mathbf{x}^{(2)}$ *is normal with mean* $\boldsymbol{\mu}^{(1)} + \mathbf{\Sigma}_{12}\mathbf{\Sigma}_{22}^{-1}(\mathbf{x}^{(2)} - \boldsymbol{\mu}^{(2)})$ *and covariance matrix* $\mathbf{\Sigma}_{11} - \mathbf{\Sigma}_{12}\mathbf{\Sigma}_{22}^{-1}\mathbf{\Sigma}_{21}$.

As an example of the above considerations let us consider the bivariate normal distribution and find the conditional distribution of X_1 given $X_2 = x_2$. In this case $\boldsymbol{\mu}^{(1)} = \mu_1$, $\boldsymbol{\mu}^{(2)} = \mu_2$, $\mathbf{\Sigma}_{11} = \sigma_1^2$, $\mathbf{\Sigma}_{12} = \sigma_1\sigma_2\rho$, and $\mathbf{\Sigma}_{22} = \sigma_2^2$. Thus the (1×1) matrix of regression coefficients is $\mathbf{\Sigma}_{12}\mathbf{\Sigma}_{22}^{-1} = \sigma_1\rho/\sigma_2$, and the (1×1) matrix of partial covariances is

$$\mathbf{\Sigma}_{11 \cdot 2} = \mathbf{\Sigma}_{11} - \mathbf{\Sigma}_{12}\mathbf{\Sigma}_{22}^{-1}\mathbf{\Sigma}_{21} = \sigma_1^2 - \sigma_1^2\sigma_2^2\rho^2/\sigma_2^2 = \sigma_1^2(1 - \rho^2).$$

Thus the density of X_1 given x_2 is $n\left[x_1 \middle| \mu_1 + \frac{\sigma_1}{\sigma_2}\rho(x_2 - \mu_2), \sigma_1^2(1 - \rho^2)\right]$.

It will be noticed that the mean of this conditional distribution increases with x_2 when ρ is positive and decreases with x_2 when ρ is negative.

A geometrical interpretation of the theory is enlightening. The density $f(x_1, x_2)$ can be thought of as a surface $z = f(x_1, x_2)$ over the x_1, x_2-plane.

If we intersect this surface with the plane $x_2 = c$, we obtain a curve $z = f(x_1, c)$ over the line $x_2 = c$ in the x_1, x_2-plane. The ordinate of this curve is proportional to the conditional density of X_1 given $x_2 = c$; that is, it is proportional to the ordinate of the curve of a univariate normal distribution. In the more general case it is convenient to consider the ellipsoids of constant density in the p-dimensional space. The surfaces of constant density of $f(x_1, \cdots, x_q | c_{q+1}, \cdots, c_p)$ are the intersections of the surfaces of constant density of $f(x_1, \cdots, x_p)$ and the hyperplanes $x_{q+1} = c_{q+1}, \cdots, x_p = c_p$; these are again ellipsoids.

Further clarification of these ideas may be had by consideration of an actual population which is idealized by a normal distribution. Consider, for example, a population of father-son pairs. If the population is reasonably homogeneous, the heights of fathers and the heights of corresponding sons have approximately a normal distribution (over a certain range). A conditional distribution may be obtained by considering sons of all fathers whose height is, say, 5 feet, 9 inches (to the accuracy of measurement); the heights of these sons will have an approximate univariate normal distribution. The mean of this normal distribution will differ from the mean of the heights of sons whose fathers' heights are 5 feet, 4 inches, say, but the variances will be the same.

We could also consider triplets of observations, the height of a father, height of the oldest son, and height of the next oldest son. The collection of heights of two sons given that the fathers' heights are 5 feet, 9 inches is a conditional distribution of two variables; the correlation between the heights of oldest and next oldest sons is a partial correlation coefficient. Holding the fathers' heights constant eliminates the effect of heredity from fathers; however, one would expect that the partial correlation coefficient would be positive since the effect of mothers' heredity and environmental factors would tend to cause brothers' heights to vary similarly.

As we have remarked above, any conditional distribution obtained from a normal distribution is normal with the mean a linear function of the variables held fixed and the covariance matrix constant. In the case of nonnormal distributions the conditional distribution of one set of variates on another does not usually have these properties. However, one can construct nonnormal distributions such that some conditional distributions have these properties. This can be done by taking as the density of $\boldsymbol{X}$ the product $n(\boldsymbol{x}^{(1)} | \boldsymbol{\mu}^{(1)} + \boldsymbol{\beta}(\boldsymbol{x}^{(2)} - \boldsymbol{\mu}^{(2)}), \boldsymbol{\Sigma}_{11 \cdot 2}) f(\boldsymbol{x}^{(2)})$ where $f(\boldsymbol{x}^{(2)})$ is an arbitrary density.

2.5.2. The Multiple Correlation Coefficient

We again consider $\boldsymbol{X}$ partitioned into $\boldsymbol{X}^{(1)}$ and $\boldsymbol{X}^{(2)}$. We shall consider some properties of $\boldsymbol{\Sigma}_{12}\boldsymbol{\Sigma}_{22}^{-1}\boldsymbol{X}^{(2)}$. Since we are interested only in functions of the covariances we shall assume $\boldsymbol{\mu} = \mathbf{0}$ (that is, $X - \boldsymbol{\mu}$ is replaced by X). We choose a component X_i of $X^{(1)}$. Then

$$\mathscr{E}(X_i | x^{(2)}) = \boldsymbol{\beta} x^{(2)}, \tag{8}$$

where

$$\boldsymbol{\beta} = \boldsymbol{\sigma}_{(i)} \boldsymbol{\Sigma}_{22}^{-1}, \tag{9}$$

and $\boldsymbol{\sigma}_{(i)}$ is the ith row of $\boldsymbol{\Sigma}_{12}$ defined by

$$\boldsymbol{\Sigma} = \begin{pmatrix} \boldsymbol{\Sigma}_{11} & \boldsymbol{\Sigma}_{12} \\ \boldsymbol{\Sigma}_{21} & \boldsymbol{\Sigma}_{22} \end{pmatrix}. \tag{10}$$

Now consider the linear function of random variables $\boldsymbol{\beta} X^{(2)}$. Since the covariance between $X_i - \boldsymbol{\beta} X^{(2)}$ and $X^{(2)}$ is

$$\mathscr{E}(X_i - \boldsymbol{\beta} X^{(2)}) X^{(2)\prime} = \boldsymbol{\sigma}_{(i)} - \boldsymbol{\beta} \boldsymbol{\Sigma}_{22} = \boldsymbol{\sigma}_{(i)} - \boldsymbol{\sigma}_{(i)} \boldsymbol{\Sigma}_{22}^{-1} \boldsymbol{\Sigma}_{22} = \mathbf{0}, \tag{11}$$

these two variates are independent. Now let us find the linear function $\boldsymbol{\alpha} X^{(2)}$ for which $(X_i - \boldsymbol{\alpha} X^{(2)})$ has minimum variance. Since $\mathscr{E}Z^2 = \mathscr{E}ZZ'$ for scalar Z, the variance is

$$\begin{aligned} \mathscr{E}(X_i - \boldsymbol{\alpha} X^{(2)})^2 &= \mathscr{E}[(X_i - \boldsymbol{\beta} X^{(2)}) + (\boldsymbol{\beta} - \boldsymbol{\alpha}) X^{(2)}]^2 \\ &= \mathscr{E}[(X_i - \boldsymbol{\beta} X^{(2)}) + (\boldsymbol{\beta} - \boldsymbol{\alpha}) X^{(2)}][(X_i' - X^{(2)\prime} \boldsymbol{\beta}') + X^{(2)\prime}(\boldsymbol{\beta}' - \boldsymbol{\alpha}')] \\ &= \mathscr{E}(X_i - \boldsymbol{\beta} X^{(2)})(X_i' - X^{(2)\prime}\boldsymbol{\beta}') + (\boldsymbol{\beta} - \boldsymbol{\alpha}) \mathscr{E} X^{(2)} X^{(2)\prime}(\boldsymbol{\beta}' - \boldsymbol{\alpha}') \end{aligned} \tag{12}$$

by (11). But (12) is

$$(\sigma_{ii} - \boldsymbol{\sigma}_{(i)} \boldsymbol{\Sigma}_{22}^{-1} \boldsymbol{\Sigma}_{22} \boldsymbol{\Sigma}_{22}^{-1} \boldsymbol{\sigma}_{(i)}') + (\boldsymbol{\beta} - \boldsymbol{\alpha}) \boldsymbol{\Sigma}_{22} (\boldsymbol{\beta} - \boldsymbol{\alpha})'. \tag{13}$$

Since $\boldsymbol{\Sigma}_{22}$ is positive definite, the second term of (13) is nonnegative and attains its minimum (namely, zero) if $\boldsymbol{\alpha} = \boldsymbol{\beta}$. Thus the regression function is the function of $X^{(2)}$ such that $(X_i - \boldsymbol{\alpha} X^{(2)})$ has minimum variance.

Now let us show that the maximum correlation between X_i and $\boldsymbol{\alpha} X^{(2)}$ is given by $\boldsymbol{\alpha} = \boldsymbol{\beta}$. We know that

$$\mathscr{E}(X_i - \boldsymbol{\beta} X^{(2)})^2 \leq \mathscr{E}(X_i - c\boldsymbol{\alpha} X^{(2)})^2 \tag{14}$$

for any c and $\boldsymbol{\alpha}$. Hence,

$$\sigma_{ii} + \mathscr{E}(\boldsymbol{\beta} X^{(2)})^2 - 2\mathscr{E} X_i \boldsymbol{\beta} X^{(2)} \leq \sigma_{ii} + c^2 \mathscr{E}(\boldsymbol{\alpha} X^{(2)})^2 - 2c\mathscr{E} X_i \boldsymbol{\alpha} X^{(2)}. \tag{15}$$

Thus

$$\begin{aligned} 2 \frac{\mathscr{E} X_i \boldsymbol{\beta} X^{(2)}}{\sqrt{\sigma_{ii}} \sqrt{\mathscr{E}(\boldsymbol{\beta} X^{(2)})^2}} &- \frac{\mathscr{E}(\boldsymbol{\beta} X^{(2)})^2}{\sqrt{\sigma_{ii}} \sqrt{\mathscr{E}(\boldsymbol{\beta} X^{(2)})^2}} \\ &\geq 2c \frac{\mathscr{E} X_i \boldsymbol{\alpha} X^{(2)}}{\sqrt{\sigma_{ii}} \sqrt{\mathscr{E}(\boldsymbol{\beta} X^{(2)})^2}} - c^2 \frac{\mathscr{E}(\boldsymbol{\alpha} X^{(2)})^2}{\sqrt{\sigma_{ii}} \sqrt{\mathscr{E}(\boldsymbol{\beta} X^{(2)})^2}}. \end{aligned} \tag{16}$$

Now choose

$$c^2 = \frac{\mathscr{E}(\boldsymbol{\beta} X^{(2)})^2}{\mathscr{E}(\boldsymbol{\alpha} X^{(2)})^2}. \tag{17}$$

Then (16) becomes

$$\frac{\mathscr{E} X_i \boldsymbol{\beta} X^{(2)}}{\sqrt{\sigma_{ii}}\sqrt{\mathscr{E}(\boldsymbol{\beta} X^{(2)})^2}} \geq \frac{\mathscr{E} X_i \boldsymbol{\alpha} X^{(2)}}{\sqrt{\sigma_{ii}}\sqrt{\mathscr{E}(\boldsymbol{\alpha} X^{(2)})^2}}. \tag{18}$$

This finishes the proof of the following theorem:

THEOREM 2.5.2. *Let X be distributed according to $N(\boldsymbol{\mu}, \boldsymbol{\Sigma})$. Let $X' = (X^{(1)\prime} X^{(2)\prime})$ and $\boldsymbol{\Sigma} = \begin{pmatrix} \boldsymbol{\Sigma}_{11} & \boldsymbol{\Sigma}_{12} \\ \boldsymbol{\Sigma}_{21} & \boldsymbol{\Sigma}_{22} \end{pmatrix}$, and let $\boldsymbol{\beta}$ be the ith row of $\boldsymbol{\Sigma}_{12}\boldsymbol{\Sigma}_{22}^{-1}$ $(i = 1, \cdots, q)$. Of all linear combinations $\boldsymbol{\alpha} X^{(2)}$, the combination that minimizes the variance of $X_i - \boldsymbol{\alpha} X^{(2)}$ and that maximizes the correlation between X_i and $\boldsymbol{\alpha} X^{(2)}$ is the linear combination $\boldsymbol{\beta} X^{(2)}$.*

DEFINITION 2.5.3. *The maximum correlation between X_i and the linear combination $\boldsymbol{\alpha} X^{(2)}$ is called the multiple correlation coefficient between X_i and $X^{(2)}$.*

It follows that this is

$$\begin{aligned} \bar{R}_{i \cdot q+1, \cdots, p} &= \frac{\mathscr{E} \boldsymbol{\beta} X^{(2)} X_i}{\sqrt{\sigma_{ii}}\sqrt{\mathscr{E} \boldsymbol{\beta} X^{(2)} X^{(2)\prime} \boldsymbol{\beta}'}} = \frac{\boldsymbol{\sigma}_{(i)} \boldsymbol{\Sigma}_{22}^{-1} \boldsymbol{\sigma}_{(i)}'}{\sqrt{\sigma_{ii}}\sqrt{\boldsymbol{\sigma}_{(i)} \boldsymbol{\Sigma}_{22}^{-1} \boldsymbol{\sigma}_{(i)}'}} \\ &= \frac{\sqrt{\boldsymbol{\sigma}_{(i)} \boldsymbol{\Sigma}_{22}^{-1} \boldsymbol{\sigma}_{(i)}'}}{\sqrt{\sigma_{ii}}}. \end{aligned} \tag{19}$$

A useful formula is

$$1 - \bar{R}^2_{i \cdot q+1, \cdots, p} = \frac{\sigma_{ii} - \boldsymbol{\sigma}_{(i)} \boldsymbol{\Sigma}_{22}^{-1} \boldsymbol{\sigma}_{(i)}'}{\sigma_{ii}} = \frac{|\boldsymbol{\Sigma}^*|}{\sigma_{ii} |\boldsymbol{\Sigma}_{22}|}, \tag{20}$$

where

$$\boldsymbol{\Sigma}^* = \begin{pmatrix} \sigma_{ii} & \boldsymbol{\sigma}_{(i)} \\ \boldsymbol{\sigma}_{(i)}' & \boldsymbol{\Sigma}_{22} \end{pmatrix}. \tag{21}$$

Since

$$\sigma_{ii \cdot q+1, \cdots, p} = \sigma_{ii} - \boldsymbol{\sigma}_{(i)} \boldsymbol{\Sigma}_{22}^{-1} \boldsymbol{\sigma}_{(i)}', \tag{22}$$

it follows that

$$\sigma_{ii \cdot q+1, \cdots, p} = (1 - \bar{R}^2_{i \cdot q+1, \cdots, p}) \sigma_{ii}. \tag{23}$$

This shows incidentally that any conditional variance of a component of X cannot be greater than the variance. In fact, the larger $\bar{R}_{i \cdot q+1, \cdots, p}$ is, the more is the reduction in variance by going to the conditional distribution.

2.5.3. Some Formulas for Partial Correlations

We now consider relations between several conditional distributions obtained by holding several different sets of variates fixed. These relations are useful because they enable us to compute one set of conditional parameters from another set. A very special case is

$$\rho_{12\cdot 3} = \frac{\rho_{12} - \rho_{13}\rho_{23}}{\sqrt{1-\rho_{23}^2}\,\sqrt{1-\rho_{13}^2}}; \tag{24}$$

this follows from (7) when $p = 3$ and $q = 2$. We shall now find a generalization of this result. The derivation is tedious, but is given here for completeness.

Let

$$X = \begin{pmatrix} X^{(1)} \\ X^{(2)} \\ X^{(3)} \end{pmatrix}, \tag{25}$$

where $X^{(1)}$ is of p_1 components, $X^{(2)}$ of p_2 components, and $X^{(3)}$ of p_3 components. Suppose we have the conditional distribution of $X^{(1)}$ and $X^{(2)}$ given $X^{(3)} = x^{(3)}$; how do we find the conditional distribution of $X^{(1)}$ given $X^{(2)} = x^{(2)}$ and $X^{(3)} = x^{(3)}$? We know that the conditional mean and covariance of $\begin{pmatrix} X^{(1)} \\ X^{(2)} \end{pmatrix}$ are

$$\begin{aligned} \mathscr{E}\left[\begin{pmatrix} X^{(1)} \\ X^{(2)} \end{pmatrix} \middle| x^{(3)}\right] &= \begin{pmatrix} \mu^{(1)} \\ \mu^{(2)} \end{pmatrix} + \begin{pmatrix} \Sigma_{13}\Sigma_{33}^{-1} \\ \Sigma_{23}\Sigma_{33}^{-1} \end{pmatrix}(x^{(3)} - \mu^{(3)}) \\ &= \begin{pmatrix} \mu^{(1)} \\ \mu^{(2)} \end{pmatrix} + \mathrm{B}_{\cdot 3}(x^{(3)} - \mu^{(3)}), \end{aligned} \tag{26}$$

and

$$\begin{aligned} \mathscr{C}\left[\begin{pmatrix} X^{(1)} \\ X^{(2)} \end{pmatrix}, \begin{pmatrix} X^{(1)} \\ X^{(2)} \end{pmatrix}' \middle| x^{(3)}\right] &= \begin{pmatrix} \Sigma_{11} & \Sigma_{12} \\ \Sigma_{21} & \Sigma_{22} \end{pmatrix} - \begin{pmatrix} \Sigma_{13} \\ \Sigma_{23} \end{pmatrix} \Sigma_{33}^{-1}(\Sigma_{31} \quad \Sigma_{32}) \\ &= \Sigma_{\cdot 3} = \begin{pmatrix} \Sigma_{11\cdot 3} & \Sigma_{12\cdot 3} \\ \Sigma_{21\cdot 3} & \Sigma_{22\cdot 3} \end{pmatrix}, \end{aligned} \tag{27}$$

say. Similarly,

$$\mathscr{E}\left[X^{(1)} \middle| \begin{pmatrix} x^{(2)} \\ x^{(3)} \end{pmatrix}\right] = \mu^{(1)} + (\Sigma_{12} \quad \Sigma_{13}) \begin{pmatrix} \Sigma_{22} & \Sigma_{23} \\ \Sigma_{32} & \Sigma_{33} \end{pmatrix}^{-1} \begin{pmatrix} x^{(2)} - \mu^{(2)} \\ x^{(3)} - \mu^{(3)} \end{pmatrix}, \tag{28}$$

$$\begin{aligned} \mathscr{C}\left[X^{(1)}, X^{(1)\prime} \middle| \begin{pmatrix} x^{(2)} \\ x^{(3)} \end{pmatrix}\right] &= \Sigma_{11} - (\Sigma_{12} \quad \Sigma_{13}) \begin{pmatrix} \Sigma_{22} & \Sigma_{23} \\ \Sigma_{32} & \Sigma_{33} \end{pmatrix}^{-1} \begin{pmatrix} \Sigma_{21} \\ \Sigma_{31} \end{pmatrix} \\ &= \Sigma_{11\cdot 23}. \end{aligned} \tag{29}$$

We want to express (28) and (29) in terms of (26) and (27). We treat the distribution of $\begin{pmatrix} X^{(1)} \\ X^{(2)} \end{pmatrix}$ given $x^{(3)}$ as a multivariate normal distribution. Then we know that

$$(30)\quad \begin{aligned} \mathscr{E}(X^{(1)}|x^{(2)}, x^{(3)}) &= \mathscr{E}(X^{(1)}|x^{(3)}) + \Sigma_{12\cdot 3}\Sigma_{22\cdot 3}^{-1}[x^{(2)} - \mathscr{E}(X^{(2)}|x^{(3)})] \\ &= \mu^{(1)} + \Sigma_{13}\Sigma_{33}^{-1}(x^{(3)} - \mu^{(3)}) + (\Sigma_{12} - \Sigma_{13}\Sigma_{33}^{-1}\Sigma_{32}) \\ &\qquad (\Sigma_{22} - \Sigma_{23}\Sigma_{33}^{-1}\Sigma_{32})^{-1}[x^{(2)} - \mu^{(2)} - \Sigma_{23}\Sigma_{33}^{-1}(x^{(3)} - \mu^{(3)})] \end{aligned}$$

and

$$(31)\quad \begin{aligned} \mathscr{C}[X^{(1)}, X^{(1)\prime}|x^{(2)}, x^{(3)}] &= \Sigma_{11\cdot 3} - \Sigma_{12\cdot 3}(\Sigma_{22\cdot 3})^{-1}\Sigma_{21\cdot 3} \\ &= (\Sigma_{11} - \Sigma_{13}\Sigma_{33}^{-1}\Sigma_{31}) \\ &\quad - (\Sigma_{12} - \Sigma_{13}\Sigma_{33}^{-1}\Sigma_{32})(\Sigma_{22} - \Sigma_{23}\Sigma_{33}^{-1}\Sigma_{32})^{-1}(\Sigma_{21} - \Sigma_{23}\Sigma_{33}^{-1}\Sigma_{31}). \end{aligned}$$

We know that (28) and (30) are the same and (29) and (31) are the same. This could also be checked algebraically.

In particular, for $p_1 = q$, $p_2 = 1$, and $p_3 = p - q - 1$, we obtain

$$(32)\quad \sigma_{ij\cdot q+1,\cdots,p} = \sigma_{ij\cdot q+2,\cdots,p} - \frac{\sigma_{i,q+1\cdot q+2,\cdots,p}\,\sigma_{j,q+1\cdot q+2,\cdots,p}}{\sigma_{q+1,q+1\cdot q+2,\cdots,p}}, \qquad i, j = 1, \cdots, q.$$

Since

$$(33)\quad \sigma_{ii\cdot q+1,\cdots,p} = \sigma_{ii\cdot q+2,\cdots,p}(1 - \rho^2_{i,q+1\cdot q+2,\cdots,p}) .$$

we obtain

$$(34)\quad \rho_{ij\cdot q+1,\cdots,p} = \frac{\rho_{ij\cdot q+2,\cdots,p} - \rho_{i,q+1\cdot q+2,\cdots,p}\,\rho_{j,q+1\cdot q+2,\cdots,p}}{\sqrt{1 - \rho^2_{i,q+1\cdot q+2,\cdots,p}}\sqrt{1 - \rho^2_{j,q+1\cdot q+2,\cdots,p}}}.$$

This is a useful recursion formula to compute from $\{\rho_{ij}\}$ in succession $\{\rho_{ij\cdot p}\}$, $\{\rho_{ij\cdot p-1,p}\}$, $\cdots$, $\rho_{12\cdot 3,\cdots,p}$.

2.6. THE CHARACTERISTIC FUNCTION; MOMENTS

2.6.1. The Characteristic Function

The characteristic function of a multivariate normal distribution has a form similar to the density function. From the characteristic function moments and semi-invariants can be found easily.

DEFINITION 2.6.1. *The characteristic function of a random vector* X *is*

$$(1)\quad \phi(t) = \mathscr{E}e^{it'X}$$

defined for every real vector t.

To make this definition meaningful we need to define the expected value of a complex-valued function of a random vector.

DEFINITION 2.6.2. *Let the complex-valued function $g(\boldsymbol{x})$ be written as $g(\boldsymbol{x}) = g_1(\boldsymbol{x}) + ig_2(\boldsymbol{x})$, where $g_1(\boldsymbol{x})$ and $g_2(\boldsymbol{x})$ are real-valued. Then the expected value of $g(\boldsymbol{X})$ is*

$$\mathscr{E}g(\boldsymbol{X}) = \mathscr{E}g_1(\boldsymbol{X}) + i\mathscr{E}g_2(\boldsymbol{X}). \tag{2}$$

In particular,

$$\mathscr{E}e^{i\boldsymbol{t}'\boldsymbol{X}} = \mathscr{E}\cos \boldsymbol{t}'\boldsymbol{X} + i\mathscr{E}\sin \boldsymbol{t}'\boldsymbol{X}. \tag{3}$$

To evaluate the characteristic function of a vector $\boldsymbol{X}$ it is often convenient to use the following lemma:

LEMMA 2.6.1. *Let $\boldsymbol{X}' = (\boldsymbol{X}^{(1)\prime} \quad \boldsymbol{X}^{(2)\prime})$. If $\boldsymbol{X}^{(1)}$ and $\boldsymbol{X}^{(2)}$ are independent and $g(\boldsymbol{x}) = g^{(1)}(\boldsymbol{x}^{(1)})g^{(2)}(\boldsymbol{x}^{(2)})$, then*

$$\mathscr{E}g(\boldsymbol{X}) = \mathscr{E}g^{(1)}(\boldsymbol{X}^{(1)})\mathscr{E}g^{(2)}(\boldsymbol{X}^{(2)}). \tag{4}$$

Proof. If $g(\boldsymbol{x})$ is real-valued and $\boldsymbol{X}$ has a density,

$$\begin{aligned}
\mathscr{E}g(\boldsymbol{X}) &= \int_{-\infty}^{\infty} \cdots \int_{-\infty}^{\infty} g(\boldsymbol{x})f(\boldsymbol{x})\,dx_1 \cdots dx_p \\
&= \int_{-\infty}^{\infty} \cdots \int_{-\infty}^{\infty} g^{(1)}(\boldsymbol{x}^{(1)})g^{(2)}(\boldsymbol{x}^{(2)})f^{(1)}(\boldsymbol{x}^{(1)})f^{(2)}(\boldsymbol{x}^{(2)})\,dx_1 \cdots dx_p \\
&= \int_{-\infty}^{\infty} \cdots \int_{-\infty}^{\infty} g^{(1)}(\boldsymbol{x}^{(1)})f^{(1)}(\boldsymbol{x}^{(1)})\,dx_1 \cdots dx_q \\
&\qquad \int_{-\infty}^{\infty} \cdots \int_{-\infty}^{\infty} g^{(2)}(\boldsymbol{x}^{(2)})f^{(2)}(\boldsymbol{x}^{(2)})\,dx_{q+1} \cdots dx_p \\
&= \mathscr{E}g^{(1)}(\boldsymbol{X}^{(1)})\mathscr{E}g^{(2)}(\boldsymbol{X}^{(2)}).
\end{aligned} \tag{5}$$

If $g(\boldsymbol{x})$ is complex-valued,

$$\begin{aligned}
g(\boldsymbol{x}) &= [g_1^{(1)}(\boldsymbol{x}^{(1)}) + ig_2^{(1)}(\boldsymbol{x}^{(1)})][g_1^{(2)}(\boldsymbol{x}^{(2)}) + ig_2^{(2)}(\boldsymbol{x}^{(2)})] \\
&= g_1^{(1)}(\boldsymbol{x}^{(1)})g_1^{(2)}(\boldsymbol{x}^{(2)}) - g_2^{(1)}(\boldsymbol{x}^{(1)})g_2^{(2)}(\boldsymbol{x}^{(2)}) \\
&\qquad + i[g_2^{(1)}(\boldsymbol{x}^{(1)})g_1^{(2)}(\boldsymbol{x}^{(2)}) + g_1^{(1)}(\boldsymbol{x}^{(1)})g_2^{(2)}(\boldsymbol{x}^{(2)})].
\end{aligned} \tag{6}$$

Then

$$\begin{aligned}
\mathscr{E}g(\boldsymbol{X}) &= \mathscr{E}[g_1^{(1)}(\boldsymbol{X}^{(1)})g_1^{(2)}(\boldsymbol{X}^{(2)}) - g_2^{(1)}(\boldsymbol{X}^{(1)})g_2^{(2)}(\boldsymbol{X}^{(2)})] \\
&\qquad + i\mathscr{E}[g_2^{(1)}(\boldsymbol{X}^{(1)})g_1^{(2)}(\boldsymbol{X}^{(2)}) + g_1^{(1)}(\boldsymbol{X}^{(1)})g_2^{(2)}(\boldsymbol{X}^{(2)})] \\
&= \mathscr{E}g_1^{(1)}(\boldsymbol{X}^{(1)})\mathscr{E}g_1^{(2)}(\boldsymbol{X}^{(2)}) - \mathscr{E}g_2^{(1)}(\boldsymbol{X}^{(1)})\mathscr{E}g_2^{(2)}(\boldsymbol{X}^{(2)}) \\
&\qquad + i[\mathscr{E}g_2^{(1)}(\boldsymbol{X}^{(1)})\mathscr{E}g_1^{(2)}(\boldsymbol{X}^{(2)}) + \mathscr{E}g_1^{(1)}(\boldsymbol{X}^{(1)})\mathscr{E}g_2^{(2)}(\boldsymbol{X}^{(2)})] \\
&= [\mathscr{E}g_1^{(1)}(\boldsymbol{X}^{(1)}) + i\mathscr{E}g_2^{(1)}(\boldsymbol{X}^{(1)})][\mathscr{E}g_1^{(2)}(\boldsymbol{X}^{(2)}) + i\mathscr{E}g_2^{(2)}(\boldsymbol{X}^{(2)})] \\
&= \mathscr{E}g^{(1)}(\boldsymbol{X}^{(1)})\mathscr{E}g^{(2)}(\boldsymbol{X}^{(2)}).
\end{aligned} \tag{7}$$

By applying Lemma 2.6.1 successively to $g(\mathbf{x}) = e^{i\mathbf{t}'\mathbf{x}}$, we derive

LEMMA 2.6.2. *If the components of* $\mathbf{X}$ *are independently distributed*

$$\mathscr{E}e^{i\mathbf{t}'\mathbf{X}} = \prod_{j=1}^{p} \mathscr{E}e^{it_jX_j}. \tag{8}$$

We now find the characteristic function of a random vector with a normal distribution.

THEOREM 2.6.1. *The characteristic function of* $\mathbf{X}$ *which is distributed according to* $N(\boldsymbol{\mu}, \boldsymbol{\Sigma})$ is

$$\phi(\mathbf{t}) = \mathscr{E}e^{i\mathbf{t}'\mathbf{X}} = e^{i\mathbf{t}'\boldsymbol{\mu} - \frac{1}{2}\mathbf{t}'\boldsymbol{\Sigma}\mathbf{t}} \tag{9}$$

for every real vector $\mathbf{t}$.

Proof. From Corollary 4 of Appendix 1 we know there is a nonsingular matrix $\mathbf{C}$ such that

$$\mathbf{C}'\boldsymbol{\Sigma}^{-1}\mathbf{C} = \mathbf{I}. \tag{10}$$

Thus

$$\begin{aligned} \boldsymbol{\Sigma}^{-1} &= \mathbf{C}'^{-1}\mathbf{C}^{-1} \\ &= (\mathbf{C}\mathbf{C}')^{-1}. \end{aligned} \tag{11}$$

Let

$$\mathbf{X} - \boldsymbol{\mu} = \mathbf{C}\mathbf{Y}. \tag{12}$$

Then $\mathbf{Y}$ is distributed according to $N(\mathbf{0}, \mathbf{I})$.

Now the characteristic function of $\mathbf{Y}$ is

$$\Psi(\mathbf{u}) = \mathscr{E}e^{i\mathbf{u}'\mathbf{Y}} = \prod_{j=1}^{p} \mathscr{E}e^{iu_jY_j}. \tag{13}$$

Since Y_j is distributed according to $N(0, 1)$,

$$\Psi(\mathbf{u}) = \prod_{j=1}^{p} e^{-\frac{1}{2}u_j^2} = e^{-\frac{1}{2}\mathbf{u}'\mathbf{u}}. \tag{14}$$

Thus

$$\begin{aligned} \phi(\mathbf{t}) &= \mathscr{E}e^{i\mathbf{t}'\mathbf{X}} = \mathscr{E}e^{i\mathbf{t}'(\mathbf{C}\mathbf{Y}+\boldsymbol{\mu})} \\ &= e^{i\mathbf{t}'\boldsymbol{\mu}}\mathscr{E}e^{i\mathbf{t}'\mathbf{C}\mathbf{Y}} \\ &= e^{i\mathbf{t}'\boldsymbol{\mu}}e^{-\frac{1}{2}(\mathbf{t}'\mathbf{C})(\mathbf{t}'\mathbf{C})'} \end{aligned} \tag{15}$$

for $\mathbf{t}'\mathbf{C} = \mathbf{u}'$; the third equality is easily verified by writing both sides of this equality as integrals. But this is

$$\begin{aligned} \phi(\mathbf{t}) &= e^{i\mathbf{t}'\boldsymbol{\mu}}e^{-\frac{1}{2}\mathbf{t}'\mathbf{C}\mathbf{C}'\mathbf{t}} \\ &= e^{i\mathbf{t}'\boldsymbol{\mu} - \frac{1}{2}\mathbf{t}'\boldsymbol{\Sigma}\mathbf{t}} \end{aligned} \tag{16}$$

by (11). This proves the theorem.

The characteristic function of the normal distribution is very useful. It is clear that we can use this method of proof to demonstrate the results of Section 2.4. If $\boldsymbol{Z} = \boldsymbol{D}\boldsymbol{X}$, then the characteristic function of $\boldsymbol{Z}$ is

$$(17)\qquad \begin{aligned} \mathscr{E} e^{it'Z} &= \mathscr{E} e^{it'DX} = \mathscr{E} e^{i(D't)'X} \\ &= e^{i(D't)'\mu - \frac{1}{2}(D't)'\Sigma(D't)} \\ &= e^{it'(D\mu) - \frac{1}{2}t'(D\Sigma D')t}, \end{aligned}$$

which is the characteristic function of $N(\boldsymbol{D}\boldsymbol{\mu}, \boldsymbol{D}\boldsymbol{\Sigma}\boldsymbol{D}')$ (by Theorem 2.6.1).

It is interesting to use the characteristic function to show that it is only the multivariate normal distribution that has the property that every linear combination of variates is normally distributed. Consider a vector $\boldsymbol{Y}$ of p components with density $f(\boldsymbol{y})$ and characteristic function

$$(18)\qquad \Psi(\boldsymbol{u}) = \mathscr{E} e^{iu'Y} = \int_{-\infty}^{\infty} \cdots \int_{-\infty}^{\infty} e^{iu'y} f(\boldsymbol{y})\, dy_1 \cdots dy_p,$$

and suppose the mean of $\boldsymbol{Y}$ is $\boldsymbol{\mu}$ and the covariance matrix is $\boldsymbol{\Sigma}$. Suppose $\boldsymbol{u}'\boldsymbol{Y}$ is normally distributed for every $\boldsymbol{u}$. Then the characteristic function of such a linear combination is

$$(19)\qquad \mathscr{E} e^{itu'Y} = e^{itu'\mu - \frac{1}{2}t^2 u'\Sigma u}.$$

Now set $t = 1$. Since the right-hand side is then the characteristic function of $N(\boldsymbol{\mu}, \boldsymbol{\Sigma})$, the result is proved (by Theorems 2.6.1 and 2.6.3).

THEOREM 2.6.2. *If every linear combination of the components of a vector* $\boldsymbol{Y}$ *is normally distributed, then* $\boldsymbol{Y}$ *is normally distributed.*

It might be pointed out in passing that it is essential that *every* linear combination be normally distributed for Theorem 2.6.2 to hold. For instance, if $\boldsymbol{Y}' = (Y_1, Y_2)$ and Y_1 and Y_2 are not independent, then Y_1 and Y_2 can each have a marginal normal distribution. An example is most easily given geometrically. Let X_1, X_2 have a joint normal distribution with means 0. Move the same mass in adjoining Figure 1 from rectangle A to C and from B to D. It will be seen that the resulting distribution of $\boldsymbol{Y}$ is such that the marginal distributions of Y_1 and Y_2 are the same as of X_1 and X_2 respectively, which are normal, and yet the joint distribution of Y_1 and Y_2 is not normal.

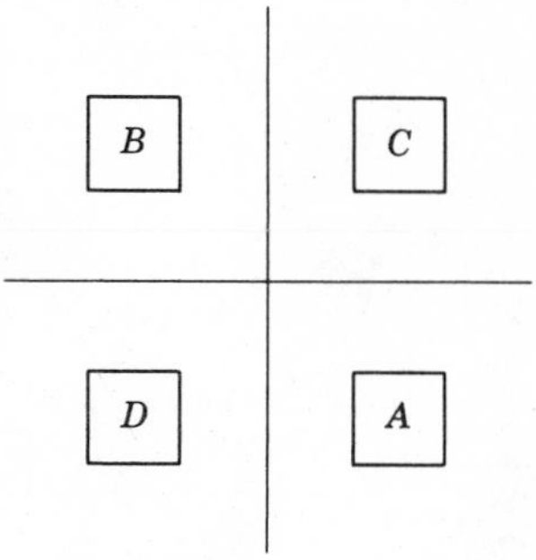

FIGURE 1

This example can be used also to demonstrate that two variables, Y_1 and Y_2, can be uncorrelated and the marginal distribution of each may be normal, but the pair need not have a joint normal distribution and need not be independent. This is done by choosing the rectangles so that for the resultant distribution the expected value of $Y_1 Y_2$ is zero. It is clear geometrically that this can be done.

For future reference we state two useful theorems concerning characteristic functions.

THEOREM 2.6.3. *If the random vector* $\boldsymbol{X}$ *has the density* $f(\boldsymbol{x})$ *and the characteristic function* $\phi(\boldsymbol{t})$, *then*

$$f(\boldsymbol{x}) = \frac{1}{(2\pi)^p} \int_{-\infty}^{\infty} \cdots \int_{-\infty}^{\infty} e^{-i\boldsymbol{t}'\boldsymbol{x}} \phi(\boldsymbol{t})\, dt_1 \cdots dt_p. \tag{20}$$

This shows that the characteristic function determines the density function uniquely. If $\boldsymbol{X}$ does not have a density, the characteristic function uniquely defines the probability of any "continuity interval." In the univariate case a continuity interval is an interval such that the cdf does not have a discontinuity at an endpoint of the interval.

THEOREM 2.6.4. *Let* $\{F_j(\boldsymbol{x})\}$ *be a sequence of cdf's and let* $\{\phi_j(\boldsymbol{t})\}$ *be the sequence of corresponding characteristic functions. A necessary and sufficient condition that* $F_j(\boldsymbol{x})$ *converge to a distribution* $F(\boldsymbol{x})$ *is that, for every* $\boldsymbol{t}$, $\phi_j(\boldsymbol{t})$ *converge to a limit* $\phi(\boldsymbol{t})$ *which is continuous at* $\boldsymbol{t} = \boldsymbol{0}$. *When this condition is satisfied, the limit* $\phi(\boldsymbol{t})$ *is identical with the characteristic function of the limiting distribution* $F(\boldsymbol{x})$.

For the proofs of these two theorems, the reader is referred to Cramér (1946), Sections 10.6, 10.7.

2.6.2. The Moments and Semi-invariants

The moments of $X_1, \cdots, X_p$ with a joint normal distribution can be obtained from the characteristic function (9). The mean is

$$\begin{aligned} \mathscr{E} X_h &= \frac{1}{i} \frac{\partial \phi}{\partial t_h}\bigg|_{\boldsymbol{t}=\boldsymbol{0}} \\ &= \frac{1}{i}\{-\sum_j \sigma_{hj} t_j + i\mu_h\}\phi(\boldsymbol{t})\bigg|_{\boldsymbol{t}=\boldsymbol{0}} \\ &= \mu_h. \end{aligned} \tag{21}$$

The second moment is

$$\begin{aligned} \mathscr{E} X_h X_j &= \frac{1}{i^2} \frac{\partial^2 \phi}{\partial t_h \partial t_j}\bigg|_{\boldsymbol{t}=\boldsymbol{0}} \\ &= \frac{1}{i^2}\{(-\sum_k \sigma_{hk} t_k + i\mu_h)(-\sum_k \sigma_{kj} t_k + i\mu_j) - \sigma_{hj}\}\phi(\boldsymbol{t})\bigg|_{\boldsymbol{t}=\boldsymbol{0}} \\ &= \sigma_{hj} + \mu_h \mu_j. \end{aligned} \tag{22}$$

Thus

(23) $$\text{Variance}\,(X_i) = \mathscr{E}(X_i - \mu_i)^2 = \sigma_{ii},$$

(24) $$\text{Covariance}\,(X_i, X_j) = \mathscr{E}(X_i - \mu_i)(X_j - \mu_j) = \sigma_{ij}.$$

Any third moment about the mean is

(25) $$\mathscr{E}(X_i - \mu_i)(X_j - \mu_j)(X_k - \mu_k) = 0.$$

The fourth moment about the mean is

(26) $$\mathscr{E}(X_i - \mu_i)(X_j - \mu_j)(X_k - \mu_k)(X_l - \mu_l) = \sigma_{ij}\sigma_{kl} + \sigma_{ik}\sigma_{jl} + \sigma_{il}\sigma_{jk}.$$

DEFINITION 2.6.3. *If all the moments of a distribution exist, then the semi-invariants are the coefficients* κ *in*

(27) $$\log \phi(t) = \sum_{s_1, \cdots, s_p = 0}^{\infty} \kappa_{s_1 \cdots s_p} \frac{(it_1)^{s_1} \cdots (it_p)^{s_p}}{s_1! \cdots s_p!}.$$

In the case of the multivariate normal distribution $\kappa_{10 \cdots 0} = \mu_1, \cdots,$ $\kappa_{0 \cdots 01} = \mu_p,$ $\kappa_{20 \cdots 0} = \sigma_{11}, \cdots, \kappa_{0 \cdots 02} = \sigma_{pp}, \kappa_{110 \cdots 0} = \sigma_{12}, \cdots.$ The semi-invariants for which $\sum s_i > 2$ are 0.

REFERENCES

Section 2.2. Cramér (1946), 260–270, 291–297; Kendall (1943), 19–22, 79–81, 104–105; Kolmogorov (1950); Mood (1950), 74–86, 102–103; Wilks (1943), 5–40.

Section 2.3. Cadwell (1951); Cramér (1946), 287–290, 310–312; David (1953); Kendall (1941), (1943), 22, 79–80, 89, 133–134, 376–378; Mood (1950), 165–170, 176–180; K. Pearson (1930), (1931); Plackett (1954); Polya (1949); Wilks (1943), 59–68.

Section 2.4. Cramér (1946), 312–313; Mood (1950), 181; Wilks (1943), 68–71.

Section 2.5. Cramér (1946), 305–308, 314–316; Kendall (1943), 334–335, 368–376, 380–381; Mood (1950), 181–184; K. Pearson (1896); Wilks (1943), 40–46, 71; Yule (1897a), (1897b).

Section 2.6. Cook (1951); Cramér (1946), 100–103, 310–311, 376–378; Kendall (1943), 79–80; Mood (1950), 184–186.

Chapter 2. Z. W. Birnbaum (1950); Cansado (1951); Fieller, Lewis, and Pearson (1955); Gnedenko (1948); Johnson (1949); Kamat (1953); McFadden (1955); Moran (1956); Nabeya (1951); Oberg (1947); Skitovic (1953); Vaswani (1950).

PROBLEMS

1. (Sec. 2.2) Let $f(x, y) = 1,\ 0 \le x \le 1,\ 0 \le y \le 1,$
$= 0$, otherwise.

Find:

(*a*) $F(x, y)$.

(*b*) $F(x)$.

(*c*) $f(x)$.

(*d*) $f(x|y)$. [*Note:* $f(x_0|y_0) = 0$ if $f(x_0, y_0) = 0$.]

(*e*) $\mathscr{E}X^nY^m$.

(*f*) Prove X and Y are independent.

2. (Sec. 2.2) Let $f(x, y) = 2,\ 0 \leq y \leq x \leq 1,$
$= 0,$ otherwise.

Find:

(*a*) $F(x, y)$.
(*b*) $F(x)$.
(*c*) $f(x)$.
(*d*) $G(y)$.
(*e*) $g(y)$.
(*f*) $f(x|y)$.
(*g*) $f(y|x)$.
(*h*) $\mathscr{E}X^n Y^m$.
(*i*) Are X and Y independent?

3. (Sec. 2.3) Sketch the ellipses $f(x, y) = 0.06$, where $f(x, y)$ is the bivariate normal density with

(*a*) $\mu_x = 1,\ \mu_y = 2,\ \sigma_x^2 = 1,\ \sigma_y^2 = 1,\ \rho_{xy} = 0.$
(*b*) $\mu_x = 0,\ \mu_y = 0,\ \sigma_x^2 = 1,\ \sigma_y^2 = 1,\ \rho_{xy} = 0.$
(*c*) $\mu_x = 0,\ \mu_y = 0,\ \sigma_x^2 = 1,\ \sigma_y^2 = 1,\ \rho_{xy} = 0.2.$
(*d*) $\mu_x = 0,\ \mu_y = 0,\ \sigma_x^2 = 1,\ \sigma_y^2 = 1,\ \rho_{xy} = 0.8.$
(*e*) $\mu_x = 0,\ \mu_y = 0,\ \sigma_x^2 = 4,\ \sigma_y^2 = 1,\ \rho_{xy} = 0.8.$

4. (Sec. 2.3) Find $\boldsymbol{b}$ and $\boldsymbol{A}$ so that the following densities can be written in the form of (23). Also find μ_x, μ_y, σ_x, σ_y and ρ_{xy}.

(*a*) $\dfrac{1}{2\pi} \exp \{-\tfrac{1}{2}[(x-1)^2 + (y-2)^2]\}.$

(*b*) $\dfrac{1}{2.4\pi} \exp \left(- \dfrac{\dfrac{x^2}{4} - 1.6 \dfrac{xy}{2} + y^2}{0.72} \right).$

(*c*) $\dfrac{1}{2\pi} \exp [-\tfrac{1}{2}(x^2 + y^2 + 4x - 6y + 13)].$

(*d*) $\dfrac{1}{2\pi} \exp [-\tfrac{1}{2}(2x^2 + y^2 + 2xy - 22x - 14y + 65)].$

5. (Sec. 2.3) Which densities in Problem 4 define distributions in which X and Y are independent?

6. (Sec. 2.3) For each matrix $\boldsymbol{A}$ in Problem 4 find $\boldsymbol{C}$ so that $\boldsymbol{C}'\boldsymbol{A}\boldsymbol{C} = \boldsymbol{I}$.

7. (Sec. 2.3) Let $\boldsymbol{b} = \boldsymbol{0}$,

$$\boldsymbol{A} = \begin{pmatrix} 7 & 3 & 2 \\ 3 & 4 & 1 \\ 2 & 1 & 2 \end{pmatrix}.$$

(*a*) Write the density (23).
(*b*) Find $\boldsymbol{\Sigma}$.

8. (Sec. 2.4) (*a*) Write the marginal density of X for each case in Problem 3.

(*b*) Indicate the marginal distribution of X for each case in Problem 4 by the notation $N(a, b)$.

(*c*) Write the marginal density of X_1 and X_2 in Problem 7.

9. (Sec. 2.4) What is the distribution of $Z = X - Y$ when X and Y have each density in Problem 3?

10. (Sec. 2.4) What is the distribution of $X_1 + 2X_2 - 3X_3$ when X_1, X_2, X_3 have the distribution defined in Problem 7?

11. (Sec. 2.4) Let X_i be independently distributed, each according to $N(\mu, \sigma^2)$.

(*a*) What is the joint distribution of $\boldsymbol{X} = \begin{pmatrix} X_1 \\ \cdot \\ \cdot \\ \cdot \\ X_N \end{pmatrix}$? Find the vector of means and the covariance matrix.

(*b*) Using Theorem 2.4.4 find the marginal distribution of $\bar{X} = \frac{1}{N}\sum X_i$.

12. (Sec. 2.4) Let the X_i be independent with X_i having distribution $N(\beta + \gamma z_i, \sigma^2)$, where z_i is a given number ($i = 1, \cdots, N$) and $\sum_i z_i = 0$.

(*a*) Find the joint distribution of $\begin{pmatrix} X_1 \\ \cdot \\ \cdot \\ \cdot \\ X_N \end{pmatrix}$.

(*b*) Find the joint distribution of $\bar{X}$ and $g = \sum X_i z_i / \sum z_i^2$ for $\sum z_i^2 > 0$.

13. (Sec. 2.4) Let the $\begin{pmatrix} X_i \\ Y_i \end{pmatrix}$ be independently distributed, each according to

$$N\left[\begin{pmatrix} \mu \\ \nu \end{pmatrix}, \begin{pmatrix} \sigma_{xx} & \sigma_{xy} \\ \sigma_{xy} & \sigma_{yy} \end{pmatrix}\right], \qquad i = 1, 2, 3.$$

(*a*) Find the joint distribution of the six variables.

(*b*) Find the joint distribution of $\begin{pmatrix} \bar{X} \\ \bar{Y} \end{pmatrix}$.

14. (Sec. 2.4) Let $\boldsymbol{X}$ have a (singular) normal distribution with mean $\boldsymbol{0}$ and covariance matrix

$$\boldsymbol{\Sigma} = \begin{pmatrix} 4 & 2 \\ 2 & 1 \end{pmatrix}.$$

(*a*) Prove $\boldsymbol{\Sigma}$ is of rank 1.

(*b*) Find $\boldsymbol{A}$ so $\boldsymbol{X} = \boldsymbol{A}Y$ and Y has a nonsingular normal distribution, and give the density of Y.

15. (Sec. 2.4) Let

$$\boldsymbol{\Sigma} = \begin{pmatrix} 2 & -1 & 3 \\ -1 & 5 & -3 \\ 3 & -3 & 5 \end{pmatrix}.$$

(*a*) Find a vector $\boldsymbol{u} \neq \boldsymbol{0}$ so $\boldsymbol{\Sigma u} = \boldsymbol{0}$. (*Hint:* Take cofactors of any column.)

(*b*) Show that any matrix of the form $\boldsymbol{G} = (\boldsymbol{H}\ \boldsymbol{u})$, where $\boldsymbol{H}$ is 3×2, has the property

$$\boldsymbol{G}'\boldsymbol{\Sigma G} = \begin{pmatrix} \boldsymbol{H}'\boldsymbol{\Sigma H} & \boldsymbol{0} \\ \boldsymbol{0} & 0 \end{pmatrix}.$$

(*c*) Using (*a*) and (*b*), find $\boldsymbol{B}$ to satisfy (34).

(*d*) Find $\boldsymbol{B}^{-1}$ and partition according to (37).

(*e*) Verify that $\boldsymbol{CC}' = \boldsymbol{\Sigma}$.

16. (Sec. 2.5) In each part of Problem 3, find the conditional distribution of X given $Y = y$, find the conditional distribution of Y given $X = x$, and plot each regression line on the appropriate graph in Problem 3.

17. (Sec. 2.5) Let $\boldsymbol{\mu} = \mathbf{0}$ and

$$\boldsymbol{\Sigma} = \begin{pmatrix} 1. & 0.80 & -0.40 \\ 0.80 & 1. & -0.56 \\ -0.40 & -0.56 & 1. \end{pmatrix}.$$

(*a*) Find the conditional distribution of X_1 and X_3, given $X_2 = x_2$.

(*b*) What is the partial correlation between X_1 and X_3 given X_2?

18. (Sec. 2.5) Let $\boldsymbol{A} = \boldsymbol{\Sigma}^{-1}$ and $\boldsymbol{\Sigma}$ be partitioned similarly as

$$\boldsymbol{A} = \begin{pmatrix} \boldsymbol{A}_{11} & \boldsymbol{A}_{12} \\ \boldsymbol{A}_{21} & \boldsymbol{A}_{22} \end{pmatrix}, \quad \boldsymbol{\Sigma} = \begin{pmatrix} \boldsymbol{\Sigma}_{11} & \boldsymbol{\Sigma}_{12} \\ \boldsymbol{\Sigma}_{21} & \boldsymbol{\Sigma}_{22} \end{pmatrix}.$$

By solving $\boldsymbol{A\Sigma} = \boldsymbol{I}$ according to the partitioning, prove

(*a*) $\boldsymbol{\Sigma}_{12}\boldsymbol{\Sigma}_{22}^{-1} = -\boldsymbol{A}_{11}^{-1}\boldsymbol{A}_{12}$.

(*b*) $\boldsymbol{\Sigma}_{11} - \boldsymbol{\Sigma}_{12}\boldsymbol{\Sigma}_{22}^{-1}\boldsymbol{\Sigma}_{21} = \boldsymbol{A}_{11}^{-1}$.

19. (Sec. 2.5) In Problem 7, find the conditional distribution of X_1 and X_2 given $X_3 = x_3$.

20. (Sec. 2.5) Verify (24) of Section 2.5 directly from Theorem 2.5.1.

21. (Sec. 2.5) (*a*) Show that finding $\boldsymbol{\alpha}$ to maximize the correlation between X_i and $\boldsymbol{\alpha X}^{(2)}$ is equivalent to maximizing $(\boldsymbol{\sigma}_{(i)}\boldsymbol{\alpha}')^2$ subject to $\boldsymbol{\alpha\Sigma}_{22}\boldsymbol{\alpha}'$ constant.

(*b*) Find $\boldsymbol{\alpha}$ by maximizing $(\boldsymbol{\sigma}_{(i)}\boldsymbol{\alpha}')^2 - \lambda(\boldsymbol{\alpha\Sigma}_{22}\boldsymbol{\alpha}' - c)$ where c is a constant and λ is a Lagrange multiplier.

22. (Sec. 2.5) Prove that $\bar{R}_{i\cdot q+1,\cdots,p}$ is invariant with respect to scalar transformations of the components of X (that is, that if X_j is replaced by c_jX_j, $\bar{R}_{i\cdot q+1,\cdots,p}$ is not changed, $c_j \neq 0$).

23. (Sec. 2.5) Prove that $1 - \bar{R}^2_{i\cdot q+1,\cdots,p} = \begin{vmatrix} 1 & \rho_{ij} \\ \rho_{ki} & \rho_{kj} \end{vmatrix} \Big/ |\rho_{kj}|$, $k, j = q+1, \cdots, p$.

24. (Sec. 2.5) Prove that $\bar{R}_{i\cdot q+1,\cdots,p}$ is invariant with respect to linear transformations of $\boldsymbol{X}^{(2)}$ (that is, that if $\boldsymbol{X}^{(2)}$ is replaced by $\boldsymbol{HX}^{(2)}$, where $\boldsymbol{H}$ is nonsingular, then $\bar{R}_{i\cdot q+1,\cdots,p}$ is not changed).

25. (Sec. 2.5) Find the multiple correlation coefficient between X_1 and $(X_2\ X_3)$ in Problem 17.

26. (Sec. 2.5) Prove explicitly that if $\boldsymbol{\Sigma}$ is positive definite

$$|\boldsymbol{\Sigma}| = |\boldsymbol{\Sigma}_{11} - \boldsymbol{\Sigma}_{12}\boldsymbol{\Sigma}_{22}^{-1}\boldsymbol{\Sigma}_{21}| \cdot |\boldsymbol{\Sigma}_{22}|.$$

27. (Sec. 2.5) Prove

$$|\boldsymbol{\Sigma}| \leq \prod_{i=1}^{p} \sigma_{ii}.$$

[*Hint:* Using Problem 26, prove $|\boldsymbol{\Sigma}| \leq \sigma_{11}|\boldsymbol{\Sigma}_{22}|$ where $\boldsymbol{\Sigma}_{22}$ is $(p-1) \times (p-1)$ and apply induction.]

28. (Sec. 2.5) Prove $\beta_{12\cdot 3} = \sigma_{12\cdot 3}/\sigma_{22\cdot 3}$ and $\beta_{13\cdot 2} = \sigma_{13\cdot 2}/\sigma_{33\cdot 2}$.

29. (Sec. 2.3) Let $\boldsymbol{X}$ of two components have the covariance matrix (48) of Section 2.3. Let $Y_i = X_i/\sigma_i$, $i = 1, 2$. Prove $\operatorname{Var}(Y_1 - Y_2) = 2(1 - \rho)$. How does this suggest that ρ is a measure of association between X_1 and X_2?

30. (Sec. 2.3) Prove that the principal axes of (52) of Section 2.3 are along the 45° and 135° lines with lengths $2\sqrt{c(1+\rho)}$ and $2\sqrt{c(1-\rho)}$ respectively, by transforming according to $y_1 = (z_1 + z_2)/\sqrt{2}$, $y_2 = (z_1 - z_2)/\sqrt{2}$.

31. (Sec. 2.3) Prove that ρ_{12} is an invariant characteristic of the bivariate normal distribution $N(\boldsymbol{\mu}, \boldsymbol{\Sigma})$ under transformations $x_i^* = b_i x_i + c_i$ $(i = 1, 2)$ and $b_i > 0$ and that every function of the parameters that is invariant is a function of ρ_{12}.

32. (Sec. 2.5) Let $\boldsymbol{X}_\alpha = (X_{1\alpha}, X_{2\alpha})'$ be independently distributed according to $N(\boldsymbol{0}, \boldsymbol{\Sigma})$, $\alpha = 1, 2$. What is the conditional distribution of $X_{11}, X_{12}, X_{21}, X_{22}$ given that $X_{11} = X_{12}$?

33. (Sec. 2.3) Suppose the scalar random variables $X_1, \cdots, X_n$ are independent and have a density which is a function only of $x_1^2 + \cdots + x_n^2$. Prove that the X_i are normally distributed with mean 0 and common variance. Indicate the mildest conditions on the density for your proof.

34. (Sec. 2.5) Let (X_1, X_2) have the density $n(\boldsymbol{x}|\boldsymbol{0}, \boldsymbol{\Sigma}) = f(x_1, x_2)$. Let the density of X_2 given $X_1 = x_1$ be $f(x_2|x_1)$. Let the joint density of X_1, X_2, X_3 be $f(x_1, x_2)f(x_3|x_1)$. Find the covariance matrix of X_1, X_2, X_3 and the partial correlation between X_2 and X_3 for given X_1.

35. (Sec. 2.5) Prove $1 - \bar{R}_{1\cdot 23}^2 = (1 - \rho_{12}^2)(1 - \rho_{13\cdot 2}^2)$. [*Hint:* Use the fact that the variance of X_1 in the conditional distribution given x_2 and x_3 is $(1 - \bar{R}_{1\cdot 23}^2)\sigma_{11}$.]

36. (Sec. 2.3) Show that if $\Pr\{X \geq 0, Y \geq 0\} = \alpha$ for the distribution

$$N\left[\begin{pmatrix}0\\0\end{pmatrix}, \begin{pmatrix}\sigma_x^2 & \sigma_x\sigma_y\rho\\ \sigma_x\sigma_y\rho & \sigma_y^2\end{pmatrix}\right],$$

then $\rho = \cos(1 - 2\alpha)\pi$.

37. (Sec. 2.6) Let $\boldsymbol{Y}$ be distributed according to $N(\boldsymbol{0}, \boldsymbol{\Sigma})$. Differentiating the characteristic function, verify (25) and (26).

38. (Sec. 2.3) Prove that if $\rho_{ij} = \rho$ $(i \neq j;\ i, j = 1, \cdots, p)$, then $\rho \geq -1/(p-1)$.

39. (Sec. 2.5) Prove

$$\beta_{ij\cdot q+1,\cdots,j-1,j+1,\cdots,p} = \sigma_{ij\cdot q+1,\cdots,j-1,j+1,\cdots,p}/\sigma_{jj\cdot q+1,\cdots,j-1,j+1,\cdots,p}$$

$(i = 1, \cdots, q;\ j = q+1, \cdots, p)$. [*Hint:* Prove this for the special case $j = q + 1$ by considering (30) with $p_1 = q$, $p_2 = 1$. $p_3 = p - q - 1$.]

40. (Sec. 2.5) If $p = 2$, can there be a difference between the ordinary correlation between X_1 and X_2 and the multiple correlation between X_1 and $\boldsymbol{X}^{(2)} = X_2$? Explain.

41. (Sec. 2.4) Prove that if the joint (marginal) distribution of X_1 and X_2 is singular (that is, degenerate), then the joint distribution of X_1, X_2, and X_3 is singular.

42. (Sec. 2.5) Give a necessary and sufficient condition for $\bar{R}_{i\cdot q+1,\cdots,p} = 0$ in terms of $\sigma_{i,q+1}, \cdots, \sigma_{ip}$.

43. (Sec. 2.2) Let $f(x, y) = C$ for $x^2 + y^2 \leq k^2$ and 0 elsewhere. Prove $C = 1/(\pi k^2)$, $\mathscr{E}X = \mathscr{E}Y = 0$, $\mathscr{E}X^2 = \mathscr{E}Y^2 = k^2/4$, and $\mathscr{E}XY = 0$. Are X and Y independent?

44. (Sec. 2.3) Let the density of the p-component $\boldsymbol{Y}$ be $f(\boldsymbol{y}) = \Gamma(\frac{1}{2}p + 1)/[(p+2)\pi]^{\frac{1}{2}p}$ for $\boldsymbol{y}'\boldsymbol{y} \leq p + 2$ and 0 elsewhere. Then $\mathscr{E}\boldsymbol{Y} = \boldsymbol{0}$ and $\mathscr{E}\boldsymbol{Y}\boldsymbol{Y}' = \boldsymbol{I}$. From this result prove that if the density of $\boldsymbol{X}$ is $g(\boldsymbol{x}) = \sqrt{|\boldsymbol{A}|}\,\Gamma(\frac{1}{2}p + 1)/[(p+2)\pi]^{\frac{1}{2}p}$ for $(\boldsymbol{x} - \boldsymbol{\mu})'\boldsymbol{A}(\boldsymbol{x} - \boldsymbol{\mu}) \leq p + 2$ and 0 elsewhere then $\mathscr{E}\boldsymbol{X} = \boldsymbol{\mu}$ and $\mathscr{E}(\boldsymbol{X} - \boldsymbol{\mu})(\boldsymbol{X} - \boldsymbol{\mu})' = \boldsymbol{A}^{-1}$.

45. (Sec. 2.2) Let $F(x_1, x_2)$ be the joint cdf of X_1, X_2 and let $F_i(x_i)$ be the marginal cdf of $X_i (i = 1, 2)$. Prove that if $F_i(x_i)$ is continuous $(i = 1, 2)$, then $F(x_1, x_2)$ is continuous.

CHAPTER 3

Estimation of the Mean Vector and the Covariance Matrix

3.1. INTRODUCTION

The normal distribution is specified completely by the mean vector $\boldsymbol{\mu}$ and the covariance matrix $\boldsymbol{\Sigma}$. The first statistical problem is how to estimate these parameters on the basis of a sample of observations. In Section 3.2 it is shown that the maximum likelihood estimate of $\boldsymbol{\mu}$ is the sample mean; the maximum likelihood estimate of $\boldsymbol{\Sigma}$ is proportional to the matrix of sample variances and covariances. A sample variance is a sum of squares of deviations of observations from the sample mean divided by one less than the number of observations in the sample; a sample covariance is similarly defined in terms of cross products. The sample covariance matrix is an unbiased estimate of $\boldsymbol{\Sigma}$.

The distribution of the sample mean vector is given in Section 3.3, and it is shown how one can test the hypothesis that $\boldsymbol{\mu}$ is a given vector when $\boldsymbol{\Sigma}$ is known. The case of $\boldsymbol{\Sigma}$ unknown will be treated in Chapter 5.

3.2. THE MAXIMUM LIKELIHOOD ESTIMATES OF THE MEAN VECTOR AND THE COVARIANCE MATRIX

Given a sample of (vector) observations from a p-variate (nondegenerate) normal distribution, we ask for estimates of the mean vector $\boldsymbol{\mu}$ and the covariance matrix $\boldsymbol{\Sigma}$ of the distribution. We shall deduce the maximum likelihood estimates.

It turns out that the method of maximum likelihood is very useful in various estimation and hypothesis testing problems concerning the multivariate normal distribution. The maximum likelihood estimates or modifications of them usually have some optimum properties. In the particular case studied here, the estimates are asymptotically efficient [Cramér (1946), Sec. 33.3].

Suppose our sample of N observations on $\boldsymbol{X}$ distributed according to $N(\boldsymbol{\mu}, \boldsymbol{\Sigma})$ is $\boldsymbol{x}_1, \cdots, \boldsymbol{x}_N$, where $N > p$. The likelihood function is

$$(1) \qquad L = \frac{1}{(2\pi)^{\frac{1}{2}pN}|\boldsymbol{\Sigma}|^{\frac{1}{2}N}} \exp\left[-\tfrac{1}{2}\sum_{\alpha=1}^{N}(\boldsymbol{x}_\alpha - \boldsymbol{\mu})'\boldsymbol{\Sigma}^{-1}(\boldsymbol{x}_\alpha - \boldsymbol{\mu})\right].$$

Since the exponent is written in terms of $\boldsymbol{\Sigma}^{-1}$, we shall first find the maximum likelihood estimates of $\boldsymbol{\mu}$ and $\boldsymbol{\Sigma}^{-1} = \boldsymbol{\Psi}$, say. In the likelihood function the vectors $\boldsymbol{x}_1, \cdots, \boldsymbol{x}_N$ are fixed at the sample values and L is a function of $\boldsymbol{\mu}$ and $\boldsymbol{\Psi}$. To emphasize that these quantities are variables (and not parameters) we shall denote them by $\boldsymbol{\mu}^*$ and $\boldsymbol{\Psi}^*$. Then the logarithm of the likelihood function is

$$(2) \quad \log L = -\tfrac{1}{2}pN\log(2\pi) + \tfrac{1}{2}N\log|\boldsymbol{\Psi}^*| - \tfrac{1}{2}\sum_{\alpha=1}^{N}(\boldsymbol{x}_\alpha - \boldsymbol{\mu}^*)'\boldsymbol{\Psi}^*(\boldsymbol{x}_\alpha - \boldsymbol{\mu}^*).$$

Since $\log L$ is an increasing function of L, its maximum is at the same point in the space of $\boldsymbol{\mu}^*, \boldsymbol{\Psi}^*$ as the maximum of L.

Let the sample mean be

$$(3) \qquad \bar{\boldsymbol{x}} = \frac{1}{N}\sum_{\alpha=1}^{N}\boldsymbol{x}_\alpha = \begin{pmatrix} \frac{1}{N}\sum_{\alpha=1}^{N}x_{1\alpha} \\ \cdot \\ \cdot \\ \cdot \\ \frac{1}{N}\sum_{\alpha=1}^{N}x_{p\alpha} \end{pmatrix} = \begin{pmatrix} \bar{x}_1 \\ \cdot \\ \cdot \\ \cdot \\ \bar{x}_p \end{pmatrix},$$

and let the matrix of sums of squares and cross products of deviations about the mean be

$$(4) \qquad A = \sum_{\alpha=1}^{N}(\boldsymbol{x}_\alpha - \bar{\boldsymbol{x}})(\boldsymbol{x}_\alpha - \bar{\boldsymbol{x}})'$$

$$= \left[\sum_{\alpha=1}^{N}(x_{i\alpha} - \bar{x}_i)(x_{j\alpha} - \bar{x}_j)\right], \qquad i, j = 1, \cdots, p.$$

It will be convenient to use the following lemma:

LEMMA 3.2.1. *Let $\boldsymbol{x}_1, \cdots, \boldsymbol{x}_N$ be N (p-component) vectors and let $\bar{\boldsymbol{x}}$ be defined by* (3). *Then for any vector* $\boldsymbol{b}$

$$(5) \quad \sum_{\alpha=1}^{N}(\boldsymbol{x}_\alpha - \boldsymbol{b})(\boldsymbol{x}_\alpha - \boldsymbol{b})' = \sum_{\alpha=1}^{N}(\boldsymbol{x}_\alpha - \bar{\boldsymbol{x}})(\boldsymbol{x}_\alpha - \bar{\boldsymbol{x}})' + N(\bar{\boldsymbol{x}} - \boldsymbol{b})(\bar{\boldsymbol{x}} - \boldsymbol{b})'.$$

Proof.

$$(6)\quad \sum_{\alpha=1}^{N}(\boldsymbol{x}_\alpha - \boldsymbol{b})(\boldsymbol{x}_\alpha - \boldsymbol{b})' = \sum_{\alpha=1}^{N}[(\boldsymbol{x}_\alpha - \bar{\boldsymbol{x}}) + (\bar{\boldsymbol{x}} - \boldsymbol{b})][(\boldsymbol{x}_\alpha - \bar{\boldsymbol{x}}) + (\bar{\boldsymbol{x}} - \boldsymbol{b})]'$$

$$= \sum_{\alpha=1}^{N}[(\boldsymbol{x}_\alpha - \bar{\boldsymbol{x}})(\boldsymbol{x}_\alpha - \bar{\boldsymbol{x}})' + (\boldsymbol{x}_\alpha - \bar{\boldsymbol{x}})(\bar{\boldsymbol{x}} - \boldsymbol{b})' + (\bar{\boldsymbol{x}} - \boldsymbol{b})(\boldsymbol{x}_\alpha - \bar{\boldsymbol{x}})' + (\bar{\boldsymbol{x}} - \boldsymbol{b})(\bar{\boldsymbol{x}} - \boldsymbol{b})']$$

$$= \sum_{\alpha=1}^{N}(\boldsymbol{x}_\alpha - \bar{\boldsymbol{x}})(\boldsymbol{x}_\alpha - \bar{\boldsymbol{x}})' + \left[\sum_{\alpha=1}^{N}(\boldsymbol{x}_\alpha - \bar{\boldsymbol{x}})\right](\bar{\boldsymbol{x}} - \boldsymbol{b})' + (\bar{\boldsymbol{x}} - \boldsymbol{b})\sum_{\alpha=1}^{N}(\boldsymbol{x}_\alpha - \bar{\boldsymbol{x}})' + N(\bar{\boldsymbol{x}} - \boldsymbol{b})(\bar{\boldsymbol{x}} - \boldsymbol{b})'.$$

The second and third terms are $\mathbf{0}$ because $\sum(\boldsymbol{x}_\alpha - \bar{\boldsymbol{x}}) = \sum \boldsymbol{x}_\alpha - N\bar{\boldsymbol{x}} = \mathbf{0}$ by (3). Q.E.D.

When we let $\boldsymbol{b} = \boldsymbol{\mu}^*$, we have

$$(7)\quad \sum_{\alpha=1}^{N}(\boldsymbol{x}_\alpha - \boldsymbol{\mu}^*)(\boldsymbol{x}_\alpha - \boldsymbol{\mu}^*)' = \sum_{\alpha=1}^{N}(\boldsymbol{x}_\alpha - \bar{\boldsymbol{x}})(\boldsymbol{x}_\alpha - \bar{\boldsymbol{x}})' + N(\bar{\boldsymbol{x}} - \boldsymbol{\mu}^*)(\bar{\boldsymbol{x}} - \boldsymbol{\mu}^*)'$$

$$= \boldsymbol{A} + N(\bar{\boldsymbol{x}} - \boldsymbol{\mu}^*)(\bar{\boldsymbol{x}} - \boldsymbol{\mu}^*)'.$$

Using this result and the properties of the trace of a matrix ($\operatorname{tr} \boldsymbol{CD} = \sum c_{ij}d_{ji} = \operatorname{tr} \boldsymbol{DC}$) we have

$$(8)\quad \sum_{\alpha=1}^{N}(\boldsymbol{x}_\alpha - \boldsymbol{\mu}^*)'\boldsymbol{\Psi}^*(\boldsymbol{x}_\alpha - \boldsymbol{\mu}^*) = \operatorname{tr}\sum_{\alpha=1}^{N}(\boldsymbol{x}_\alpha - \boldsymbol{\mu}^*)'\boldsymbol{\Psi}^*(\boldsymbol{x}_\alpha - \boldsymbol{\mu}^*)$$

$$= \operatorname{tr}\sum_{\alpha=1}^{N}\boldsymbol{\Psi}^*(\boldsymbol{x}_\alpha - \boldsymbol{\mu}^*)(\boldsymbol{x}_\alpha - \boldsymbol{\mu}^*)'$$

$$= \operatorname{tr}\boldsymbol{\Psi}^*\boldsymbol{A} + \operatorname{tr}\boldsymbol{\Psi}^*N(\bar{\boldsymbol{x}} - \boldsymbol{\mu}^*)(\bar{\boldsymbol{x}} - \boldsymbol{\mu}^*)'$$

$$= \operatorname{tr}\boldsymbol{\Psi}^*\boldsymbol{A} + N(\bar{\boldsymbol{x}} - \boldsymbol{\mu}^*)'\boldsymbol{\Psi}^*(\bar{\boldsymbol{x}} - \boldsymbol{\mu}^*).$$

Thus we can write (2) as

$$(9)\quad \log L = -\tfrac{1}{2}pN\log(2\pi) + \tfrac{1}{2}N\log|\boldsymbol{\Psi}^*| - \tfrac{1}{2}\operatorname{tr}\boldsymbol{\Psi}^*\boldsymbol{A} - \tfrac{1}{2}N(\bar{\boldsymbol{x}} - \boldsymbol{\mu}^*)'\boldsymbol{\Psi}^*(\bar{\boldsymbol{x}} - \boldsymbol{\mu}^*).$$

Since $\boldsymbol{\Psi}^*$ is positive semidefinite, $N(\bar{\boldsymbol{x}} - \boldsymbol{\mu}^*)'\boldsymbol{\Psi}^*(\bar{\boldsymbol{x}} - \boldsymbol{\mu}^*) \geq 0$ and is 0 if $\boldsymbol{\mu}^* = \bar{\boldsymbol{x}}$. To maximize the second and third terms of (9) we use the following lemma (which is also used in later chapters):

LEMMA 3.2.2. *Let*

$$(10)\quad f(\boldsymbol{C}) = \tfrac{1}{2}N\log|\boldsymbol{C}| - \tfrac{1}{2}\sum_{i,j=1}^{p}c_{ij}d_{ij},$$

where $\boldsymbol{C} = (c_{ij})$ *is positive semidefinite and where* $\boldsymbol{D} = (d_{ij})$ *is positive* *definite. Then the maximum of $f(C)$ is taken on at $C = ND^{-1}$ and the maximum is*

$$f(ND^{-1}) = \tfrac{1}{2}pN \log N - \tfrac{1}{2}N \log |D| - \tfrac{1}{2}pN. \tag{11}$$

Proof. We note that $f(C) \to -\infty$ if C approaches a singular matrix or if one or more elements of C approach ∞ and/or $-\infty$ (nondiagonal elements)†. Thus maxima of $f(C)$ are defined by setting equal to zero the derivatives with respect to the elements of C. Using Theorem 7 of Appendix 1, we find

$$\frac{\partial f}{\partial c_{kk}} = \frac{1}{2}\frac{N}{|C|}\frac{\partial |C|}{\partial c_{kk}} - \frac{1}{2}d_{kk} = \frac{1}{2}N\frac{\text{cof } c_{kk}}{|C|} - \frac{1}{2}d_{kk}, \tag{12}$$

where "cof c_{kk}" denotes the cofactor of c_{kk} in C. For $k \neq l$,

$$\frac{\partial f}{\partial c_{kl}} = N\frac{\text{cof } c_{kl}}{|C|} - d_{kl}, \tag{13}$$

since $c_{kl} = c_{lk}$. Setting $2\partial f/\partial c_{kk}$ and $\partial f/\partial c_{kl}$ equal to 0 and using the fact that $\text{cof } c_{kl}/|C|$ is the l, kth element of C^{-1}, we obtain $NC^{-1} = D$. Thus $C = ND^{-1}$. The value of the maximum is

$$\begin{aligned} f(ND^{-1}) &= \tfrac{1}{2}N \log |ND^{-1}| - \tfrac{1}{2} \operatorname{tr} ND^{-1}D \\ &= \tfrac{1}{2}N \log N^p|D^{-1}| - \tfrac{1}{2} \operatorname{tr} NI \\ &= \tfrac{1}{2}Np \log N - \tfrac{1}{2}N \log |D| - \tfrac{1}{2}Np. \end{aligned} \tag{14}$$

Applying this lemma to (9) with the last term zero, we find that the maximum occurs at

$$\Psi^* = \left(\frac{1}{N}A\right)^{-1}. \tag{15}$$

We assume that A is nonsingular; we shall see in Chapter 7 that the probability is 0 of drawing a sample ($N > p$) such that A is singular. Thus A^{-1} exists, and Ψ^* is positive definite. Therefore $\mu^* = \bar{x}$ is the only value of μ^* to make the last term of (9) zero. Thus the maximum likelihood estimates of μ and Ψ are $\hat{\mu} = \bar{x}$ and $\hat{\Psi} = NA^{-1}$.

To find the maximum likelihood estimate of Σ, we need the following lemma:

LEMMA 3.2.3. *Let $f(\theta)$ be a real-valued function defined on a certain set S and let ϕ be a single-valued function, with a single-valued inverse,*

† Let $D = EE'$, $C = E'^{-1}C^*E^{-1}$. Then $f = k + \frac{1}{2}N \log |C^*| - \frac{1}{2}\sum c_{ii}^*$ since $\operatorname{tr} CD = \operatorname{tr} (E'^{-1}C^*E^{-1}EE') = \operatorname{tr} C^*$. If one or more elements of C approach ∞ and/or $-\infty$, one or more diagonal elements of C^* approach ∞. But by Problem 27 of Chapter 2, $|C^*| \leq \Pi c_{ii}^*$. Thus, $f = k + \frac{1}{2}N \log |C^*| - \frac{1}{2}\sum c_{ii}^* \leq k + \frac{1}{2}N \sum \log c_{ii}^* - \frac{1}{2}\sum c_{ii}^*$ which approaches $-\infty$ if one or more c_{ij} become infinite.

on S to some other set S^*; that is, to each $\theta \in S$ there corresponds a unique $\theta^* \in S^*$ and conversely to each $\theta^* \in S^*$ there corresponds a unique $\theta \in S$. Let

$$g(\theta^*) = f[\phi^{-1}(\theta^*)]. \tag{16}$$

Then if $f(\theta)$ attains a maximum at $\theta = \theta_0$, $g(\theta^*)$ attains a maximum at $\theta^* = \theta_0^* = \phi(\theta_0)$. If the maximum of $f(\theta)$ at θ_0 is unique, so is the maximum of $g(\theta^*)$ at θ_0^*.

Proof. By hypothesis

$$f(\theta_0) \geq f(\theta) \tag{17}$$

for all $\theta \in S$. Then for any $\theta^* \in S^*$

$$\begin{aligned} g(\theta^*) &= f[\phi^{-1}(\theta^*)] = f(\theta) \leq f(\theta_0) = g[\phi(\theta_0)] \\ &= g(\theta_0^*). \end{aligned} \tag{18}$$

Thus $g(\theta^*)$ attains a maximum at θ_0^*. If the maximum of $f(\theta)$ at θ_0 is unique, there is strict inequality above for $\theta \neq \theta_0$, and the maximum of $g(\theta^*)$ is unique. We have the following corollary:

COROLLARY 3.2.1. *If on the basis of a given sample $\hat{\theta}_1, \cdots, \hat{\theta}_m$ are maximum likelihood estimates of the parameters $\theta_1, \cdots, \theta_m$ of a distribution, then $\phi_1(\hat{\theta}_1, \cdots, \hat{\theta}_m), \cdots, \phi_m(\hat{\theta}_1, \cdots, \hat{\theta}_m)$ are maximum likelihood estimates of $\phi_1(\theta_1, \cdots, \theta_m), \cdots, \phi_m(\theta_1, \cdots, \theta_m)$ if the transformation from $\theta_1, \cdots, \theta_m$ to $\phi_1, \cdots, \phi_m$ is one-to-one. If the estimates of $\theta_1, \cdots, \theta_m$ are unique, then the estimates of $\phi_1, \cdots, \phi_m$ are unique.*

It follows from the corollary that the maximum likelihood estimate of $\boldsymbol{\Sigma}$ is $\hat{\boldsymbol{\Sigma}} = \hat{\boldsymbol{\Psi}}^{-1} = (1/N)A$. We summarize these results as follows:

THEOREM 3.2.1. *If $\boldsymbol{x}_1, \cdots, \boldsymbol{x}_N$ constitute a sample from $N(\boldsymbol{\mu}, \boldsymbol{\Sigma})$ $(p < N)$, the maximum likelihood estimates of $\boldsymbol{\mu}$ and $\boldsymbol{\Sigma}$ are $\hat{\boldsymbol{\mu}} = \bar{\boldsymbol{x}} = (1/N)\sum_\alpha \boldsymbol{x}_\alpha$ and $\hat{\boldsymbol{\Sigma}} = (1/N)\sum_\alpha(\boldsymbol{x}_\alpha - \bar{\boldsymbol{x}})(\boldsymbol{x}_\alpha - \bar{\boldsymbol{x}})'$.*

Computation of the estimates of $\boldsymbol{\Sigma}$ is made easier because by the specialization of Lemma 3.2.1.

$$\sum_{\alpha=1}^{N} (\boldsymbol{x}_\alpha - \bar{\boldsymbol{x}})(\boldsymbol{x}_\alpha - \bar{\boldsymbol{x}})' = \sum_{\alpha=1}^{N} \boldsymbol{x}_\alpha \boldsymbol{x}_\alpha' - N\bar{\boldsymbol{x}}\bar{\boldsymbol{x}}'. \tag{19}$$

An element of $\sum_{\alpha=1}^N \boldsymbol{x}_\alpha \boldsymbol{x}_\alpha'$ is computed as $\sum_{\alpha=1}^N x_{i\alpha} x_{j\alpha}$, and an element of $N\bar{\boldsymbol{x}}\bar{\boldsymbol{x}}'$ is computed as $N\bar{x}_i\bar{x}_j$ or $(\sum_{\alpha=1}^N x_{i\alpha})(\sum_{\alpha=1}^N x_{j\alpha})/N$.

COROLLARY 3.2.2. *If $\boldsymbol{x}_1, \cdots, \boldsymbol{x}_N$ constitutes a sample from $N(\boldsymbol{\mu}, \boldsymbol{\Sigma})$, where $\sigma_{ij} = \sigma_i \sigma_j \rho_{ij}$ $(\rho_{ii} = 1)$, the maximum likelihood estimate of $\boldsymbol{\mu}$ is $\hat{\boldsymbol{\mu}} = \bar{\boldsymbol{x}} = (1/N)\sum_\alpha \boldsymbol{x}_\alpha$, the maximum likelihood estimate of σ_i^2 is $\hat{\sigma}_i^2 = (1/N)\sum_\alpha (x_{i\alpha} - \bar{x}_i)^2 = (1/N)(\sum_\alpha x_{i\alpha}^2 - N\bar{x}_i^2)$, where $x_{i\alpha}$ is the ith component of*

x_α and $\bar{x}_i$ is the ith component of $\bar{x}$, and the maximum likelihood estimate of ρ_{ij} is

$$\hat{\rho}_{ij} = \frac{\sum_\alpha (x_{i\alpha} - \bar{x}_i)(x_{j\alpha} - \bar{x}_j)}{\sqrt{\sum_\alpha (x_{i\alpha} - \bar{x}_i)^2}\sqrt{\sum_\alpha (x_{j\alpha} - \bar{x}_j)^2}} = \frac{\sum_\alpha x_{i\alpha} x_{j\alpha} - N\bar{x}_i\bar{x}_j}{\sqrt{\sum_\alpha x_{i\alpha}^2 - N\bar{x}_i^2}\sqrt{\sum_\alpha x_{j\alpha}^2 - N\bar{x}_j^2}}. \tag{20}$$

Proof. The set of parameters $\mu_i = \mu_i$, $\sigma_i^2 = \sigma_{ii}$, and $\rho_{ij} = \sigma_{ij}/\sqrt{\sigma_{ii}\sigma_{jj}}$ is a one-to-one transform of the set of parameters μ_i and σ_{ij}. Therefore by Corollary 3.2.1. the estimate of μ_i is $\hat{\mu}_i$, of σ_i^2 is $\hat{\sigma}_{ii}$, and of ρ_{ij} is.

$$\hat{\rho}_{ij} = \frac{\hat{\sigma}_{ij}}{\sqrt{\hat{\sigma}_{ii}\hat{\sigma}_{jj}}}. \tag{21}$$

Pearson (1896) gave a justification for this estimate of ρ_{ij}, and (20) is sometimes called the Pearson correlation coefficient. It is usually denoted by r_{ij}.

A convenient geometrical interpretation of this sample $(x_1\, x_2 \cdots x_N) = X$ is in terms of the rows of X. Let

$$X = \begin{pmatrix} x_{11} & \cdots & x_{1N} \\ \cdot & & \cdot \\ \cdot & & \cdot \\ \cdot & & \cdot \\ x_{p1} & \cdots & x_{pN} \end{pmatrix} = \begin{pmatrix} y_1 \\ \cdot \\ \cdot \\ \cdot \\ y_p \end{pmatrix}; \tag{22}$$

that is, y_i is the ith row of X. The vector y_i can be considered as a vector in an N-dimensional space with the αth coordinate of one end-point being $x_{i\alpha}$ and the other end point at the origin. Thus the sample is represented by p vectors in N-dimensional Euclidean space. By definition of Euclidean metric the squared length of y_i (that is, the squared distance of one end-point from the other) is $y_i y_i' = \sum_\alpha x_{i\alpha}^2$.

Now let us show that the cosine of the angle between y_i and y_j is $y_i y_j'/\sqrt{y_i y_i' y_j y_j'} = \sum_\alpha x_{i\alpha} x_{j\alpha}/\sqrt{\sum_\alpha x_{i\alpha}^2 \sum_\alpha x_{j\alpha}^2}$. Choose the scalar d so the vector dy_j is orthogonal to $y_i - dy_j$; that is, $0 = dy_j(y_i - dy_j)' = d(y_j y_i' - d y_j y_j')$. Therefore $d = y_j y_i'/y_j y_j'$. We decompose y_i into $y_i - dy_j$ and dy_j ($y_i = (y_i - dy_j) + dy_j$) as indicated in Figure 1. The absolute value of the cosine of the angle between y_i and y_j is the length of dy_j divided by the length of y_i; that is, it is $\sqrt{dy_j(dy_j)'/y_i y_i'} = \sqrt{dy_j y_i' d/y_i y_i'}$; the cosine is $y_i y_j'/\sqrt{y_i y_i' y_j y_j'}$. This proves the desired result.

To give a geometric interpretation of a_{ii} and $a_{ij}/\sqrt{a_{ii}a_{jj}}$ we introduce the equiangular line which is the line going through the origin and the point $(1, 1, \cdots, 1)$. See Figure 2. The projection of y_i on the vector

$\boldsymbol{\varepsilon} = (1, 1, \cdots, 1)$ is $(\boldsymbol{\varepsilon} \boldsymbol{y}_i'/\boldsymbol{\varepsilon}\boldsymbol{\varepsilon}')\boldsymbol{\varepsilon} = (\sum_\alpha x_{i\alpha}/\sum_\alpha 1)\boldsymbol{\varepsilon} = \bar{x}_i \boldsymbol{\varepsilon} = (\bar{x}_i, \bar{x}_i, \cdots, \bar{x}_i)$. Then we decompose $\boldsymbol{y}_i$ into $\bar{x}_i\boldsymbol{\varepsilon}$, the projection on the equiangular line, and $\boldsymbol{y}_i - \bar{x}_i\boldsymbol{\varepsilon}$, the projection of $\boldsymbol{y}_i$ on the plane perpendicular to the equiangular

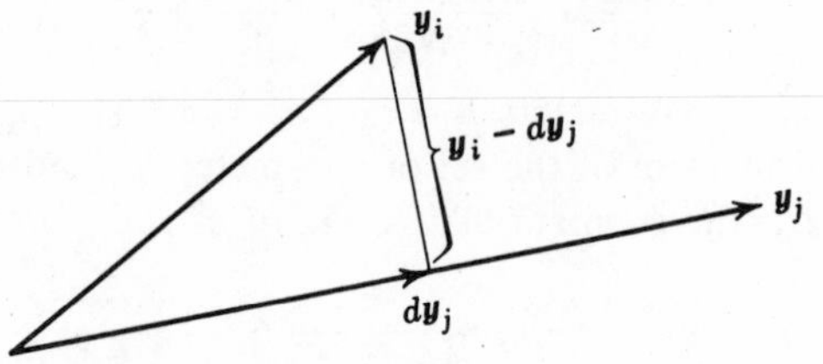

FIGURE 1

line. The squared length of $\boldsymbol{y}_i - \bar{x}_i\boldsymbol{\varepsilon}$ is $(\boldsymbol{y}_i - \bar{x}_i\boldsymbol{\varepsilon})(\boldsymbol{y}_i - \bar{x}_i\boldsymbol{\varepsilon})' = \sum_\alpha (x_{i\alpha} - \bar{x}_i)^2$; this is $N\hat{\sigma}_{ii} = a_{ii}$. Translate $\boldsymbol{y}_i - \bar{x}_i\boldsymbol{\varepsilon}$ and $\boldsymbol{y}_j - \bar{x}_j\boldsymbol{\varepsilon}$ so that each vector has an end point at the origin; the αth coordinate of the first vector is

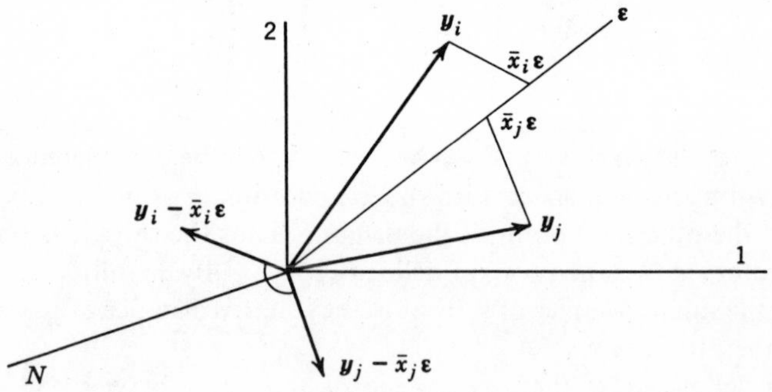

FIGURE 2

$x_{i\alpha} - \bar{x}_i$ and of the second is $x_{j\alpha} - \bar{x}_j$. The cosine of the angle between these two vectors is

$$(23) \qquad \frac{(\boldsymbol{y}_i - \bar{x}_i\boldsymbol{\varepsilon})(\boldsymbol{y}_j - \bar{x}_j\boldsymbol{\varepsilon})'}{\sqrt{(\boldsymbol{y}_i - \bar{x}_i\boldsymbol{\varepsilon})(\boldsymbol{y}_i - \bar{x}_i\boldsymbol{\varepsilon})'(\boldsymbol{y}_j - \bar{x}_j\boldsymbol{\varepsilon})(\boldsymbol{y}_j - \bar{x}_j\boldsymbol{\varepsilon})'}}$$

$$= \frac{\sum (x_{i\alpha} - \bar{x}_i)(x_{j\alpha} - \bar{x}_j)}{\sqrt{\sum (x_{i\alpha} - \bar{x}_i)^2 \sum (x_{j\alpha} - \bar{x}_j)^2}}.$$

As an example of the calculations consider the data in Table 1, taken from Student (1908).

TABLE 1

Patient	Drug A x_1	Drug B x_2
1	1.9	0.7
2	0.8	−1.6
3	1.1	−0.2
4	0.1	−1.2
5	−0.1	−0.1
6	4.4	3.4
7	5.5	3.7
8	1.6	0.8
9	4.6	0.0
10	3.4	2.0

The measurement $x_{11} = 1.9$ on the first patient is the number of hours' increase in his sleep due to the use of the soporific drug A, and $x_{21} = 0.7$ is the number of hours' increase due to drug B, etc. Assuming that each pair (that is, each row in the table) is an observation from $N(\boldsymbol{\mu}, \boldsymbol{\Sigma})$, we find that

$$\hat{\boldsymbol{\mu}} = \bar{x} = \begin{pmatrix} 2.33 \\ 0.75 \end{pmatrix},$$

$$\hat{\boldsymbol{\Sigma}} = \begin{pmatrix} 3.61 & 2.56 \\ 2.56 & 2.88 \end{pmatrix}, \tag{24}$$

$$S = \begin{pmatrix} 4.01 & 2.85 \\ 2.85 & 3.20 \end{pmatrix},$$

and $\hat{\rho}_{12} = r_{12} = 0.7952$. ($S$ will be defined later.)

3.3. THE DISTRIBUTION OF THE SAMPLE MEAN VECTOR; INFERENCE CONCERNING THE MEAN WHEN THE COVARIANCE MATRIX IS KNOWN

3.3.1. Distribution Theory

In the univariate case the mean of a sample is distributed normally and independently of the sample variance. Similarly, the sample mean $\bar{X}$ defined in Section 3.2 is distributed normally and independently of $\hat{\boldsymbol{\Sigma}}$.

To prove this result we shall make a transformation of the set of observation vectors. Because this kind of transformation is used several times in this book we first prove a more general theorem.

THEOREM 3.3.1. *Suppose* $X_1, \cdots, X_N$ *are independent, where* X_α *is distributed according to* $N(\boldsymbol{\mu}_\alpha, \boldsymbol{\Sigma})$. *Let* $C = (c_{\alpha\beta})$ *be an orthogonal matrix.*

Then $Y_\alpha = \sum_{\beta=1}^N c_{\alpha\beta} X_\beta$ *is distributed according to* $N(\nu_\alpha, \Sigma)$, *where* $\nu_\alpha = \sum_{\beta=1}^N c_{\alpha\beta}\mu_\beta$ *and* $Y_1, \cdots, Y_N$ *are independent.*

Proof. The set of vectors $\{Y_\alpha\}$ have a joint normal distribution because the entire set of components are a set of linear combinations of the components of $\{X_\alpha\}$ which have a joint normal distribution. The expected value of Y_α is

$$(1) \qquad \mathscr{E}Y_\alpha = \mathscr{E}\sum_{\beta=1}^N c_{\alpha\beta}X_\beta = \sum_{\beta=1}^N c_{\alpha\beta}\mathscr{E}X_\beta = \sum_{\beta=1}^N c_{\alpha\beta}\mu_\beta = \nu_\alpha.$$

The covariance matrix between Y_α and Y_γ is

$$(2) \qquad \begin{aligned} \mathscr{C}(Y_\alpha, Y_\gamma') &= \mathscr{E}(Y_\alpha - \nu_\alpha)(Y_\gamma - \nu_\gamma)' \\ &= \mathscr{E}\left[\sum_{\beta=1}^N c_{\alpha\beta}(X_\beta - \mu_\beta)\right]\left[\sum_{\varepsilon=1}^N c_{\gamma\varepsilon}(X_\varepsilon - \mu_\varepsilon)'\right] \\ &= \sum_{\beta,\varepsilon=1}^N c_{\alpha\beta}c_{\gamma\varepsilon}\mathscr{E}(X_\beta - \mu_\beta)(X_\varepsilon - \mu_\varepsilon)' \\ &= \sum_{\beta,\varepsilon=1}^N c_{\alpha\beta}c_{\gamma\varepsilon}\delta_{\beta\varepsilon}\Sigma \\ &= \sum_{\beta=1}^N c_{\alpha\beta}c_{\gamma\beta}\Sigma \\ &= \delta_{\alpha\gamma}\Sigma, \end{aligned}$$

where $\delta_{\alpha\gamma}$ is the Kronecker delta ($= 1$ if $\alpha = \gamma$ and $= 0$ if $\alpha \neq \gamma$). This shows that Y_α is independent of Y_γ and Y_α has the covariance matrix Σ.

We also use the following general lemma:

LEMMA 3.3.1. *If* $C = (c_{\alpha\beta})$ *is orthogonal, then* $\sum_{\alpha=1}^N X_\alpha X_\alpha' = \sum_{\alpha=1}^N Y_\alpha Y_\alpha'$, *where* $Y_\alpha = \sum_{\beta=1}^N c_{\alpha\beta} X_\beta$.

Proof.

$$(3) \qquad \begin{aligned} \sum_\alpha Y_\alpha Y_\alpha' &= \sum_\alpha \sum_\beta c_{\alpha\beta}X_\beta \sum_\gamma c_{\alpha\gamma}X_\gamma' \\ &= \sum_{\beta,\gamma}\left(\sum_\alpha c_{\alpha\beta}c_{\alpha\gamma}\right)X_\beta X_\gamma' \\ &= \sum_{\beta,\gamma}\delta_{\beta\gamma}X_\beta X_\gamma' \\ &= \sum_\beta X_\beta X_\beta'. \end{aligned}$$

Let $X_1, \cdots, X_N$ be independent, each distributed according to $N(\mu, \Sigma)$. There exists an $N \times N$ orthogonal matrix $B = (b_{\alpha\beta})$ with the last row

$$(4) \qquad (1/\sqrt{N}, \cdots, 1/\sqrt{N}).$$

This transformation is a rotation in the N-dimensional space described in Section 3.2 with the equiangular line going into the Nth coordinate axis. Let $A = N\hat{\mathbf{\Sigma}}$, defined in Section 3.2, and let

$$Z_\alpha = \sum_\beta b_{\alpha\beta} X_\beta. \tag{5}$$

Then

$$\begin{aligned} Z_N &= \sum_\beta b_{N\beta} X_\beta = \sum_\beta \frac{1}{\sqrt{N}} X_\beta \\ &= \sqrt{N}\bar{X}. \end{aligned} \tag{6}$$

By Lemma 3.3.1, we have

$$\begin{aligned} A &= \sum_{\alpha=1}^{N} X_\alpha X_\alpha' - N\bar{X}\bar{X}' \\ &= \sum_{\alpha=1}^{N} Z_\alpha Z_\alpha' - Z_N Z_N' \\ &= \sum_{\alpha=1}^{N-1} Z_\alpha Z_\alpha'. \end{aligned} \tag{7}$$

Since Z_N is independent of $Z_1, \cdots, Z_{N-1}$, $\bar{X}$ is independent of A. Since

$$\mathscr{E} Z_N = \sum_\beta b_{N\beta} \mathscr{E} X_\beta = \sum_\beta \frac{1}{\sqrt{N}} \mu = \sqrt{N}\mu, \tag{8}$$

Z_N is distributed according to $N(\sqrt{N}\mu, \Sigma)$ and $\bar{X} = (1/\sqrt{N})Z_N$ is distributed according to $N[\mu, (1/N)\Sigma]$. We note

$$\begin{aligned} \mathscr{E} Z_\alpha &= \sum_\beta b_{\alpha\beta} \mathscr{E} X_\beta = \sum_\beta b_{\alpha\beta} \mu \\ &= \sum_\beta b_{\alpha\beta} b_{N\beta} \sqrt{N} \mu \\ &= 0, \qquad \alpha \neq N. \end{aligned} \tag{9}$$

THEOREM 3.3.2. *The mean of a sample of N from $N(\mu, \Sigma)$ is distributed according to $N[\mu, (1/N)\Sigma]$ and independently of $\hat{\Sigma}$, the maximum likelihood estimate of Σ. $N\hat{\Sigma}$ is distributed as $\sum_{\alpha=1}^{N-1} Z_\alpha Z_\alpha'$, where Z_α is distributed according to $N(\mathbf{0}, \Sigma)$ independently of $Z_\beta (\alpha \neq \beta)$.*

We note that

$$\mathscr{E}\hat{\Sigma} = \frac{1}{N} \mathscr{E} \sum_{\alpha=1}^{N-1} Z_\alpha Z_\alpha' = \frac{N-1}{N} \Sigma. \tag{10}$$

Thus $\hat{\Sigma}$ is a biased estimate of Σ. We shall, therefore, define

$$S = \frac{1}{N-1} A = \frac{1}{N-1} \sum_{\alpha=1}^{N} (x_\alpha - \bar{x})(x_\alpha - \bar{x})' \tag{11}$$

as the *sample covariance matrix*. It is an unbiased estimate of $\mathbf{\Sigma}$ and the diagonal elements are the usual (unbiased) sample variances of the components of $\boldsymbol{X}$.

3.3.2. Tests and Confidence Regions for $\boldsymbol{\mu}$ when $\mathbf{\Sigma}$ Is Known

A statistical problem of considerable importance is that of testing the hypothesis that the mean vector of a normal distribution is a given vector, and a related problem is that of giving a confidence region for the unknown vector of means. We now go on to study these problems under the assumption that the covariance matrix $\mathbf{\Sigma}$ is known. In Chapter 5 we consider these problems when the covariance matrix is unknown.

In the univariate case one bases a test or a confidence interval on the fact that the difference between the sample mean and the population mean is normally distributed with a zero mean and known variance; then tables of the normal distribution can be used to set up significance points or to compute confidence intervals. In the multivariate case one uses the fact that the difference between the sample mean vector and the population mean vector is normally distributed with mean vector zero and known covariance matrix. One could set up limits for each component on the basis of the distribution, but this procedure has the disadvantages that the choice of limits is somewhat arbitrary and in the case of tests leads to tests that may be very poor against some alternatives, and, moreover, such limits are difficult to compute because tables are available only for the bivariate case. The procedures given below, however, are easily computed and furthermore can be given general intuitive and theoretical justifications.

The procedures are based on the following theorem:

THEOREM 3.3.3. *If the m-component vector $\boldsymbol{Y}$ is distributed according to $N(\mathbf{0}, \boldsymbol{T})$ (nonsingular), then $\boldsymbol{Y}'\boldsymbol{T}^{-1}\boldsymbol{Y}$ is distributed according to the χ^2-distribution with m degrees of freedom.*

Proof. Let $\boldsymbol{C}$ be a nonsingular matrix such that $\boldsymbol{CTC}' = \boldsymbol{I}$ and define $\boldsymbol{Z} = \boldsymbol{CY}$. Then $\boldsymbol{Z}$ is normally distributed with mean $\mathscr{E}\boldsymbol{Z} = \boldsymbol{C}\mathscr{E}\boldsymbol{Y} = \mathbf{0}$ and covariance matrix $\mathscr{E}\boldsymbol{ZZ}' = \mathscr{E}\boldsymbol{CYY}'\boldsymbol{C}' = \boldsymbol{CTC}' = \boldsymbol{I}$. Then $\boldsymbol{Y}'\boldsymbol{T}^{-1}\boldsymbol{Y} = \boldsymbol{Z}'(\boldsymbol{C}')^{-1}\boldsymbol{T}^{-1}\boldsymbol{C}^{-1}\boldsymbol{Z} = \boldsymbol{Z}'(\boldsymbol{CTC}')^{-1}\boldsymbol{Z} = \boldsymbol{Z}'\boldsymbol{Z}$, which is the sum of squares of the components of $\boldsymbol{Z}$. Since the components of $\boldsymbol{Z}$ are independently distributed according to $N(0, 1)$, $\boldsymbol{Z}'\boldsymbol{Z} = \boldsymbol{Y}'\boldsymbol{T}^{-1}\boldsymbol{Y}$ is a χ^2-variable with m degrees of freedom. (See Problem 5 of Chapter 7.)

Since $\sqrt{N}(\bar{\boldsymbol{X}} - \boldsymbol{\mu})$ is distributed according to $N(\mathbf{0}, \mathbf{\Sigma})$, it follows from the theorem that

$$N(\bar{\boldsymbol{X}} - \boldsymbol{\mu})'\mathbf{\Sigma}^{-1}(\bar{\boldsymbol{X}} - \boldsymbol{\mu}) \tag{12}$$

has a χ^2-distribution with p degrees of freedom. This is the fundamental fact we use in setting up tests and confidence regions concerning $\boldsymbol{\mu}$.

Let $\chi_p^2(\alpha)$ be the number such that

$$\Pr\{\chi_p^2 \geq \chi_p^2(\alpha)\} = \alpha. \tag{13}$$

Thus

$$\Pr\{N(\bar{\mathbf{X}} - \boldsymbol{\mu})'\boldsymbol{\Sigma}^{-1}(\bar{\mathbf{X}} - \boldsymbol{\mu}) \geq \chi_p^2(\alpha)\} = \alpha. \tag{14}$$

To test the hypothesis that $\boldsymbol{\mu} = \boldsymbol{\mu}_0$, where $\boldsymbol{\mu}_0$ is a specified vector, we use as our critical region

$$N(\bar{\mathbf{x}} - \boldsymbol{\mu}_0)'\boldsymbol{\Sigma}^{-1}(\bar{\mathbf{x}} - \boldsymbol{\mu}_0) \geq \chi_p^2(\alpha). \tag{15}$$

If we obtain a sample such that (15) is satisfied, we reject the null hypothesis. It can be seen intuitively that the probability is greater than α of rejecting the hypothesis if $\boldsymbol{\mu}$ is very much different from $\boldsymbol{\mu}_0$, since in the space of $\bar{\mathbf{x}}$ (15) defines an ellipse with a center at $\boldsymbol{\mu}_0$, and when $\boldsymbol{\mu}$ is far from $\boldsymbol{\mu}_0$ the density of $\bar{\mathbf{x}}$ will be concentrated at a point near the edge or outside of the ellipse. The proof of Theorem 3.3.3 can be extended to show that $N(\bar{\mathbf{X}} - \boldsymbol{\mu}_0)'\boldsymbol{\Sigma}^{-1}(\bar{\mathbf{X}} - \boldsymbol{\mu}_0)$ is distributed as a noncentral χ^2 with p degrees of freedom and noncentral parameter $N(\boldsymbol{\mu} - \boldsymbol{\mu}_0)'\boldsymbol{\Sigma}^{-1}(\boldsymbol{\mu} - \boldsymbol{\mu}_0)$ when $\bar{\mathbf{X}}$ is the mean of a sample of N from $N(\boldsymbol{\mu}, \boldsymbol{\Sigma})$ [given by Bose (1936a), (1936b)]. Pearson (1900) first proved Theorem 3.3.3. (See Section 5.4.)

Now consider the following statement made on the basis of a sample with mean $\bar{\mathbf{x}}$: "The mean of the distribution satisfies

$$N(\bar{\mathbf{x}} - \boldsymbol{\mu}^*)'\boldsymbol{\Sigma}^{-1}(\bar{\mathbf{x}} - \boldsymbol{\mu}^*) \leq \chi_p^2(\alpha) \tag{16}$$

as an inequality on $\boldsymbol{\mu}^*$." We see from (14) that the probability that a sample be drawn such that the above statement is true is $1 - \alpha$ because the event in (14) is equivalent to the statement being false. Thus, the set of $\boldsymbol{\mu}^*$ satisfying (16) is a confidence region for $\boldsymbol{\mu}$ of confidence $1 - \alpha$.

In the p-dimensional space of $\bar{\mathbf{x}}$, (15) is the surface and exterior of an ellipsoid with center $\boldsymbol{\mu}_0$, the shape of the ellipsoid depending on $\boldsymbol{\Sigma}^{-1}$ and the size on $(1/N)\chi_p^2(\alpha)$ for given $\boldsymbol{\Sigma}^{-1}$. In the p-dimensional space of $\boldsymbol{\mu}^*$, (16) is the surface and interior of an ellipsoid with its center at $\bar{\mathbf{x}}$. If $\boldsymbol{\Sigma}^{-1} = I$, then (14) says that the probability is α that the distance between $\bar{\mathbf{x}}$ and $\boldsymbol{\mu}$ is greater than $\sqrt{\chi_p^2(\alpha)/N}$.

THEOREM 3.3.4. *If $\bar{\mathbf{x}}$ is the mean of a sample of N drawn from $N(\boldsymbol{\mu}, \boldsymbol{\Sigma})$ and $\boldsymbol{\Sigma}$ is known, then* (15) *gives a critical region of size α for testing the hypothesis $\boldsymbol{\mu} = \boldsymbol{\mu}_0$ and* (16) *gives a confidence region for $\boldsymbol{\mu}$ of confidence $1 - \alpha$. $\chi_p^2(\alpha)$ is chosen to satisfy* (13).

The same technique can be used for the corresponding two sample problems. Suppose we have a sample $\{\mathbf{x}_\alpha^{(1)}\}(\alpha = 1, \cdots, N_1)$ from the distribution $N(\boldsymbol{\mu}^{(1)}, \boldsymbol{\Sigma})$ and a sample $\{\mathbf{x}_\alpha^{(2)}\}(\alpha = 1, \cdots, N_2)$ from a second normal population $N(\boldsymbol{\mu}^{(2)}, \boldsymbol{\Sigma})$ with the same covariance matrix. Then the two sample means

$$\bar{\boldsymbol{x}}^{(1)} = \frac{1}{N_1} \sum_{\alpha=1}^{N_1} \boldsymbol{x}_\alpha^{(1)}, \tag{17}$$

$$\bar{\boldsymbol{x}}^{(2)} = \frac{1}{N_2} \sum_{\alpha=1}^{N_2} \boldsymbol{x}_\alpha^{(2)}$$

are distributed independently according to $N[\boldsymbol{\mu}^{(1)}, (1/N_1)\boldsymbol{\Sigma}]$ and $N[\boldsymbol{\mu}^{(2)}, (1/N_2)\boldsymbol{\Sigma}]$ respectively. The difference of the two sample means $\boldsymbol{y} = \bar{\boldsymbol{x}}^{(1)} - \bar{\boldsymbol{x}}^{(2)}$ is distributed according to $N\{\boldsymbol{\nu}, [(1/N_1) + (1/N_2)]\boldsymbol{\Sigma}\}$, where $\boldsymbol{\nu} = \boldsymbol{\mu}^{(1)} - \boldsymbol{\mu}^{(2)}$. Thus

$$\frac{N_1 N_2}{N_1 + N_2} (\boldsymbol{y} - \boldsymbol{\nu})' \boldsymbol{\Sigma}^{-1} (\boldsymbol{y} - \boldsymbol{\nu}) \leq \chi_p^2(\alpha) \tag{18}$$

is a confidence region for the difference $\boldsymbol{\nu}$ of the two mean vectors, and a critical region for testing the hypothesis $\boldsymbol{\mu}^{(1)} = \boldsymbol{\mu}^{(2)}$ is given by

$$\frac{N_1 N_2}{N_1 + N_2} (\bar{\boldsymbol{x}}^{(1)} - \bar{\boldsymbol{x}}^{(2)})' \boldsymbol{\Sigma}^{-1} (\bar{\boldsymbol{x}}^{(1)} - \bar{\boldsymbol{x}}^{(2)}) \geq \chi_p^2(\alpha). \tag{19}$$

Mahalanobis (1930) suggested $(\boldsymbol{\mu}^{(1)} - \boldsymbol{\mu}^{(2)})' \boldsymbol{\Sigma}^{-1} (\boldsymbol{\mu}^{(1)} - \boldsymbol{\mu}^{(2)})$ as a measure of the distance between two populations.

3.3.3. Sufficient Statistics for $\boldsymbol{\mu}$ and $\boldsymbol{\Sigma}$

It has been shown that

$$\sum_\alpha (\boldsymbol{x}_\alpha - \boldsymbol{\mu})(\boldsymbol{x}_\alpha - \boldsymbol{\mu})' = \boldsymbol{A} + N(\bar{\boldsymbol{x}} - \boldsymbol{\mu})(\bar{\boldsymbol{x}} - \boldsymbol{\mu})', \tag{20}$$

$$\sum_\alpha (\boldsymbol{x}_\alpha - \boldsymbol{\mu})' \boldsymbol{\Sigma}^{-1} (\boldsymbol{x}_\alpha - \boldsymbol{\mu}) = \operatorname{tr}(\boldsymbol{\Sigma}^{-1}\boldsymbol{A}) + N(\bar{\boldsymbol{x}} - \boldsymbol{\mu})' \boldsymbol{\Sigma}^{-1} (\bar{\boldsymbol{x}} - \boldsymbol{\mu}). \tag{21}$$

Thus the density of $\boldsymbol{X}_1, \cdots, \boldsymbol{X}_N$ can be written

$$\begin{aligned} &K \exp\{-\tfrac{1}{2}[N(\bar{\boldsymbol{x}} - \boldsymbol{\mu})' \boldsymbol{\Sigma}^{-1} (\bar{\boldsymbol{x}} - \boldsymbol{\mu}) + \operatorname{tr}(\boldsymbol{\Sigma}^{-1}\boldsymbol{A})]\} \\ &\quad = K_1 \exp[-\tfrac{1}{2} N(\bar{\boldsymbol{x}} - \boldsymbol{\mu})' \boldsymbol{\Sigma}^{-1} (\bar{\boldsymbol{x}} - \boldsymbol{\mu})] K_2 \exp[-\tfrac{1}{2}\operatorname{tr}(\boldsymbol{\Sigma}^{-1}\boldsymbol{A})]. \end{aligned} \tag{22}$$

Thus $\bar{\boldsymbol{x}}$ and $(1/N)\boldsymbol{A}$ form a sufficient set of statistics for $\boldsymbol{\mu}$ and $\boldsymbol{\Sigma}$. If $\boldsymbol{\Sigma}$ is known, $\bar{\boldsymbol{x}}$ is a sufficient statistic for $\boldsymbol{\mu}$. However, if $\boldsymbol{\mu}$ is known, $(1/N)\boldsymbol{A}$ is not a sufficient statistic for $\boldsymbol{\Sigma}$, but $(1/N)\sum_{\alpha=1}^N (\boldsymbol{x}_\alpha - \boldsymbol{\mu})(\boldsymbol{x}_\alpha - \boldsymbol{\mu})'$ is a sufficient statistic for $\boldsymbol{\Sigma}$. The reader is reminded that $\boldsymbol{t}$ is a sufficient statistic for $\boldsymbol{\theta}$ if [Cramér (1946), p. 488]

$$\prod_{\alpha=1}^N f(\boldsymbol{x}_\alpha; \boldsymbol{\theta}) = g(\boldsymbol{t}; \boldsymbol{\theta}) h(\boldsymbol{x}_1, \cdots, \boldsymbol{x}_N), \tag{23}$$

where $f(\boldsymbol{x}_\alpha; \boldsymbol{\theta})$ is the density of the αth observation; $g(\boldsymbol{t}; \boldsymbol{\theta})$ is the density of $\boldsymbol{t}$ and $h(\boldsymbol{x}_1, \cdots, \boldsymbol{x}_N)$ does not depend on $\boldsymbol{\theta}$.

If a q-component random vector Y has mean vector $\mathscr{E}Y = \boldsymbol{\nu}$ and covariance matrix $\mathscr{E}(Y - \boldsymbol{\nu})(Y - \boldsymbol{\nu})' = \boldsymbol{\Psi}$, then

$$(y - \boldsymbol{\nu})'\boldsymbol{\Psi}^{-1}(y - \boldsymbol{\nu}) = q + 2 \tag{24}$$

is called the concentration ellipsoid of Y [see Cramér (1946), p. 300]. The density defined by a uniform distribution over the interior of this ellipsoid has the same mean vector and covariance matrix as Y (see Problem 44 of Chapter 2). Let $\boldsymbol{\theta}$ be a vector of q parameters in a distribution and let $\boldsymbol{t}$ be a vector of unbiased estimates (that is, $\mathscr{E}\boldsymbol{t} = \boldsymbol{\theta}$) based on N observations from that distribution with covariance matrix $\boldsymbol{\Psi}$. Then the ellipsoid

$$N(\boldsymbol{t} - \boldsymbol{\theta})'\mathscr{E}\left(\frac{\partial \log f}{\partial \boldsymbol{\theta}}\right)\left(\frac{\partial \log f}{\partial \boldsymbol{\theta}}\right)'(\boldsymbol{t} - \boldsymbol{\theta}) = q + 2 \tag{25}$$

lies entirely within the ellipsoid of concentration of $\boldsymbol{t}$; $\partial \log f/\partial\boldsymbol{\theta}$ denotes the column vector of derivatives of the density (or probability function) with respect to the components of $\boldsymbol{\theta}$. The discussion by Cramér (1946, p. 495) is in terms of scalar observations, but it is clear that it holds true for vector observations. If (25) is the ellipsoid of concentration of $\boldsymbol{t}$, then $\boldsymbol{t}$ is said to be efficient. In general, the square of the ratio of the volume of (25) to that of the ellipsoid of concentration is defined as the efficiency of $\boldsymbol{t}$. In the case of the multivariate normal distribution, if $\boldsymbol{\theta} = \boldsymbol{\mu}$, the vector of means, then $\bar{\boldsymbol{x}}$ is efficient. If $\boldsymbol{\theta}$ includes both $\boldsymbol{\mu}$ and $\boldsymbol{\Sigma}$, then $\bar{\boldsymbol{x}}$ and S have efficiency $[(N - 1)/N]^{p(p+1)/2}$.

REFERENCES

Section 3.2. Aitken (1948b); Chown and Moran (1951); Cramér (1946), 394–398, 403–404; Des Raj (1953a), (1953b); Dwyer (1949); Fisher (1947b); Frets (1921); Hotelling (1948); Hughes (1949); Kendall (1943), 329–334, 337–339; Mood (1950), 186–188; K. Pearson (1896); Student (1908); Votaw, Rafferty, and Deemer (1950); Yule (1897b), (1907).

Section 3.3. R. C. Bose (1936a), (1936b); S. N. Bose (1936), (1937); Cramér (1946), 313; Das (1948); Mahalanobis (1930), (1936); K. Pearson (1900), (1926), (1928); Wilks (1932b), (1943), 100–101, 103–105, 120–121.

PROBLEMS

1. (Sec. 3.2) Find $\hat{\boldsymbol{\mu}}$, $\hat{\boldsymbol{\Sigma}}$, and $(\hat{\rho}_{ij})$ for the data given in Table 2, taken from Frets (1921).

2. (Sec. 3.2) Verify the numerical results of (24).

3. (Sec. 3.2) Compute $\hat{\boldsymbol{\mu}}$, $\hat{\boldsymbol{\Sigma}}$, S, and $\hat{\rho}$ for the following pairs of observations: (34, 55), (12, 29), (33, 75), (44, 89), (89, 62), (59, 69), (50, 41), (88, 67).

TABLE 2

Head Length, First Son	Head Breadth, First Son	Head Length, Second Son	Head Breadth, Second Son
x_1	x_2	x_3	x_4
191	155	179	145
195	149	201	152
181	148	185	149
183	153	188	149
176	144	171	142
208	157	192	152
189	150	190	149
197	159	189	152
188	152	197	159
192	150	187	151
179	158	186	148
183	147	174	147
174	150	185	152
190	159	195	157
188	151	187	158
163	137	161	130
195	155	183	158
186	153	173	148
181	145	182	146
175	140	165	137
192	154	185	152
174	143	178	147
176	139	176	143
197	167	200	158
190	163	187	150

4. (Sec. 3.2) Use the facts that $|C^*| = \prod \lambda_i$, tr $C^* = \sum \lambda_i$, and $C^* = I$ if $\lambda_1 = \cdots = \lambda_p = 1$, where $\lambda_1, \cdots, \lambda_p$ are the characteristic roots of C^*, to prove Lemma 3.2.2. [*Hint:* Use f as given in the footnote to the proof of Lemma 3.2.2.]

5. (Sec. 3.2) Prove that $\hat{\rho}_{ij}$ is invariant with respect to location and scale transformations (that is, that $\hat{\rho}_{ij} = \hat{\rho}^*_{ij}$, where $\hat{\rho}^*_{ij}$ is computed on the basis of $x^*_{i\alpha} = c_i x_{i\alpha} + d_i$ with $c_i > 0$).

6. (Sec. 3.3) Let X_α be distributed according to $N(\gamma c_\alpha, \Sigma)$, $\alpha = 1, \cdots, N$, where $\sum c_\alpha^2 > 0$. Show that the distribution of $g = (1/\sum c_\alpha^2)\sum c_\alpha X_\alpha$ is $N[\gamma, (1/\sum c_\alpha^2)\Sigma]$. Show that $E = \sum_\alpha (X_\alpha - gc_\alpha)(X_\alpha - gc_\alpha)'$ is independently distributed as $\sum_{\alpha=1}^{N-1} Z_\alpha Z_\alpha'$, where the Z_α are independent, each with distribution $N(\mathbf{0}, \Sigma)$. [*Hint:* Let $Z_\alpha = \sum b_{\alpha\beta} X_\beta$, where $b_{N\beta} = c_\beta / \sqrt{\sum c_\alpha^2}$ and B is orthogonal.]

7. (Sec. 3.3) Let the m-component vector Y be distributed according to $N(\nu, T)$. Prove $Y'T^{-1}Y$ is distributed as $\sum_{i=1}^m Z_i^2$, where the Z_i are independently normally distributed with variances one and $\mathscr{E}Z_1 = \sqrt{\nu' T^{-1} \nu}$ and $\mathscr{E}Z_i = 0$, $i > 1$.

8. (Sec. 3.3) Prove that the power of the test in (19) is a function only of p and $[N_1N_2/(N_1 + N_2)](\boldsymbol{\mu}^{(1)} - \boldsymbol{\mu}^{(2)})'\boldsymbol{\Sigma}^{-1}(\boldsymbol{\mu}^{(1)} - \boldsymbol{\mu}^{(2)})$, given α.

9. (Sec. 3.2) Let x_1 be the body weight (in kilograms) of a cat and x_2 the heart weight (in grams). [Data from Fisher (1947b).]

(*a*) In a sample of 47 female cats the relevant data are

$$\sum \boldsymbol{x}_\alpha = \begin{pmatrix} 110.9 \\ 432.5 \end{pmatrix}, \qquad \sum \boldsymbol{x}_\alpha \boldsymbol{x}_\alpha' = \begin{pmatrix} 265.13 & 1029.62 \\ 1029.62 & 4064.71 \end{pmatrix}.$$

Find $\hat{\boldsymbol{\mu}}$, $\hat{\boldsymbol{\Sigma}}$, $\boldsymbol{S}$, and $\hat{\rho}$.

(*b*) In a sample of 97 male cats the relevant data are

$$\sum \boldsymbol{x}_\alpha = \begin{pmatrix} 281.3 \\ 1098.3 \end{pmatrix}, \qquad \sum \boldsymbol{x}_\alpha \boldsymbol{x}_\alpha' = \begin{pmatrix} 836.75 & 3275.55 \\ 3275.55 & 13056.17 \end{pmatrix}.$$

Find $\hat{\boldsymbol{\mu}}$, $\hat{\boldsymbol{\Sigma}}$, $\boldsymbol{S}$, and $\hat{\rho}$.

10. (Sec. 3.3) Prove that $\bar{\boldsymbol{x}}$ is efficient for estimating $\boldsymbol{\mu}$.

11. (Sec. 3.3) Prove that $\bar{\boldsymbol{x}}$ and $\boldsymbol{S}$ have efficiency $[(N-1)/N]^{p(p+1)/2}$ for estimating $\boldsymbol{\mu}$ and $\boldsymbol{\Sigma}$.

12. Let $\boldsymbol{Z}(k) = (Z_{ij}(k))$, where $i = 1, \cdots, p$, $j = 1, \cdots, q$, and $k = 1, 2, \cdots$, be a sequence of random matrices. Let one norm of a matrix $\boldsymbol{A}$ be $N_1(\boldsymbol{A}) = \max_{i,j} \bmod (a_{ij})$ and another be $N_2(\boldsymbol{A}) = \sum_{i,j} a_{ij}^2 = \operatorname{tr} \boldsymbol{AA}'$. Some alternative ways of defining stochastic convergence of $\boldsymbol{Z}(k)$ to $\boldsymbol{B}$ $(p \times q)$ are

(*a*) $N_1(\boldsymbol{Z}(k) - \boldsymbol{B})$ converges stochastically to 0,

(*b*) $N_2(\boldsymbol{Z}(k) - \boldsymbol{B})$ converges stochastically to 0, and

(*c*) $Z_{ij}(k) - b_{ij}$ converges stochastically to 0, $i = 1, \cdots, p$; $j = 1, \cdots, q$.

Prove that these three definitions are equivalent. Note that the definition of $X(k)$ converging stochastically to a is that for every arbitrary positive δ and ε, we can find K large enough so that for $k > K$

$$\Pr\{|X(k) - a| < \delta\} > 1 - \varepsilon.$$

13. (Sec. 3.2) Prove that $\bar{\boldsymbol{x}}$ is a consistent estimate of $\boldsymbol{\mu}$ and $\boldsymbol{S}$ is a consistent estimate of $\boldsymbol{\Sigma}$; that is, that $\bar{\boldsymbol{x}}$ and $\boldsymbol{S}$ converge stochastically to $\boldsymbol{\mu}$ and $\boldsymbol{\Sigma}$ respectively.

14. Let the q-component vector $\boldsymbol{t}$ be a sufficient statistic for the r-component vector $\boldsymbol{\theta}$. $\boldsymbol{t}$ is said to be complete if $\mathscr{E}_{\boldsymbol{\theta}} f(\boldsymbol{t}) = 0$, for every $\boldsymbol{\theta}$ implies $f(\boldsymbol{t}) = 0$ for every $\boldsymbol{t}$ except in a set of probability 0 for every $\boldsymbol{\theta}$. Prove $\bar{\boldsymbol{x}}$ is complete (for given $\boldsymbol{\Sigma}$). [*Hint:* For $\boldsymbol{\Sigma} = \boldsymbol{I}$,

$$\mathscr{E}_{\boldsymbol{\mu}} f(\bar{\boldsymbol{x}}) = k e^{-N\boldsymbol{\mu}'\boldsymbol{\mu}} \int \cdots \int f(\bar{\boldsymbol{x}})\, e^{N\bar{\boldsymbol{x}}'\boldsymbol{\mu}} e^{-\frac{1}{2}N\bar{\boldsymbol{x}}'\bar{\boldsymbol{x}}}\, d\bar{x}_1 \cdots d\bar{x}_p$$

and use the fact that the integral is the Laplace transform of $f(\bar{\boldsymbol{x}})e^{-\frac{1}{2}N\bar{\boldsymbol{x}}'\bar{\boldsymbol{x}}}$ (Wilks, 1943, p. 39).]

CHAPTER 4

The Distributions and Uses of Sample Correlation Coefficients

4.1. INTRODUCTION

In Chapter 2, in which the multivariate normal distribution was introduced, it was shown that a measure of dependence between two normal variates is the correlation coefficient $\rho_{ij} = \sigma_{ij}/\sqrt{\sigma_{ii}\sigma_{jj}}$. In a conditional distribution of $X_1, \cdots, X_q$, given $X_{q+1} = x_{q+1}, \cdots, X_p = x_p$, the partial correlation $\rho_{ij \cdot q+1, \cdots, p}$ measures the dependence between X_i and X_j. The third kind of correlation discussed was the multiple correlation which measures the relationship between one variate and a set of others. In this chapter we treat the sample equivalents of these quantities; they are point estimates of the population quantities. The distributions of the sample correlations are found. Tests of hypotheses and confidence intervals are treated.

In the cases of joint normal distributions these correlation coefficients are the natural measures of dependence. In the population they are the only parameters except for location (means) and scale (variances) parameters. In the sample the correlation coefficients are derived as the reasonable estimates of the population correlations. Since the sample means and variances are location and scale estimates, the sample correlations (that is, standardized sample second moments) give all possible information about the population correlations. The sample correlations are the functions of the sufficient statistics that are invariant with respect to location and scale transformations; the population correlations are the functions of the parameters that are invariant with respect to these transformations.

In "regression theory" or least squares, one variable is considered random or "dependent" and the others fixed or "independent." In correlation theory we consider several variables as random and treat them symmetrically. If we start with a joint normal distribution and hold all variables fixed except one, we obtain the least squares model because the expected value of the random variable in the conditional distribution is a linear function of the variables held fixed. The sample regression coefficients obtained in least squares are functions of the sample variances and correlations.

In testing independence we will see that we arrive at the same tests in either case (that is, joint normal distribution or in the conditional distribution of least squares). The probability theory under the null hypothesis is the same. The distribution of the test criterion when the null hypothesis is not true differs in the two cases. If all variables may be considered random one uses "correlation" theory as given here; if only one variable is random one uses "least squares" theory (which is considered in some generality in Chapter 8).

4.2. CORRELATION COEFFICIENT OF A BIVARIATE SAMPLE

4.2.1. The Distribution when the Population Correlation Is Zero; Tests of the Hypothesis of Lack of Correlation

In Section 3.2 it was shown that if one has a sample (of p-component vectors) $\boldsymbol{x}_1, \cdots, \boldsymbol{x}_N$ from a normal distribution the maximum likelihood estimate of the correlation between X_i and X_j (two components of the random vector $\boldsymbol{X}$) is

$$r_{ij} = \frac{\sum_{\alpha=1}^{N} (x_{i\alpha} - \bar{x}_i)(x_{j\alpha} - \bar{x}_j)}{\sqrt{\sum_{\alpha=1}^{N} (x_{i\alpha} - \bar{x}_i)^2} \sqrt{\sum_{\alpha=1}^{N} (x_{j\alpha} - \bar{x}_j)^2}} \tag{1}$$

where $x_{i\alpha}$ is the ith component of $\boldsymbol{x}_\alpha$ and

$$\bar{x}_i = \frac{1}{N} \sum_{\alpha=1}^{N} x_{i\alpha}. \tag{2}$$

In this section we shall find the distribution of r_{ij} when the population correlation between X_i and X_j is zero, and we shall see how to use the sample correlation coefficient to test the hypothesis that the population coefficient is zero.

For convenience we shall treat r_{12}; it is clear that the same theory holds for each r_{ij}. Since r_{12} depends only on the first two coordinates of each $\boldsymbol{x}_\alpha$, it is also obvious that to find the distribution of r_{12} we need only consider the joint distribution of (x_{11}, x_{21}), $(x_{12}, x_{22}), \cdots, (x_{1N}, x_{2N})$. We can reformulate the problems to be considered here, therefore, in terms of a bivariate normal distribution. Let $\boldsymbol{x}_1^*, \cdots, \boldsymbol{x}_N^*$ be observation vectors from

$$N\left[\begin{pmatrix} \mu_1 \\ \mu_2 \end{pmatrix}, \begin{pmatrix} \sigma_1^2 & \sigma_1\sigma_2\rho \\ \sigma_2\sigma_1\rho & \sigma_2^2 \end{pmatrix}\right]. \tag{3}$$

We shall consider

$$r = \frac{a_{12}}{\sqrt{a_{11}}\sqrt{a_{22}}}, \tag{4}$$

$$a_{ij} = \sum_{\alpha=1}^{N} (x_{i\alpha} - \bar{x}_i)(x_{j\alpha} - \bar{x}_j) \qquad (i, j = 1, 2) \tag{5}$$

and $\bar{x}_i$ is defined by (2) ($x_{i\alpha}$ being the ith component of $\boldsymbol{x}_\alpha^*$).

From Section 3.3 we see that a_{11}, a_{12}, and a_{22} are distributed like

$$a_{ij} = \sum_{\alpha=1}^{n} z_{i\alpha} z_{j\alpha} \qquad (i, j = 1, 2), \tag{6}$$

where $n = N - 1$ and $(z_{1\alpha}, z_{2\alpha})$ is distributed according to

$$N\left[\begin{pmatrix} 0 \\ 0 \end{pmatrix}, \begin{pmatrix} \sigma_1^2 & \sigma_1\sigma_2\rho \\ \sigma_2\sigma_1\rho & \sigma_2^2 \end{pmatrix}\right]. \tag{7}$$

independently of $(z_{1\beta}, z_{2\beta})$, $\alpha \neq \beta$.

Let $\boldsymbol{z}_i' = (z_{i1}, \cdots, z_{in})$, $i = 1, 2$. These two vectors can be represented in an n-dimensional space; see Figure 1. The correlation coefficient is

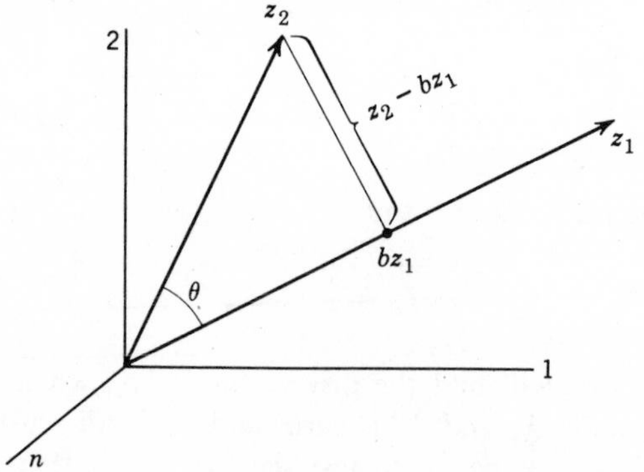

FIGURE 1

the cosine of the angle, say θ, between $\boldsymbol{z}_1$ and $\boldsymbol{z}_2$. To find the distribution of $\cos\theta$ we shall start out to find the distribution of $\cot\theta$. Since $\boldsymbol{z}_2 = (\boldsymbol{z}_2 - b\boldsymbol{z}_1) + b\boldsymbol{z}_1$, we shall let the scalar b be a function of $\boldsymbol{z}_1$ and $\boldsymbol{z}_2$ such that $\boldsymbol{z}_2 - b\boldsymbol{z}_1$ is orthogonal to $b\boldsymbol{z}_1$. Then

$$\cot\theta = b\sqrt{\boldsymbol{z}_1'\boldsymbol{z}_1/(\boldsymbol{z}_2 - b\boldsymbol{z}_1)'(\boldsymbol{z}_2 - b\boldsymbol{z}_1)}.$$

If $\boldsymbol{z}_1$ is fixed we can rotate coordinate axes so that the first coordinate axis lies along $\boldsymbol{z}_1$. Then $b\boldsymbol{z}_1$ has only the first coordinate different from zero and $\mathbf{z}_2 - b\mathbf{z}_1$ has this first coordinate equal to zero. We shall show that $\cot\theta$ is proportional to a t-variable when $\rho = 0$.

The conditional distribution of $Z_{2\alpha}$ (using the capital letter to denote a random variable) given $Z_{1\alpha} = z_{1\alpha}$ is $N(\beta z_{1\alpha}, \sigma^2)$, where $\beta = \rho\sigma_2/\sigma_1$ and $\sigma^2 = \sigma_2^2(1-\rho^2)$ (see Section 2.5). The joint distribution of $\boldsymbol{Z}_2$ given $\boldsymbol{Z}_1 = \mathbf{z}_1$ is $N(\beta\mathbf{z}_1, \sigma^2\boldsymbol{I})$ since the $Z_{2\alpha}$ are independent. More precisely, the joint density of $\boldsymbol{Z}_1$, $\boldsymbol{Z}_2$ is

$$\prod_{\alpha=1}^{n} n\left[\begin{pmatrix} z_{1\alpha} \\ z_{2\alpha} \end{pmatrix} \middle| \begin{pmatrix} 0 \\ 0 \end{pmatrix}, \quad \begin{pmatrix} \sigma_1^2 & \sigma_1\sigma_2\rho \\ \sigma_1\sigma_2\rho & \sigma_2^2 \end{pmatrix}\right];$$

the marginal density of $\boldsymbol{Z}_1$ is

$$\prod_{\alpha=1}^{n} n(z_{1\alpha}|0, \sigma_1^2) = n(\mathbf{z}_1|\mathbf{0}, \sigma_1^2\boldsymbol{I});$$

hence the conditional density of $\boldsymbol{Z}_2$ given $\boldsymbol{Z}_1 = \mathbf{z}_1$ is the joint density of $\boldsymbol{Z}_1$ and $\boldsymbol{Z}_2$ divided by the marginal density of $\boldsymbol{Z}_1$ (at $\mathbf{z}_1$), which is

$$\prod_{\alpha=1}^{n}\left\{n\left[\begin{pmatrix} z_{1\alpha} \\ z_{2\alpha} \end{pmatrix} \middle| \begin{pmatrix} 0 \\ 0 \end{pmatrix}, \begin{pmatrix} \sigma_1^2 & \sigma_1\sigma_2\rho \\ \sigma_1\sigma_2\rho & \sigma_2^2 \end{pmatrix}\right] \middle/ n(z_{1\alpha}|0, \sigma_1^2)\right\} = \prod_{\alpha=1}^{n} n(z_{2\alpha}|\beta z_{1\alpha}, \sigma^2).$$

Let $b = \boldsymbol{Z}_2'\mathbf{z}_1/\mathbf{z}_1'\mathbf{z}_1$ $(= a_{21}/a_{11})$ so that $b\mathbf{z}_1'(\boldsymbol{Z}_2 - b\mathbf{z}_1) = 0$ and let $V = (\boldsymbol{Z}_2 - b\mathbf{z}_1)'(\boldsymbol{Z}_2 - b\mathbf{z}_1) = \boldsymbol{Z}_2'\boldsymbol{Z}_2 - b^2\mathbf{z}_1'\mathbf{z}_1$ $(= a_{22} - a_{12}^2/a_{11})$. Then $\cot\theta = b\sqrt{a_{11}/V}$. The rotation of coordinate axes involves choosing an $(n \times n)$ orthogonal matrix $\boldsymbol{C}$ with first row $(1/c)\mathbf{z}_1'$, where $c^2 = \mathbf{z}_1'\mathbf{z}_1$.

We now apply Theorem 3.3.1 with $\boldsymbol{X}_\alpha = Z_{2\alpha}$. Let $Y_\alpha = \sum_\beta c_{\alpha\beta}Z_{2\beta}$. Then the $\{Y_\alpha\}$ are independently normally distributed with variance σ^2 and means

$$\mathscr{E}Y_1 = \sum_{\gamma=1}^{n} c_{1\gamma}\beta z_{1\gamma} = \frac{\beta}{c}\sum_{\gamma=1}^{n} z_{1\gamma}^2 = \beta c, \tag{8}$$

$$\mathscr{E}Y_\alpha = \sum_{\gamma=1}^{n} c_{\alpha\gamma}\beta z_{1\gamma} = \beta c\sum_{\gamma=1}^{n} c_{\alpha\gamma}c_{1\gamma} = 0, \qquad \alpha \neq 1. \tag{9}$$

We have $b = \sum_\alpha Z_{2\alpha}z_{1\alpha}/\sum_\alpha z_{1\alpha}^2 = c\sum_\alpha Z_{2\alpha}c_{1\alpha}/c^2 = Y_1/c$ and, from Lemma 3.3.1,

$$\begin{aligned} V &= \sum_{\alpha=1}^{n} Z_{2\alpha}^2 - b^2\sum_{\alpha=1}^{n} z_{1\alpha}^2 = \sum_{\alpha=1}^{n} Y_\alpha^2 - Y_1^2 \\ &= \sum_{\alpha=2}^{n} Y_\alpha^2, \end{aligned} \tag{10}$$

which is independent of b.

LEMMA 4.2.1. *If* $(Z_{1\alpha}, Z_{2\alpha}), \alpha = 1, \cdots, n$, *are independent, each pair with distribution* (7), *then the conditional distribution of* $b = \sum_\alpha Z_{2\alpha} Z_{1\alpha} / \sum_\alpha Z_{1\alpha}^2$ *and* $V/\sigma^2 = \sum_\alpha (Z_{2\alpha} - bZ_{1\alpha})^2/\sigma^2$ *given* $Z_{1\alpha} = z_{1\alpha}$ $(\alpha = 1, \cdots, n)$ *is* $N(\beta, \sigma^2/c^2)$ $(c^2 = \sum_\alpha z_{1\alpha}^2)$ *and that of* χ^2 *with* $n - 1$ *degrees of freedom respectively, and* b *and* V *are independent.*

If $\rho = 0$, then $\beta = 0$; then b is distributed conditionally according to $N(0, \sigma^2/c^2)$, and

$$\frac{cb/\sigma}{\sqrt{\dfrac{V/\sigma^2}{n-1}}} = \frac{cb}{\sqrt{\dfrac{V}{n-1}}} \tag{11}$$

has a conditional t-distribution with $n - 1$ degrees of freedom. However, this random variable is

$$\begin{aligned} \sqrt{n-1}\,\frac{\sqrt{a_{11}}\,a_{12}/a_{11}}{\sqrt{a_{22} - a_{12}^2/a_{11}}} &= \sqrt{n-1}\,\frac{a_{12}/\sqrt{a_{11}a_{22}}}{\sqrt{1 - [a_{12}^2/(a_{11}a_{22})]}} \\ &= \sqrt{n-1}\,\frac{r}{\sqrt{1-r^2}}. \end{aligned} \tag{12}$$

Thus $\sqrt{n-1}\, r/\sqrt{1-r^2}$ has a conditional t-distribution with $n - 1$ degrees of freedom. The density of t is

$$\frac{\Gamma(\frac{1}{2}n)}{\sqrt{n-1}\,\Gamma[\frac{1}{2}(n-1)]\sqrt{\pi}} \left(1 + \frac{t^2}{n-1}\right)^{-\frac{1}{2}n}, \tag{13}$$

and the density of $W = r/\sqrt{1-r^2}$ is

$$\frac{\Gamma(\frac{1}{2}n)}{\Gamma[\frac{1}{2}(n-1)]\sqrt{\pi}} (1 + w^2)^{-\frac{1}{2}n}. \tag{14}$$

Since $w = r(1 - r^2)^{-\frac{1}{2}}$, $dw/dr = (1 - r^2)^{-\frac{3}{2}}$. Therefore the density of r is (replacing n by $N - 1$)

$$\frac{\Gamma[\frac{1}{2}(N-1)]}{\Gamma[\frac{1}{2}(N-2)]\sqrt{\pi}} (1 - r^2)^{\frac{1}{2}(N-4)}. \tag{15}$$

It should be noted that (15) is the conditional density of r for z_1 fixed. However, since (15) does not depend on z_1, it is also the marginal density of r.

THEOREM 4.2.1. *Let* $\mathbf{X}_1, \cdots, \mathbf{X}_N$ *be independent, each with distribution* $N(\boldsymbol{\mu}, \boldsymbol{\Sigma})$. *If* $\rho_{ij} = 0$, *the density of* r_{ij} *defined by* (1) *is* (15).

From (15) we see that the density is symmetric about the origin. For

$N > 4$, it has a mode at $r = 0$ and its order of contact with the r axis at ± 1 is $\frac{1}{2}(N - 5)$ for N odd and $\frac{1}{2}N - 3$ for N even. Since the density is even the odd moments are zero; in particular the mean is zero. The even moments are found by integration (letting $x = r^2$ and using the definition of the β-function). It is left to the reader to verify that $\mathscr{E}r^{2m} = \Gamma[\frac{1}{2}(N-1)]\Gamma(m+\frac{1}{2})/\{\sqrt{\pi}\Gamma[\frac{1}{2}(N-1)+m]\}$ and in particular that the variance is $1/(N-1)$.

The most important use of Theorem 4.2.1 is to find significance points for testing the hypothesis that a pair of variables are not correlated. Consider the hypothesis

$$H\colon \rho_{ij} = 0 \tag{16}$$

for some particular pair (i, j). It would seem reasonable to reject this hypothesis if the corresponding sample correlation coefficient were very different from zero. Now how do we decide what we mean by "very different"?

Let us suppose we are interested in testing H against the alternative hypotheses $\rho_{ij} > 0$. Then we reject H if the sample correlation coefficient r_{ij} is greater than some number r_0. The probability of rejecting H when H is true is

$$\int_{r_0}^{1} k_N(r)\, dr, \tag{17}$$

where $k_N(r)$ is (15), the density of a correlation coefficient based on N observations. We choose r_0 so (17) is the desired significance level. If we test H against alternatives $\rho_{ij} < 0$, we reject H when $r_{ij} < -r_0$.

Now suppose we are interested in alternatives $\rho_{ij} \neq 0$; that is, ρ_{ij} may be either positive or negative. Then we reject the hypothesis H if $r_{ij} > r_1$ or $r_{ij} < -r_1$. The probability of rejection when H is true is

$$\int_{-1}^{-r_1} k_N(r)\, dr + \int_{r_1}^{1} k_N(r)\, dr. \tag{18}$$

The number r_1 is chosen so that (18) is the desired significance level.

The significance points r_1 are given in many books including Table VI of Fisher and Yates (1942); the index n in Table VI is equal to our $N - 2$. Since $\sqrt{N-2}\,r/\sqrt{1-r^2}$ has the t-distribution with $N - 2$ degrees of freedom, t-tables can also be used. Against alternatives $\rho_{ij} \neq 0$, reject H if

$$\sqrt{N-2}\,\frac{|r_{ij}|}{\sqrt{1-r_{ij}^2}} > t_{N-2}(\alpha), \tag{19}$$

where $t_{N-2}(\alpha)$ is the two-tailed significance point of the t-statistic with $N-2$ degrees of freedom for significance level α. Against alternatives $\rho_{ij} > 0$, reject H if

$$\sqrt{N-2}\,\frac{r_{ij}}{\sqrt{1-r_{ij}^2}} > \bar{t}_{N-2}(\alpha), \tag{20}$$

where $\bar{t}_{N-2}(\alpha) = t_{N-2}(2\alpha)$ is the one-tailed significance point.

From (11) and (12) we see that $\sqrt{N-2}\,r/\sqrt{1-r^2}$ is the proper statistic for testing the hypothesis that the regression of Z_2 on z_1 is zero. In terms of the original observations $\{x_{i\alpha}\}$, we have

$$\sqrt{N-2}\,\frac{r}{\sqrt{1-r^2}} = \frac{b\sqrt{\sum_\alpha (x_{1\alpha}-\bar{x}_1)^2}}{\sqrt{\sum_\alpha [x_{2\alpha}-\bar{x}_2-b(x_{1\alpha}-\bar{x}_1)]^2/(N-2)}}, \tag{21}$$

where $b = \sum_\alpha (x_{2\alpha}-\bar{x}_2)(x_{1\alpha}-\bar{x}_1)/\sum_\alpha (x_{1\alpha}-\bar{x}_1)^2$ is the least squares regression coefficient of $x_{2\alpha}$ on $x_{1\alpha}$. It is seen that the test of $\rho_{12} = 0$ is equivalent to the test that the regression of X_2 on x_1 is zero (that is, that $\rho_{12}\sigma_2/\sigma_1 = 0$).

To illustrate this procedure we consider the example given in Section 3.2. Let us test the null hypothesis that the effects of the two drugs are uncorrelated against the alternative that they are positively correlated. We shall use the 5% level of significance. For $N = 10$, the 5% significance point (r_0) is 0.5494. Our observed correlation coefficient of 0.7952 is significant; we reject the hypothesis that the effects of the two drugs are independent.

4.2.2. The Distribution when the Population Coefficient Is Nonzero; Tests of Hypotheses and Confidence Regions

To find the distribution of the sample correlation coefficient when the population coefficient is different from zero, we shall first derive the joint density of a_{11}, a_{12}, and a_{22}. In Section 4.2.1 we saw that, conditional on $\mathbf{z}_1$ held fixed, the random variables $b = a_{12}/a_{11}$ and $V/\sigma^2 = (a_{22} - a_{12}^2/a_{11})/\sigma^2$ are distributed independently according to $N(\beta, \sigma^2/c^2)$ and the χ^2-distribution with $n-1$ degrees of freedom, respectively. Denoting the density of the χ^2-distribution by $g_{n-1}(v)$, we write the conditional density of b and V as $n(b|\beta, \sigma^2/a_{11})g_{n-1}(v/\sigma^2)/\sigma^2$. The joint density of $\mathbf{Z}_1$, b, and V is $n(\mathbf{z}_1|\mathbf{0}, \sigma_1^2\mathbf{I})n(b|\beta, \sigma^2/a_{11})g_{n-1}(v/\sigma^2)/\sigma^2$. The marginal density of $\mathbf{Z}_1'\mathbf{Z}_1/\sigma_1^2 = a_{11}/\sigma_1^2$ is $g_n(v)$; that is, the density of a_{11} is

$$\frac{1}{\sigma_1^2}g_n\left(\frac{a_{11}}{\sigma_1^2}\right) = \int \cdots \int_{\mathbf{z}_1'\mathbf{z}_1 = a_{11}} n(\mathbf{z}_1|\mathbf{0}, \sigma_1^2\mathbf{I})\,dW, \tag{22}$$

where dW is the proper volume element.

The integration is over the sphere $z_1'z_1 = a_{11}$; thus, dW is an element of area on this sphere. (See Problem 1 of Chapter 7 for the use of angular coordinates in defining dW.) Thus the joint density of b, V, and a_{11} is

$$(23)\quad \int_{z_1'z_1=a_{11}}\cdots\int n(b|\beta, \sigma^2/a_{11})g_{n-1}(v/\sigma^2)\frac{1}{\sigma^2}n(z_1|\mathbf{0}, \sigma_1^2 I)\,dW$$

$$= g_n(a_{11}/\sigma_1^2)n(b|\beta, \sigma^2/a_{11})g_{n-1}(v/\sigma^2)/(\sigma_1^2\sigma^2)$$

$$= \frac{(a_{11})^{\frac{1}{2}n-1}}{(2\sigma_1^2)^{\frac{1}{2}n}\Gamma(\frac{1}{2}n)}\exp\left(-\frac{1}{2\sigma_1^2}a_{11}\right)\frac{\sqrt{a_{11}}}{\sqrt{2\pi\sigma^2}}\exp\left[-\frac{a_{11}}{2\sigma^2}(b-\beta)^2\right]$$

$$\cdot\frac{1}{(2\sigma^2)^{\frac{1}{2}(n-1)}\Gamma[\frac{1}{2}(n-1)]}v^{\frac{1}{2}(n-3)}\exp\left(-\frac{1}{2\sigma^2}v\right).$$

Now let $b = a_{12}/a_{11}$, $V = a_{22} - a_{12}^2/a_{11}$. The Jacobian is

$$(24)\quad \left|\frac{\partial(b, v)}{\partial(a_{12}, a_{22})}\right| = \begin{vmatrix} \dfrac{1}{a_{11}} & 0 \\ -2\dfrac{a_{12}}{a_{11}} & 1 \end{vmatrix} = \frac{1}{a_{11}}.$$

Thus the density of a_{11}, a_{12}, and a_{22} is

$$(25)\quad \frac{a_{11}^{\frac{1}{2}(n-3)}\left(\dfrac{a_{11}a_{22}-a_{12}^2}{a_{11}}\right)^{\frac{1}{2}(n-3)}e^{-\frac{1}{2}Q}}{2^n\sigma_1^n\left(\dfrac{\sigma_1^2\sigma_2^2-\rho^2\sigma_1^2\sigma_2^2}{\sigma_1^2}\right)^{\frac{1}{2}n}\sqrt{\pi}\Gamma(\frac{1}{2}n)\Gamma[\frac{1}{2}(n-1)]},$$

where

$$(26)\quad Q = \frac{a_{11}}{\sigma_1^2} + \frac{a_{11}}{\sigma^2}\left(\frac{a_{12}^2}{a_{11}^2} - 2\rho\frac{\sigma_1\sigma_2}{\sigma_1^2}\frac{a_{12}}{a_{11}} + \frac{\rho^2\sigma_1^2\sigma_2^2}{\sigma_1^4}\right) + \frac{1}{\sigma^2}\left(a_{22} - \frac{a_{12}^2}{a_{11}}\right)$$

$$= a_{11}\left[\frac{1}{\sigma_1^2} + \frac{\rho^2\sigma_1^2\sigma_2^2}{\sigma_1^4\sigma_2^2(1-\rho^2)}\right] - 2a_{12}\frac{\rho\sigma_2}{\sigma_1\sigma_2^2(1-\rho^2)} + \frac{a_{22}}{\sigma_2^2(1-\rho^2)}$$

$$= \frac{1}{1-\rho^2}\left(\frac{a_{11}}{\sigma_1^2} - 2\rho\frac{a_{12}}{\sigma_1\sigma_2} + \frac{a_{22}}{\sigma_2^2}\right).$$

The density can be written

$$(27)\quad \frac{|A|^{\frac{1}{2}(n-3)}e^{-\frac{1}{2}Q}}{2^n|\mathbf{\Sigma}|^{\frac{1}{2}n}\sqrt{\pi}\Gamma(\frac{1}{2}n)\Gamma[\frac{1}{2}(n-1)]}.$$

This is a special case of the Wishart distribution derived in Chapter 7.

The density of a_{11}, a_{22}, and $r = a_{12}/\sqrt{a_{11}a_{22}}$ ($da_{12} = dr\sqrt{a_{11}a_{22}}$) is

$$\frac{a_{11}^{\frac{1}{2}n-1}a_{22}^{\frac{1}{2}n-1}(1-r^2)^{\frac{1}{2}(n-3)}e^{-\frac{1}{2}Q}}{2^n[\sigma_1^2\sigma_2^2(1-\rho^2)]^{\frac{1}{2}n}\sqrt{\pi}\,\Gamma(\frac{1}{2}n)\Gamma[\frac{1}{2}(n-1)]}, \tag{28}$$

where

$$Q = \frac{1}{(1-\rho^2)}\left(\frac{a_{11}}{\sigma_1^2} - 2\rho r\frac{\sqrt{a_{11}}\sqrt{a_{22}}}{\sigma_1\sigma_2} + \frac{a_{22}}{\sigma_2^2}\right). \tag{29}$$

To find the density of r, we must integrate (28) with respect to a_{11} and a_{22} over the range 0 to ∞. There are various ways of carrying out the integration, which result in different expressions for the density. The method we shall indicate here is straightforward. We expand part of the exponential

$$\exp\left[\frac{\rho r\sqrt{a_{11}}\sqrt{a_{22}}}{(1-\rho^2)\sigma_1\sigma_2}\right] = \sum_{\alpha=0}^{\infty}\frac{(\rho r\sqrt{a_{11}}\sqrt{a_{22}})^\alpha}{\alpha![\sigma_2\sigma_1(1-\rho^2)]^\alpha}. \tag{30}$$

Then the density (28) is

$$\frac{(1-r^2)^{\frac{1}{2}(n-3)}}{\sigma_1^n\sigma_2^n(1-\rho^2)^{\frac{1}{2}n}2^n\sqrt{\pi}\,\Gamma(\frac{1}{2}n)\Gamma[\frac{1}{2}(n-1)]}\sum_{\alpha=0}^{\infty}\frac{(\rho r)^\alpha}{\alpha!(1-\rho^2)^\alpha\sigma_1^\alpha\sigma_2^\alpha} \tag{31}$$

$$\left\{\exp\left[-\frac{a_{11}}{2(1-\rho^2)\sigma_1^2}\right]a_{11}^{\frac{1}{2}(n+\alpha)-1}\right\}\left\{\exp\left[-\frac{a_{22}}{2(1-\rho^2)\sigma_2^2}\right]a_{22}^{\frac{1}{2}(n+\alpha)-1}\right\}.$$

Since

$$\int_0^\infty a_{11}^{\frac{1}{2}(n+\alpha)-1}\exp\left[-\frac{a_{11}}{2(1-\rho^2)\sigma_1^2}\right]da_{11} \tag{32}$$
$$= \Gamma[\tfrac{1}{2}(n+\alpha)][2\sigma_1^2(1-\rho^2)]^{\frac{1}{2}(n+\alpha)},$$

the integral of (31) (term by term integration is permissible) is

$$\frac{(1-r^2)^{\frac{1}{2}(n-3)}}{\sigma_1^n\sigma_2^n(1-\rho^2)^{\frac{1}{2}n}2^n\sqrt{\pi}\,\Gamma(\frac{1}{2}n)\Gamma[\frac{1}{2}(n-1)]}\sum_{\alpha=0}^{\infty}\frac{(\rho r)^\alpha}{\alpha!(1-\rho^2)^\alpha\sigma_1^\alpha\sigma_2^\alpha} \tag{33}$$
$$\cdot\,\Gamma^2[\tfrac{1}{2}(n+\alpha)]2^{n+\alpha}\sigma_1^{n+\alpha}\sigma_2^{n+\alpha}(1-\rho^2)^{n+\alpha}$$
$$= \frac{(1-\rho^2)^{\frac{1}{2}n}(1-r^2)^{\frac{1}{2}(n-3)}}{\sqrt{\pi}\,\Gamma(\frac{1}{2}n)\Gamma[\frac{1}{2}(n-1)]}\sum_{\alpha=0}^{\infty}\frac{(2\rho r)^\alpha}{\alpha!}\Gamma^2[\tfrac{1}{2}(n+\alpha)].$$

If we use the duplication formula $\Gamma(z)\Gamma(z+\frac{1}{2}) = \sqrt{\pi}\,\Gamma(2z)/2^{2z-1}$ we can modify the constant.

THEOREM 4.2.2. *The correlation coefficient in a sample of N from a bivariate normal distribution with correlation ρ is distributed with density*

$$\frac{2^{n-2}(1-\rho^2)^{\frac{1}{2}n}(1-r^2)^{\frac{1}{2}(n-3)}}{(n-2)!\pi}\sum_{\alpha=0}^{\infty}\frac{(2\rho r)^\alpha}{\alpha!}\Gamma^2[\tfrac{1}{2}(n+\alpha)], \tag{34}$$

where $n = N - 1$.

The distribution of r was first found by Fisher (1915). He also gave as another form of the density

$$\frac{(1-\rho^2)^{\frac{1}{2}n}(1-r^2)^{\frac{1}{2}(n-3)}}{\pi(n-2)!}\left[\frac{d^{n-1}}{dx^{n-1}}\left\{\frac{\cos^{-1}(-x)}{\sqrt{1-x^2}}\right\}\bigg|_{x=r\rho}\right]. \tag{35}$$

This is obtained from (28) by letting $a_{11} = ue^{-v}$ and $a_{22} = ue^{v}$.

Hotelling (1953) has made an exhaustive study of the distribution of r. He has recommended the following form, which is derived from (28) by the preceding transformation:

$$\frac{n-1}{\sqrt{2\pi}}\frac{\Gamma(n)}{\Gamma(n+\frac{1}{2})}(1-\rho^2)^{\frac{1}{2}n}(1-r^2)^{\frac{1}{2}(n-3)} (1-\rho r)^{-n+\frac{1}{2}}F\left(\tfrac{1}{2},\tfrac{1}{2};n+\tfrac{1}{2};\frac{1+\rho r}{2}\right), \tag{36}$$

where

$$F(a,b;c;x) = \sum_{j=0}^{\infty}\frac{\Gamma(a+j)}{\Gamma(a)}\frac{\Gamma(b+j)}{\Gamma(b)}\frac{\Gamma(c)}{\Gamma(c+j)}\frac{x^j}{j!} \tag{37}$$

is a hypergeometric function. The series in (36) converges more rapidly than the one in (34). Hotelling discusses methods of integrating the density and also calculates moments of r.

The cumulative distribution of r,

$$\Pr\{r \leq r^*\} = F(r^*|N,\rho), \tag{38}$$

has been tabulated by F. N. David (1938) for* $\rho = 0(.1).9$, $N = 3(1)25$, 50, 100, 200, 400, and $r^* = -1(.05)1$. (It should be noted that David's n is our N.) It is clear from the density (34) that $F(r^*|N,\rho) = 1 - F(-r^*|N,-\rho)$ because the density for r, ρ is equal to the density for $-r, -\rho$. These tables can be used for a number of statistical procedures.

First, we consider the problem of using a sample to test the hypothesis

$$H: \rho = \rho_0. \tag{39}$$

* $\rho = 0(.1).9$ means $\rho = 0, 0.1, 0.2, \cdots, 0.9$.

If the alternatives are $\rho > \rho_0$, we reject the hypothesis if the sample correlation coefficient is greater than r_0 where r_0 is chosen so $1 - F(r_0|N,\rho_0) = \alpha$, the significance level. If the alternatives are $\rho < \rho_0$, we reject the hypothesis if the sample correlation coefficient is less than r_0', where r_0' is chosen so $F(r_0'|N, \rho_0) = \alpha$. If the alternatives are $\rho \neq \rho_0$, the region of rejection is $r > r_1$ and $r < r_1'$ where r_1 and r_1' are chosen so $[1 - F(r_1|N, \rho_0)] + F(r_1'|N, \rho_0) = \alpha$. David suggests that r_1 and r_1' be chosen so $1 - F(r_1|N, \rho_0) = F(r_1'|N, \rho_0) = \frac{1}{2}\alpha$. She has shown (1937) that for $N \geq 10$, $|\rho| \leq 0.8$ this critical region is nearly the region of an unbiased test of H, that is, a test whose power function has its minimum at ρ_0.

It should be pointed out that any test based on r is invariant under transformations $x_{i\alpha}^* = c_i x_{i\alpha} + d_i (\alpha = 1, \cdots, N; i = 1, 2, c_i > 0)$ and r essentially is the only invariant of the sufficient statistics (Problem 19). The above procedure for testing $H : \rho = \rho_0$ against alternatives $\rho > \rho_0$ is uniformly most powerful among all invariant tests. (See Problems 29, 30, and 31.)

As an example suppose one wishes to test the hypothesis that $\rho = 0.5$ against alternatives $\rho \neq 0.5$ at the 5% level of significance using the correlation observed in a sample of 15. In David's tables we find (by interpolation) that $F(0.027|15, 0.5) = 0.025$ and $F(0.805|15, 0.5) = 0.975$. Hence, we reject the hypothesis if our sample r is less than 0.027 or greater than 0.805.

Secondly, we can use David's tables to compute the power function of a test of correlation. If the region of rejection of H is $r > r_1$ and $r < r_1'$, the power of the test is a function of the true correlation ρ, $[1 - F(r_1|N, \rho)] + [F(r_1'|N, \rho)]$; this is the probability of rejecting the null hypothesis when the population correlation is ρ.

As an example consider finding the power function of the test for $\rho = 0$ considered in the preceding section. The rejection region (one-sided) is $r \geq 0.5494$ at the 5% significance level. The probabilities of rejection are

TABLE 1

ρ	Prob.
−1.0	0.0000
−0.8	0.0000
−0.6	0.0004
−0.4	0.0032
−0.2	0.0147
0.0	0.0500
0.2	0.1376
0.4	0.3215
0.6	0.6235
0.8	0.9279
1.0	1.0000

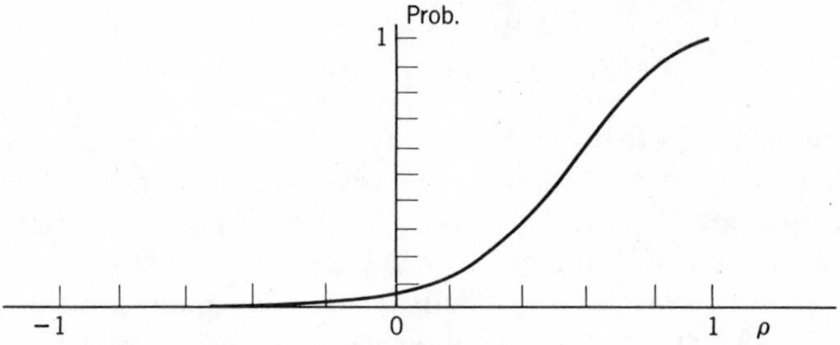

FIGURE 2

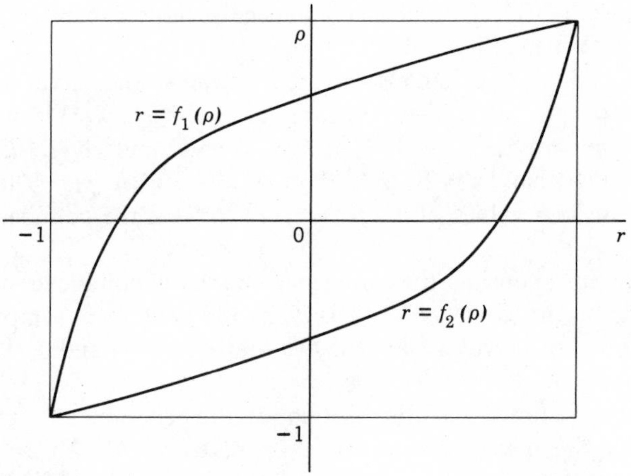

FIGURE 3

given in Table 1. The graph of the power function is illustrated in Figure 2.

Thirdly, David's computations lead to confidence regions for ρ. For given N, r_1' (defining a significance point) is a function of ρ, say, $f_1(\rho)$, and r_1 is another function of ρ, say, $f_2(\rho)$, such that

$$\Pr\{f_1(\rho) < r < f_2(\rho)|\rho\} = 1 - \alpha. \tag{40}$$

Clearly, $f_1(\rho)$ and $f_2(\rho)$ are monotonically increasing functions of ρ if r_1 and r_1' are chosen so $1 - F(r_1|N, \rho) = \frac{1}{2}\alpha = F(r_1'|N, \rho)$. If $\rho = f_i^{-1}(r)$ is the inverse of $r = f_i(\rho)$ $(i = 1, 2)$, then the inequality $f_1(\rho) < r$ is equivalent to†

† The point $(f_1(\rho), \rho)$ on the first curve is to the left of (r, ρ), and the point $(r, f_1^{-1}(r))$ is above (r, ρ).

$\rho < f_1^{-1}(r)$ and $r < f_2(\rho)$ is equivalent to $f_2^{-1}(r) < \rho$. Thus (40) can be written

$$\Pr\{f_2^{-1}(r) < \rho < f_1^{-1}(r)|\rho\} = 1 - \alpha. \tag{41}$$

This equation says that the probability is $1 - \alpha$ that we draw a sample such that the interval $(f_2^{-1}(r), f_1^{-1}(r))$ cover the parameter ρ. Thus this interval is a confidence interval for ρ with confidence coefficient $1 - \alpha$. For a given N and α the curves $r = f_1(\rho)$ and $r = f_2(\rho)$ appear as in Figure 3. In testing the hypothesis $\rho = \rho_0$, the intersection of the line $\rho = \rho_0$ and two curves gives the significance points r_1 and r_1'. In setting up a confidence region for ρ on the basis of a sample correlation r^* we find the limits $f_2^{-1}(r^*)$ and $f_1^{-1}(r^*)$ by the intersection of the line $r = r^*$ with the two curves. David gives these curves for $\alpha = 0.1, 0.05, 0.02$, and 0.01 for various values of N. One-sided confidence regions can be obtained by using only one inequality above.

The tables of $F(r|N, \rho)$ can also be used instead of the curves for finding the confidence interval. Given the sample value r^*, $f_1^{-1}(r^*)$ is the value of ρ such that $\frac{1}{2}\alpha = \Pr\{r \leq r^*|\rho\} = F(r^*|N, \rho)$ and similarly $f_2^{-1}(r^*)$ is the value of ρ such that $\frac{1}{2}\alpha = \Pr\{r \geq r^*|\rho\} = 1 - F(r^*|N, \rho)$. The interval between these two values of ρ $(f_2^{-1}(r^*), f_1^{-1}(r^*))$ is the confidence interval.

As an example, consider the confidence interval of confidence coefficient 0.95 based on the correlation of 0.7952 observed in a sample of 10. Using Graph II of David we find the two limits are 0.34 and 0.94. Hence we state that $0.34 < \rho < 0.94$.

It is of interest to see what the likelihood ratio criterion is for testing the hypothesis $\rho = \rho_0$ given a sample $\boldsymbol{x}_1, \cdots, \boldsymbol{x}_N$ from $N(\boldsymbol{\mu}, \boldsymbol{\Sigma})$, where $\boldsymbol{\mu}' = (\mu_1, \mu_2)$ and $\sigma_{11} = \sigma_1^2$, $\sigma_{12} = \sigma_{21} = \rho\sigma_1\sigma_2$ and $\sigma_{22} = \sigma_2^2$. The likelihood function can be written

$$K\sigma_1^{-N}\sigma_2^{-N}(1 - \rho^2)^{-\frac{1}{2}N} \exp\left\{-\frac{1}{2(1-\rho^2)}\left[\frac{a_{11}}{\sigma_1^2} - 2\rho\frac{a_{12}}{\sigma_1\sigma_2} + \frac{a_{22}}{\sigma_2^2} + N\frac{(\bar{x}_1 - \mu_1)^2}{\sigma_1^2} - 2\rho N\frac{(\bar{x}_1 - \mu_1)(\bar{x}_2 - \mu_2)}{\sigma_1\sigma_2} + N\frac{(\bar{x}_2 - \mu_2)^2}{\sigma_2^2}\right]\right\}. \tag{42}$$

When (42) is maximized with respect to variation of the parameters in Ω (the parameter space restricted only by $\sigma_1^2 > 0$, $\sigma_2^2 > 0$, $\rho^2 < 1$), we obtain $\hat{\sigma}_{1\Omega}^2 = a_{11}/N$, $\hat{\sigma}_{2\Omega}^2 = a_{22}/N$, $\hat{\rho}_\Omega = a_{12}/\sqrt{a_{11}a_{22}}$, $\hat{\mu}_{1\Omega} = \bar{x}_1$, $\hat{\mu}_{2\Omega} = \bar{x}_2$. When (42) is maximized with respect to variation of the parameters in ω (Ω restricted by $\rho = \rho_0$), $\hat{\mu}_{1\omega} = \bar{x}_1$ and $\hat{\mu}_{2\omega} = \bar{x}_2$ since the quadratic form of $\bar{x}_1 - \mu_1$ and $\bar{x}_2 - \mu_2$ in the exponent is negative definite (and hence its maximum value is 0). Then we can maximize the logarithm of (42), which is

$$\text{(43)}\quad \log K - N\log\sigma_1 - N\log\sigma_2 - \tfrac{1}{2}N\log(1-\rho_0^2) - \frac{1}{2(1-\rho_0^2)}\left(\frac{a_{11}}{\sigma_1^2} - 2\rho_0\frac{a_{12}}{\sigma_1\sigma_2} + \frac{a_{22}}{\sigma_2^2}\right),$$

with respect to σ_1 and σ_2. Setting the derivatives equal to 0, we obtain

$$\text{(44)}\quad -\frac{N}{\hat\sigma_{i\omega}} - \frac{1}{2(1-\rho_0^2)}\left(-2\frac{a_{ii}}{\hat\sigma_{i\omega}^3} + 2\rho_0\frac{a_{ij}}{\hat\sigma_{i\omega}^2\hat\sigma_{j\omega}}\right) = 0, \qquad i \neq j,\ i,j = 1,2.$$

This is

$$\text{(45)}\quad \frac{a_{ii}}{\hat\sigma_{i\omega}^2} - \rho_0\frac{a_{ij}}{\hat\sigma_{i\omega}\hat\sigma_{j\omega}} = N(1-\rho_0^2).$$

Adding (45) for $i = 1, j = 2$, and $i = 2, j = 1$, we obtain

$$\text{(46)}\quad \frac{a_{11}}{\hat\sigma_{1\omega}^2} - 2\rho_0\frac{a_{12}}{\hat\sigma_{1\omega}\hat\sigma_{2\omega}} + \frac{a_{22}}{\hat\sigma_{2\omega}^2} = 2N(1-\rho_0^2).$$

Subtracting (45) for $i = 1,\ j = 2$, from (45) for $i = 2,\ j = 1$ gives $a_{11}/\hat\sigma_{1\omega}^2 = a_{22}/\hat\sigma_{2\omega}^2 = a^2/\hat\sigma^2$, say. Then

$$\text{(47)}\quad \frac{a^2}{\hat\sigma^2} - \frac{\rho_0 r a^2}{\hat\sigma^2} = N(1-\rho_0^2).$$

Thus

$$\text{(48)}\quad \frac{N(1-\rho_0^2)}{1-\rho_0 r} = \frac{a^2}{\hat\sigma^2} = \frac{\sqrt{a_{11}a_{22}}}{\hat\sigma_{1\omega}\hat\sigma_{2\omega}}$$

and

$$\text{(49)}\quad |\hat{\mathbf{\Sigma}}_\omega| = (1-\rho_0^2)\hat\sigma_{1\omega}^2\hat\sigma_{2\omega}^2 = \frac{(1-\rho_0 r)^2}{1-\rho_0^2}\frac{a_{11}}{N}\frac{a_{22}}{N}.$$

The maximized likelihood function is

$$\text{(50)}\quad \max_\omega L = \frac{(1-\rho_0^2)^{\frac{1}{2}N}N^N}{(2\pi)^N(1-\rho_0 r)^N a_{11}^{\frac{1}{2}N}a_{22}^{\frac{1}{2}N}}e^{-N}.$$

For variations in Ω, the maximized likelihood function is

$$\text{(51)}\quad \max_\Omega L = \frac{N^N}{(2\pi)^N(1-r^2)^{\frac{1}{2}N}a_{11}^{\frac{1}{2}N}a_{22}^{\frac{1}{2}N}}e^{-N}.$$

The likelihood ratio criterion is, therefore,

$$\frac{\max_{\omega} L}{\max_{\Omega} L} = \frac{(1-\rho_0^2)^{\frac{1}{2}N}(1-r^2)^{\frac{1}{2}N}}{(1-\rho_0 r)^N} = \left[\frac{(1-\rho_0^2)(1-r^2)}{(1-\rho_0 r)^2}\right]^{\frac{1}{2}N}. \tag{52}$$

The likelihood ratio test is $(1-\rho_0^2)(1-r^2)(1-\rho_0 r)^{-2} < c$, where c is chosen so the probability of the inequality when samples are drawn from normal populations with correlation ρ_0 is the prescribed significance level. The critical region can be written equivalently.

$$(\rho_0^2 c - \rho_0^2 + 1)r^2 - 2\rho_0 c r + c - 1 + \rho_0^2 > 0, \tag{53}$$

or

$$r > \frac{\rho_0 c + (1-\rho_0^2)\sqrt{1-c}}{\rho_0^2 c + 1 - \rho_0^2}, \tag{54}$$

$$r < \frac{\rho_0 c - (1-\rho_0^2)\sqrt{1-c}}{\rho_0^2 c + 1 - \rho_0^2}.$$

Thus the likelihood ratio test of $H: \rho = \rho_0$ against alternatives $\rho \neq \rho_0$ has a rejection region of the form $r > r_1$ and $r < r_1'$; but r_1 and r_1' are not chosen so that the probability of each inequality is $\alpha/2$ when H is true, but they are taken to be of the form given in (54) where c is chosen so that the probability of the two inequalities is α.

4.2.3. The Asymptotic Distribution of a Sample Correlation Coefficient and Fisher's z

In this section we shall show that as the sample size increases, a sample correlation coefficient tends to be normally distributed. The distribution of a particular function of a sample correlation, Fisher's z [Fisher (1921)], which has a variance approximately independent of the population correlation, tends to normality faster.

First we prove a multivariate central limit theorem:

THEOREM 4.2.3. *Let the m-component vectors* $\mathbf{Y}_1, \mathbf{Y}_2, \cdots$ *be independently and identically distributed with means* $\mathscr{E}\mathbf{Y}_\alpha = \boldsymbol{\nu}$ *and covariance matrices* $\mathscr{E}(\mathbf{Y}_\alpha - \boldsymbol{\nu})(\mathbf{Y}_\alpha - \boldsymbol{\nu})' = \mathbf{T}$. *Then the limiting distribution of* $(1/\sqrt{n})\sum_{\alpha=1}^{n}(\mathbf{Y}_\alpha - \boldsymbol{\nu})$ *as* $n \to \infty$ *is* $N(\mathbf{0}, \mathbf{T})$.

Proof. Let

$$\phi_n(\mathbf{t}, u) = \mathscr{E} \exp\left[iu\mathbf{t}' \frac{1}{\sqrt{n}} \sum_{\alpha=1}^{n} (\mathbf{Y}_\alpha - \boldsymbol{\nu})\right], \tag{55}$$

where u is a scalar and $\mathbf{t}$ an m-component vector. For fixed $\mathbf{t}$, $\phi_n(\mathbf{t}, u)$ can be considered as the characteristic function of $(1/\sqrt{n})\sum_{\alpha=1}^{n}(\mathbf{t}'\mathbf{Y}_\alpha - \mathscr{E}\mathbf{t}'\mathbf{Y}_\alpha)$.

By the univariate central limit theorem [Cramér (1946), p. 215], the limiting distribution is $N(0, \boldsymbol{t}'\mathbf{T}\boldsymbol{t})$. Therefore (Theorem 2.6.4)

$$\lim_{n\to\infty} \phi_n(\boldsymbol{t}, u) = e^{-\frac{1}{2}u^2\boldsymbol{t}'\mathbf{T}\boldsymbol{t}} \tag{56}$$

for every u and $\boldsymbol{t}$ (for $\boldsymbol{t} = \mathbf{0}$ a special and obvious argument is used). Let $u = 1$ to obtain

$$\lim_{n\to\infty} \mathscr{E} \exp\left[i\boldsymbol{t}' \frac{1}{\sqrt{n}} \sum_{\alpha=1}^{n} (\boldsymbol{Y}_\alpha - \boldsymbol{\nu})\right] = e^{-\frac{1}{2}\boldsymbol{t}'\mathbf{T}\boldsymbol{t}} \tag{57}$$

for every $\boldsymbol{t}$. Since $e^{-\frac{1}{2}\boldsymbol{t}'\mathbf{T}\boldsymbol{t}}$ is continuous at $\boldsymbol{t} = \mathbf{0}$, the convergence is uniform in some neighborhood of $\boldsymbol{t} = \mathbf{0}$. The theorem follows.

Now we wish to show that the sample covariance matrix is asymptotically normally distributed as the sample size increases.

THEOREM 4.2.4. *Let* $A(n) = \sum_{\alpha=1}^{N}(\boldsymbol{X}_\alpha - \bar{\boldsymbol{X}}_N)(\boldsymbol{X}_\alpha - \bar{\boldsymbol{X}}_N)'$, *where* $\boldsymbol{X}_1, \boldsymbol{X}_2, \cdots$ *are independently distributed according to* $N(\boldsymbol{\mu}, \boldsymbol{\Sigma})$ *and* $n = N - 1$. *Then the asymptotic distribution of* $\boldsymbol{B}(n) = (1/\sqrt{n})[A(n) - n\boldsymbol{\Sigma}]$ *is normal with mean* $\mathbf{0}$ *and covariances*

$$\mathscr{E} b_{ij}(n) b_{kl}(n) = \sigma_{ik}\sigma_{jl} + \sigma_{il}\sigma_{jk}. \tag{58}$$

Proof. As shown earlier, $A(n)$ is distributed as $A(n) = \sum_{\alpha=1}^{n} \boldsymbol{Z}_\alpha \boldsymbol{Z}_\alpha'$, where $\boldsymbol{Z}_1, \boldsymbol{Z}_2, \cdots$ are distributed independently according to $N(\mathbf{0}, \boldsymbol{\Sigma})$. We arrange the elements of $\boldsymbol{Z}_\alpha \boldsymbol{Z}_\alpha'$ in a vector such as

$$\boldsymbol{Y}_\alpha = \begin{pmatrix} Z_{1\alpha}^2 \\ Z_{1\alpha}Z_{2\alpha} \\ \cdot \\ \cdot \\ \cdot \\ Z_{2\alpha}^2 \\ \cdot \\ \cdot \\ \cdot \\ Z_{p\alpha}^2 \end{pmatrix}. \tag{59}$$

The moments of $\boldsymbol{Y}_\alpha$ can be deduced from the moments of $\boldsymbol{Z}_\alpha$ as given in Section 2.6. We have $\mathscr{E} Z_{i\alpha}Z_{j\alpha} = \sigma_{ij}$, $\mathscr{E} Z_{i\alpha}Z_{j\alpha}Z_{k\alpha}Z_{l\alpha} = \sigma_{ij}\sigma_{kl} + \sigma_{ik}\sigma_{jl} + \sigma_{il}\sigma_{jk}$, $\mathscr{E}(Z_{i\alpha}Z_{j\alpha} - \sigma_{ij})(Z_{k\alpha}Z_{l\alpha} - \sigma_{kl}) = \sigma_{ik}\sigma_{jl} + \sigma_{il}\sigma_{jk}$. Thus the vectors $\boldsymbol{Y}_\alpha$ defined by (59) satisfy the conditions of Theorem 4.2.3 with the elements of $\boldsymbol{\nu}$ being the elements of $\boldsymbol{\Sigma}$ arranged in vector form similar to (59) and the elements of $\mathbf{T}$ being given above. If the elements of $A(n)$ are arranged in vector form similar to (59), say the vector $\boldsymbol{W}(n)$, then $\boldsymbol{W}(n) - n\boldsymbol{\nu} = \sum_{\alpha=1}^{n}(\boldsymbol{Y}_\alpha - \boldsymbol{\nu})$. By Theorem 4.2.3 $(1/\sqrt{n})[\boldsymbol{W}(n) - n\boldsymbol{\nu}]$ is asymptotically normally distributed with mean $\mathbf{0}$ and the covariance matrix of $\boldsymbol{Y}_\alpha$. Q.E.D.

We are particularly interested in the sample correlation

$$r(n) = \frac{A_{ij}(n)}{\sqrt{A_{ii}(n)A_{jj}(n)}} \tag{60}$$

for some i and j $(i \neq j)$. This can also be written

$$r(n) = \frac{C_{ij}(n)}{\sqrt{C_{ii}(n)C_{jj}(n)}}, \tag{61}$$

where $C_{gh}(n) = A_{gh}(n)/\sqrt{\sigma_{gg}\sigma_{hh}}$. The set $C_{ii}(n)$, $C_{jj}(n)$, and $C_{ij}(n)$ is distributed like

$$\sum_{\alpha=1}^{n} \begin{pmatrix} Z_{i\alpha}^* \\ Z_{j\alpha}^* \end{pmatrix} (Z_{i\alpha}^* \; Z_{j\alpha}^*) = \sum_{\alpha=1}^{n} \begin{pmatrix} Z_{i\alpha}/\sqrt{\sigma_{ii}} \\ Z_{j\alpha}/\sqrt{\sigma_{jj}} \end{pmatrix} (Z_{i\alpha}/\sqrt{\sigma_{ii}}, Z_{j\alpha}/\sqrt{\sigma_{jj}}), \tag{62}$$

where the $(Z_{i\alpha}^*, Z_{j\alpha}^*)$ are independent, each with distribution $N\left[\begin{pmatrix} 0 \\ 0 \end{pmatrix}, \begin{pmatrix} 1 & \rho \\ \rho & 1 \end{pmatrix}\right]$ and $\rho = \sigma_{ij}/\sqrt{\sigma_{ii}\sigma_{jj}}$. Let

$$U(n) = \frac{1}{n}\begin{pmatrix} C_{ii}(n) \\ C_{jj}(n) \\ C_{ij}(n) \end{pmatrix}, \tag{63}$$

$$\boldsymbol{b} = \begin{pmatrix} 1 \\ 1 \\ \rho \end{pmatrix}. \tag{64}$$

Then $\sqrt{n}(\boldsymbol{U}(n) - \boldsymbol{b})$ is asymptotically normally distributed with mean $\mathbf{0}$ and covariance matrix

$$\begin{pmatrix} 2 & 2\rho^2 & 2\rho \\ 2\rho^2 & 2 & 2\rho \\ 2\rho & 2\rho & 1+\rho^2 \end{pmatrix}. \tag{65}$$

Now we need the general theorem:

THEOREM 4.2.5. *Let* $\boldsymbol{U}(n)$ *be an m-component random vector and* $\boldsymbol{b}$ *a fixed vector. Assume* $\sqrt{n}(\boldsymbol{U}(n) - \boldsymbol{b})$ *is asymptotically distributed*† *according to* $N(\mathbf{0}, \mathbf{T})$. *Let* $w = f(\boldsymbol{u})$ *be a function of a*

† Then plim $\boldsymbol{U}(n) = \boldsymbol{b}$; that is, $\boldsymbol{U}(n)$ converges stochastically to $\boldsymbol{b}$; see Problem 12 of Chapter 3.

vector $\boldsymbol{u}$ *with first and second derivatives existing in a neighborhood of* $\boldsymbol{u} = \boldsymbol{b}$. *Let* $\left.\dfrac{\partial f(\boldsymbol{u})}{\partial u_i}\right|_{\boldsymbol{u}=\boldsymbol{b}}$ *be the ith component of* $\boldsymbol{\phi}_b$. *Then the limiting distribution of* $\sqrt{n}[f(\boldsymbol{U}(n)) - f(\boldsymbol{b})]$ *is*

$$N(0, \boldsymbol{\phi}_b'\mathbf{T}\boldsymbol{\phi}_b). \tag{66}$$

This theorem is essentially proved in Cramér (1946), p. 366.

It is clear that $\boldsymbol{U}(n)$ defined by (63) with $\boldsymbol{b}$ and $\mathbf{T}$ defined by (64) and (65), respectively, satisfy the conditions of the theorem. The function

$$r = \frac{u_3}{\sqrt{u_1 u_2}} = u_3 u_1^{-\frac{1}{2}} u_2^{-\frac{1}{2}} \tag{67}$$

satisfies the conditions; the elements of $\boldsymbol{\phi}_b$ are

$$\begin{aligned}
\left.\frac{\partial r}{\partial u_1}\right|_{\boldsymbol{u}=\boldsymbol{b}} &= -\tfrac{1}{2}u_3 u_1^{-\frac{3}{2}} u_2^{-\frac{1}{2}}\Big|_{\boldsymbol{u}=\boldsymbol{b}} = -\tfrac{1}{2}\rho, \\
\left.\frac{\partial r}{\partial u_2}\right|_{\boldsymbol{u}=\boldsymbol{b}} &= -\tfrac{1}{2}u_3 u_1^{-\frac{1}{2}} u_2^{-\frac{3}{2}}\Big|_{\boldsymbol{u}=\boldsymbol{b}} = -\tfrac{1}{2}\rho, \\
\left.\frac{\partial r}{\partial u_3}\right|_{\boldsymbol{u}=\boldsymbol{b}} &= u_1^{-\frac{1}{2}} u_2^{-\frac{1}{2}}\Big|_{\boldsymbol{u}=\boldsymbol{b}} = 1,
\end{aligned} \tag{68}$$

and $f(\boldsymbol{b}) = \rho$. The asymptotic variance of $\sqrt{n}(r(n) - \rho)$ is

$$\begin{aligned}
&(-\tfrac{1}{2}\rho, -\tfrac{1}{2}\rho, 1)\begin{pmatrix} 2 & 2\rho^2 & 2\rho \\ 2\rho^2 & 2 & 2\rho \\ 2\rho & 2\rho & 1+\rho^2 \end{pmatrix}\begin{pmatrix} -\tfrac{1}{2}\rho \\ -\tfrac{1}{2}\rho \\ 1 \end{pmatrix} \\
&\qquad = (\rho - \rho^3, \rho - \rho^3, 1 - \rho^2)\begin{pmatrix} -\tfrac{1}{2}\rho \\ -\tfrac{1}{2}\rho \\ 1 \end{pmatrix} \\
&\qquad = 1 - 2\rho^2 + \rho^4 \\
&\qquad = (1 - \rho^2)^2.
\end{aligned} \tag{69}$$

Thus we obtain the following:

THEOREM 4.2.6. *If* $r(n)$ *is the sample correlation coefficient of a sample of* $N(= n + 1)$ *from a normal distribution with correlation* ρ, *then* $\sqrt{n}(r(n) - \rho)/(1 - \rho^2)$ [*or* $\sqrt{N}(r(n) - \rho)/(1 - \rho^2)$] *is asymptotically distributed according to* $N(0, 1)$.

It is clear from Theorem 4.2.5 that if $f(x)$ is a function with first and second derivatives at $x = \rho$, then $\sqrt{n}[f(r) - f(\rho)]$ is asymptotically normally distributed with mean zero and variance $\left(\left.\dfrac{\partial f}{\partial x}\right|_{x=\rho}\right)^2 (1 - \rho^2)^2$.

A useful function to consider is one whose asymptotic variance is constant (here unity) independent of the parameter ρ. This function satisfies the equation

$$f'(\rho) = \frac{1}{1-\rho^2} = \frac{1}{2}\left(\frac{1}{1+\rho} + \frac{1}{1-\rho}\right). \tag{70}$$

Thus $f(\rho)$ can be taken as $\frac{1}{2}[\log(1+\rho) - \log(1-\rho)] = \frac{1}{2}\log[(1+\rho)/(1-\rho)]$. The so-called "Fisher's z" is

$$z = \tfrac{1}{2}\log\frac{1+r}{1-r}. \tag{71}$$

Let

$$\zeta = \tfrac{1}{2}\log\frac{1+\rho}{1-\rho}. \tag{72}$$

THEOREM 4.2.7. *Let z be defined by* (71) *where r is the correlation coefficient of a sample of $N(= n + 1)$ from a bivariate normal distribution with correlation ρ; let ζ be defined by* (72). *Then $\sqrt{n}(z - \zeta)$ is asymptotically normally distributed with mean* 0 *and variance* 1.

It can be shown that to a closer approximation

$$\mathscr{E}z \sim \zeta + \frac{\rho}{2n}, \tag{73}$$

$$\mathscr{E}(z-\zeta)^2 \sim \frac{1}{n-2} \sim \mathscr{E}\left(z - \zeta - \frac{\rho}{2n}\right)^2. \tag{74}$$

The latter follows from an expression

$$\mathscr{E}(z-\zeta)^2 = \frac{1}{n} + \frac{8-\rho^2}{4n^2} + \cdots \tag{75}$$

and holds good for ρ^2/n^2 small. Hotelling (1953) gives moments of z to order n^{-3}. An important property of Fisher's z is that the approach to normality is much more rapid than for r. David (1938) makes some comparisons between the tabulated probabilities and the probabilities computed by assuming z is normally distributed. She recommends that for $N > 25$ one take z as normally distributed with mean and variance given by (73) and (74).

We shall now indicate how Theorem 4.2.7 can be used.

(*a*) Suppose we wish to test the hypothesis $\rho = \rho_0$ on the basis of a sample of N against the alternatives $\rho \neq \rho_0$. We compute r and then z by (71). Let

$$\zeta_0 = \tfrac{1}{2}\log\frac{1+\rho_0}{1-\rho_0}. \tag{76}$$

Then a region of rejection at the 5% significance level is

$$\sqrt{N-3}\,|z - \zeta_0| > 1.96. \tag{77}$$

A better region is

$$\sqrt{N-3}\,|z - \zeta_0 - \tfrac{1}{2}\rho_0/(N-1)| > 1.96. \tag{78}$$

(*b*) Suppose we have a sample of N_1 from one population and a sample of N_2 from a second population. How do we test the hypothesis that the two correlation coefficients are equal, $\rho_1 = \rho_2$? From Theorem 4.2.7 we know that if the null hypothesis is true $z_1 - z_2$ [where z_1 and z_2 are defined by (71) for the two sample correlation coefficients] is asymptotically normally distributed with mean 0 and variance $1/(N_1 - 3) + 1/(N_2 - 3)$. As a critical region of size 5%, we use

$$\frac{|z_1 - z_2|}{\sqrt{1/(N_1-3) + 1/(N_2-3)}} > 1.96. \tag{79}$$

(*c*) Under the conditions of (*b*) assume that $\rho_1 = \rho_2 = \rho$. How do we use the results of both samples to give a joint estimate of ρ? Since z_1 and z_2 have variances $1/(N_1 - 3)$ and $1/(N_2 - 3)$, respectively, we can estimate ζ by

$$\frac{(N_1 - 3)z_1 + (N_2 - 3)z_2}{N_1 + N_2 - 6} \tag{80}$$

and convert this to an estimate of ρ by the inverse of (71).

(*d*) Let r be the sample correlation from N observations. How do we obtain a confidence interval for ρ? We know that approximately

$$\Pr\{-1.96 < \sqrt{N-3}(z - \zeta) < 1.96\} = 0.95. \tag{81}$$

From this we deduce that $[-1.96/\sqrt{N-3} + z,\ 1.96/\sqrt{N-3} + z]$ is a confidence region for ζ. From this we obtain the region for ρ using the fact $\rho = \tanh \zeta = (e^{\zeta} - e^{-\zeta})/(e^{\zeta} + e^{-\zeta})$ which is a monotonic transformation. Thus the confidence region is

$$\tanh(z - 1.96/\sqrt{N-3}) < \rho < \tanh(z + 1.96/\sqrt{N-3}). \tag{82}$$

4.3. PARTIAL CORRELATION COEFFICIENTS

4.3.1. Estimation of Partial Correlations

Partial correlation coefficients are correlation coefficients in conditional distributions. It was shown in Section 2.5 that if $\boldsymbol{X}$ is distributed according to $N(\boldsymbol{\mu}, \boldsymbol{\Sigma})$ then the conditional distribution of a subvector $\boldsymbol{X}^{(1)}$ given

$\boldsymbol{X}^{(2)} = \boldsymbol{x}^{(2)}$ [where $\boldsymbol{X}' = (\boldsymbol{X}^{(1)\prime}\boldsymbol{X}^{(2)\prime})$] is $N[\boldsymbol{\mu}^{(1)} + \boldsymbol{\beta}(\boldsymbol{x}^{(2)} - \boldsymbol{\mu}^{(2)}), \boldsymbol{\Sigma}_{11\cdot 2}]$, where

$$\boldsymbol{\beta} = \boldsymbol{\Sigma}_{12}\boldsymbol{\Sigma}_{22}^{-1}, \tag{1}$$

$$\boldsymbol{\Sigma}_{11\cdot 2} = \boldsymbol{\Sigma}_{11} - \boldsymbol{\Sigma}_{12}\boldsymbol{\Sigma}_{22}^{-1}\boldsymbol{\Sigma}_{21}. \tag{2}$$

The partial correlations of $\boldsymbol{X}^{(1)}$ given $\boldsymbol{x}^{(2)}$ are the correlations calculated in the usual way from $\boldsymbol{\Sigma}_{11\cdot 2}$. In this section we are interested in statistical problems concerning these correlation coefficients.

First we consider the problem of estimation. Suppose we have a sample of N from $N(\boldsymbol{\mu}, \boldsymbol{\Sigma})$. What are the maximum likelihood estimates of the partial correlations of $\boldsymbol{X}^{(1)}$ (of q components), $\rho_{ij\cdot q+1,\cdots,p}$? We know that the maximum likelihood estimate of $\boldsymbol{\Sigma}$ is

$$\hat{\boldsymbol{\Sigma}} = \frac{1}{N}\sum_{\alpha=1}^{N}(\boldsymbol{x}_\alpha - \bar{\boldsymbol{x}})(\boldsymbol{x}_\alpha - \bar{\boldsymbol{x}})', \tag{3}$$

where $\bar{\boldsymbol{x}} = (1/N)\sum \boldsymbol{x}_\alpha$. The correspondence between $\boldsymbol{\Sigma}$ and $\boldsymbol{\Sigma}_{11\cdot 2}$, $\boldsymbol{\beta}$, and $\boldsymbol{\Sigma}_{22}$ is one-to-one by virtue of (1) and (2) and

$$\boldsymbol{\Sigma}_{12} = \boldsymbol{\beta}\boldsymbol{\Sigma}_{22}, \tag{4}$$

$$\boldsymbol{\Sigma}_{11} = \boldsymbol{\Sigma}_{11\cdot 2} + \boldsymbol{\beta}\boldsymbol{\Sigma}_{22}\boldsymbol{\beta}'. \tag{5}$$

It follows from Corollary 3.2.1 that the maximum likelihood estimates of $\boldsymbol{\Sigma}_{11\cdot 2}$, $\boldsymbol{\beta}$, and $\boldsymbol{\Sigma}_{22}$ are $\hat{\boldsymbol{\Sigma}}_{11\cdot 2} = \hat{\boldsymbol{\Sigma}}_{11} - \hat{\boldsymbol{\Sigma}}_{12}\hat{\boldsymbol{\Sigma}}_{22}^{-1}\hat{\boldsymbol{\Sigma}}_{21}$, $\hat{\boldsymbol{\beta}} = \hat{\boldsymbol{\Sigma}}_{12}\hat{\boldsymbol{\Sigma}}_{22}^{-1}$, and $\hat{\boldsymbol{\Sigma}}_{22}$. Furthermore, it follows that the maximum likelihood estimates of the partial correlations are

$$\hat{\rho}_{ij\cdot q+1,\cdots,p} = \frac{\hat{\sigma}_{ij\cdot q+1,\cdots,p}}{\sqrt{\hat{\sigma}_{ii\cdot q+1,\cdots,p}\hat{\sigma}_{jj\cdot q+1,\cdots,p}}}, \qquad i,j = 1,\cdots,q, \tag{6}$$

where $\hat{\sigma}_{ij\cdot q+1,\cdots,p}$ is the i,jth element of $\hat{\boldsymbol{\Sigma}}_{11\cdot 2}$.

THEOREM 4.3.1. *Let $\boldsymbol{x}_1,\cdots,\boldsymbol{x}_N$ be a sample of N from $N(\boldsymbol{\mu}, \boldsymbol{\Sigma})$. The maximum likelihood estimates of $\rho_{ij\cdot q+1,\cdots,p}$, the partial correlations of the first q components conditional on the last $p - q$ components, are given by*

$$\hat{\rho}_{ij\cdot q+1,\cdots,p} = \frac{a_{ij\cdot q+1,\cdots,p}}{\sqrt{a_{ii\cdot q+1,\cdots,p}a_{jj\cdot q+1,\cdots,p}}}, \tag{7}$$

where

$$(a_{ij\cdot q+1,\cdots,p}) = A_{11} - A_{12}A_{22}^{-1}A_{21} = A_{11\cdot 2}, \tag{8}$$

$$\begin{pmatrix} A_{11} & A_{12} \\ A_{21} & A_{22} \end{pmatrix} = A = \sum_{\alpha=1}^{N}(\boldsymbol{x}_\alpha - \bar{\boldsymbol{x}})(\boldsymbol{x}_\alpha - \bar{\boldsymbol{x}})'. \tag{9}$$

The estimate $\hat{\rho}_{ij\cdot q+1,\cdots,p}$, denoted by $r_{ij\cdot q+1,\cdots,p}$, is called the *sample partial correlation coefficient between X_i and X_j holding $X_{q+1},\cdots,X_p$ fixed.*

Two geometric interpretations of the above theory can be given. In p-dimensional space, $\boldsymbol{x}_1, \cdots, \boldsymbol{x}_N$ represent N points. The sample regression function

(10) $$\boldsymbol{x}^{(1)} = \bar{\boldsymbol{x}}^{(1)} + \hat{\boldsymbol{\beta}}(\boldsymbol{x}^{(2)} - \bar{\boldsymbol{x}}^{(2)})$$

is a $(p - q)$-dimensional hyperplane which is the intersection of q $(p - 1)$-dimensional hyperplanes,

(11) $$x_i = \bar{x}_i + \sum_{j=q+1}^{p} \hat{\beta}_{ij}(x_j - \bar{x}_j), \qquad i = 1, \cdots, q,$$

where x_i, x_j are running variables. Here $\hat{\beta}_{ij}$ is an element of $\hat{\boldsymbol{\beta}} = \hat{\boldsymbol{\Sigma}}_{12}\hat{\boldsymbol{\Sigma}}_{22}^{-1} = \boldsymbol{A}_{12}\boldsymbol{A}_{22}^{-1}$. The ith row of $\hat{\boldsymbol{\beta}}$ is $(\hat{\beta}_{i,q+1}, \cdots, \hat{\beta}_{ip})$. Each right-hand side of (11) is the least squares "regression" function of x_i on $x_{q+1}, \cdots, x_p$; that is, if we project the points $\boldsymbol{x}_1, \cdots, \boldsymbol{x}_N$ on the coordinate hyperplane of $x_i, x_{q+1}, \cdots, x_p$, then (11) is the regression plane. The point with coordinates

(12) $$\begin{aligned} x_i &= \bar{x}_i + \sum_{j=q+1}^{p} \hat{\beta}_{ij}(x_{j\alpha} - \bar{x}_j), & i &= 1, \cdots, q, \\ x_j &= x_{j\alpha}, & j &= q + 1, \cdots, p, \end{aligned}$$

is on the hyperplane (11). The difference in the ith coordinate of $\boldsymbol{x}_\alpha$ and the point (12) is $y_{i\alpha} = x_{i\alpha} - [\bar{x}_i + \sum_{j=q+1}^{p}\hat{\beta}_{ij}(x_{j\alpha} - \bar{x}_j)]$ for $i = 1, \cdots, q$ and 0 for the other coordinates. Let

(13) $$\boldsymbol{y}_\alpha = \begin{pmatrix} y_{1\alpha} \\ \cdot \\ \cdot \\ \cdot \\ y_{q\alpha} \end{pmatrix}.$$

These points can be represented as N points in a q-dimensional space. Then $\boldsymbol{A}_{11\cdot 2} = \sum_{\alpha=1}^{N} \boldsymbol{y}_\alpha \boldsymbol{y}_\alpha'$.

We can also interpret the sample as p points in N-space. Let $\boldsymbol{z}_j = (x_{j1}, \cdots, x_{jN})$ be the jth point, and let $\boldsymbol{z}_{p+1} = (1, \cdots, 1)$ be a $(p + 1)$st point. The point with coordinates $\bar{x}_i, \cdots, \bar{x}_i$ is $\bar{x}_i \boldsymbol{z}_{p+1}$. The projection of $\boldsymbol{z}_i$ on the hyperplane spanned by $\boldsymbol{z}_{q+1}, \cdots, \boldsymbol{z}_{p+1}$ is

(14) $$\boldsymbol{z}_i^* = \bar{x}_i \boldsymbol{z}_{p+1} + \sum_{j=q+1}^{p} \hat{\beta}_{ij}(\boldsymbol{z}_j - \bar{x}_j \boldsymbol{z}_{p+1});$$

this is the point on the hyperplane that is at a minimum distance from $\boldsymbol{z}_i$. Let $\tilde{\boldsymbol{z}}_i$ be the vector from $\boldsymbol{z}_i^*$ to $\boldsymbol{z}_i$, that is, $\boldsymbol{z}_i - \boldsymbol{z}_i^*$, or equivalently this vector translated so that one end point is at the origin. The set of vectors $\tilde{\boldsymbol{z}}_1, \cdots, \tilde{\boldsymbol{z}}_q$ are the projections of $\boldsymbol{z}_1, \cdots, \boldsymbol{z}_q$ on the hyperplane, orthogonal to $\mathbf{z}_{q+1}, \cdots, \mathbf{z}_{p+1}$. Then $\tilde{\mathbf{z}}_i'\tilde{\mathbf{z}}_i = a_{ii \cdot q+1, \cdots, p}$, the length squared of $\tilde{\mathbf{z}}_i$ (that is, the distance squared of $\mathbf{z}_i$ from $\mathbf{z}_i^*$). $\tilde{\mathbf{z}}_i'\tilde{\mathbf{z}}_j/\sqrt{\tilde{\mathbf{z}}_i'\tilde{\mathbf{z}}_i\tilde{\mathbf{z}}_j'\tilde{\mathbf{z}}_j} = r_{ij \cdot q+1, \cdots, p}$ is the cosine of the angle between $\tilde{\mathbf{z}}_i$ and $\tilde{\mathbf{z}}_j$.

As an example of partial correlation study we consider some data [Hooker (1907)] on yield of hay (X_1) in hundredweights per acre, spring rainfall (X_2) in inches, and accumulated termperature above 42° F in the spring (X_3) for an English area over 20 years. The estimates of μ_i, σ_i $(= \sqrt{\sigma_{ii}})$, and ρ_{ij} are

$$\hat{\boldsymbol{\mu}} = \bar{x} = \begin{pmatrix} 28.02 \\ 4.91 \\ 594 \end{pmatrix},$$

$$\begin{pmatrix} \hat{\sigma}_1 \\ \hat{\sigma}_2 \\ \hat{\sigma}_3 \end{pmatrix} = \begin{pmatrix} 4.42 \\ 1.10 \\ 85 \end{pmatrix}, \tag{15}$$

$$\begin{pmatrix} 1 & \hat{\rho}_{12} & \hat{\rho}_{13} \\ \hat{\rho}_{21} & 1 & \hat{\rho}_{23} \\ \hat{\rho}_{31} & \hat{\rho}_{32} & 1 \end{pmatrix} = \begin{pmatrix} 1.00 & 0.80 & -0.40 \\ 0.80 & 1.00 & -0.56 \\ -0.40 & -0.56 & 1.00 \end{pmatrix}.$$

From the correlations we observe that yield and rainfall are positively related, yield and temperature are negatively related, and rainfall and temperature are negatively related. What interpretation is to be given to the apparent negative relation between yield and temperature? Does high temperature tend to cause low yield or is high temperature associated with low rainfall and hence with low yield? To answer this question we consider the correlation between yield and temperature when rainfall is held fixed; that is, we use the data given above to estimate the partial correlation between X_1 and X_3. It is*

$$\frac{\hat{\sigma}_{13 \cdot 2}}{\sqrt{\hat{\sigma}_{11 \cdot 2}\hat{\sigma}_{33 \cdot 2}}} = 0.097. \tag{16}$$

Thus, if the effect of rainfall is removed, yield and temperature are positively correlated. The conclusion is that both high rainfall and high temperature increase hay yield, but in most years high rainfall occurs with low temperature and vice versa.

4.3.2. The Distribution of the Sample Partial Correlation Coefficient

The partial correlations are computed from $A_{11 \cdot 2} = A_{11} - A_{12}A_{22}^{-1}A_{21}$ (as indicated in Theorem 4.3.1) in the same way that correlations are

* We compute with $\hat{\boldsymbol{\Sigma}}$ as if it were $\boldsymbol{\Sigma}$.

computed from A. To obtain the distribution of the correlations we showed that A was distributed as $\sum_{\alpha=1}^{N-1} Z_\alpha Z'_\alpha$, where the Z_α are independent, each with distribution $N(\mathbf{0},\mathbf{\Sigma})$. To find the distribution of a sample partial correlation we shall show a similar result. This will follow from the following theorem which is stated in more general form so that it can be used more easily later.

THEOREM 4.3.2. *Suppose* $Y_1, \cdots, Y_m$ *are independent with* Y_α *distributed according to* $N(\mathbf{\Gamma} w_\alpha, \mathbf{\Phi})$, *where* w_α *is an r-component vector. Let* $G = \sum_\alpha Y_\alpha w'_\alpha H^{-1}$, *where* $H = \sum_\alpha w_\alpha w'_\alpha$ *and is nonsingular. Then* $\sum_{\alpha=1}^{m} Y_\alpha Y'_\alpha - GHG'$ *is distributed as* $\sum_{\alpha=1}^{m-r} U_\alpha U'_\alpha$, *where the* U_α *are independently distributed, each according to* $N(\mathbf{0}, \mathbf{\Phi})$, *and independently of* G.

Proof. Let $W = (w_1, \cdots, w_m)$ and let F be a square matrix such that (see Theorem 6 of Appendix 1) $FHF' = I$; then $F'^{-1}H^{-1}F^{-1} = I$. Let $E_2 = FW$; thus $W = F^{-1}E_2$. Then

$$E_2E_2' = FWW'F' = F\sum_{\alpha=1}^{m} w_\alpha w'_\alpha F' = FHF' = I. \tag{17}$$

Thus the m-component rows of E_2 are orthogonal and of unit length. It is possible to find an $(m - r) \times m$ matrix E_1 such that

$$E = \begin{pmatrix} E_1 \\ E_2 \end{pmatrix} \tag{18}$$

is orthogonal (see Appendix 1, Lemma 2). Now let $Y = (Y_1, \cdots, Y_m) = UE$ or $U = YE'$ (that is, $U_\alpha = \Sigma_\beta e_{\alpha\beta} Y_\beta$). By Theorem 3.3.1 the columns of U, say U_α, are independently and normally distributed, each with covariance matrix $\mathbf{\Phi}$. The means are given by

$$\begin{aligned} \mathscr{E}U = \mathscr{E}YE' &= \mathbf{\Gamma} WE' \\ &= \mathbf{\Gamma} F^{-1}E_2(E_1'\ E_2') \\ &= (\mathbf{0}\ \mathbf{\Gamma} F^{-1}). \end{aligned} \tag{19}$$

To complete the proof of Theorem 4.3.2 we need to show that

$$\sum Y_\alpha Y'_\alpha - GHG' = \sum_{\alpha=1}^{m-r} U_\alpha U'_\alpha. \tag{20}$$

From Theorem 3.3.2 we have

$$\sum_{\alpha=1}^{m} Y_\alpha Y'_\alpha = \sum_{\alpha=1}^{m} U_\alpha U'_\alpha.$$

Also

$$GHG' = (YW'H^{-1})H(H^{-1}WY') \tag{21}$$
$$= UEE_2'(F^{-1})'H^{-1}F^{-1}E_2E'U'$$
$$= U\begin{pmatrix}E_1\\E_2\end{pmatrix}E_2'E_2(E_1'\ E_2')U'$$
$$= U\begin{pmatrix}0\\I\end{pmatrix}(0\ I)U'$$
$$= \sum_{\alpha=m-r+1}^{m} U_\alpha U_\alpha'.$$

Thus

$$\sum_{\alpha=1}^{m} Y_\alpha Y_\alpha' - GHG' = \sum_{\alpha=1}^{m} U_\alpha U_\alpha' - \sum_{\alpha=m-r+1}^{m} U_\alpha U_\alpha' = \sum_{\alpha=1}^{m-r} U_\alpha U_\alpha'. \tag{22}$$

This proves the theorem.

It follows from the above considerations that when $\mathbf{\Gamma} = \mathbf{0}$, $\mathscr{E}U = \mathbf{0}$, and we obtain the following:

COROLLARY 4.3.1. *If* $\mathbf{\Gamma} = \mathbf{0}$, *the matrix* GHG' *defined in Theorem* 4.3.2 *is distributed as* $\sum_{\alpha=m-r+1}^{m} U_\alpha U_\alpha'$, *where the* U_α *are independently distributed, each according to* $N(\mathbf{0},\mathbf{\Phi})$.

We now find the distribution of $A_{11\cdot2}$ in the same form. It was shown in Theorem 3.3.1 that A is distributed as $\sum_{\alpha=1}^{N-1} Z_\alpha Z_\alpha'$, where the Z_α are independent, each with distribution $N(\mathbf{0}, \mathbf{\Sigma})$. Let Z_α be partitioned into two subvectors of q and $p-q$ components respectively,

$$Z_\alpha = \begin{pmatrix}Z_\alpha^{(1)}\\Z_\alpha^{(2)}\end{pmatrix}. \tag{23}$$

Then $A_{ij} = \sum_{\alpha=1}^{N-1} Z_\alpha^{(i)} Z_\alpha^{(j)\prime}$. The conditional density of $Z_1^{(1)}, \cdots, Z_{N-1}^{(1)}$ given $Z_1^{(2)} = z_1^{(2)}, \cdots, Z_{N-1}^{(2)} = z_{N-1}^{(2)}$ is

$$\frac{\prod_{\alpha=1}^{N-1} n(z_\alpha|\mathbf{0}, \mathbf{\Sigma})}{\prod_{\alpha=1}^{N-1} n(z_\alpha^{(2)}|\mathbf{0}, \mathbf{\Sigma}_{22})} = \prod_{\alpha=1}^{N-1} \frac{n(z_\alpha|\mathbf{0}, \mathbf{\Sigma})}{n(z_\alpha^{(2)}|\mathbf{0}, \mathbf{\Sigma}_{22})} \tag{24}$$
$$= \prod_{\alpha=1}^{N-1} n(z_\alpha^{(1)}|\mathbf{\beta} z_\alpha^{(2)}, \mathbf{\Sigma}_{11\cdot2}).$$

where $\mathbf{\beta} = \mathbf{\Sigma}_{12}\mathbf{\Sigma}_{22}^{-1}$ and $\mathbf{\Sigma}_{11\cdot2} = \mathbf{\Sigma}_{11} - \mathbf{\Sigma}_{12}\mathbf{\Sigma}_{22}^{-1}\mathbf{\Sigma}_{21}$. Now we apply Theorem 4.3.2 with $Z_\alpha^{(1)} = Y_\alpha$, $z_\alpha^{(2)} = w_\alpha$, $N-1 = m$, $p-q = r$, $\mathbf{\beta} = \mathbf{\Gamma}$, $\mathbf{\Sigma}_{11\cdot2} =$

$\boldsymbol{\Phi}$, $A_{11} = \sum Y_\alpha Y'_\alpha$, $A_{12}A_{22}^{-1} = G$, $A_{22} = H$. We find that the conditional distribution of $A_{11} - (A_{12}A_{22}^{-1})A_{22}(A_{22}^{-1}A'_{12}) = A_{11\cdot 2}$ given $Z_\alpha^{(2)} = z_\alpha^{(2)}$ is that of $\sum_{\alpha=1}^{N-1-(p-q)} U_\alpha U'_\alpha$, where the U_α are independent, each with distribution $N(\mathbf{0}, \boldsymbol{\Sigma}_{11\cdot 2})$. Since this distribution does not depend on $\{z_\alpha^{(2)}\}$ we obtain

THEOREM 4.3.3. *The matrix* $\mathbf{A}_{11\cdot 2} = \mathbf{A}_{11} - \mathbf{A}_{12}\mathbf{A}_{22}^{-1}\mathbf{A}_{21}$ *is distributed as* $\sum_{\alpha=1}^{N-1-(p-q)} \mathbf{U}_\alpha \mathbf{U}'_\alpha$, *where the* $\mathbf{U}_\alpha$ *are independently distributed, each according to* $N(\mathbf{0}, \boldsymbol{\Sigma}_{11\cdot 2})$.

As a corollary we obtain

COROLLARY 4.3.2. *If* $\boldsymbol{\Sigma}_{12} = \mathbf{0}(\boldsymbol{\beta} = \mathbf{0})$, *then* $\mathbf{A}_{11\cdot 2}$ *is distributed as* $\sum_{\alpha=1}^{N-1-(p-q)} \mathbf{U}_\alpha \mathbf{U}'_\alpha$ *and* $\mathbf{A}_{12}\mathbf{A}_{22}^{-1}\mathbf{A}_{21}$ *is distributed as* $\sum_{\alpha=N-(p-q)}^{N-1} \mathbf{U}_\alpha \mathbf{U}'_\alpha$, *where the* $\mathbf{U}_\alpha$ *are independently distributed, each according to* $N(\mathbf{0}, \boldsymbol{\Sigma}_{11\cdot 2})$.

Now it follows that the distribution of $r_{ij\cdot q+1,\cdots,p}$ based on N observations is the same as an ordinary correlation coefficient based on $N - (p - q)$ observations with a corresponding population correlation value of $\rho_{ij\cdot q+1,\cdots,p}$.

THEOREM 4.3.4. *If the cdf of* r_{ij} *based on a sample of* N *from a normal distribution with correlation* ρ_{ij} *is denoted by* $F(r|N, \rho_{ij})$, *then the cdf of the sample partial correlation* $r_{ij\cdot q+1,\cdots,p}$ *based on a sample of* N *from a normal distribution is* $F[r|N - (p - q), \rho_{ij\cdot q+1,\cdots,p}]$.

This distribution was derived by Fisher in 1924.

4.3.3. Tests of Hypotheses and Confidence Regions for Partial Correlation Coefficients

Since the distribution of a sample partial correlation, $r_{ij\cdot q+1,\cdots,p}$ based on a sample of N from a distribution with population correlation $\rho_{ij\cdot q+1,\cdots,p}$ equal to a certain value, ρ, say, is the same as the distribution of an ordinary correlation r based on a sample of size $N - (p - q)$ from a distribution with the corresponding population correlation of ρ, all statistical inference procedures for the ordinary population correlation can be used for the partial correlation. The procedure for the partial correlation is exactly the same except that N is replaced by $N - (p - q)$. To illustrate this rule we give two examples.

Example 1. Suppose that on the basis of a sample of size N we wish to obtain a confidence interval for $\rho_{ij\cdot q+1,\cdots,p}$. The sample partial correlation is $r_{ij\cdot q+1,\cdots,p}$. The procedure is to use David's charts for $N - (p - q)$. In the example at the end of Section 4.3.1, we might want to find a confidence interval for $\rho_{12\cdot 3}$ with confidence coefficient 0.95. The sample partial correlation is $r_{12\cdot 3} = 0.790$. We use the chart (or table) for $N - (p - q) = 20 - 1 = 19$. The interval is $0.52 < \rho_{12\cdot 3} < 0.92$.

Example 2. Suppose that on the basis of a sample of size N we use

Fisher's z for an approximate significance test of $\rho_{ij\cdot q+1,\cdots,p} = \rho_0$ against two-sided alternatives. We let

$$z = \tfrac{1}{2}\log\frac{1 + r_{ij\cdot q+1,\cdots,p}}{1 - r_{ij\cdot q+1,\cdots,p}},$$

(25)

$$\zeta_0 = \tfrac{1}{2}\log\frac{1 + \rho_0}{1 - \rho_0}.$$

Then $\sqrt{N - (p - q) - 3}\,(z - \zeta_0)$ is compared with the significance points of the standardized normal distribution. In the example at the end of Section 4.3.1, we might wish to test the hypothesis $\rho_{13\cdot 2} = 0$ at the 0.05 level. Then $\zeta_0 = 0$ and $\sqrt{20 - 1 - 3}\,(0.0973) = 0.3892$. This value is clearly nonsignificant ($|0.3892| < 1.96$), and hence the data do not indicate rejection of the null hypothesis.

4.4. THE MULTIPLE CORRELATION COEFFICIENT

4.4.1. Estimation of Multiple Correlation

The population multiple correlation between one variate and a set of variates was defined in Section 2.5. For sake of convenience in this section we will treat the case of the multiple correlation between X_1 and the set $X_2, \cdots, X_p$; we shall not need subscripts on R. The variables can always be numbered so that the desired multiple correlation is this one (any irrelevant variables being omitted). Then the multiple correlation in the population is

$$(1)\qquad \bar{R} = \frac{\boldsymbol{\beta}\boldsymbol{\Sigma}_{22}\boldsymbol{\beta}'}{\sqrt{\sigma_{11}\boldsymbol{\beta}\boldsymbol{\Sigma}_{22}\boldsymbol{\beta}'}} = \sqrt{\frac{\boldsymbol{\beta}\boldsymbol{\Sigma}_{22}\boldsymbol{\beta}'}{\sigma_{11}}} = \sqrt{\frac{\boldsymbol{\sigma}_{(1)}\boldsymbol{\Sigma}_{22}^{-1}\boldsymbol{\sigma}_{(1)}'}{\sigma_{11}}},$$

where $\boldsymbol{\beta}$, $\boldsymbol{\sigma}_{(1)}$, and $\boldsymbol{\Sigma}_{22}$ are defined by

$$(2)\qquad \boldsymbol{\Sigma} = \begin{pmatrix} \sigma_{11} & \boldsymbol{\sigma}_{(1)} \\ \boldsymbol{\sigma}_{(1)}' & \boldsymbol{\Sigma}_{22} \end{pmatrix},$$

$$(3)\qquad \boldsymbol{\beta} = \boldsymbol{\sigma}_{(1)}\boldsymbol{\Sigma}_{22}^{-1}.$$

Given a sample $\boldsymbol{x}_1, \cdots, \boldsymbol{x}_N$ $(N > p)$ we estimate $\boldsymbol{\Sigma}$ by $S = [N/(N-1)]\hat{\boldsymbol{\Sigma}}$ or

$$(4)\qquad \hat{\boldsymbol{\Sigma}} = \frac{1}{N}A = \frac{1}{N}\sum_{\alpha=1}^{N}(\boldsymbol{x}_\alpha - \bar{\boldsymbol{x}})(\boldsymbol{x}_\alpha - \bar{\boldsymbol{x}})' = \begin{pmatrix} \hat{\sigma}_{11} & \hat{\boldsymbol{\sigma}}_{(1)} \\ \hat{\boldsymbol{\sigma}}_{(1)}' & \hat{\boldsymbol{\Sigma}}_{22} \end{pmatrix},$$

and we estimate $\boldsymbol{\beta}$ by $\hat{\boldsymbol{\beta}} = \hat{\boldsymbol{\sigma}}_{(1)}\hat{\boldsymbol{\Sigma}}_{22}^{-1} = \boldsymbol{a}_{(1)}A_{22}^{-1}$. We define the sample multiple correlation coefficient by

$$(5)\qquad R = \sqrt{\frac{\hat{\boldsymbol{\beta}}\hat{\boldsymbol{\Sigma}}_{22}\hat{\boldsymbol{\beta}}'}{\hat{\sigma}_{11}}} = \sqrt{\frac{\hat{\boldsymbol{\sigma}}_{(1)}\hat{\boldsymbol{\Sigma}}_{22}^{-1}\hat{\boldsymbol{\sigma}}_{(1)}'}{\hat{\sigma}_{11}}} = \sqrt{\frac{\boldsymbol{a}_{(1)}A_{22}^{-1}\boldsymbol{a}_{(1)}'}{a_{11}}}.$$

That this is the maximum likelihood estimate of $\bar{R}$ is justified by Corollary 3.2.1 since we can define $\bar{R}$, $\boldsymbol{\sigma}_{(1)}$, $\boldsymbol{\Sigma}_{22}$ as a one-to-one transformation of $\boldsymbol{\Sigma}$. Another expression for R [see (20) of Section 2.5] is

$$(6)\qquad 1 - R^2 = \frac{|\hat{\boldsymbol{\Sigma}}|}{\hat{\sigma}_{11}|\hat{\boldsymbol{\Sigma}}_{22}|} = \frac{|A|}{a_{11}|A_{22}|}.$$

R and $\hat{\boldsymbol{\beta}}$ have the same properties in the sample that $\bar{R}$ and $\boldsymbol{\beta}$ have in the population. For example, of all [$(p-1)$-component row] vectors $\boldsymbol{d}$ defining linear combinations $\boldsymbol{d}\boldsymbol{x}_\alpha^{(2)}$ of the components of $\boldsymbol{x}_\alpha^{(2)}$, the vector $\boldsymbol{d} = \hat{\boldsymbol{\beta}}$ is the one that minimizes $\sum_{\alpha=1}^N [(x_{1\alpha} - \bar{x}_1) - \boldsymbol{d}(\boldsymbol{x}_\alpha^{(2)} - \bar{\boldsymbol{x}}^{(2)})]^2$. First we observe that

$$(7)\qquad \sum_{\alpha=1}^N [(x_{1\alpha} - \bar{x}_1) - \hat{\boldsymbol{\beta}}(\boldsymbol{x}_\alpha^{(2)} - \bar{\boldsymbol{x}}^{(2)})](\boldsymbol{x}_\alpha^{(2)} - \bar{\boldsymbol{x}}^{(2)})' = \boldsymbol{a}_{(1)} - \hat{\boldsymbol{\beta}}A_{22} = \boldsymbol{0},$$

since $\hat{\boldsymbol{\beta}} = \boldsymbol{a}_{(1)}A_{22}^{-1}$. Thus

$$(8)\qquad \begin{aligned} &\sum_{\alpha=1}^N [(x_{1\alpha} - \bar{x}_1) - \boldsymbol{d}(\boldsymbol{x}_\alpha^{(2)} - \bar{\boldsymbol{x}}^{(2)})]^2 \\ &\quad = \sum_{\alpha=1}^N \{[(x_{1\alpha} - \bar{x}_1) - \hat{\boldsymbol{\beta}}(\boldsymbol{x}_\alpha^{(2)} - \bar{\boldsymbol{x}}^{(2)})] + (\hat{\boldsymbol{\beta}} - \boldsymbol{d})(\boldsymbol{x}_\alpha^{(2)} - \bar{\boldsymbol{x}}^{(2)})\}^2 \\ &\quad = \sum_{\alpha=1}^N [(x_{1\alpha} - \bar{x}_1) - \hat{\boldsymbol{\beta}}(\boldsymbol{x}_\alpha^{(2)} - \bar{\boldsymbol{x}}^{(2)})]^2 + (\hat{\boldsymbol{\beta}} - \boldsymbol{d})A_{22}(\hat{\boldsymbol{\beta}} - \boldsymbol{d})'. \end{aligned}$$

Since A_{22} is positive definite (except for samples occurring with probability 0), the minimum occurs when $\boldsymbol{d} - \hat{\boldsymbol{\beta}} = \boldsymbol{0}$. This proves the result. The minimum is

$$(9)\qquad \begin{aligned} \sum_{\alpha=1}^N [(x_{1\alpha} - \bar{x}_1) - \hat{\boldsymbol{\beta}}(\boldsymbol{x}_\alpha^{(2)} - \bar{\boldsymbol{x}}^{(2)})]^2 &= a_{11} - \hat{\boldsymbol{\beta}}A_{22}\hat{\boldsymbol{\beta}}' \\ &= a_{11} - \boldsymbol{a}_{(1)}A_{22}^{-1}\boldsymbol{a}_{(1)}' \\ &= a_{11\cdot 2} \end{aligned}$$

(as defined in Section 4.3, with $q = 1$).

This result can be given an interesting geometric interpretation. The N-component vector with αth component $x_{i\alpha} - \bar{x}_i$ is the projection of the vector with αth component $x_{i\alpha}$ on the plane orthogonal to the equiangular line. We have p such vectors. $\boldsymbol{d}(\boldsymbol{x}_\alpha^{(2)} - \bar{\boldsymbol{x}}^{(2)})$ is the αth component of a vector in the hyperplane spanned by the last $p - 1$ vectors. Since (8) is the distance between the first vector and the linear combination of the last $p - 1$ vectors, $\hat{\boldsymbol{\beta}}(\boldsymbol{x}_\alpha^{(2)} - \bar{\boldsymbol{x}}^{(2)})$ is a component of the vector which minimizes this distance. The interpretation of (7) is that the vector with component $(x_{1\alpha} - \bar{x}_1) - \hat{\beta}(x_\alpha^{(2)} - \bar{x}^{(2)})$ is orthogonal to each of the last $p - 1$ vectors. Thus the vector with component $\hat{\beta}(x_\alpha^{(2)} - \bar{x}^{(2)})$ is the projection of the first vector on the hyperplane. See Figure 4. The length squared of the projection vector is

$$\sum_{\alpha=1}^{N} [\hat{\beta}(x_\alpha^{(2)} - \bar{x}^{(2)})]^2 = \hat{\beta} A_{22} \hat{\beta}' = a_{(1)} A_{22}^{-1} a_{(1)}'$$

and the length of the first vector is $\sum_\alpha (x_{1\alpha} - \bar{x}_1)^2 = a_{11}$. Thus R is the cosine of the angle between the first vector and its projection.

In Section 3.2 we saw that the ordinary correlation coefficient is the cosine of the angle between the two vectors involved (in the plane orthogonal to the equiangular line). Another property of R is that it is the

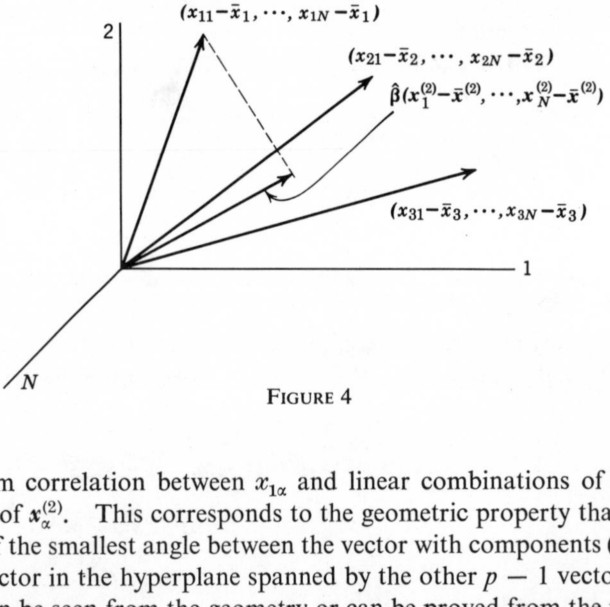

FIGURE 4

maximum correlation between $x_{1\alpha}$ and linear combinations of the components of $x_\alpha^{(2)}$. This corresponds to the geometric property that R is the cosine of the smallest angle between the vector with components $(x_{1\alpha} - \bar{x}_1)$ and a vector in the hyperplane spanned by the other $p - 1$ vectors. This result can be seen from the geometry or can be proved from the preceding result here in a manner analogous to the proof used for $\bar{R}$.

The geometric interpretations are entirely in terms of the vectors in the $(N - 1)$-dimensional hyperplane orthogonal to the equiangular line. It was shown in Section 3.3 that the vector $(x_{i1} - \bar{x}_i, \cdots, x_{iN} - \bar{x}_i)$ in this hyperplane can be designated as $(z_{i1}, \cdots, z_{i,N-1})$, where the $z_{i\alpha}$ are the coordinates referred to an $(N - 1)$-dimensional coordinate system in the hyperplane. It was shown that the new coordinates are obtained from the old by the transformation $x_{i\alpha} = \sum_{\beta=1}^{N} z_{i\beta} b_{\beta\alpha}$, where B is an orthogonal matrix with last row $(1/\sqrt{N}, \cdots, 1/\sqrt{N})$. Then

$$a_{ij} = \sum_{\alpha=1}^{N} (x_{i\alpha} - \bar{x}_i)(x_{j\alpha} - \bar{x}_j) = \sum_{\alpha=1}^{N-1} z_{i\alpha} z_{j\alpha}. \tag{10}$$

It will be convenient to refer to the multiple correlation defined in terms of $z_{i\alpha}$ as the "multiple correlation without subtracting the means."

The computation of R involves taking the square root of the ratio of $\boldsymbol{a}_{(1)} A_{22}^{-1} \boldsymbol{a}'_{(1)}$ to a_{11}. Since A is computed directly from the observations, only the calculation of $\boldsymbol{a}_{(1)} A_{22}^{-1} \boldsymbol{a}'_{(1)}$ requires special techniques. This kind of computation is considered in Section 5.3.1.

4.4.2. Distribution of the Sample Multiple Correlation Coefficient when the Population Multiple Correlation Coefficient Is Zero

From (5) we have

$$R^2 = \frac{\boldsymbol{a}_{(1)} A_{22}^{-1} \boldsymbol{a}'_{(1)}}{a_{11}}; \tag{11}$$

then

$$1 - R^2 = 1 - \frac{\boldsymbol{a}_{(1)} A_{22}^{-1} \boldsymbol{a}'_{(1)}}{a_{11}} = \frac{a_{11} - \boldsymbol{a}_{(1)} A_{22}^{-1} \boldsymbol{a}'_{(1)}}{a_{11}} \tag{12}$$

$$= \frac{a_{11\cdot 2}}{a_{11}},$$

and

$$\frac{R^2}{1 - R^2} = \frac{\boldsymbol{a}_{(1)} A_{22}^{-1} \boldsymbol{a}'_{(1)}}{a_{11\cdot 2}}. \tag{13}$$

For $q = 1$, Corollary 4.3.2 states that when $\boldsymbol{\beta} = \boldsymbol{0}$, that is, when $\bar{R} = 0$, $a_{11\cdot 2}$ is distributed as $\sum_{\alpha=1}^{N-p} V_\alpha^2$ and $\boldsymbol{a}_{(1)} A_{22}^{-1} \boldsymbol{a}'_{(1)}$ is distributed as $\sum_{\alpha=N-p+1}^{N-1} V_\alpha^2$, where V_α are independent, each with distribution $N(0, \sigma_{11\cdot 2})$. Then $a_{11\cdot 2}/\sigma_{11\cdot 2}$ and $\boldsymbol{a}_{(1)} A_{22}^{-1} \boldsymbol{a}'_{(1)}/\sigma_{11\cdot 2}$ are distributed independently as χ^2-variables with $N - p$ and $p - 1$ degrees of freedom respectively. Thus

$$\frac{R^2}{1 - R^2} \cdot \frac{N - p}{p - 1} = \frac{\boldsymbol{a}_{(1)} A_{22}^{-1} \boldsymbol{a}'_{(1)}/\sigma_{11\cdot 2}}{a_{11\cdot 2}/\sigma_{11\cdot 2}} \cdot \frac{N - p}{p - 1} \tag{14}$$

$$= \frac{\chi_{p-1}^2}{\chi_{N-p}^2} \cdot \frac{N - p}{p - 1}$$

$$= F_{p-1, N-p}$$

has the F-distribution with $p-1$ and $N-p$ degrees of freedom. The density of F is

$$\frac{\Gamma[\frac{1}{2}(N-1)]}{\Gamma[\frac{1}{2}(p-1)]\Gamma[\frac{1}{2}(N-p)]}\left(\frac{p-1}{N-p}\right)^{\frac{1}{2}(p-1)} f^{\frac{1}{2}(p-1)-1}\left(1+\frac{p-1}{N-p}f\right)^{-\frac{1}{2}(N-1)} \tag{15}$$

Thus the density of

$$R=\sqrt{\frac{\dfrac{p-1}{N-p}F_{p-1,\,N-p}}{1+\dfrac{p-1}{N-p}F_{p-1,\,N-p}}} \tag{16}$$

is

$$2\frac{\Gamma[\frac{1}{2}(N-1)]}{\Gamma[\frac{1}{2}(p-1)]\Gamma[\frac{1}{2}(N-p)]}R^{p-2}(1-R^2)^{\frac{1}{2}(N-p)-1}. \tag{17}$$

THEOREM 4.4.1. *Let R be the sample multiple correlation coefficient [defined by (5)] between X_1 and $\boldsymbol{X}^{(2)\prime}=(X_2,\cdots,X_p)$ based on a sample of N from $N(\boldsymbol{\mu},\boldsymbol{\Sigma})$. If $\bar{R}=0$ (that is, if $(\sigma_{12},\cdots,\sigma_{1p})=\boldsymbol{\sigma}_{(1)}=\boldsymbol{0}=\boldsymbol{\beta}$), then $[R^2/(1-R^2)]\cdot[(N-p)/(p-1)]$ is distributed as F with $p-1$ and $N-p$ degrees of freedom.*

It should be noticed that $p-1$ is the number of components of $\boldsymbol{X}^{(2)}$ and $N-p=N-(p-1)-1$. If the multiple correlation is between a component X_i and q other components, the numbers are q and $N-q-1$.

It might be observed that $R^2/(1-R^2)$ is the quantity that arises in regression (or least squares) theory for testing the hypothesis that the regression of X_1 on $X_2,\cdots,X_p$ is zero.

If $\bar{R}\neq 0$, the distribution of R is much more difficult to derive. This distribution will be obtained in Section 4.4.3.

Now let us consider the statistical problem of testing the hypothesis $H\colon \bar{R}=0$ on the basis of a sample of N from $N(\boldsymbol{\mu},\boldsymbol{\Sigma})$. [$\bar{R}$ is the population multiple correlation between X_1 and $(X_2,\cdots,X_p)$.] Since $\bar{R}\geq 0$, the alternatives considered are $\bar{R}>0$.

Let us derive the likelihood ratio test of this hypothesis. The likelihood function is

$$L(\boldsymbol{\mu}^*,\boldsymbol{\Sigma}^*)=\frac{1}{(2\pi)^{\frac{1}{2}pN}|\boldsymbol{\Sigma}^*|^{\frac{1}{2}N}}\exp\left[-\tfrac{1}{2}\sum_{\alpha}(\boldsymbol{x}_\alpha-\boldsymbol{\mu}^*)'\boldsymbol{\Sigma}^{*-1}(\boldsymbol{x}_\alpha-\boldsymbol{\mu}^*)\right]. \tag{18}$$

The observations are given; L is a function of the indeterminates $\boldsymbol{\mu}^*,\boldsymbol{\Sigma}^*$. Let ω be the region in the parameter space Ω specified by the null hypothesis. The likelihood ratio criterion is

$$\lambda = \frac{\max_{\boldsymbol{\mu}^*,\, \boldsymbol{\Sigma}^* \epsilon \omega} L(\boldsymbol{\mu}^*, \boldsymbol{\Sigma}^*)}{\max_{\boldsymbol{\mu}^*,\, \boldsymbol{\Sigma}^* \epsilon \Omega} L(\boldsymbol{\mu}^*, \boldsymbol{\Sigma}^*)}. \tag{19}$$

The likelihood ratio test consists in rejecting the null hypothesis if λ is less than some predetermined constant. Intuitively, one rejects the null hypothesis if the density of the observations under the most favorable choice of parameters in the null hypothesis is much less than the density under the most favorable unrestricted choice of parameters. Likelihood ratio tests have some desirable asymptotic properties [Wald (1943)]. In most problems concerning the multivariate normal distribution, the likelihood ratio tests are optimal, or at least reasonable.

Here Ω is the space of $\boldsymbol{\mu}^*, \boldsymbol{\Sigma}^*$ positive definite and ω is the region in this space where $\bar{R} = \sqrt{\boldsymbol{\sigma}_{(1)}\boldsymbol{\Sigma}_{22}^{-1}\boldsymbol{\sigma}_{(1)}'}/\sqrt{\sigma_{11}} = 0$; that is, where $\boldsymbol{\sigma}_{(1)}\boldsymbol{\Sigma}_{22}^{-1}\boldsymbol{\sigma}_{(1)}' = 0$. Because $\boldsymbol{\Sigma}_{22}^{-1}$ is positive definite, this condition is equivalent to $\boldsymbol{\sigma}_{(1)} = \mathbf{0}$. The maximum of $L(\boldsymbol{\mu}^*, \boldsymbol{\Sigma}^*)$ over Ω occurs at $\boldsymbol{\mu}^* = \hat{\boldsymbol{\mu}} = \bar{\boldsymbol{x}}$ and $\boldsymbol{\Sigma}^* = \hat{\boldsymbol{\Sigma}} = (1/N)A = (1/N)\sum_{\alpha=1}^{N}(\boldsymbol{x}_\alpha - \bar{\boldsymbol{x}})(\boldsymbol{x}_\alpha - \bar{\boldsymbol{x}})'$ and is

$$\max_{\boldsymbol{\mu}^*,\, \boldsymbol{\Sigma}^* \epsilon \Omega} L(\boldsymbol{\mu}^*, \boldsymbol{\Sigma}^*) = \frac{N^{\frac{1}{2}pN} e^{-\frac{1}{2}pN}}{(2\pi)^{\frac{1}{2}pN} |A|^{\frac{1}{2}N}}. \tag{20}$$

In ω, the likelihood function is

$$L(\boldsymbol{\mu}^*, \boldsymbol{\Sigma}^* | \boldsymbol{\sigma}_{(1)}^* = \mathbf{0}) = \frac{1}{(2\pi)^{\frac{1}{2}N} \sigma_{11}^{*\frac{1}{2}N}} \exp\left[-\tfrac{1}{2}\sum(x_{1\alpha} - \mu_1^*)^2/\sigma_{11}^*\right] \tag{21}$$

$$\cdot \frac{1}{(2\pi)^{\frac{1}{2}(p-1)N} |\boldsymbol{\Sigma}_{22}^*|^{\frac{1}{2}N}} \exp\left[-\tfrac{1}{2}\sum(\boldsymbol{x}_\alpha^{(2)} - \boldsymbol{\mu}^{(2)*})'\boldsymbol{\Sigma}_{22}^{*-1}(\boldsymbol{x}_\alpha^{(2)} - \boldsymbol{\mu}^{(2)*})\right].$$

The first factor is maximized at $\mu_1^* = \hat{\mu}_1 = \bar{x}_1$ and $\sigma_{11}^* = \hat{\sigma}_{11} = (1/N)a_{11}$, and the second factor is maximized at $\boldsymbol{\mu}^{(2)*} = \hat{\boldsymbol{\mu}}^{(2)} = \bar{\boldsymbol{x}}^{(2)}$ and $\boldsymbol{\Sigma}_{22}^* = \hat{\boldsymbol{\Sigma}}_{22} = (1/N)A_{22}$. The value of the maximized function is

$$\max_{\boldsymbol{\mu}^*,\, \boldsymbol{\Sigma}^* \epsilon \omega} L(\boldsymbol{\mu}^*, \boldsymbol{\Sigma}^*) = \frac{N^{\frac{1}{2}N} e^{-\frac{1}{2}N}}{(2\pi)^{\frac{1}{2}N} a_{11}^{\frac{1}{2}N}} \cdot \frac{N^{\frac{1}{2}(p-1)N} e^{-\frac{1}{2}(p-1)N}}{(2\pi)^{\frac{1}{2}(p-1)N} |A_{22}|^{\frac{1}{2}N}}. \tag{22}$$

Thus the likelihood ratio criterion is [see (6)]

$$\lambda = \frac{|A|^{\frac{1}{2}N}}{a_{11}^{\frac{1}{2}N} |A_{22}|^{\frac{1}{2}N}} = (1 - R^2)^{\frac{1}{2}N}. \tag{23}$$

The likelihood ratio test consists of the critical region $\lambda < \lambda_0$, where λ_0 is chosen so the probability of the above inequality when $\bar{R} = 0$ is the significance level α. It is clear that an equivalent test is

$$1 - \lambda^{2/N} = R^2 > 1 - \lambda_0^{2/N}. \tag{24}$$

Since $[R^2/(1 - R^2)][(N - p)/(p - 1)]$ is a monotonic function of R, an equivalent test involves this ratio being larger than a constant. When $\bar{R} = 0$, this ratio has an $F_{p-1, N-p}$-distribution. Hence, the test region is

$$\frac{R^2}{1 - R^2} \cdot \frac{N - p}{p - 1} > F_{p-1, N-p}(\alpha), \tag{25}$$

where $F_{p-1, N-p}(\alpha)$ is the (upper) significance point corresponding to the α significance level.

THEOREM 4.4.2. *Given a sample $\boldsymbol{x}_1, \cdots, \boldsymbol{x}_N$ from $N(\boldsymbol{\mu}, \boldsymbol{\Sigma})$, the likelihood ratio test at significance level α for the hypothesis $\bar{R} = 0$, where $\bar{R}$ is the population multiple correlation coefficient between X_1 and $(X_2, \cdots, X_p)$, is given by* (25), *where R is the sample multiple correlation coefficient defined by* (5).

From the fact that the density of R is monotonically increasing in $\bar{R}$ (see Section 4.4.3) one can assert that the test (25) is uniformly most powerful for testing $\bar{R} = 0$ in the class of tests depending on R. Since R is invariant under transformations $x_{1\alpha}^* = cx_{1\alpha} + d$ and $\boldsymbol{x}_\alpha^{(2)*} = \boldsymbol{C}\boldsymbol{x}_\alpha^{(2)} + \boldsymbol{e}$ and is the only function of the sufficient statistics that is invariant (see Problem 20), one can state that the test (25) is the uniformly most powerful invariant test.

As an example consider the data given at the end of Section 4.3.1. The sample multiple correlation coefficient is given by

$$1 - R^2 = \frac{\begin{vmatrix} 1 & r_{12} & r_{13} \\ r_{12} & 1 & r_{23} \\ r_{13} & r_{23} & 1 \end{vmatrix}}{\begin{vmatrix} 1 & r_{23} \\ r_{23} & 1 \end{vmatrix}} = \frac{\begin{vmatrix} 1.00 & 0.80 & -0.40 \\ 0.80 & 1.00 & -0.56 \\ -0.40 & -0.56 & 1.00 \end{vmatrix}}{\begin{vmatrix} 1.00 & -0.56 \\ -0.56 & 1.00 \end{vmatrix}} = 0.357. \tag{26}$$

Thus R is 0.802. If we wish to test the hypothesis at the 0.01 level that hay yield is independent of spring rainfall and temperature, we compare $[R^2/(1 - R^2)][(20 - 3)/(3 - 1)] = 15.3$ with $F_{2,17}(0.01) = 6.11$ and find the result significant; that is, we reject the null hypothesis.

It should be noted that the test of independence between X_1 and $(X_2, \cdots, X_p) = \boldsymbol{X}^{(2)\prime}$ is equivalent to the test that if the regression of X_1 on $\boldsymbol{x}^{(2)}$ (that is, the conditional expected value of X_1 given $X_2 = x_2, \cdots, X_p = x_p$) is $\mu_1 + \boldsymbol{\beta}(\boldsymbol{x}^{(2)} - \boldsymbol{\mu}^{(2)})$, the vector of regression coefficients is $\boldsymbol{0}$. $\hat{\boldsymbol{\beta}} = \boldsymbol{a}_{(1)}\boldsymbol{A}_{22}^{-1}$ is the usual least squares estimate of $\boldsymbol{\beta}$ with expected value $\boldsymbol{\beta}$ and covariance matrix $\sigma_{11\cdot 2}\boldsymbol{A}_{22}^{-1}$ (when the $\boldsymbol{X}_\alpha^{(2)}$ are fixed) and $a_{11\cdot 2}/(N - p)$ is the usual estimate of $\sigma_{11\cdot 2}$. Thus [see (13)]

$$\frac{R^2}{1 - R^2} \cdot \frac{N - p}{p - 1} = \frac{\hat{\boldsymbol{\beta}}\boldsymbol{A}_{22}\hat{\boldsymbol{\beta}}'}{a_{11\cdot 2}} \cdot \frac{N - p}{p - 1} \tag{27}$$

is the usual F-statistic for testing the hypothesis that the regression of X_1 on $x_2, \cdots, x_p$ is 0.

It is of interest to note that $\bar{R}$ is the only function of the parameters $\boldsymbol{\mu}$ and $\boldsymbol{\Sigma}$ that is invariant under changes of location of the variables, changes in scale of X_1, and nonsingular linear transformations of $\boldsymbol{X}^{(2)}$. Similarly, R is the only function of $\bar{\boldsymbol{x}}$ and $\hat{\boldsymbol{\Sigma}}$, a sufficient set of statistics for $\boldsymbol{\mu}$ and $\boldsymbol{\Sigma}$, which is invariant under similar transformations.

4.4.3. Distribution of the Sample Multiple Correlation Coefficient when the Population Multiple Correlation Coefficient Is not Zero

In this section we shall find the distribution of R when the null hypothesis is not true. We shall find that the distribution depends only on the population multiple correlation coefficient $\bar{R}$.

First let us consider the conditional distribution of $R^2/(1 - R^2) = \boldsymbol{a}_{(1)}A_{22}^{-1}\boldsymbol{a}_{(1)}'/a_{11\cdot 2}$ given $\boldsymbol{Z}_\alpha^{(2)} = \boldsymbol{z}_\alpha^{(2)}$, $\alpha = 1, \cdots, n$. Under these conditions $Z_{1\alpha}$ are independently distributed, $Z_{1\alpha}$ according to $N(\boldsymbol{\beta}\boldsymbol{z}_\alpha^{(2)}, \sigma_{11\cdot 2})$, where $\boldsymbol{\beta} = \boldsymbol{\sigma}_{(1)}\boldsymbol{\Sigma}_{22}^{-1}$ and $\sigma_{11\cdot 2} = \sigma_{11} - \boldsymbol{\sigma}_{(1)}\boldsymbol{\Sigma}_{22}^{-1}\boldsymbol{\sigma}_{(1)}'$. The conditions are those of Theorem 4.3.2 with $Y_\alpha = Z_{1\alpha}$, $\boldsymbol{\Gamma} = \boldsymbol{\beta}$, $\boldsymbol{w}_\alpha = \boldsymbol{z}_\alpha^{(2)} (r = p - 1)$, $\boldsymbol{\Phi} = \sigma_{11\cdot 2}$, $m = n$. Then $a_{11\cdot 2} = a_{11} - \boldsymbol{a}_{(1)}A_{22}^{-1}\boldsymbol{a}_{(1)}'$ corresponds to $\sum_1^m Y_\alpha Y_\alpha' - \boldsymbol{GHG}'$, and $a_{11\cdot 2}/\sigma_{11\cdot 2}$ has a χ^2-distribution with $n - (p - 1)$ degrees of freedom. $\boldsymbol{a}_{(1)}A_{22}^{-1}\boldsymbol{a}_{(1)}' = (\boldsymbol{a}_{(1)}A_{22}^{-1})A_{22}(\boldsymbol{a}_{(1)}A_{22}^{-1})'$ corresponds to $\boldsymbol{GHG}'$ and is distributed as $\sum U_\alpha^2 (\alpha = n - (p - 1) + 1, \cdots, n)$ where $\operatorname{Var}(U_\alpha) = \sigma_{11\cdot 2}$ and

$$\mathscr{E}(U_{n-p+2}, \cdots, U_n) = \boldsymbol{\Gamma F}^{-1}, \tag{28}$$

where $\boldsymbol{FHF}' = \boldsymbol{I}\,[\boldsymbol{H} = \boldsymbol{F}^{-1}(\boldsymbol{F}')^{-1}]$. Then $\boldsymbol{a}_{(1)}A_{22}^{-1}\boldsymbol{a}_{(1)}'/\sigma_{11\cdot 2}$ is distributed as $\sum_\alpha (U_\alpha/\sqrt{\sigma_{11\cdot 2}})^2$, where $\operatorname{Var}(U_\alpha/\sqrt{\sigma_{11\cdot 2}}) = 1$ and

$$\begin{aligned}\sum_\alpha \left(\frac{\mathscr{E} U_\alpha}{\sqrt{\sigma_{11\cdot 2}}}\right)^2 &= \frac{1}{\sigma_{11\cdot 2}} \boldsymbol{\Gamma F}^{-1}(\boldsymbol{\Gamma F}^{-1})' = \frac{\boldsymbol{\Gamma H \Gamma}'}{\sigma_{11\cdot 2}} \\ &= \frac{\boldsymbol{\beta} A_{22} \boldsymbol{\beta}'}{\sigma_{11\cdot 2}}.\end{aligned} \tag{29}$$

Thus (conditionally) $\boldsymbol{a}_{(1)}A_{22}^{-1}\boldsymbol{a}_{(1)}'/\sigma_{11\cdot 2}$ has a noncentral χ^2-distribution with $p - 1$ degrees of freedom and noncentrality parameter $\boldsymbol{\beta}A_{22}\boldsymbol{\beta}'/\sigma_{11\cdot 2}$ (see Theorem 5.4.1). We are led to the following theorem:

THEOREM 4.4.3. *Let R be the sample multiple correlation coefficient between X_1 and $\boldsymbol{X}^{(2)\prime} = (X_2, \cdots, X_p)$ based on N observations $(x_{11}, \boldsymbol{x}_1^{(2)}), \cdots, (x_{1N}, \boldsymbol{x}_N^{(2)})$. The conditional distribution of $[R^2/(1 - R^2)][(N - p)/(p - 1)]$ given $\boldsymbol{x}_\alpha^{(2)}$ fixed is noncentral F with $p - 1$ and $N - p$ degrees of freedom and noncentrality parameter $\boldsymbol{\beta}A_{22}\boldsymbol{\beta}'/\sigma_{11\cdot 2}$.*

The conditional density (from Theorem 5.4.2) of $F = [R^2/(1 - R^2)][(N - p)/(p - 1)]$ is

$$(30)\quad \frac{(p-1)e^{-\frac{1}{2}\beta A_{22}\beta'/\sigma_{11\cdot 2}}}{(N-p)\Gamma[\frac{1}{2}(N-p)]} \sum_{\alpha=0}^{\infty} \frac{\left(\dfrac{\beta A_{22}\beta'}{2\sigma_{11\cdot 2}}\right)^{\alpha} \left[\dfrac{(p-1)f}{N-p}\right]^{\frac{1}{2}(p-1)+\alpha-1} \Gamma[\frac{1}{2}(N-1)+\alpha]}{\alpha!\,\Gamma[\frac{1}{2}(p-1)+\alpha]\left[1+\dfrac{(p-1)f}{N-p}\right]^{\frac{1}{2}(N-1)+\alpha}},$$

and the conditional density of $W = R^2$ is $(df = [(N - p)/(p - 1)](1 - w)^{-2}\, dw)$

$$(31)\quad \frac{e^{-\frac{1}{2}\beta A_{22}\beta'/\sigma_{11\cdot 2}}}{\Gamma[\frac{1}{2}(N-p)]}(1-w)^{\frac{1}{2}(N-p)-1} \sum_{\alpha=0}^{\infty} \frac{\left(\dfrac{\beta A_{22}\beta'}{2\sigma_{11\cdot 2}}\right)^{\alpha} w^{\frac{1}{2}(p-1)+\alpha-1}\Gamma[\frac{1}{2}(N-1)+\alpha]}{\alpha!\,\Gamma[\frac{1}{2}(p-1)+\alpha]}.$$

To get the unconditional density we need to multiply (31) by the density of $Z_1^{(2)}, \cdots, Z_n^{(2)}$ to get the joint density of W and $Z_1^{(2)}, \cdots, Z_n^{(2)}$ and then integrate with respect to the latter set to obtain the marginal density of W. We have

$$(32)\quad \frac{\beta A_{22}\beta'}{\sigma_{11\cdot 2}} = \frac{\beta \sum_{\alpha=1}^{n} z_\alpha^{(2)} z_\alpha^{(2)\prime}\beta'}{\sigma_{11\cdot 2}} = \sum_{\alpha=1}^{n}\left(\frac{\beta z_\alpha^{(2)}}{\sqrt{\sigma_{11\cdot 2}}}\right)^2.$$

Since the distribution of $Z_\alpha^{(2)}$ is $N(\mathbf{0}, \Sigma_{22})$, the distribution of $\beta Z_\alpha^{(2)}/\sqrt{\sigma_{11\cdot 2}}$ is normal with mean zero and variance

$$(33)\quad \mathscr{E}\left(\frac{\beta Z_\alpha^{(2)}}{\sqrt{\sigma_{11\cdot 2}}}\right)^2 = \frac{\mathscr{E}\beta Z_\alpha^{(2)} Z_\alpha^{(2)\prime}\beta'}{\sigma_{11\cdot 2}} = \frac{\beta\Sigma_{22}\beta'}{\sigma_{11} - \beta\Sigma_{22}\beta'} = \frac{\beta\Sigma_{22}\beta'/\sigma_{11}}{1 - \beta\Sigma_{22}\beta'/\sigma_{11}} = \frac{\bar{R}^2}{1-\bar{R}^2}.$$

Thus $(\beta A_{22}\beta'/\sigma_{11\cdot 2})/[\bar{R}^2/(1 - \bar{R}^2)]$ has a χ^2-distribution with n degrees of freedom. Let $\bar{R}^2/(1-\bar{R}^2)=\phi$. Then $\boldsymbol{\beta} A_{22}\boldsymbol{\beta}'/\sigma_{11\cdot 2}=\phi\chi_n^2$. We compute

$$(34)\quad \mathscr{E}e^{-\frac{1}{2}\phi\chi_n^2}\left(\frac{\phi\chi_n^2}{2}\right)^\alpha$$

$$=\frac{\phi^\alpha}{2^\alpha}\int_0^\infty u^\alpha e^{-\frac{1}{2}\phi u}\frac{1}{2^{\frac{1}{2}n}\Gamma(\frac{1}{2}n)}u^{\frac{1}{2}n-1}e^{-\frac{1}{2}u}\,du$$

$$=\frac{\phi^\alpha}{2^\alpha}\int_0^\infty \frac{1}{2^{\frac{1}{2}n}\Gamma(\frac{1}{2}n)}u^{\frac{1}{2}n+\alpha-1}e^{-\frac{1}{2}(1+\phi)u}\,du$$

$$=\frac{\phi^\alpha}{(1+\phi)^{\frac{1}{2}n+\alpha}}\frac{\Gamma(\frac{1}{2}n+\alpha)}{\Gamma(\frac{1}{2}n)}\int_0^\infty\frac{1}{2^{\frac{1}{2}n+\alpha}\Gamma(\frac{1}{2}n+\alpha)}v^{\frac{1}{2}n+\alpha-1}e^{-\frac{1}{2}v}\,dv$$

$$=\frac{\phi^\alpha}{(1+\phi)^{\frac{1}{2}n+\alpha}}\frac{\Gamma(\frac{1}{2}n+\alpha)}{\Gamma(\frac{1}{2}n)}.$$

Applying this result to (31) we obtain as the density of R^2

$$(35)\quad \frac{(1-R^2)^{\frac{1}{2}(n-p-1)}(1-\bar{R}^2)^{\frac{1}{2}n}}{\Gamma[\frac{1}{2}(n-p+1)]\Gamma(\frac{1}{2}n)}\sum_{\mu=0}^\infty\frac{(\bar{R}^2)^\mu(R^2)^{\frac{1}{2}(p-1)+\mu-1}\Gamma^2(\frac{1}{2}n+\mu)}{\mu!\Gamma[\frac{1}{2}(p-1)+\mu]}.$$

Fisher found this distribution in 1928.

It is easily verified that another way of writing the density is

$$(36)\quad \frac{\Gamma(\frac{1}{2}n)(1-\bar{R}^2)^{\frac{1}{2}n}}{\Gamma[\frac{1}{2}(n-p+1)]\Gamma[\frac{1}{2}(p-1)]}(R^2)^{\frac{1}{2}(p-3)}(1-R^2)^{\frac{1}{2}(n-p-1)}$$
$$F[\tfrac{1}{2}n,\tfrac{1}{2}n;\ \tfrac{1}{2}(p-1);\ R^2\bar{R}^2],$$

where F is the hypergeometric function.

Another form of the density can be obtained when $n-p+1$ is even. We have

$$(37)\quad \sum_{\mu=0}^\infty\frac{(R^2\bar{R}^2)^\mu}{\mu!}\frac{\Gamma^2(\frac{1}{2}n+\mu)}{\Gamma[\frac{1}{2}(p-1)+\mu]}$$

$$=\sum_{\mu=0}^\infty\frac{(R^2\bar{R}^2)^\mu}{\mu!}\Gamma(\tfrac{1}{2}n+\mu)\left(\frac{\partial}{\partial t}\right)^{\frac{1}{2}(n-p+1)}t^{\frac{1}{2}n+\mu-1}\bigg]_{t=1}$$

$$=\left(\frac{\partial}{\partial t}\right)^{\frac{1}{2}(n-p+1)}t^{\frac{1}{2}n-1}\sum_{\mu=0}^\infty\frac{(t\bar{R}^2R^2)^\mu}{\mu!}\frac{\Gamma(\frac{1}{2}n+\mu)}{\Gamma(\frac{1}{2}n)}\bigg]_{t=1}\Gamma(\tfrac{1}{2}n)$$

$$=\Gamma(\tfrac{1}{2}n)\left(\frac{\partial}{\partial t}\right)^{\frac{1}{2}(n-p+1)}t^{\frac{1}{2}n-1}(1-tR^2\bar{R}^2)^{-\frac{1}{2}n}\bigg]_{t=1}.$$

The density is therefore

$$(38) \qquad \frac{(1-\bar{R}^2)^{\frac{1}{2}n}(R^2)^{\frac{1}{2}(p-3)}(1-R^2)^{\frac{1}{2}(n-p-1)}}{\Gamma[\frac{1}{2}(n-p+1)]} \left(\frac{\partial}{\partial t}\right)^{\frac{1}{2}(n-p+1)} t^{\frac{1}{2}n-1}(1-tR^2\bar{R}^2)^{-\frac{1}{2}n}\Bigg]_{t=1}.$$

THEOREM 4.4.4. *The density of the square of the multiple correlation coefficient, R^2, between X_1 and $X_2, \cdots, X_p$ based on a sample of $N = n + 1$ is given by* (35) *or* (36) [*or* (38) *in the case of $n - p + 1$ even*], *where $\bar{R}^2$ is the corresponding population multiple correlation coefficient.*

The moments of R are

$$(39) \qquad \mathscr{E}R^h = \frac{(1-\bar{R}^2)^{\frac{1}{2}n}}{\Gamma[\frac{1}{2}(n-p+1)]\Gamma(\frac{1}{2}n)} \sum_{\mu=0}^{\infty} \frac{(\bar{R}^2)^{\mu}\Gamma^2(\frac{1}{2}n+\mu)}{\Gamma[\frac{1}{2}(p-1)+\mu]\mu!} \int_0^{\infty}(1-R^2)^{\frac{1}{2}(n-p+1)-1}(R^2)^{\frac{1}{2}(p+h-1)+\mu-1}d(R^2)$$

$$= \frac{(1-\bar{R}^2)^{\frac{1}{2}n}}{\Gamma(\frac{1}{2}n)} \sum_{\mu=0}^{\infty} \frac{(\bar{R}^2)^{\mu}\Gamma^2(\frac{1}{2}n+\mu)\Gamma[\frac{1}{2}(p+h-1)+\mu]}{\mu!\Gamma[\frac{1}{2}(p-1)+\mu]\Gamma[\frac{1}{2}(n+h)+\mu]}.$$

REFERENCES

Section 4.2. Cramér (1946), 394–401; David (1937), (1938); Fisher (1915), (1921); Fisher and Yates (1942); Garwood (1933); Gayen (1951); Harley (1954); Hotelling (1953); Kendall (1943), 324–347; Kono (1952); K. Pearson (1929); Sato (1951); Soper, Young, Cave, Lee, and Pearson (1917); Wilks (1943), 116–120.

Section 4.3. Cramér (1946), 410–413; Fisher (1924); Hall (1927); Hooker (1907); Isserlis (1914), (1916); Kelley (1916); Kendall (1943), 368–379; Miner (1922).

Section 4.4. Banerjee (1952); Cramér (1946), 413–415; Fisher (1928); Hall (1927); Isserlis (1917); Kendall (1943), 380–385; Moran (1950); S. N. Roy and Bose (1940); Wilks (1932c), (1943), 244–245; Wishart (1931).

Chapter 4. Bartlett (1933); Elfving (1947); Ezekiel (1930); Frisch (1929); Kelley (1928); Maritz (1953); Simonsen (1944–45).

PROBLEMS

1. (Sec. 4.2.1) Sketch $k_N(r) = \dfrac{\Gamma[\frac{1}{2}(N-1)]}{\Gamma(\frac{1}{2}N-1)\sqrt{\pi}}(1-r^2)^{\frac{1}{2}(N-4)}$ for (*a*) $N = 3$, (*b*) $N = 4$, (*c*) $N = 5$, and (*d*) $N = 10$.

2. (Sec. 4.2.1) Using the data of Problem 1 of Chapter 3, test the hypothesis that X_1 and X_2 are independent against all alternatives of dependence at significance level 0.01.

3. (Sec. 4.2.1) Suppose a sample correlation of 0.65 is observed in a sample of 10. Test the hypothesis of independence against the alternatives of positive correlation at significance level 0.05.

4. (Sec. 4.2.2) Suppose a sample correlation of 0.65 is observed in a sample of 20. Test the hypothesis that the population correlation is 0.4 against the alternatives that the population correlation is greater than 0.4 at significance level 0.05.

5. (Sec. 4.2.1) Find the significance points for testing $\rho = 0$ at the 0.01 level with $N = 15$ observations against alternatives (*a*) $\rho \neq 0$, (*b*) $\rho > 0$, and (*c*) $\rho < 0$.

6. (Sec. 4.2.2) Find significance points for testing $\rho = 0.6$ at the 0.01 level with $N = 20$ observations against alternatives (*a*) $\rho \neq 0.6$, (*b*) $\rho > 0.6$, and (*c*) $\rho < 0.6$.

7. (Sec. 4.2.2) Table the power function at $\rho = -1(0.2)1$ for the tests in Problem 5. Sketch the graph of each power function.

8. (Sec. 4.2.2) Table the power function at $\rho = -1(0.2)1$ for the tests in Problem 6. Sketch the graph of each power function.

9. (Sec. 4.2.2) Using the data of Problem 1 of Chapter 3 find a (two-sided) confidence interval for ρ_{12} with confidence coefficient 0.99.

10. (Sec. 4.2.2) Suppose $N = 10$, $r = 0.795$. Find a one-sided confidence interval for ρ [of the form $(r_0, 1)$] with confidence coefficient 0.95.

11. (Sec. 4.2.3) Use Fisher's z to test the hypothesis $\rho = 0.7$ against alternatives $\rho \neq 0.7$ at the 0.05 level with $r = 0.5$ and $N = 50$.

12. (Sec. 4.2.3) Use Fisher's z to test the hypothesis $\rho_1 = \rho_2$ against the alternatives $\rho_1 \neq \rho_2$ at the 0.01 level with $r_1 = 0.5$, $N_1 = 40$, $r_2 = 0.6$, $N_2 = 40$.

13. (Sec. 4.2.3) Use Fisher's z to estimate ρ based on sample correlations of -0.7 ($N = 30$) and of -0.6 ($N = 40$).

14. (Sec. 4.2.3) Use Fisher's z to obtain a confidence interval for ρ with confidence 0.95 based on a sample correlation of 0.65 and a sample size of 25.

15. (Sec. 4.3.2) Find a confidence interval for $\rho_{13\cdot 2}$ with confidence 0.95 based on $r_{13\cdot 2} = 0.097$ and $N = 20$.

16. (Sec. 4.3.2) Use Fisher's z to test the hypothesis $\rho_{12\cdot 34} = 0$ against alternatives $\rho_{12\cdot 34} \neq 0$ at significance level 0.01 with $r_{12\cdot 34} = 0.14$ and $N = 40$.

17. The estimates of $\boldsymbol{\mu}$ and $\boldsymbol{\Sigma}$ in Problem 1 of Chapter 3 are

$$\bar{\mathbf{x}}' = (185.72 \quad 151.12 \quad 183.84 \quad 149.24)$$

$$\mathbf{S} = \left(\begin{array}{cc|cc} 95.2933 & 52.8683 & 69.6617 & 46.1117 \\ 52.8683 & 54.3600 & 51.3117 & 35.0533 \\ \hline 69.6617 & 51.3117 & 100.8067 & 56.5400 \\ 46.1117 & 35.0533 & 56.5400 & 45.0233 \end{array}\right).$$

(*a*) Find the estimates of the parameters of the conditional distribution of (x_3, x_4) given (x_1, x_2); that is, find $\mathbf{S}_{21}\mathbf{S}_{11}^{-1}$ and $\mathbf{S}_{22\cdot 1} = \mathbf{S}_{22} - \mathbf{S}_{21}\mathbf{S}_{11}^{-1}\mathbf{S}_{12}$.

(*b*) Find the partial correlation $r_{34\cdot 12}$.

(*c*) Use Fisher's z to find a confidence interval for $\rho_{34\cdot 12}$ with confidence 0.95.

(*d*) Find the sample multiple correlation coefficients between x_3 and (x_1, x_2) and between x_4 and (x_1, x_2).

(*e*) Test the hypotheses that x_3 is independent of (x_1, x_2) and x_4 is independent of (x_1, x_2) at significance level 0.05.

18. Let the components of $\boldsymbol{X}$ correspond to scores on tests in arithmetic speed (X_1), arithmetic power (X_2), memory for words (X_3), memory for meaningful symbols (X_4),

and memory for meaningless symbols (X_5). The observed correlations in a sample of 140 are [Kelley (1928)]

$$\begin{pmatrix} 1.0000 & 0.4248 & 0.0420 & 0.0215 & 0.0573 \\ 0.4248 & 1.0000 & 0.1487 & 0.2489 & 0.2843 \\ 0.0420 & 0.1487 & 1.0000 & 0.6693 & 0.4662 \\ 0.0215 & 0.2489 & 0.6693 & 1.0000 & 0.6915 \\ 0.0573 & 0.2843 & 0.4662 & 0.6915 & 1.0000 \end{pmatrix}$$

(*a*) Find the partial correlation between X_4 and X_5, holding X_3 fixed.

(*b*) Find the partial correlation between X_1 and X_2, holding X_3, X_4, and X_5 fixed.

(*c*) Find the multiple correlation between X_1 and the set X_3, X_4, and X_5.

(*d*) Test the hypothesis at the 1% significance level that arithmetic speed is independent of the three memory scores.

19. (Sec. 4.2) In the bivariate case prove that any function of the sufficient statistics $\bar{x}$ and $\hat{\mathbf{\Sigma}}$ invariant under changes of location and scale (that is, under transformations $x^*_{i\alpha} = c_i x_{i\alpha} + d_i$, $i = 1, 2$, $c_i > 0$) is a function of r_{12}.

20. (Sec. 4.4) Prove the statement at the end of Section 4.4.2 that any function of the sufficient statistics $\bar{x}$ and $\hat{\mathbf{\Sigma}}$ that is invariant under changes of location and scale of $x_{1\alpha}$ and nonsingular linear transformations of $\mathbf{x}^{(2)}_\alpha$ (that is, $x^*_{1\alpha} = c_1 x_{1\alpha} + d_1$, $\mathbf{x}^{*(2)}_\alpha = \mathbf{C}\mathbf{x}^{(2)}_\alpha + \mathbf{d}$) is a function of R.

21. Prove that if $\rho_{ij\cdot q+1,\cdots,p} = 0$, then $\sqrt{N-2-(p-q)}\, r_{ij\cdot q+1,\cdots,p}/\sqrt{1-r^2_{ij\cdot q+1,\cdots,p}}$ is distributed according to the t-distribution with $N-2-(p-q)$ degrees of freedom.

22. Let $X' = (X_1, X_2, \mathbf{X}^{(2)\prime})$ have the distribution $N(\boldsymbol{\mu}, \mathbf{\Sigma})$. The conditional distribution of X_1 given $X_2 = x_2$ and $\mathbf{X}^{(2)} = \mathbf{x}^{(2)}$ is

$$N[\mu_1 + \gamma_2(x_2 - \mu_2) + \boldsymbol{\gamma}'(\mathbf{x}^{(2)} - \boldsymbol{\mu}^{(2)}), \sigma_{11\cdot 2,\cdots,p}],$$

where

$$\begin{pmatrix} \sigma_{22} & \boldsymbol{\sigma}_{(2)} \\ \boldsymbol{\sigma}'_{(2)} & \mathbf{\Sigma}_{22} \end{pmatrix}\begin{pmatrix} \gamma_2 \\ \boldsymbol{\gamma} \end{pmatrix} = \begin{pmatrix} \sigma_{12} \\ \boldsymbol{\sigma}'_{(1)} \end{pmatrix}.$$

The estimate of γ_2 and $\boldsymbol{\gamma}$ is given by

$$\begin{pmatrix} a_{22} & \mathbf{a}_{(2)} \\ \mathbf{a}'_{(2)} & \mathbf{A}_{22} \end{pmatrix}\begin{pmatrix} c_2 \\ \mathbf{c} \end{pmatrix} = \begin{pmatrix} a_{12} \\ \mathbf{a}'_{(1)} \end{pmatrix}.$$

Show $c_2 = a_{12\cdot 3,\cdots,p}/a_{22\cdot 3,\cdots,p}$. [*Hint:* Solve for $\mathbf{c}$ in terms of c_2 and the a's and substitute.]

23. In the notation of Problem 22 prove

$$\begin{aligned} a_{11\cdot 2,\cdots,p} &= a_{11} - \mathbf{a}_{(1)}\mathbf{A}_{22}^{-1}\mathbf{a}'_{(1)} - c_2^2(a_{22} - \mathbf{a}_{(2)}\mathbf{A}_{22}^{-1}\mathbf{a}'_{(2)}) \\ &= a_{11\cdot 3,\cdots,p} - c_2^2 a_{22\cdot 3,\cdots,p}. \end{aligned}$$

Hint: Use

$$a_{11\cdot 2,\cdots,p} = a_{11} - (c_2\ \mathbf{c}')\begin{pmatrix} a_{22} & \mathbf{a}_{(2)} \\ \mathbf{a}'_{(2)} & \mathbf{A}_{22} \end{pmatrix}\begin{pmatrix} c_2 \\ \mathbf{c} \end{pmatrix}.$$

24. Prove that $1/a_{22\cdot 3,\cdots,p}$ is the element in the upper left-hand corner of

$$\begin{pmatrix} a_{22} & \mathbf{a}_{(2)} \\ \mathbf{a}'_{(2)} & \mathbf{A}_{22} \end{pmatrix}^{-1}.$$

25. Using the results in Problems 21, 22, 23, and 24 prove that the test for $\rho_{12\cdot 3,\ldots,p} = 0$ is equivalent to the usual t-test that $\gamma_2 = 0$.

26. (Sec. 4.2.2) Prove that when $N = 2$ and $\rho = 0$, $\Pr\{r = 1\} = \Pr\{r = -1\} = \frac{1}{2}$.

27. Let $\boldsymbol{X}' = (\boldsymbol{Y}'\ \boldsymbol{Z}')$, where $\boldsymbol{Y}$ has p components and $\boldsymbol{Z}$ has q components, be distributed according to $N(\boldsymbol{\mu}, \boldsymbol{\Sigma})$, where

$$\boldsymbol{\mu} = \begin{pmatrix} \boldsymbol{\mu}_y \\ \boldsymbol{\mu}_z \end{pmatrix}, \qquad \boldsymbol{\Sigma} = \begin{pmatrix} \boldsymbol{\Sigma}_{yy} & \boldsymbol{\Sigma}_{yz} \\ \boldsymbol{\Sigma}_{zy} & \boldsymbol{\Sigma}_{zz} \end{pmatrix}.$$

Let M observations be made on $\boldsymbol{X}$ and $N - M$ additional observations be made on $\boldsymbol{Y}$. Find the maximum likelihood estimates of $\boldsymbol{\mu}$ and $\boldsymbol{\Sigma}$. [*Hint:* Express the likelihood function in terms of the marginal density of $\boldsymbol{Y}$ and the conditional density of $\boldsymbol{Z}$ given $\boldsymbol{Y}$.]

28. Suppose $\boldsymbol{X}$ is distributed according to $N(\boldsymbol{0}, \boldsymbol{\Sigma})$, where

$$\boldsymbol{\Sigma} = \begin{pmatrix} 1 & \rho & \rho^2 \\ \rho & 1 & \rho \\ \rho^2 & \rho & 1 \end{pmatrix}.$$

Show that on the basis of one observation, $\boldsymbol{x}' = (x_1, x_2, x_3)$, we can obtain a confidence interval for ρ (with confidence coefficient $1 - \epsilon$) by using as end points of the interval the solutions in t of

$$(x_2^2 + K_\epsilon)t^2 - 2(x_1x_2 + x_2x_3)t + x_1^2 + x_2^2 + x_3^3 - K_\epsilon = 0,$$

where K_ϵ is the significance point of the χ^2-distribution with three degrees of freedom at significance level ϵ.

29. (Sec. 4.2) Let $k_N(r, \rho)$ be the density of the sample correlation coefficient r for a given value of ρ and N. Prove that r has a monotone likelihood ratio; that is, show that if $\rho_1 > \rho_2$, then $k_N(r, \rho_1)/k_N(r, \rho_2)$ is monotonically increasing in r. [*Hint:* Using (36), prove that if

$$F[\tfrac{1}{2}, \tfrac{1}{2}; n + \tfrac{1}{2}; \tfrac{1}{2}(1 + \rho r)] = \sum_{\alpha=0}^{\infty} c_\alpha (1 + \rho r)^\alpha = g(r, \rho)$$

has a monotone ratio, then $k_N(r, \rho)$ does. Show

$$\frac{\partial^2}{\partial\rho\,\partial r} \log g(r, \rho) = \frac{\sum\limits_{\alpha,\beta=0}^{\infty} c_\alpha c_\beta [(\alpha - \beta)^2 r\rho + (\alpha + \beta)](1 + r\rho)^{\alpha+\beta-2}}{2\left[\sum\limits_{\alpha=0}^{\infty} c_\alpha (1 + r\rho)^\alpha\right]^2};$$

if $(\partial^2/\partial\rho\,\partial r)\log g(r, \rho) > 0$, then $g(r, \rho)$ has a monotone ratio. Show the numerator of the above expression is positive by showing that for each α the sum on β is positive; use the fact that $c_{\alpha+1} < \frac{1}{2}c_\alpha$.]

30. (Sec. 4.2) Show that of all tests of ρ_0 against a specific $\rho_1 (> \rho_0)$ based on r, the procedures for which $r > c$ implies rejection are the best. [*Hint:* This follows from Problem 29.]

31. (Sec. 4.2) Show that of all tests of $\rho = \rho_0$ against $\rho > \rho_0$ based on r, a procedure for which $r > c$ implies rejection is uniformly most powerful.

32. (Sec. 4.2) Prove r has a monotone likelihood ratio for $r > 0$, $\rho > 0$ by proving $h(r) = k_N(r, \rho_1)/k_N(r, \rho_2)$ is monotonically increasing for $\rho_1 > \rho_2$. $h(r)$ is a constant times $(\sum_{\alpha=0}^{\infty} c_\alpha \rho_1^\alpha r^\alpha)/(\sum_{\alpha=0}^{\infty} c_\alpha \rho_2^\alpha r^\alpha)$. In the numerator of $h'(r)$, show that the coefficient of r^β is positive.

33. (Sec. 4.4) Prove that conditional on $Z_{1\alpha} = z_{1\alpha}(\alpha = 1, \cdots, n) R^2/(1 - R^2)$ is distributed like $T^2/(N^* - 1)$, where $T^2 = N^* \bar{x}' S^{-1} \bar{x}$ based on $N^* = n$ observations on a vector $\boldsymbol{X}$ with $p^* = (p - 1)$ components, with mean vector $(c/\sigma_{11})\boldsymbol{\sigma}'_{(1)}$ $(nc^2 = \Sigma z_{1\alpha}^2)$ and covariance matrix $\boldsymbol{\Sigma}_{22\cdot 1} = \boldsymbol{\Sigma}_{22} - (1/\sigma_{11})\boldsymbol{\sigma}'_{(1)}\boldsymbol{\sigma}_{(1)}$. [*Hint:* The conditional distribution of $\boldsymbol{Z}_\alpha^{(2)}$ given $Z_{1\alpha} = z_{1\alpha}$ is $N[(1/\sigma_{11})\boldsymbol{\sigma}'_{(1)} z_{1\alpha}, \boldsymbol{\Sigma}_{22\cdot 1}]$. There is an $n \times n$ orthogonal matrix $\boldsymbol{B}$ which carries $(z_{11}, \cdots, z_{1n})$ into $(c, \cdots, c)$ and $(Z_{i1}, \cdots, Z_{in})$ into $(Y_{i1}, \cdots, Y_{in})$, $i = 2, \cdots, p$. Let the new $\boldsymbol{X}'_\alpha$ be $(Y_{2\alpha}, \cdots, Y_{p\alpha})$.]

34. (Sec. 4.4) Prove that the noncentrality parameter in the distribution in Problem 33 is $(a_{11}/\sigma_{11})\bar{R}^2/(1 - \bar{R}^2)$.

35. (Sec. 4.4) Find the distribution of $R^2/(1 - R^2)$ by multiplying the density of Problem 33 by the density of a_{11} and integrating with respect to a_{11}.

36. (Sec. 4.2) Prove that if $\boldsymbol{\Sigma}$ is diagonal the sets r_{ij} and a_{ii} are independently distributed. [*Hint:* Use the facts that r_{ij} is invariant under scale transformations and that the density of the observations depends only on the a_{ii}.]

37. (Sec. 4.2.1) Prove that if $\rho = 0$

$$\mathscr{E} r^{2m} = \frac{\Gamma[\tfrac{1}{2}(N-1)]\Gamma(m + \tfrac{1}{2})}{\sqrt{\pi}\,\Gamma[\tfrac{1}{2}(N-1) + m]}$$

38. (Sec. 4.2.2) Prove $f_1(\rho)$ and $f_2(\rho)$ are monotonically increasing functions of ρ.

39. (Sec. 4.2.2) Prove that the density of the sample correlation r [given by (34)] is

$$\frac{n-1}{\pi}(1 - \rho^2)^{\frac{1}{2}n}(1 - r^2)^{\frac{1}{2}(n-3)} \int_0^1 \frac{x^{n-1}\,dx}{(1 - \rho r x)^n \sqrt{1 - x^2}}.$$

[*Hint:* Expand $(1 - \rho r x)^{-n}$ in a power series, integrate, and use the duplication formula for the gamma function.]

CHAPTER 5

The Generalized T^2-Statistic

5.1. INTRODUCTION

One of the most important groups of problems in univariate statistics relates to questions concerning the mean of a given distribution when the variance of the distribution is unknown. On the basis of a sample one may wish to decide whether the mean is equal to a number specified in advance, or one may wish to give an interval within which the mean lies. The statistic usually used in univariate statistics is the difference between the mean of the sample, $\bar{x}$, and the hypothetical population mean divided by the sample standard deviation, s. If the distribution sampled is $N(\mu, \sigma^2)$, then

$$t = \sqrt{N}\,\frac{\bar{x} - \mu}{s} \tag{1}$$

has the well-known t-distribution with $N - 1$ degrees of freedom, where N is the number of observations in the sample. On the basis of this fact, one can set up a test of the hypothesis $\mu = \mu_0$, where μ_0 is specified, or one can set up a confidence interval for the unknown parameter μ.

The multivariate analogue of the square of t given in (1) is

$$T^2 = N(\bar{\mathbf{x}} - \boldsymbol{\mu})' \mathbf{S}^{-1}(\bar{\mathbf{x}} - \boldsymbol{\mu}), \tag{2}$$

where $\bar{\mathbf{x}}$ is the mean vector of a sample of N and $\mathbf{S}$ is the sample covariance matrix. It will be shown how this statistic can be used for testing hypotheses about the mean vector $\boldsymbol{\mu}$ of the population and for obtaining confidence regions for the unknown $\boldsymbol{\mu}$. Other uses of this statistical procedure will be indicated. Hotelling (1931) proposed the T^2-statistic for two samples and derived the distribution under the null hypothesis.

5.2. DERIVATION OF THE GENERALIZED T^2-STATISTIC AND ITS DISTRIBUTION

5.2.1. Derivation of the T^2-Statistic as a Function of the Likelihood Ratio Criterion

Although the T^2-statistic has many uses, we shall begin our discussion by showing that the likelihood ratio test of the hypothesis $H\colon \boldsymbol{\mu} = \boldsymbol{\mu}_0$ on the basis of a sample from $N(\boldsymbol{\mu}, \boldsymbol{\Sigma})$ is based on the T^2-statistic given in (2) of Section 5.1. Suppose we have N observations $\boldsymbol{x}_1, \cdots, \boldsymbol{x}_N (N > p)$. The likelihood function is

$$L(\boldsymbol{\mu}, \boldsymbol{\Sigma}^{-1}) = \frac{|\boldsymbol{\Sigma}^{-1}|^{\frac{1}{2}N}}{(2\pi)^{\frac{1}{2}pN}} \exp\left[-\tfrac{1}{2}\sum_{\alpha=1}^{N}(\boldsymbol{x}_\alpha - \boldsymbol{\mu})'\boldsymbol{\Sigma}^{-1}(\boldsymbol{x}_\alpha - \boldsymbol{\mu})\right]. \tag{1}$$

The observations are given; L is a function of the indeterminates $\boldsymbol{\mu}, \boldsymbol{\Sigma}^{-1}$. (We shall not distinguish in notation between the indeterminates and the parameters.) The likelihood ratio criterion is

$$\lambda = \frac{\max\limits_{\boldsymbol{\Sigma}^{-1}} L(\boldsymbol{\mu}_0, \boldsymbol{\Sigma}^{-1})}{\max\limits_{\boldsymbol{\mu}, \boldsymbol{\Sigma}^{-1}} L(\boldsymbol{\mu}, \boldsymbol{\Sigma}^{-1})}, \tag{2}$$

that is, the numerator is the maximum of the likelihood function for $\boldsymbol{\mu}, \boldsymbol{\Sigma}^{-1}$ in the parameter space restricted by the null hypothesis ($\boldsymbol{\mu} = \boldsymbol{\mu}_0$, $\boldsymbol{\Sigma}^{-1}$ positive definite), and the denominator is the maximum over the entire parameter space ($\boldsymbol{\Sigma}^{-1}$ positive definite). When the parameters are unrestricted the maximum occurs when $\boldsymbol{\mu}, \boldsymbol{\Sigma}^{-1}$ are defined by the maximum likelihood estimates (Section 3.2) of $\boldsymbol{\mu}$ and $\boldsymbol{\Sigma}$

$$\hat{\boldsymbol{\mu}}_\Omega = \bar{\boldsymbol{x}}, \tag{3}$$

$$\hat{\boldsymbol{\Sigma}}_\Omega = \frac{1}{N}\sum_\alpha (\boldsymbol{x}_\alpha - \bar{\boldsymbol{x}})(\boldsymbol{x}_\alpha - \bar{\boldsymbol{x}})'. \tag{4}$$

When $\boldsymbol{\mu} = \boldsymbol{\mu}_0$, the likelihood function is maximized at

$$\hat{\boldsymbol{\Sigma}}_\omega = \frac{1}{N}\sum_\alpha (\boldsymbol{x}_\alpha - \boldsymbol{\mu}_0)(\boldsymbol{x}_\alpha - \boldsymbol{\mu}_0)' \tag{5}$$

by Lemma 3.2.2. Furthermore, by Lemma 3.2.2

$$\max_{\boldsymbol{\Sigma}^{-1}, \boldsymbol{\mu}} L(\boldsymbol{\mu}, \boldsymbol{\Sigma}^{-1}) = \frac{1}{(2\pi)^{\frac{1}{2}pN}|\hat{\boldsymbol{\Sigma}}_\Omega|^{\frac{1}{2}N}} e^{-\frac{1}{2}pN}, \tag{6}$$

$$\max_{\boldsymbol{\Sigma}^{-1}} L(\boldsymbol{\mu}_0, \boldsymbol{\Sigma}^{-1}) = \frac{1}{(2\pi)^{\frac{1}{2}pN}|\hat{\boldsymbol{\Sigma}}_\omega|^{\frac{1}{2}N}} e^{-\frac{1}{2}pN}. \tag{7}$$

Thus the likelihood ratio criterion is

$$\begin{aligned}\lambda &= \frac{|\hat{\boldsymbol{\Sigma}}_\Omega|^{\frac{1}{2}N}}{|\hat{\boldsymbol{\Sigma}}_\omega|^{\frac{1}{2}N}} = \frac{|\sum(\boldsymbol{x}_\alpha - \bar{\boldsymbol{x}})(\boldsymbol{x}_\alpha - \bar{\boldsymbol{x}})'|^{\frac{1}{2}N}}{|\sum(\boldsymbol{x}_\alpha - \boldsymbol{\mu}_0)(\boldsymbol{x}_\alpha - \boldsymbol{\mu}_0)'|^{\frac{1}{2}N}} \\ &= \frac{|A|^{\frac{1}{2}N}}{|A + N(\bar{\boldsymbol{x}} - \boldsymbol{\mu}_0)(\bar{\boldsymbol{x}} - \boldsymbol{\mu}_0)'|^{\frac{1}{2}N}},\end{aligned} \tag{8}$$

where

$$A = \sum_{\alpha} (\mathbf{x}_\alpha - \bar{\mathbf{x}})(\mathbf{x}_\alpha - \bar{\mathbf{x}})' = (N-1)S. \tag{9}$$

For $|B| \neq 0$ and by (62) of Appendix 1

$$\begin{aligned} \begin{vmatrix} B & C \\ D & E \end{vmatrix} &= \begin{vmatrix} B & C \\ D & E \end{vmatrix} \cdot \begin{vmatrix} I & -B^{-1}C \\ 0 & I \end{vmatrix} \\ &= \begin{vmatrix} B & 0 \\ D & E - DB^{-1}C \end{vmatrix} \\ &= |B| \cdot |E - DB^{-1}C|. \end{aligned} \tag{10}$$

Use of this result twice shows

$$\begin{aligned} \lambda^{2/N} &= \frac{|A|}{|A + [\sqrt{N}(\bar{\mathbf{x}} - \boldsymbol{\mu}_0)][\sqrt{N}(\bar{\mathbf{x}} - \boldsymbol{\mu}_0)]'|} \\ &= \frac{|A|}{\begin{vmatrix} 1 & \sqrt{N}(\bar{\mathbf{x}} - \boldsymbol{\mu}_0)' \\ -\sqrt{N}(\bar{\mathbf{x}} - \boldsymbol{\mu}_0) & A \end{vmatrix}} \\ &= \frac{1}{1 + N(\bar{\mathbf{x}} - \boldsymbol{\mu}_0)'A^{-1}(\bar{\mathbf{x}} - \boldsymbol{\mu}_0)} \\ &= \frac{1}{1 + T^2/(N-1)}, \end{aligned} \tag{11}$$

where

$$T^2 = N(\bar{\mathbf{x}} - \boldsymbol{\mu}_0)'S^{-1}(\bar{\mathbf{x}} - \boldsymbol{\mu}_0) = (N-1)N(\bar{\mathbf{x}} - \boldsymbol{\mu}_0)'A^{-1}(\bar{\mathbf{x}} - \boldsymbol{\mu}_0). \tag{12}$$

The likelihood ratio test is defined by the critical region (region of rejection)

$$\lambda \leq \lambda_0, \tag{13}$$

where λ_0 is chosen so that the probability of (13) when the null hypothesis is true is equal to the significance level. If we take the $N/2$th root of both sides of (13) and invert, subtract 1, and multiply by $(N-1)$, we obtain

$$T^2 \geq T_0^2, \tag{14}$$

where

$$T_0^2 = (N-1)(\lambda_0^{-2/N} - 1). \tag{15}$$

THEOREM 5.2.1. *The likelihood ratio test of the hypothesis* $\boldsymbol{\mu} = \boldsymbol{\mu}_0$ *for the distribution* $N(\boldsymbol{\mu}, \boldsymbol{\Sigma})$ *is given by* (14), *where* T^2 *is defined by* (12), $\bar{\mathbf{x}}$ *is the mean of a sample of* N *from* $N(\boldsymbol{\mu}, \boldsymbol{\Sigma})$, S *is the covariance matrix of the* *sample, and T_0^2 is chosen so that the probability of* (14) *under the null hypothesis is equal to the chosen significance level.*

The Student t-test has the property that when testing $\mu = 0$ it is invariant with respect to scale transformations. If the scalar random variable X is distributed according to $N(\mu, \sigma^2)$, then $X^* = cX$ is distributed according to $N(c\mu, c^2\sigma^2)$ which is in the same class, and the hypothesis $\mathscr{E}X = 0$ is equivalent to $\mathscr{E}X^* = \mathscr{E}cX = 0$. If the observations x_α are transformed similarly, $x_\alpha^* = cx_\alpha$, then, for $c > 0$, t^* computed from x_α^* is the same as t computed from x_α. Thus, whatever the unit of measurement the statistical result is the same.

The generalized T^2-test has a similar property. If the vector random variable $\boldsymbol{X}$ is distributed according to $N(\boldsymbol{\mu}, \boldsymbol{\Sigma})$, then $\boldsymbol{X}^* = \boldsymbol{C}\boldsymbol{X}$ (for $|\boldsymbol{C}| \neq 0$) is distributed according to $N(\boldsymbol{C}\boldsymbol{\mu}, \boldsymbol{C}\boldsymbol{\Sigma}\boldsymbol{C}')$ which is in the same class. The hypothesis $\mathscr{E}\boldsymbol{X} = \boldsymbol{0}$ is equivalent to the hypothesis $\mathscr{E}\boldsymbol{X}^* = \mathscr{E}\boldsymbol{C}\boldsymbol{X} = \boldsymbol{0}$. If the observations $\boldsymbol{x}_\alpha$ are transformed in the same way, $\boldsymbol{x}_\alpha^* = \boldsymbol{C}\boldsymbol{x}_\alpha$, then T^{*2} computed on the basis of $\boldsymbol{x}_\alpha^*$ is the same as T^2 computed on the basis of $\boldsymbol{x}_\alpha$. This follows from the facts that $\bar{\boldsymbol{x}}^* = \boldsymbol{C}\bar{\boldsymbol{x}}$ and $\boldsymbol{A}^* = \boldsymbol{C}\boldsymbol{A}\boldsymbol{C}'$ and the following lemma:

LEMMA 5.2.1. *For any $p \times p$ nonsingular matrices $\boldsymbol{C}$ and $\boldsymbol{H}$ and any vector $\boldsymbol{k}$,*

$$\boldsymbol{k}'\boldsymbol{H}^{-1}\boldsymbol{k} = (\boldsymbol{C}\boldsymbol{k})'(\boldsymbol{C}\boldsymbol{H}\boldsymbol{C}')^{-1}(\boldsymbol{C}\boldsymbol{k}). \tag{16}$$

Proof.

$$\begin{aligned}(\boldsymbol{C}\boldsymbol{k})'(\boldsymbol{C}\boldsymbol{H}\boldsymbol{C}')^{-1}(\boldsymbol{C}\boldsymbol{k}) &= \boldsymbol{k}'\boldsymbol{C}'(\boldsymbol{C}')^{-1}\boldsymbol{H}^{-1}\boldsymbol{C}^{-1}\boldsymbol{C}\boldsymbol{k} \\ &= \boldsymbol{k}'\boldsymbol{H}^{-1}\boldsymbol{k}.\end{aligned} \tag{17}$$

We will show in Section 5.5 that of all tests invariant with regard to such transformations, (14) is the uniformly most powerful.

We can give a geometric interpretation of the $2/N$th root of the likelihood ratio criterion

$$\lambda^{2/N} = \frac{\left|\sum_{\alpha=1}^{N}(\boldsymbol{x}_\alpha - \bar{\boldsymbol{x}})(\boldsymbol{x}_\alpha - \bar{\boldsymbol{x}})'\right|}{\left|\sum_{\alpha=1}^{N}(\boldsymbol{x}_\alpha - \boldsymbol{\mu}_0)(\boldsymbol{x}_\alpha - \boldsymbol{\mu}_0)'\right|} \tag{18}$$

in terms of parallelotopes (see Section 7.5). In the p-dimensional representation the numerator of $\lambda^{2/N}$ is the sum of squares of volumes of all parallelotopes with principal edges p vectors, each with one end point at $\bar{\boldsymbol{x}}$ and the other at an $\boldsymbol{x}_\alpha$. The denominator is the sum of squares of volumes of all parallelotopes with principal edges p vectors, each with one end point at $\boldsymbol{\mu}_0$ and the other at $\boldsymbol{x}_\alpha$. If the sum of squared volumes involving vectors emanating from $\bar{\boldsymbol{x}}$, the "center" of the $\boldsymbol{x}_\alpha$, is much less than that involving vectors emanating from $\boldsymbol{\mu}_0$, then we reject the hypothesis that $\boldsymbol{\mu}_0$ is the mean of the distribution.

There is also an interpretation in the N-dimensional representation. Let $\boldsymbol{y}_i = (x_{i1}, \cdots, x_{iN})$ be the ith vector. Then

$$\sqrt{N}\,\bar{x}_i = \sum_{\alpha=1}^{N} \frac{1}{\sqrt{N}} x_{i\alpha} \tag{19}$$

is the distance from the origin of the projection of $\boldsymbol{y}_i$ on the equiangular line (with direction cosines $1/\sqrt{N}, \cdots, 1/\sqrt{N}$). The coordinates of the projection are $(\bar{x}_i, \cdots, \bar{x}_i)$. Then $(x_{i1} - \bar{x}_i, \cdots, x_{iN} - \bar{x}_i)$ is the projection of $\boldsymbol{y}_i$ on the plane through the origin perpendicular to the equiangular line. The numerator of $\lambda^{2/N}$ is the square of the p-dimensional volume of the parallelotope with principal edges, the vectors $(x_{i1} - \bar{x}_i, \cdots, x_{iN} - \bar{x}_i)$. A point $(x_{i1} - \mu_{0i}, \cdots, x_{iN} - \mu_{0i})$ is obtained from $\boldsymbol{y}_i$ by translation parallel to the equiangular line (by a distance $\sqrt{N}\mu_{0i}$). The denominator of $\lambda^{2/N}$ is the square of the volume of the parallelotope with principal edges these vectors. Then $\lambda^{2/N}$ is the ratio of these squared volumes.

5.2.2. The Distribution of T^2

In this section we will find the distribution of T^2 under general conditions, including the case when the null hypothesis is not true. Let $T^2 = \boldsymbol{Y}'\boldsymbol{S}^{-1}\boldsymbol{Y}$ where $\boldsymbol{Y}$ is distributed according to $N(\boldsymbol{\nu}, \boldsymbol{\Sigma})$ and $n\boldsymbol{S}$ is distributed independently as $\sum_{\alpha=1}^{n} \boldsymbol{Z}_\alpha \boldsymbol{Z}_\alpha'$ with the $\boldsymbol{Z}_\alpha$ independent, each with distribution $N(\boldsymbol{0}, \boldsymbol{\Sigma})$. T^2 defined in Section 5.2.1 is a special case of this with $\boldsymbol{Y} = \sqrt{N}(\bar{\boldsymbol{x}} - \boldsymbol{\mu}_0)$ and $\boldsymbol{\nu} = \sqrt{N}(\boldsymbol{\mu} - \boldsymbol{\mu}_0)$ and $n = N - 1$. Let $\boldsymbol{D}$ be a nonsingular matrix such that $\boldsymbol{D\Sigma D}' = \boldsymbol{I}$, and define

$$\begin{aligned} \boldsymbol{Y}^* &= \boldsymbol{DY}, \\ \boldsymbol{S}^* &= \boldsymbol{DSD}', \\ \boldsymbol{\nu}^* &= \boldsymbol{D\nu}. \end{aligned} \tag{20}$$

Then $T^2 = \boldsymbol{Y}^{*\prime}\boldsymbol{S}^{*-1}\boldsymbol{Y}^*$ (by Lemma 5.2.1) where $\boldsymbol{Y}^*$ is distributed according to $N(\boldsymbol{\nu}^*, \boldsymbol{I})$ and $n\boldsymbol{S}^*$ is distributed independently as $\sum_{\alpha=1}^{n}\boldsymbol{Z}_\alpha^*\boldsymbol{Z}_\alpha^{*\prime} = \sum_{\alpha=1}^{n}\boldsymbol{DZ}_\alpha(\boldsymbol{DZ}_\alpha)'$ with the $\boldsymbol{Z}_\alpha^* = \boldsymbol{DZ}_\alpha$ independent, each with distribution $N(\boldsymbol{0}, \boldsymbol{I})$. We note $\boldsymbol{\nu}'\boldsymbol{\Sigma}^{-1}\boldsymbol{\nu} = \boldsymbol{\nu}^{*\prime}(\boldsymbol{I})^{-1}\boldsymbol{\nu}^* = \boldsymbol{\nu}^{*\prime}\boldsymbol{\nu}^*$.

Let the first row of a $p \times p$ orthogonal matrix $\boldsymbol{Q}$ be defined by

$$q_{1i} = \frac{Y_i^*}{\sqrt{\boldsymbol{Y}^{*\prime}\boldsymbol{Y}^*}}, \qquad i = 1, \cdots, p; \tag{21}$$

this is permissible since $\sum_i q_{1i}^2 = 1$. The other $p - 1$ rows can be defined by some arbitrary rule by Lemma 2 of Appendix 1. Since Q depends on Y^* it is a random matrix. Now let

$$U = QY^*, \quad B = QnS^*Q'. \tag{22}$$

By the way Q was defined

$$\begin{aligned} U_1 &= \sum q_{1i} Y_i^* = \sqrt{Y^{*\prime}Y^*}, \\ U_j &= \sum q_{ji} Y_i^* \\ &= \sqrt{Y^{*\prime}Y^*}\sum q_{ji}q_{1i} \\ &= 0, \qquad j \neq 1. \end{aligned} \tag{23}$$

Then

$$\frac{T^2}{n} = U'B^{-1}U = (U_1 0 \cdots 0)\begin{pmatrix} b^{11} & b^{12} \cdots & b^{1p} \\ b^{21} & b^{22} \cdots & b^{2p} \\ \cdot & \cdot & \cdot \\ \cdot & \cdot & \cdot \\ \cdot & \cdot & \cdot \\ b^{p1} & b^{p2} \cdots & b^{pp} \end{pmatrix}\begin{pmatrix} U_1 \\ 0 \\ \cdot \\ \cdot \\ \cdot \\ 0 \end{pmatrix} = U_1^2 b^{11}, \tag{24}$$

where $(b^{ij}) = B^{-1}$. By Problem 18 of Chapter 2, $1/b^{11} = b_{11} - b_{(1)}B_{22}^{-1}b'_{(1)} = b_{11\cdot 2,\cdots,p}$, where

$$B = \begin{pmatrix} b_{11} & b_{(1)} \\ b'_{(1)} & B_{22} \end{pmatrix}, \tag{25}$$

and $T^2/n = U_1^2/b_{11\cdot 2,\cdots,p} = Y^{*\prime}Y^*/b_{11\cdot 2,\cdots,p}$. The conditional distribution of B given Q is that of $\sum_{\alpha=1}^n \mathbf{V}_\alpha \mathbf{V}'_\alpha$, where conditionally the $\mathbf{V}_\alpha = QZ_\alpha^*$ are independent, each with distribution $N(\mathbf{0}, I)$. By Theorem 4.3.3 $b_{11\cdot 2,\cdots,p}$ is conditionally distributed as $\sum_{\alpha=1}^{n-(p-1)} W_\alpha^2$ where conditionally the W_α are independent, each with the distribution $N(0, 1)$; that is, $b_{11\cdot 2,\cdots,p}$ is conditionally distributed as χ^2 with $n - (p - 1)$ degrees of freedom. Since the conditional distribution of $b_{11\cdot 2,\cdots,p}$ does not depend on Q, it is unconditionally distributed as χ^2. The quantity $Y^{*\prime}Y^*$ has a noncentral χ^2-distribution with p degrees of freedom and noncentrality parameter $\boldsymbol{\nu}^{*\prime}\boldsymbol{\nu}^* = \boldsymbol{\nu}'\boldsymbol{\Sigma}^{-1}\boldsymbol{\nu}$. Then T^2/n is distributed as the ratio of a noncentral χ^2 and an independent χ^2.

THEOREM 5.2.2. *Let $T^2 = Y'S^{-1}Y$, where Y is distributed according to $N(\boldsymbol{\nu}, \boldsymbol{\Sigma})$ and nS is independently distributed as $\sum_{\alpha=1}^n Z_\alpha Z'_\alpha$ with the Z_α independent, each with distribution $N(\mathbf{0}, \boldsymbol{\Sigma})$. Then $(T^2/n)[(n - p + 1)/p]$ is distributed as a noncentral F with p and $n - p + 1$ degrees of freedom and noncentrality parameter $\boldsymbol{\nu}'\boldsymbol{\Sigma}^{-1}\boldsymbol{\nu}$. If $\boldsymbol{\nu} = \mathbf{0}$, the distribution is the central F-distribution.*

We shall call this the T^2-distribution with n degrees of freedom.

COROLLARY 5.2.1. *Let* $x_1, \cdots, x_N$ *be a sample from* $N(\mu, \Sigma)$, *and let* $T^2 = N(\bar{x} - \mu_0)'S^{-1}(\bar{x} - \mu_0)$. *The distribution of* $[T^2/(N-1)][(N-p)/p]$ *is noncentral F with p and N − p degrees of freedom and noncentrality parameter* $N(\mu - \mu_0)'\Sigma^{-1}(\mu - \mu_0)$. *If* $\mu = \mu_0$, *then the F-distribution is central.*

The above derivation of the T^2-distribution is due to A. H. Bowker (personal communication). The noncentral F-density and tables of the distribution are discussed in Section 5.4.

5.3. USES OF THE T^2-STATISTIC

5.3.1. Testing the Hypothesis that the Mean Vector Is a Given Vector; Computation of T^2

The likelihood ratio test of the hypothesis $\mu = \mu_0$ on the basis of a sample of N from $N(\mu, \Sigma)$ is equivalent to

$$T^2 \geq T_0^2 \tag{1}$$

as given in Section 5.2.1. If the significance level is α, then the $100\alpha\%$ point of the F-distribution is taken, that is,

$$T_0^2 = \frac{(N-1)p}{N-p} F_{p, N-p}(\alpha). \tag{2}$$

The choice of significance level may depend on the power of the test. We shall discuss this in Section 5.4.

The statistic T^2 is easily computed from the sample. Let

$$A^{-1}(\bar{x} - \mu_0) = b. \tag{3}$$

This vector is the solution of

$$Ab = (\bar{x} - \mu_0). \tag{4}$$

If the vector b is obtained by solving (4), then

$$\frac{T^2}{N-1} = N(\bar{x} - \mu_0)'b. \tag{5}$$

Thus, it is unnecessary to compute A^{-1} or S^{-1}. In fact, if one uses the Doolittle method for the solution of (4), one need only proceed through the forward solution. In Section 8.2 it is shown that the forward solution amounts to multiplying (4) on the left by a matrix F (such that FA is triangular) and this result by D^{-1}, where D is a diagonal matrix with diagonal elements equal to the corresponding diagonal elements of FA. On the right-hand side one has $F(\bar{x} - \mu_0)$ and $D^{-1}F(\bar{x} - \mu_0)$ respectively.

Then $[F(\bar{x} - \mu_0)]'[D^{-1}F(\bar{x} - \mu_0)] = (\bar{x} - \mu_0)'F'D^{-1}F(\bar{x} - \mu_0) = (\bar{x} - \mu_0)' A^{-1}(\bar{x} - \mu_0)$ because $F'D^{-1}F = A^{-1}$. The details of the procedure and proof are given in Section 8.2.

It is of interest to note that $T^2/(N - 1)$ is the nonzero root of

$$|N(\bar{x} - \mu_0)(\bar{x} - \mu_0)' - \lambda A| = 0. \tag{6}$$

LEMMA 5.3.1. *If ν is a vector of p components and if B is a nonsingular pth order matrix, then $\nu'B^{-1}\nu$ is the nonzero root of*

$$|\nu\nu' - \lambda B| = 0. \tag{7}$$

Proof. The nonzero root, say λ_1, of (7) is associated with a characteristic vector β satisfying

$$\nu\nu'\beta = \lambda_1 B\beta. \tag{8}$$

Since $\lambda_1 \neq 0$, $\nu'\beta \neq 0$. Multiplying on the left by $\nu'B^{-1}$, we obtain

$$(\nu'B^{-1}\nu)(\nu'\beta) = \lambda_1(\nu'\beta); \tag{9}$$

this proves the lemma. In the case above $\nu = \sqrt{N}(\bar{x} - \mu_0)$ and $B = A$.

5.3.2. A Confidence Region for the Mean Vector

If μ is the mean of $N(\mu, \Sigma)$, we know that the probability is $1 - \alpha$ of drawing a sample of N with mean $\bar{x}$ and covariance matrix S such that

$$N(\bar{x} - \mu)'S^{-1}(\bar{x} - \mu) \leq T_0^2(\alpha). \tag{10}$$

Thus, if we compute (10) for a particular sample, we have confidence $1 - \alpha$ that (10) is a true statement concerning μ. The inequality

$$N(\bar{x} - m)'S^{-1}(\bar{x} - m) \leq T_0^2(\alpha) \tag{11}$$

is the interior and boundary of an ellipsoid in the p-dimensional space of m with center at $\bar{x}$ and with size and shape depending on S^{-1} and α. See Figure 1. We state that μ lies within this ellipsoid. Over random samples (11) is a random ellipsoid.

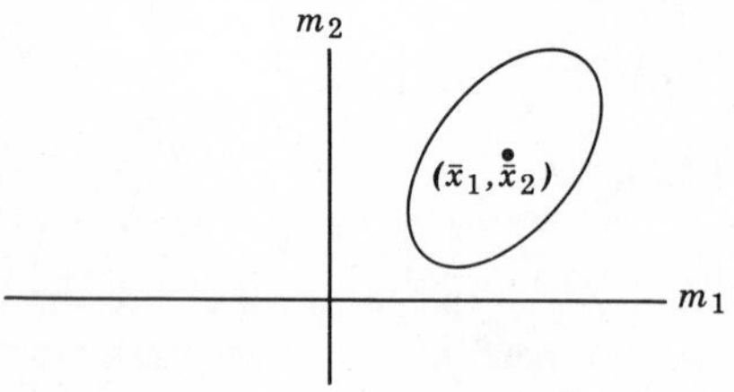

FIGURE 1

5.3.3. A Two-Sample Problem

Another situation in which the T^2-statistic is used is that in which the null hypothesis is that the mean of one normal population is equal to the mean of the other where the covariance matrices are assumed equal but unknown. Suppose $y_1^{(i)}, \cdots, y_{N_i}^{(i)}$ is a sample from $N(\mu^{(i)}, \Sigma)$,

$i = 1, 2$. We wish to test the null hypothesis $\boldsymbol{\mu}^{(1)} = \boldsymbol{\mu}^{(2)}$. $\bar{\mathbf{y}}^{(i)}$ is distributed according to $N[\boldsymbol{\mu}^{(i)}, (1/N_i)\boldsymbol{\Sigma}]$. Consequently, $\sqrt{N_1 N_2/(N_1 + N_2)}(\bar{\mathbf{y}}^{(1)} - \bar{\mathbf{y}}^{(2)})$ is distributed according to $N(\mathbf{0}, \boldsymbol{\Sigma})$ under the null hypothesis. If we let

$$(12) \qquad S = \frac{1}{N_1 + N_2 - 2}\left\{\sum_{\alpha=1}^{N_1}(\mathbf{y}_\alpha^{(1)} - \bar{\mathbf{y}}^{(1)})(\mathbf{y}_\alpha^{(1)} - \bar{\mathbf{y}}^{(1)})' + \sum_{\alpha=1}^{N_2}(\mathbf{y}_\alpha^{(2)} - \bar{\mathbf{y}}^{(2)})(\mathbf{y}_\alpha^{(2)} - \bar{\mathbf{y}}^{(2)})'\right\},$$

then $(N_1 + N_2 - 2)S$ is distributed as $\sum_{\alpha=1}^{N_1+N_2-2} \mathbf{Z}_\alpha \mathbf{Z}_\alpha'$, where $\mathbf{Z}_\alpha$ is distributed according to $N(\mathbf{0}, \boldsymbol{\Sigma})$. Thus

$$(13) \qquad T^2 = \frac{N_1 N_2}{N_1 + N_2}(\bar{\mathbf{y}}^{(1)} - \bar{\mathbf{y}}^{(2)})' S^{-1}(\bar{\mathbf{y}}^{(1)} - \bar{\mathbf{y}}^{(2)})$$

is distributed as T^2 with $N_1 + N_2 - 2$ degrees of freedom. The critical region is

$$T^2 \geq \frac{(N_1 + N_2 - 2)p}{N_1 + N_2 - p - 1} F_{p, N_1+N_2-p-1}(\alpha)$$

with significance level α. We can obviously set up a confidence region for $\boldsymbol{\mu}^{(2)} - \boldsymbol{\mu}^{(1)}$ based on the T^2-statistic.

An example may be given from Fisher (1936). Let x_1 = sepal length, x_2 = sepal width, x_3 = petal length, x_4 = petal width. Fifty observations are taken from the population Iris versicolor (1) and 50 from the population Iris setosa (2). The data may be summarized (in centimeters) as

$$(14) \qquad \bar{\mathbf{x}}^{(1)} = \begin{pmatrix} 5.936 \\ 2.770 \\ 4.260 \\ 1.326 \end{pmatrix},$$

$$(15) \qquad \bar{\mathbf{x}}^{(2)} = \begin{pmatrix} 5.006 \\ 3.428 \\ 1.462 \\ 0.246 \end{pmatrix},$$

$$(16) \qquad 98S = \begin{pmatrix} 19.1434 & 9.0356 & 9.7634 & 3.2394 \\ 9.0356 & 11.8658 & 4.6232 & 2.4746 \\ 9.7634 & 4.6232 & 12.2978 & 3.8794 \\ 3.2394 & 2.4746 & 3.8794 & 2.4604 \end{pmatrix}.$$

The value of $T^2/98$ is 26.334 and $T^2/98 \times 95/4 = 625.5$. This value is highly significant (compared to the F-point for 4 and 95 degrees of freedom).

5.3.4. A q-Sample Problem

After considering the above example, Fisher considers a third sample drawn from a population assumed to have the same covariance matrix. He makes the same measurements on 50 Iris virginica. There is a theoretical reason for believing the gene structures of these three species to be such that the mean vectors of the three populations are related as

$$3\boldsymbol{\mu}^{(1)} = \boldsymbol{\mu}^{(3)} + 2\boldsymbol{\mu}^{(2)}, \tag{17}$$

where $\boldsymbol{\mu}^{(3)}$ is the mean vector of the third population.

This is a special case of the following general problem. Let $\{\boldsymbol{x}_\alpha^{(i)}\}$ $(\alpha = 1, \cdots, N_i;\ i = 1, \cdots, q)$ be samples from $N(\boldsymbol{\mu}^{(i)}, \boldsymbol{\Sigma})(i = 1, \cdots, q)$ respectively. Let us test the hypothesis H:

$$\sum_{i=1}^{q} \beta_i \boldsymbol{\mu}^{(i)} = \boldsymbol{\mu}, \tag{18}$$

where $\beta_1, \cdots, \beta_q$ are given scalars and $\boldsymbol{\mu}$ is a given vector. The criterion is

$$T^2 = c\left(\sum_{i=1}^{q} \beta_i \bar{\boldsymbol{x}}^{(i)} - \boldsymbol{\mu}\right)' \boldsymbol{S}^{-1}\left(\sum_{i=1}^{q} \beta_i \bar{\boldsymbol{x}}^{(i)} - \boldsymbol{\mu}\right), \tag{19}$$

where

$$\bar{\boldsymbol{x}}^{(i)} = \frac{1}{N_i} \sum_{\alpha=1}^{N_i} \boldsymbol{x}_\alpha^{(i)}, \tag{20}$$

$$\left(\sum_{i=1}^{q} N_i - q\right) \boldsymbol{S} = \sum_{i=1}^{q} \sum_{\alpha=1}^{N_i} (\boldsymbol{x}_\alpha^{(i)} - \bar{\boldsymbol{x}}^{(i)})(\boldsymbol{x}_\alpha^{(i)} - \bar{\boldsymbol{x}}^{(i)})', \tag{21}$$

$$1/c = \sum_{i=1}^{q} \frac{\beta_i^2}{N_i}. \tag{22}$$

This T^2 has the T^2-distribution with $\sum_{i=1}^{q} N_i - q$ degrees of freedom.

Fisher actually assumes in his example that the covariance matrices of the three populations may be different. Hence, he uses the technique described in Section 5.6.

5.3.5. A Problem of Symmetry

Consider testing the hypothesis $H: \mu_1 = \mu_2 = \cdots = \mu_p$ on the basis of a sample $\boldsymbol{x}_1, \cdots, \boldsymbol{x}_N$ from $N(\boldsymbol{\mu}, \boldsymbol{\Sigma})$ where $\boldsymbol{\mu}' = (\mu_1, \cdots, \mu_p)$. Let $\boldsymbol{C}$ be any $(p-1) \times p$ matrix of rank $p - 1$ such that

$$\boldsymbol{C}\boldsymbol{\varepsilon} = \boldsymbol{0}, \tag{23}$$

where $\boldsymbol{\varepsilon}' = (1, \cdots, 1)$. Then

$$\boldsymbol{y}_\alpha = \boldsymbol{C}\boldsymbol{x}_\alpha \qquad (\alpha = 1, \cdots, N) \tag{24}$$

has mean $\boldsymbol{C\mu}$ and covariance matrix $\boldsymbol{C\Sigma C'}$. The hypothesis H is $\boldsymbol{C\mu} = \boldsymbol{0}$. The statistic to be used is

$$T^2 = N\bar{\boldsymbol{y}}'\boldsymbol{S}^{-1}\bar{\boldsymbol{y}}, \tag{25}$$

where

$$\bar{\boldsymbol{y}} = \frac{1}{N}\sum_{\alpha=1}^{N} \boldsymbol{y}_\alpha = \boldsymbol{C}\bar{\boldsymbol{x}}, \tag{26}$$

$$\begin{aligned} \boldsymbol{S} &= \frac{1}{N-1}\sum_{\alpha=1}^{N}(\boldsymbol{y}_\alpha - \bar{\boldsymbol{y}})(\boldsymbol{y}_\alpha - \bar{\boldsymbol{y}})' \\ &= \frac{1}{N-1}\boldsymbol{C}\sum_{\alpha=1}^{N}(\boldsymbol{x}_\alpha - \bar{\boldsymbol{x}})(\boldsymbol{x}_\alpha - \bar{\boldsymbol{x}})'\boldsymbol{C}'. \end{aligned} \tag{27}$$

This statistic has the T^2-distribution with $N - 1$ degrees of freedom for a $(p - 1)$-dimensional distribution. This T^2-statistic is invariant under any linear transformation in the $p - 1$ dimensions orthogonal to $\boldsymbol{\varepsilon}$. Hence, the statistic is independent of the choice of $\boldsymbol{C}$.

An example of this sort has been given by Rao (1948b). Let N be the amount of cork in a boring from the North into a cork tree; let E, S, and W be defined similarly. The set of amounts in four borings on one tree is considered as an observation from a 4-variate normal distribution. The question is whether the cork trees have the same amount of cork on each side. We make a transformation

$$\begin{aligned} y_1 &= N - E - W + S, \\ y_2 &= S - W, \\ y_3 &= N - S. \end{aligned} \tag{28}$$

The number of observations made is 28.

The vector of means is

$$\bar{\boldsymbol{y}} = \begin{pmatrix} 8.86 \\ 4.50 \\ 0.86 \end{pmatrix}; \tag{29}$$

the covariance matrix for $\boldsymbol{y}$ is

$$\boldsymbol{S} = \begin{pmatrix} 128.72 & 61.41 & -21.02 \\ 61.41 & 56.93 & -28.30 \\ -21.02 & -28.30 & 63{\cdot}53 \end{pmatrix}. \tag{30}$$

The value of $T^2/(N-1)$ is 0.768. The statistic $0.768 \times 25/3 = 6.402$ is to be compared with the F-significance point with 3 and 25 degrees of freedom. It is significant at the 1% level.

5.4. THE DISTRIBUTION OF T^2 UNDER ALTERNATIVE HYPOTHESES; THE POWER FUNCTION

In Section 5.2.2 we showed that $(T^2/n)(N-p)/p$ has a noncentral F-distribution. In this section we shall discuss the noncentral χ^2 and F-distributions, the tabulation of the latter, and applications to procedures based on T^2.

The central χ^2-distribution is the distribution of the sum of squares of independent (scalar) normal variables with means 0 and variances 1; the noncentral χ^2-distribution is the generalization of this when the means may be different from 0. Let $\boldsymbol{Y}$(of p components) be distributed according to $N(\boldsymbol{\nu}, \boldsymbol{I})$. Let $\boldsymbol{Q}$ be an orthogonal matrix with elements of the first row being

$$q_{1i} = \frac{\nu_i}{\sqrt{\boldsymbol{\nu}'\boldsymbol{\nu}}}. \tag{1}$$

Then $\boldsymbol{Z} = \boldsymbol{QY}$ is distributed according to $N(\boldsymbol{\lambda}, \boldsymbol{I})$, where

$$\boldsymbol{\lambda} = \begin{pmatrix} \tau \\ 0 \\ \cdot \\ \cdot \\ \cdot \\ 0 \end{pmatrix} \tag{2}$$

and $\tau = \sqrt{\boldsymbol{\nu}'\boldsymbol{\nu}}$ (see Section 5.2.2). Let $V = \boldsymbol{Y}'\boldsymbol{Y} = \boldsymbol{Z}'\boldsymbol{Z} = \sum_{i=1}^p Z_i^2$. Then $W = \sum_{i=2}^p Z_i^2$ has a χ^2-distribution with $p-1$ degrees of freedom (Problem 5 of Chapter 7), and Z_1 and W have as joint density

$$\begin{aligned} &\frac{1}{\sqrt{2\pi}} e^{-\frac{1}{2}(z_1-\tau)^2} \frac{1}{2^{\frac{1}{2}(p-1)}\Gamma[\frac{1}{2}(p-1)]} w^{\frac{1}{2}(p-1)-1} e^{-\frac{1}{2}w} \\ &\qquad = Ce^{-\frac{1}{2}(\tau^2+z_1{}^2+w)} w^{\frac{1}{2}(p-3)} e^{\tau z_1} \\ &\qquad = Ce^{-\frac{1}{2}(\tau^2+z_1{}^2+w)} w^{\frac{1}{2}(p-3)} \sum_{\alpha=0}^{\infty} \frac{\tau^\alpha z_1^\alpha}{\alpha!}, \end{aligned} \tag{3}$$

where $C^{-1} = 2^{\frac{1}{2}p}\sqrt{\pi}\,\Gamma[\frac{1}{2}(p-1)]$. The joint density of $V = W + Z_1^2$ and Z_1 is obtained by substituting $w = v - z_1^2$ (the Jacobian is 1),

$$Ce^{-\frac{1}{2}(\tau^2+v)}(v-z_1^2)^{\frac{1}{2}(p-3)} \sum_{\alpha=0}^{\infty} \frac{\tau^\alpha z_1^\alpha}{\alpha!}. \tag{4}$$

The joint density of V and $U = Z_1/\sqrt{V}$ is $(dz_1 = \sqrt{v}\,du)$

$$Ce^{-\frac{1}{2}(\tau^2+v)}v^{\frac{1}{2}(p-2)}(1-u^2)^{\frac{1}{2}(p-3)}\sum_{\alpha=0}^{\infty}\frac{\tau^\alpha v^{\frac{1}{2}\alpha}u^\alpha}{\alpha!}. \tag{5}$$

The admissible range of z_1 given v is $-\sqrt{v}$ to $\sqrt{v}$, and the admissible range of u is -1 to 1. When we integrate (5) with respect to u term by term, the terms for α odd integrate to 0 since such a term is an odd function of u. In the other integrations we substitute $u = \sqrt{s}(du = \frac{1}{2}ds/\sqrt{s})$ to obtain

$$\begin{aligned}\int_{-1}^{1}(1-u^2)^{\frac{1}{2}(p-3)}u^{2\beta}\,du &= 2\int_0^1(1-u^2)^{\frac{1}{2}(p-3)}u^{2\beta}\,du \\ &= \int_0^1(1-s)^{\frac{1}{2}(p-3)}s^{\beta-\frac{1}{2}}\,ds \\ &= B[\tfrac{1}{2}(p-1), \beta+\tfrac{1}{2}] \\ &= \frac{\Gamma[\frac{1}{2}(p-1)]\Gamma(\beta+\frac{1}{2})}{\Gamma(\frac{1}{2}p+\beta)},\end{aligned} \tag{6}$$

by the usual properties of the beta and gamma functions. Thus the density of V is

$$\frac{1}{2^{\frac{1}{2}p}\sqrt{\pi}}e^{-\frac{1}{2}(\tau^2+v)}v^{\frac{1}{2}p-1}\sum_{\beta=0}^{\infty}\frac{(\tau^2)^\beta v^\beta}{(2\beta)!}\frac{\Gamma(\beta+\frac{1}{2})}{\Gamma(\frac{1}{2}p+\beta)}. \tag{7}$$

This is the density of the noncentral χ^2-distribution with p degrees of freedom and noncentrality parameter τ^2.

THEOREM 5.4.1. *If Y (of p components) is distributed according to $N(\mathbf{\nu}, I)$, then $V = Y'Y$ has the density* (7), *where* $\tau = \sqrt{\mathbf{\nu}'\mathbf{\nu}}$.

Let V have the noncentral χ^2-distribution with p degrees of freedom and noncentrality parameter τ^2 and let W be independently distributed as χ^2 with m degrees of freedom. We shall find the density of $F = (V/p)/(W/m)$, which is the noncentral F with noncentrality parameter τ^2. The joint density of V and W is (7) multiplied by the density of W, which is $2^{-\frac{1}{2}m}\Gamma^{-1}(\frac{1}{2}m)w^{\frac{1}{2}m-1}e^{-\frac{1}{2}w}$. The joint density of F and W $(dv = pw\,df/m)$ is

$$\frac{e^{-\frac{1}{2}\tau^2}}{2^{\frac{1}{2}(p+m)}\sqrt{\pi}\Gamma(\frac{1}{2}m)}e^{-\frac{1}{2}w(1+pf/m)} \cdot \frac{p}{m}\sum_{\beta=0}^{\infty}\frac{(\tau^2)^\beta\Gamma(\beta+\frac{1}{2})}{(2\beta)!\Gamma(\frac{1}{2}p+\beta)}\left(\frac{pf}{m}\right)^{\frac{1}{2}p+\beta-1}w^{\frac{1}{2}(p+m)+\beta-1}. \tag{8}$$

The marginal density, obtained by integrating (8) with respect to w from 0 to ∞ (see Section 4.2.2), is

$$(9) \qquad \frac{pe^{-\frac{1}{2}\tau^2}}{m\sqrt{\pi}\,\Gamma(\frac{1}{2}m)} \sum_{\beta=0}^{\infty} \frac{(2\tau^2)^\beta \Gamma(\beta+\frac{1}{2})}{(2\beta)!\,\Gamma(\frac{1}{2}p+\beta)} \left(\frac{pf}{m}\right)^{\frac{1}{2}p+\beta-1} \frac{\Gamma[\frac{1}{2}(p+m)+\beta]}{(1+pf/m)^{\frac{1}{2}(p+m)+\beta}}.$$

If we use the gamma function duplication formula

$$(2\beta)! = \Gamma(2\beta+1) = \Gamma(\beta+\tfrac{1}{2})\Gamma(\beta+1)2^{2\beta}/\sqrt{\pi},$$

then the density is

$$(10) \qquad \frac{pe^{-\frac{1}{2}\tau^2}}{m\Gamma(\frac{1}{2}m)} \sum_{\beta=0}^{\infty} \frac{(\tau^2/2)^\beta (pf/m)^{\frac{1}{2}p+\beta-1}\Gamma[\frac{1}{2}(p+m)+\beta]}{\beta!\,\Gamma(\frac{1}{2}p+\beta)(1+pf/m)^{\frac{1}{2}(p+m)+\beta}}.$$

THEOREM 5.4.2. *If V has a noncentral χ^2-distribution with p degrees of freedom and noncentrality parameter τ^2 and W has an independent χ^2-distribution with m degrees of freedom, then $F = (V/p)/(W/m)$ has the density* (10).

If $T^2 = N(\bar{\mathbf{x}} - \boldsymbol{\mu}_0)S^{-1}(\bar{\mathbf{x}} - \boldsymbol{\mu}_0)$ is based on a sample of N from $N(\boldsymbol{\mu}, \boldsymbol{\Sigma})$, then $(T^2/n)(N-p)/p$ has the noncentral F-distribution with p and $N-p$ degrees of freedom and noncentrality parameter $N(\boldsymbol{\mu} - \boldsymbol{\mu}_0)'\boldsymbol{\Sigma}^{-1}(\boldsymbol{\mu} - \boldsymbol{\mu}_0) = \tau^2$. From (10) we find that the density of T^2 is

$$(11) \qquad \frac{e^{-\frac{1}{2}\tau^2}}{(N-1)\Gamma[\frac{1}{2}(N-p)]} \sum_{\beta=0}^{\infty} \frac{(\tau^2/2)^\beta [(t^2)/(N-1)]^{\frac{1}{2}p+\beta-1}\Gamma(\frac{1}{2}N+\beta)}{\beta!\,\Gamma(\frac{1}{2}p+\beta)[1+(t^2)/(N-1)]^{\frac{1}{2}N+\beta}}.$$

Tables have been given by Tang (1938) of the probability of accepting the null hypothesis (that is, the probability of Type II Error) for various values of τ^2 and for significance levels 0.05 and 0.01. His number of degrees of freedom f_1 is our $p[1(1)8]$, his f_2 is our $n - p + 1[2, 4(1)30, 60, \infty]$ and his noncentral parameter ϕ is related to our τ^2 by

$$(12) \qquad \phi = \frac{\tau}{\sqrt{p+1}}$$

$[1(\frac{1}{2})3(1)8]$. His accompanying tables of significance points are for $T^2/(T^2 + N - 1)$.

As an example, suppose $p = 4$, $n - p + 1 = 20$, and we are testing the null hypothesis $\boldsymbol{\mu} = \mathbf{0}$ at the 1% level of significance. We would like to know the probability, say, that we accept the null hypothesis when $\phi = 2.5$ ($\tau^2 = 31.25$). It is 0.227. If we think the disadvantage of accepting the null hypothesis when N, $\boldsymbol{\mu}$, and $\boldsymbol{\Sigma}$ are such that $\tau^2 = 31.25$ is less than the disadvantage of rejecting the null hypothesis when it is true, then we may find it reasonable to conduct the test as assumed.

However, if the disadvantage of one type of error is about equal to that of the other, it would seem reasonable to desire to bring down the probability of a Type II error. Thus, if we used a significance level of 5%, the probability of Type II error (for $\phi = 2.5$) is only 0.043.

Emma Lehmer (1944) has computed tables of ϕ for given significance level and given probability of Type II error. Her tables can be used to see what value of τ^2 is needed to make the probability of acceptance of the null hypothesis sufficiently low when $\boldsymbol{\mu} \neq \mathbf{0}$. For instance, if we want to be able to reject the hypothesis $\boldsymbol{\mu} = \mathbf{0}$ on the basis of a sample for a given $\boldsymbol{\mu}$ and $\boldsymbol{\Sigma}$, we may be able to choose N so that $N\boldsymbol{\mu}'\boldsymbol{\Sigma}^{-1}\boldsymbol{\mu} = \tau^2$ is sufficiently large. Of course, the difficulty with these considerations is that we usually do not know exactly the values of $\boldsymbol{\mu}$ and $\boldsymbol{\Sigma}$ (hence, τ^2) for which we want the probability of rejection at a certain value.

The distribution of T^2 when the null hypothesis is not true was derived by different methods by Hsu (1938) and Bose and Roy (1938a).

5.5. SOME OPTIMUM PROPERTIES OF THE T^2-TEST

In this section we shall indicate that the T^2-test is the best in certain classes of tests and sketch briefly the proofs of these results.

The hypothesis $\boldsymbol{\mu} = \mathbf{0}$ is to be tested on the basis of the N observations $\boldsymbol{x}_1, \cdots, \boldsymbol{x}_N$ from $N(\boldsymbol{\mu}, \boldsymbol{\Sigma})$. First we consider the class of tests based on the statistics $A = \sum(\boldsymbol{x}_\alpha - \bar{\boldsymbol{x}})(\boldsymbol{x}_\alpha - \bar{\boldsymbol{x}})'$ and $\bar{\boldsymbol{x}}$ which are invariant with respect to the transformations $A^* = CAC'$ and $\bar{\boldsymbol{x}}^* = C\bar{\boldsymbol{x}}$, where C is nonsingular. The transformation $\boldsymbol{x}_\alpha^* = C\boldsymbol{x}_\alpha$ leaves the problem invariant; that is, in terms of $\boldsymbol{x}_\alpha^*$ we test the hypothesis $\mathscr{E}\boldsymbol{x}_\alpha^* = \mathbf{0}$ given that $\boldsymbol{x}_1^*, \cdots, \boldsymbol{x}_N^*$ are N observations from a multivariate normal population. It seems reasonable that we require a solution that is also invariant with respect to these transformations; that is, we look for a critical region which is not changed by a nonsingular linear transformation (the definition of the region is the same in different coordinate systems).

THEOREM 5.5.1. *Given the observations* $\boldsymbol{x}_1, \cdots, \boldsymbol{x}_N$ *from* $N(\boldsymbol{\mu}, \boldsymbol{\Sigma})$, *of all tests of* $\boldsymbol{\mu} = \mathbf{0}$ *based on* $\bar{\boldsymbol{x}}$ *and* $A = \sum(\boldsymbol{x}_\alpha - \bar{\boldsymbol{x}})(\boldsymbol{x}_\alpha - \bar{\boldsymbol{x}})'$ *which are invariant with respect to transformations* $\bar{\boldsymbol{x}}^* = C\bar{\boldsymbol{x}}$, $A^* = CAC'$ (C *nonsingular*), *the* T^2-*test is uniformly most powerful.*

Proof. First, as we have seen in Section 5.2.1, any test based on T^2 is invariant. Second, this function is the only invariant, for if $f(\bar{\boldsymbol{x}}, A)$ is invariant, then $f(\bar{\boldsymbol{x}}, A) = f(\bar{\boldsymbol{x}}^*, I)$, where only the first coordinate of $\bar{\boldsymbol{x}}^*$ is different from zero and it is $\sqrt{\bar{\boldsymbol{x}}'A^{-1}\bar{\boldsymbol{x}}}$. (There is a matrix C such that $C\bar{\boldsymbol{x}} = \bar{\boldsymbol{x}}^*$ and $CAC' = I$.) Thus $f(\bar{\boldsymbol{x}}, A)$ depends only on $\bar{\boldsymbol{x}}'A^{-1}\bar{\boldsymbol{x}}$. Thus an invariant test must be based on $\bar{\boldsymbol{x}}'A^{-1}\bar{\boldsymbol{x}}$. Third, we can apply the Neyman-Pearson fundamental lemma (Lehmann) to the distribution of T^2 [(11) of Section 5.4] to find the uniformly most powerful test based on

T^2, against a simple alternative $\tau^2 = N\boldsymbol{\mu}'\boldsymbol{\Sigma}^{-1}\boldsymbol{\mu}$. The most powerful test of $\tau^2 = 0$ is based on the ratio of (11) to (11) with $\tau^2 = 0$. The test is

$$(1)\quad c < e^{-\frac{1}{2}\tau^2} \sum_{\alpha=0}^{\infty} \frac{(\tau^2/2)^\alpha (t^2/n)^{\frac{1}{2}p+\alpha-1}(1 + t^2/n)^{-\frac{1}{2}(n+1)+\alpha}\Gamma[\frac{1}{2}(n+1) - \alpha]}{\alpha!\Gamma(\frac{1}{2}p + \alpha)} \Bigg/ \frac{(t^2/n)^{\frac{1}{2}p-1}(1 + t^2/n)^{-\frac{1}{2}(n+1)}\Gamma[\frac{1}{2}(n+1)]}{\Gamma(\frac{1}{2}p)}$$

$$= \frac{\Gamma(\frac{1}{2}p)}{\Gamma[\frac{1}{2}(n+1)]} e^{-\frac{1}{2}\tau^2} \sum_{\alpha=0}^{\infty} \frac{(\tau^2/2)^\alpha \Gamma[\frac{1}{2}(n+1) + \alpha]}{\alpha!\Gamma(\frac{1}{2}p + \alpha)} \left(\frac{t^2/n}{1 + t^2/n}\right)^\alpha.$$

The right-hand side of (1) is a strictly increasing function of $\dfrac{t^2/n}{1 + t^2/n}$, hence of t^2. Thus the inequality is $t^2 > k$ for k suitably chosen. Since this does not depend on the alternative τ^2, the test is uniformly most powerful.

DEFINITION 5.5.1. *A critical function $\psi(\bar{\mathbf{x}}, \mathbf{A})$ is a function with values between* 0 *and* 1 *(inclusive) such that $\mathscr{E}\psi(\bar{\mathbf{x}}, \mathbf{A}) = \varepsilon$, the significance level, when $\boldsymbol{\mu} = \mathbf{0}$.* A randomized test consists of rejecting the hypothesis with probability $\psi(\mathbf{x}, \mathbf{B})$ when $\bar{\mathbf{x}} = \mathbf{x}$ and $\mathbf{A} = \mathbf{B}$. A nonrandomized test is defined when $\psi(\bar{\mathbf{x}}, \mathbf{A})$ takes on only the values 0 and 1. Using the form of the Neyman-Pearson lemma appropriate for critical functions we obtain the following corollary:

COROLLARY 5.5.1. *On the basis of observations $\mathbf{x}_1, \cdots, \mathbf{x}_N$ from $N(\boldsymbol{\mu}, \boldsymbol{\Sigma})$, of all randomized tests based on $\bar{\mathbf{x}}$ and $\mathbf{A}$ which are invariant with respect to transformations $\bar{\mathbf{x}}^* = \mathbf{C}\bar{\mathbf{x}}$, $\mathbf{A}^* = \mathbf{CAC}'$($\mathbf{C}$ nonsingular), the T^2-test is uniformly most powerful.*

THEOREM 5.5.2. *On the basis of observations $\mathbf{x}_1, \cdots, \mathbf{x}_N$ from $N(\boldsymbol{\mu}, \boldsymbol{\Sigma})$, of all tests of $\boldsymbol{\mu} = \mathbf{0}$ which are invariant with respect to transformations $\mathbf{x}_\alpha^* = \mathbf{C}\mathbf{x}_\alpha$ ($\mathbf{C}$ nonsingular), the T^2-test is a uniformly most powerful test; that is, the T^2-test is at least as powerful as any other invariant test.*

Proof. Let $\psi(\mathbf{x}_1, \cdots, \mathbf{x}_N)$ be the critical function of an invariant test. Then

$$(2)\qquad \mathscr{E}[\psi(\mathbf{x}_1, \cdots, \mathbf{x}_N)] = \mathscr{E}_{\bar{\mathbf{x}}, \mathbf{A}}\{\mathscr{E}[\psi(\mathbf{x}_1, \cdots, \mathbf{x}_N)|\bar{\mathbf{x}}, \mathbf{A}]\}.$$

Since $\bar{\mathbf{x}}, \mathbf{A}$ are sufficient statistics for $\boldsymbol{\mu}, \boldsymbol{\Sigma}$, $\mathscr{E}[\psi(\mathbf{x}_1, \cdots, \mathbf{x}_N)|\bar{\mathbf{x}}, \mathbf{A}]$ depends only on $\bar{\mathbf{x}}, \mathbf{A}$. It is invariant and has the same power as $\psi(\mathbf{x}_1, \cdots, \mathbf{x}_N)$. Thus each test in this larger class can be replaced by one in the smaller class (depending only on $\bar{\mathbf{x}}$ and $\mathbf{A}$) which has identical power. Corollary 5.5.1 completes the proof.

THEOREM 5.5.3. *Given observations $\mathbf{x}_1, \cdots, \mathbf{x}_N$ from $N(\boldsymbol{\mu}, \boldsymbol{\Sigma})$, of all tests of $\boldsymbol{\mu} = \mathbf{0}$ based on $\bar{\mathbf{x}}$ and $\mathbf{A} = \sum(\mathbf{x}_\alpha - \bar{\mathbf{x}})(\mathbf{x}_\alpha - \bar{\mathbf{x}})'$ with power depending only on $N\boldsymbol{\mu}'\boldsymbol{\Sigma}^{-1}\boldsymbol{\mu}$, the T^2-test is uniformly most powerful.*

Proof. We wish to reduce this theorem to Theorem 5.5.1 by identifying the class of tests with power depending on $N\boldsymbol{\mu}'\boldsymbol{\Sigma}^{-1}\boldsymbol{\mu}$ with the class of invariant tests. We need the following definition:

DEFINITION 5.5.2. *A test $\psi(\boldsymbol{x}_1, \cdots, \boldsymbol{x}_N)$ is said to be almost invariant if*

$$\psi(\boldsymbol{x}_1, \cdots, \boldsymbol{x}_N) = \psi(C\boldsymbol{x}_1, \cdots, C\boldsymbol{x}_N) \tag{3}$$

for all $\boldsymbol{x}_1, \cdots, \boldsymbol{x}_N$ except for a set of $\boldsymbol{x}_1, \cdots, \boldsymbol{x}_N$ of Lebesgue measure zero; this exceptional set may depend on C.

It is clear that Theorems 5.5.1 and 5.5.2 hold if we extend the definition of invariant test to mean that (3) holds except for a fixed set of $\boldsymbol{x}_1, \cdots, \boldsymbol{x}_N$ of measure 0 (the set not depending on C). It has been shown by Hunt and Stein (Lehmann) that in our problem almost invariance implies invariance (in the broad sense).

Now we wish to argue that if $\psi(\bar{\boldsymbol{x}}, A)$ has a power depending only on $N\boldsymbol{\mu}'\boldsymbol{\Sigma}^{-1}\boldsymbol{\mu}$, it is almost invariant. Since the power of $\psi(\bar{\boldsymbol{x}}, A)$ depends only on $N\boldsymbol{\mu}'\boldsymbol{\Sigma}^{-1}\boldsymbol{\mu}$, the power is

$$\begin{aligned} \mathscr{E}_{\boldsymbol{\mu}, \boldsymbol{\Sigma}}\psi(\bar{\boldsymbol{x}}, A) &\equiv \mathscr{E}_{C^{-1}\boldsymbol{\mu}, C^{-1}\boldsymbol{\Sigma}C^{-1\prime}}\psi(\bar{\boldsymbol{x}}, A) \\ &\equiv \mathscr{E}_{\boldsymbol{\mu}, \boldsymbol{\Sigma}}\psi(C\bar{\boldsymbol{x}}, CAC'). \end{aligned} \tag{4}$$

The second and third terms of (4) are merely different ways of writing the same integral. Thus

$$\mathscr{E}_{\boldsymbol{\mu}, \boldsymbol{\Sigma}}[\psi(\bar{\boldsymbol{x}}, A) - \psi(C\bar{\boldsymbol{x}}, CAC')] \equiv 0, \tag{5}$$

identically in $\boldsymbol{\mu}, \boldsymbol{\Sigma}$. Since $\bar{\boldsymbol{x}}, A$ are a complete sufficient set of statistics for $\boldsymbol{\mu}, \boldsymbol{\Sigma}, f(\bar{\boldsymbol{x}}, A) = \psi(\bar{\boldsymbol{x}}, A) - \psi(C\bar{\boldsymbol{x}}, CAC') = 0$ almost everywhere. To prove this last statement we consider

$$K\int \cdots \int f(\bar{\boldsymbol{x}}, A) \frac{|A|^{\frac{1}{2}(N-p-2)} \exp\{-\frac{1}{2}[\operatorname{tr} \boldsymbol{\Sigma}^{-1}A + N(\bar{\boldsymbol{x}} - \boldsymbol{\mu})'\boldsymbol{\Sigma}^{-1}(\bar{\boldsymbol{x}} - \boldsymbol{\mu})]\}}{|\boldsymbol{\Sigma}|^{\frac{1}{2}N}} d\bar{\boldsymbol{x}}\, dA \equiv 0, \tag{6}$$

where $d\bar{\boldsymbol{x}} = \prod d\bar{x}_i$, $dA = \prod da_{ij}$, and

$$K^{-1} = 2^{\frac{1}{2}pN}\pi^{\frac{1}{4}p(p-1)+\frac{1}{2}p} \prod_{i=1}^{p} \Gamma[\tfrac{1}{2}(N-i)]N^{-\frac{1}{2}p}, \tag{7}$$

and the identity (6) is for all $\boldsymbol{\mu}, \boldsymbol{\Sigma}$. The left-hand side of (6) is the integral of $f(\bar{\boldsymbol{x}}, A)$ times the density of $\bar{\boldsymbol{x}}$ times the density of A (the Wishart density of Section 7.2). In (6) replace $\boldsymbol{\Sigma}^{-1}$ by $\boldsymbol{I} - 2\boldsymbol{\theta}$, where $\boldsymbol{\theta} = \boldsymbol{\theta}'$, and $\boldsymbol{\mu}$ by $(\boldsymbol{I} - 2\boldsymbol{\theta})^{-1}\boldsymbol{t}$. Then (6) is

$$\begin{aligned} K\int \cdots \int f(\bar{\boldsymbol{x}}, A)|\boldsymbol{I} - 2\boldsymbol{\theta}|^{\frac{1}{2}N}|A|^{\frac{1}{2}(N-p-2)} \cdot \exp\{-\tfrac{1}{2}[\operatorname{tr} (\boldsymbol{I} - 2\boldsymbol{\theta}) \\ (A + N\bar{\boldsymbol{x}}\bar{\boldsymbol{x}}') - 2N\boldsymbol{t}'\bar{\boldsymbol{x}} + N\boldsymbol{t}'(\boldsymbol{I} - 2\boldsymbol{\theta})^{-1}\boldsymbol{t}]\}\, d\bar{\boldsymbol{x}}\, dA \equiv 0. \end{aligned} \tag{8}$$

Multiply (8) by $|\boldsymbol{I} - 2\boldsymbol{\theta}|^{-\frac{1}{2}N} \exp [\frac{1}{2}N\boldsymbol{t}'(\boldsymbol{I} - 2\boldsymbol{\theta})^{-1}\boldsymbol{t}]$ to obtain

$$(9)\quad K\int \cdots \int f[\bar{\boldsymbol{x}},(A + N\bar{\boldsymbol{x}}\bar{\boldsymbol{x}}') - N\bar{\boldsymbol{x}}\bar{\boldsymbol{x}}'] \cdot |A|^{\frac{1}{2}(N-p-2)} \exp [-\tfrac{1}{2} \operatorname{tr} (A + N\bar{\boldsymbol{x}}\bar{\boldsymbol{x}}') + \operatorname{tr} \boldsymbol{\theta}(A + N\bar{\boldsymbol{x}}\bar{\boldsymbol{x}}') + N\boldsymbol{t}'\bar{\boldsymbol{x}}]\, d\bar{\boldsymbol{x}}\, dA = 0.$$

This is the Laplace transform of $f[\bar{\boldsymbol{x}},(A + N\bar{\boldsymbol{x}}\bar{\boldsymbol{x}}') - N\bar{\boldsymbol{x}}\bar{\boldsymbol{x}}'] \cdot K|A|^{\frac{1}{2}(N-p-2)} \exp [-\frac{1}{2} \operatorname{tr} (A + N\bar{\boldsymbol{x}}\bar{\boldsymbol{x}}')]$ with respect to the variables $N\bar{\boldsymbol{x}}$ and $A + N\bar{\boldsymbol{x}}\bar{\boldsymbol{x}}'$. Since this is 0,

$$(10)\qquad f(\bar{\boldsymbol{x}}, A) = 0$$

except for a set of measure 0. Thus Theorem 5.5.3 follows.

As Theorem 5.5.2 follows from Theorem 5.5.1, so does the following theorem from Theorem 5.5.3:

THEOREM 5.5.4. *On the basis of observations* $\boldsymbol{x}_1, \cdots, \boldsymbol{x}_N$ *from* $N(\boldsymbol{\mu}, \boldsymbol{\Sigma})$, *of all tests of* $\boldsymbol{\mu} = \boldsymbol{0}$ *with power depending only on* $N\boldsymbol{\mu}'\boldsymbol{\Sigma}^{-1}\boldsymbol{\mu}$, *the* T^2*-test is a uniformly most powerful test.*

Theorem 5.5.4 was first proved by Simaika (1941). The results and proofs given in this section follow Lehmann (mimeographed notes). Hsu (1945) has proved an optimum property of the T^2-test that involves averaging the power over $\boldsymbol{\mu}$ and $\boldsymbol{\Sigma}$.

5.6. THE MULTIVARIATE BEHRENS-FISHER PROBLEM

We shall now give the multivariate analogue of Scheffé's solution (1943) for the Behrens-Fisher problem. Let $\{\boldsymbol{x}_\alpha^{(i)}\}(\alpha = 1, \cdots, N_i;\ i = 1, 2)$ be samples from $N(\boldsymbol{\mu}^{(i)}, \boldsymbol{\Sigma}_i)$ $(i = 1, 2)$. We wish to test the hypothesis that $\boldsymbol{\mu}^{(1)} = \boldsymbol{\mu}^{(2)}$. The mean $\bar{\boldsymbol{x}}^{(1)}$ of the first sample is normally distributed with expected value

$$(1)\qquad \mathscr{E}\bar{\boldsymbol{x}}^{(1)} = \boldsymbol{\mu}^{(1)}$$

and covariance matrix

$$(2)\qquad \mathscr{E}(\bar{\boldsymbol{x}}^{(1)} - \boldsymbol{\mu}^{(1)})(\bar{\boldsymbol{x}}^{(1)} - \boldsymbol{\mu}^{(1)})' = \frac{1}{N_1}\boldsymbol{\Sigma}_1.$$

Similarly, the mean $\bar{\boldsymbol{x}}^{(2)}$ of the second sample is normally distributed with expected value

$$(3)\qquad \mathscr{E}\bar{\boldsymbol{x}}^{(2)} = \boldsymbol{\mu}^{(2)}$$

and covariance matrix

$$(4)\qquad \mathscr{E}(\bar{\boldsymbol{x}}^{(2)} - \boldsymbol{\mu}^{(2)})(\bar{\boldsymbol{x}}^{(2)} - \boldsymbol{\mu}^{(2)})' = \frac{1}{N_2}\boldsymbol{\Sigma}_2.$$

Thus $\bar{\boldsymbol{x}}^{(1)} - \bar{\boldsymbol{x}}^{(2)}$ has mean $\boldsymbol{\mu}^{(1)} - \boldsymbol{\mu}^{(2)}$ and covariance matrix $(1/N_1)\boldsymbol{\Sigma}_1 + (1/N_2)\boldsymbol{\Sigma}_2$. We cannot use the technique of Section 5.2, however, because

$$(5)\qquad \sum_{\alpha=1}^{N_1}(\boldsymbol{x}_\alpha^{(1)} - \bar{\boldsymbol{x}}^{(1)})(\boldsymbol{x}_\alpha^{(1)} - \bar{\boldsymbol{x}}^{(1)})' + \sum_{\alpha=1}^{N_2}(\boldsymbol{x}_\alpha^{(2)} - \bar{\boldsymbol{x}}^{(2)})(\boldsymbol{x}_\alpha^{(2)} - \bar{\boldsymbol{x}}^{(2)})$$

does not have the Wishart distribution with covariance matrix a multiple of $(1/N_1)\boldsymbol{\Sigma}_1 + (1/N_2)\boldsymbol{\Sigma}_2$.

If $N_1 = N_2 = N$, say, we can use the T^2-test in an obvious way. Let $\boldsymbol{y}_\alpha = \boldsymbol{x}_\alpha^{(1)} - \boldsymbol{x}_\alpha^{(2)}$ (assuming the numbering of the observations in the two samples is independent of the observations themselves). Then $\boldsymbol{y}_\alpha$ is normally distributed with mean $\boldsymbol{\mu}^{(1)} - \boldsymbol{\mu}^{(2)}$ and covariance matrix $\boldsymbol{\Sigma}_1 + \boldsymbol{\Sigma}_2$, and independently of $\boldsymbol{y}_\beta (\beta \neq \alpha)$. Let $\bar{\boldsymbol{y}} = \sum_{\alpha=1}^{N} \boldsymbol{y}_\alpha / N = \bar{\boldsymbol{x}}^{(1)} - \bar{\boldsymbol{x}}^{(2)}$ and define $\boldsymbol{S}$ by

$$(6)\qquad \begin{aligned}(N-1)\boldsymbol{S} &= \sum_{\alpha=1}^{N}(\boldsymbol{y}_\alpha - \bar{\boldsymbol{y}})(\boldsymbol{y}_\alpha - \bar{\boldsymbol{y}})' \\ &= \sum_{\alpha=1}^{N}(\boldsymbol{x}_\alpha^{(1)} - \boldsymbol{x}_\alpha^{(2)} - \bar{\boldsymbol{x}}^{(1)} + \bar{\boldsymbol{x}}^{(2)})(\boldsymbol{x}_\alpha^{(1)} - \boldsymbol{x}_\alpha^{(2)} - \bar{\boldsymbol{x}}^{(1)} + \bar{\boldsymbol{x}}^{(2)})'.\end{aligned}$$

Then

$$(7)\qquad T^2 = N\bar{\boldsymbol{y}}'\boldsymbol{S}^{-1}\bar{\boldsymbol{y}}$$

is suitable for testing the hypothesis $\boldsymbol{\mu}^{(1)} - \boldsymbol{\mu}^{(2)} = \boldsymbol{0}$, and has the T^2-distribution with $N - 1$ degrees of freedom. It should be observed that if we had known $\boldsymbol{\Sigma}_1 = \boldsymbol{\Sigma}_2$, we would have used a T^2-statistic with $2N - 2$ degrees of freedom; thus we have lost $N - 1$ degrees of freedom in constructing a test which is independent of the two covariance matrices.

Now let us turn our attention to the case of $N_1 \neq N_2$. For convenience, let $N_1 < N_2$. Then we define

$$(8)\qquad \boldsymbol{y}_\alpha = \boldsymbol{x}_\alpha^{(1)} - \sqrt{\frac{N_1}{N_2}}\,\boldsymbol{x}_\alpha^{(2)} + \frac{1}{\sqrt{N_1N_2}}\sum_{\beta=1}^{N_1}\boldsymbol{x}_\beta^{(2)} - \frac{1}{N_2}\sum_{\gamma=1}^{N_2}\boldsymbol{x}_\gamma^{(2)}, \qquad \alpha = 1, \cdots, N_1.$$

The expected value of $\boldsymbol{y}_\alpha$ is

$$(9)\qquad \begin{aligned}\mathscr{E}\boldsymbol{y}_\alpha &= \boldsymbol{\mu}^{(1)} - \sqrt{\frac{N_1}{N_2}}\,\boldsymbol{\mu}^{(2)} + \frac{N_1}{\sqrt{N_1N_2}}\boldsymbol{\mu}^{(2)} - \frac{N_2}{N_2}\boldsymbol{\mu}^{(2)} \\ &= \boldsymbol{\mu}^{(1)} - \boldsymbol{\mu}^{(2)}.\end{aligned}$$

The covariance matrix of $\boldsymbol{y}_\alpha$ and $\boldsymbol{y}_\beta$ is

$$(10)\quad \begin{aligned} &\mathscr{E}(\boldsymbol{y}_\alpha - \mathscr{E}\boldsymbol{y}_\alpha)(\boldsymbol{y}_\beta - \mathscr{E}\boldsymbol{y}_\beta)' \\ &= \mathscr{E}\Bigg[(\boldsymbol{x}_\alpha^{(1)} - \boldsymbol{\mu}^{(1)}) - \sqrt{\frac{N_1}{N_2}}(\boldsymbol{x}_\alpha^{(2)} - \boldsymbol{\mu}^{(2)}) \\ &\qquad + \frac{1}{\sqrt{N_1N_2}}\sum_{\gamma=1}^{N_1}(\boldsymbol{x}_\gamma^{(2)} - \boldsymbol{\mu}^{(2)}) - \frac{1}{N_2}\sum_{\gamma=1}^{N_2}(\boldsymbol{x}_\gamma^{(2)} - \boldsymbol{\mu}^{(2)})\Bigg] \\ &\qquad \Bigg[(\boldsymbol{x}_\beta^{(1)} - \boldsymbol{\mu}^{(1)})' - \sqrt{\frac{N_1}{N_2}}(\boldsymbol{x}_\beta^{(2)} - \boldsymbol{\mu}^{(2)})' \\ &\qquad + \frac{1}{\sqrt{N_1N_2}}\sum_{\gamma=1}^{N_1}(\boldsymbol{x}_\gamma^{(2)} - \boldsymbol{\mu}^{(2)})' - \frac{1}{N_2}\sum_{\gamma=1}^{N_2}(\boldsymbol{x}_\gamma^{(2)} - \boldsymbol{\mu}^{(2)})'\Bigg] \\ &= \delta_{\alpha\beta}\boldsymbol{\Sigma}_1 + \frac{N_1}{N_2}\delta_{\alpha\beta}\boldsymbol{\Sigma}_2 \\ &\qquad + \boldsymbol{\Sigma}_2\left(-2\frac{1}{N_2} + \frac{2}{N_2}\sqrt{\frac{N_1}{N_2}} + \frac{N_1}{N_1N_2} - 2\frac{N_1}{\sqrt{N_1N_2}\,N_2} + \frac{N_2}{N_2^2}\right) \\ &= \delta_{\alpha\beta}\left(\boldsymbol{\Sigma}_1 + \frac{N_1}{N_2}\boldsymbol{\Sigma}_2\right). \end{aligned}$$

Thus a suitable statistic for testing $\boldsymbol{\mu}^{(1)} - \boldsymbol{\mu}^{(2)} = \boldsymbol{0}$, which has the T^2-distribution with $N_1 - 1$ degrees of freedom, is

$$(11)\qquad T^2 = N_1\bar{\boldsymbol{y}}'\boldsymbol{S}^{-1}\bar{\boldsymbol{y}},$$

where

$$(12)\qquad \bar{\boldsymbol{y}} = \frac{1}{N_1}\sum_{\alpha=1}^{N_1}\boldsymbol{y}_\alpha = \bar{\boldsymbol{x}}^{(1)} - \bar{\boldsymbol{x}}^{(2)}$$

and

$$(13)\quad \begin{aligned} (N_1 - 1)\boldsymbol{S} &= \sum_{\alpha=1}^{N_1}(\boldsymbol{y}_\alpha - \bar{\boldsymbol{y}})(\boldsymbol{y}_\alpha - \bar{\boldsymbol{y}})' \\ &= \sum_{\alpha=1}^{N_1}\left[\boldsymbol{x}_\alpha^{(1)} - \bar{\boldsymbol{x}}^{(1)} - \sqrt{\frac{N_1}{N_2}}\left(\boldsymbol{x}_\alpha^{(2)} - \frac{1}{N_1}\sum_{\beta=1}^{N_1}\boldsymbol{x}_\beta^{(2)}\right)\right] \\ &\qquad \left[\boldsymbol{x}_\alpha^{(1)} - \bar{\boldsymbol{x}}^{(1)} - \sqrt{\frac{N_1}{N_2}}\left(\boldsymbol{x}_\alpha^{(2)} - \frac{1}{N_1}\sum_{\beta=1}^{N_1}\boldsymbol{x}_\beta^{(2)}\right)\right]'. \end{aligned}$$

In terms of $\boldsymbol{u}_\alpha = \boldsymbol{x}_\alpha^{(1)} - \sqrt{N_1/N_2}\,\boldsymbol{x}_\alpha^{(2)}$ $(\alpha = 1, \cdots, N_1)$, this last equation may be written

$$(14)\qquad (N_1 - 1)\boldsymbol{S} = \sum_{\alpha=1}^{N_1}(\boldsymbol{u}_\alpha - \bar{\boldsymbol{u}})(\boldsymbol{u}_\alpha - \bar{\boldsymbol{u}})',$$

where $\bar{\boldsymbol{u}} = \sum_{\alpha=1}^{N_1}\boldsymbol{u}_\alpha/N_1$.

This procedure has been suggested by Scheffé (1943) in the univariate case. Scheffé has shown that in the univariate case this technique gives the shortest confidence intervals obtained by using the t-distribution. The advantage of the method is that $\bar{\boldsymbol{x}}^{(1)} - \bar{\boldsymbol{x}}^{(2)}$ is used, and this statistic is most relevant to $\boldsymbol{\mu}^{(1)} - \boldsymbol{\mu}^{(2)}$. The sacrifice of observations in estimating a covariance matrix is not so important. Bennett (1951) gave the extension of the procedure to the multivariate case.

This trick can be used for more general cases. Let $\{\boldsymbol{x}_\alpha^{(i)}\}$ $(\alpha = 1, \cdots, N_i;\ i = 1, \cdots, q)$ be samples from $N(\boldsymbol{\mu}^{(i)}, \boldsymbol{\Sigma}_i)$ $(i = 1, \cdots, q)$ respectively. Consider testing the hypothesis H:

$$\sum_{i=1}^{q} \beta_i \boldsymbol{\mu}^{(i)} = \boldsymbol{\mu}, \tag{15}$$

where $\beta_1, \cdots, \beta_q$ are given scalars and $\boldsymbol{\mu}$ is a given vector. If the N_i are unequal, take N_1 to be the smallest. Let

$$\boldsymbol{y}_\alpha = \beta_1 \boldsymbol{x}_\alpha^{(1)} + \sum_{i=2}^{q} \beta_i \sqrt{\frac{N_1}{N_i}} \left(\boldsymbol{x}_\alpha^{(i)} - \frac{1}{N_1} \sum_{\beta=1}^{N_1} \boldsymbol{x}_\beta^{(i)} + \frac{1}{\sqrt{N_1 N_i}} \sum_{\gamma=1}^{N_i} \boldsymbol{x}_\gamma^{(i)} \right). \tag{16}$$

Then

$$\begin{aligned} \mathscr{E}\boldsymbol{y}_\alpha &= \beta_1 \boldsymbol{\mu}^{(1)} + \sum_{i=2}^{q} \beta_i \sqrt{\frac{N_1}{N_i}} \left(\boldsymbol{\mu}^{(i)} - \frac{1}{N_1} N_1 \boldsymbol{\mu}^{(i)} + \frac{N_i}{\sqrt{N_1 N_i}} \boldsymbol{\mu}^{(i)} \right) \\ &= \sum_{i=1}^{q} \beta_i \boldsymbol{\mu}^{(i)}, \end{aligned} \tag{17}$$

$$\mathscr{E}(\boldsymbol{y}_\alpha - \mathscr{E}\boldsymbol{y}_\alpha)(\boldsymbol{y}_\beta - \mathscr{E}\boldsymbol{y}_\beta)' = \delta_{\alpha\beta} \left(\sum_{i=1}^{q} \frac{\beta_i^2 N_1}{N_i} \boldsymbol{\Sigma}_i \right). \tag{18}$$

Let $\bar{\boldsymbol{y}}$ and S be defined by

$$\begin{aligned} \bar{\boldsymbol{y}} &= \frac{1}{N_1} \sum_{\alpha=1}^{N_1} \boldsymbol{y}_\alpha \\ &= \sum_{i=1}^{q} \beta_i \bar{\boldsymbol{x}}^{(i)}, \end{aligned} \tag{19}$$

where

$$\bar{\boldsymbol{x}}^{(i)} = (1/N_i) \sum_{\beta=1}^{N_i} \boldsymbol{x}_\beta^{(i)}, \tag{20}$$

$$(N_1 - 1)S = \sum_{\alpha=1}^{N_1} (\boldsymbol{y}_\alpha - \bar{\boldsymbol{y}})(\boldsymbol{y}_\alpha - \bar{\boldsymbol{y}})'.$$

Then

$$T^2 = N_1(\bar{\boldsymbol{y}} - \boldsymbol{\mu})' S^{-1} (\bar{\boldsymbol{y}} - \boldsymbol{\mu}) \tag{21}$$

is suitable for testing H, and when the hypothesis is true this statistic has the T^2-distribution for dimension p with $N_1 - 1$ degrees of freedom. If we let

$$u_\alpha = \sum_{i=1}^{q} \beta_i \sqrt{\frac{N_1}{N_i}}\, x_\alpha^{(i)}, \qquad \alpha = 1, \cdots, N_1, \tag{22}$$

then S can be defined as

$$(N_1 - 1)S = \sum_{\alpha=1}^{N_1} (u_\alpha - \bar{u})(u_\alpha - \bar{u})'. \tag{23}$$

Another problem that is amenable to this kind of treatment is testing the hypothesis that two subvectors have equal means. Let

$$x = \begin{pmatrix} x^{(1)} \\ x^{(2)} \end{pmatrix} \tag{24}$$

be distributed normally with mean

$$\mu = \begin{pmatrix} \mu^{(1)} \\ \mu^{(2)} \end{pmatrix} \tag{25}$$

and covariance matrix

$$\Sigma = \begin{pmatrix} \Sigma_{11} & \Sigma_{12} \\ \Sigma_{21} & \Sigma_{22} \end{pmatrix}. \tag{26}$$

We assume that $x^{(1)}$ and $x^{(2)}$ are each of q components. Then $x^{(1)} - x^{(2)}$ is distributed normally with mean $\mu^{(1)} - \mu^{(2)}$ and covariance matrix

$$\begin{aligned} \mathscr{E}[(x^{(1)} - \mu^{(1)}) - (x^{(2)} - \mu^{(2)})][(x^{(1)} - \mu^{(1)}) - (x^{(2)} - \mu^{(2)})]' \\ = \Sigma_{11} - \Sigma_{21} - \Sigma_{12} + \Sigma_{22}. \end{aligned} \tag{27}$$

To test the hypothesis $\mu^{(1)} = \mu^{(2)}$ we use a T^2-statistic

$$N(\bar{x}^{(1)} - \bar{x}^{(2)})'(S_{11} - S_{21} - S_{12} + S_{22})^{-1}(\bar{x}^{(1)} - \bar{x}^{(2)}), \tag{28}$$

where the mean vector and covariance matrix of a sample are partitioned similarly to μ and Σ.

REFERENCES

Section 5.1. Hotelling (1931).

Section 5.2. R. C. Bose and S. N. Roy (1938a), (1938b); Fisher (1937); Hotelling (1931); Kendall (1946), 335–338; Rasch (1950); S. N. Roy (1939a); Wald and Wolfowitz (1944); Wilks (1943), 234–238.

Section 5.3. Cochran and Bliss (1948); Fisher (1936); Garrett (1943); Hotelling (1931); P. L. Hsu (1938); C. R. Rao (1946a), (1948b).

Section 5.4. R. C. Bose and S. N. Roy (1938a); P. K. Bose (1951a); Ferris, Grubbs, and Weaver (1946); P. L. Hsu (1938); Lehmer (1944); S. N. Roy (1939a); Tang (1938).

Section 5.5. P. L. Hsu (1945); Lehmann; Nandi (1946); Simaika (1941).

Section 5.6. Barankin (1949); Bennett (1951); G. S. James (1954); Scheffé (1943).

Chapter 5. Bhattacharyya and Narain (1939); P. K. Bose (1947b); Durand (1941); C. R. Rao (1949a); Wald (1943).

PROBLEMS

1. (Sec. 5.3) Use the data in Section 3.2 to test the hypothesis that neither drug has a soporific effect at significance level 0.01.

2. (Sec. 5.3) Using the data in Section 3.2, give a confidence region for $\boldsymbol{\mu}$ with confidence coefficient 0.95.

3. (Sec. 5.6) Use the data of Problem 17 of Chapter 4 to test the hypothesis that the mean head length and breadth of first sons is equal to those of second sons at significance level 0.01.

4. (Sec. 5.2) Let $\boldsymbol{x}_\alpha$ be distributed according to $N(\boldsymbol{\mu} + \boldsymbol{\beta}(z_\alpha - \bar{z}), \boldsymbol{\Sigma})$, $\alpha = 1, \cdots, N$, where $\bar{z} = (1/N)\sum z_\alpha$. Let $\boldsymbol{b} = [1/\sum(z_\alpha - \bar{z})^2]\sum \boldsymbol{x}_\alpha(z_\alpha - \bar{z})$, $(N-2)\boldsymbol{S} = \sum[\boldsymbol{x}_\alpha - \bar{\boldsymbol{x}} - \boldsymbol{b}(z_\alpha - \bar{z})][\boldsymbol{x}_\alpha - \bar{\boldsymbol{x}} - \boldsymbol{b}(z_\alpha - \bar{z})]'$, and $T^2 = \sum(z_\alpha - \bar{z})^2\boldsymbol{b}'\boldsymbol{S}^{-1}\boldsymbol{b}$. Show that T^2 has the T^2-distribution with $N-2$ degrees of freedom. [*Hint:* See Problem 6 of Chapter 3.]

5. Let $\bar{\boldsymbol{x}}$ and $\boldsymbol{S}$ be based on N observations from $N(\boldsymbol{\mu}, \boldsymbol{\Sigma})$ and let $\boldsymbol{x}$ be an additional observation from $N(\boldsymbol{\mu}, \boldsymbol{\Sigma})$. Show that $\boldsymbol{x} - \bar{\boldsymbol{x}}$ is distributed according to

$$N[\boldsymbol{0}, (1 + 1/N)\boldsymbol{\Sigma}].$$

Verify that $[N/(N+1)](\boldsymbol{x} - \bar{\boldsymbol{x}})'\boldsymbol{S}^{-1}(\boldsymbol{x} - \bar{\boldsymbol{x}})$ has the T^2-distribution with $N-1$ degrees of freedom. Show how this statistic can be used to give a prediction region for $\boldsymbol{x}$ based on $\bar{\boldsymbol{x}}$ and $\boldsymbol{S}$ (that is, a region such that one has a given confidence that the next observation will fall into it).

6. (Sec. 5.3) Prove the statement in Section 5.3.5 that the T^2-statistic is independent of the choice of $\boldsymbol{C}$.

7. Let $\boldsymbol{x}_\alpha^{(i)}$ be observations from $N(\boldsymbol{\mu}^{(i)}, \boldsymbol{\Sigma}_i)$, $\alpha = 1, \cdots, N_i$, $i = 1, 2$. Find the likelihood ratio criterion for testing the hypothesis $\boldsymbol{\mu}^{(1)} = \boldsymbol{\mu}^{(2)}$.

8. Prove that $\boldsymbol{\mu}'\boldsymbol{\Sigma}^{-1}\boldsymbol{\mu}$ is larger for $\boldsymbol{\mu}' = (\mu_1, \mu_2)$ than for $\boldsymbol{\mu} = (\mu_1)$ by verifying

$$\frac{1}{1-\rho^2}\left(\frac{\mu_1^2}{\sigma_1^2} - 2\rho\frac{\mu_1\mu_2}{\sigma_1\sigma_2} + \frac{\mu_2^2}{\sigma_2^2}\right) = \frac{\mu_1^2}{\sigma_1^2} + \frac{(\mu_2 - \rho\sigma_2\mu_1/\sigma_1)^2}{(1-\rho^2)\sigma_2^2}.$$

Discuss the power of the test $\mu_1 = 0$ compared to the power of the test $\mu_1 = 0$, $\mu_2 = 0$.

9. (*a*) Using the data of Section 5.3.3 test the hypothesis $\mu_1^{(1)} = \mu_1^{(2)}$.

(*b*) Test the hypothesis $\mu_1^{(1)} = \mu_1^{(2)}, \mu_2^{(1)} = \mu_2^{(2)}$.

10. Let

$$\boldsymbol{\mu} = \begin{pmatrix} \boldsymbol{\mu}^{(1)} \\ \boldsymbol{\mu}^{(2)} \end{pmatrix}, \qquad \boldsymbol{\Sigma} = \begin{pmatrix} \boldsymbol{\Sigma}_{11} & \boldsymbol{\Sigma}_{12} \\ \boldsymbol{\Sigma}_{21} & \boldsymbol{\Sigma}_{22} \end{pmatrix}.$$

Prove $\boldsymbol{\mu}'\boldsymbol{\Sigma}^{-1}\boldsymbol{\mu} \geq \boldsymbol{\mu}^{(1)\prime}\boldsymbol{\Sigma}_{11}^{-1}\boldsymbol{\mu}^{(1)}$. Give a condition for the strict inequality to hold. [*Hint:* This is the vector analogue of Problem 8.]

11. (Sec. 5.2) Using the distribution of $\bar{\boldsymbol{x}}$ and properties of $\boldsymbol{S}$, prove that T^2 under the null hypothesis has asymptotically the χ^2-distribution with p degrees of freedom.

12. Let $\boldsymbol{X}^{(i)\prime} = (\boldsymbol{Y}^{(i)\prime}, \boldsymbol{Z}^{(i)\prime})$, $i = 1, 2$, where $\boldsymbol{Y}^{(i)}$ has p components and $\boldsymbol{Z}^{(i)}$ has q components, be distributed according to $N(\boldsymbol{\mu}^{(i)}, \boldsymbol{\Sigma})$, where

$$\boldsymbol{\mu}^{(i)} = \begin{pmatrix} \boldsymbol{\mu}_y^{(i)} \\ \boldsymbol{\mu}_z^{(i)} \end{pmatrix}, \qquad \boldsymbol{\Sigma} = \begin{pmatrix} \boldsymbol{\Sigma}_{yy} & \boldsymbol{\Sigma}_{yz} \\ \boldsymbol{\Sigma}_{zy} & \boldsymbol{\Sigma}_{zz} \end{pmatrix}.$$

Find the likelihood ratio criterion (or equivalent T^2-criterion) for testing $\boldsymbol{\mu}_z^{(1)} = \boldsymbol{\mu}_z^{(2)}$ given $\boldsymbol{\mu}_y^{(1)} = \boldsymbol{\mu}_y^{(2)}$ on the basis of a sample of N_i on $\boldsymbol{X}^{(i)}$, $i = 1, 2$. [*Hint:* Express the likelihood in terms of the marginal density of $\boldsymbol{Y}^{(i)}$ and the conditional density of $\boldsymbol{Z}^{(i)}$ given $\boldsymbol{Y}^{(i)}$.]

13. Find the distribution of the criterion in the preceding problem under the null hypothesis.

14. (Sec. 5.2.2) By an argument similar to that of (11), show that

$$T^2/(N-1) = N\bar{\boldsymbol{x}}'\boldsymbol{A}^{-1}\bar{\boldsymbol{x}} = N\bar{\boldsymbol{x}}'\boldsymbol{B}^{-1}\bar{\boldsymbol{x}}/(1 - N\bar{\boldsymbol{x}}'\boldsymbol{B}^{-1}\bar{\boldsymbol{x}}),$$

where $\boldsymbol{B} = \sum_\alpha \boldsymbol{x}_\alpha \boldsymbol{x}'_\alpha$.

15. (Sec. 5.2.2) Show that $T^2/(N-1)$ in Problem 14 can be written as $R^2/(1-R^2)$ with the correspondences given in Table 1.

TABLE 1

Section 5.2	Section 4.4
$x_{0\alpha} = 1/\sqrt{N}$	$z_{1\alpha}$
$\boldsymbol{x}_\alpha$	$\boldsymbol{z}_\alpha^{(2)}$
$\sqrt{N}\bar{\boldsymbol{x}}$	$\boldsymbol{a}'_{(1)} = \sum z_{1\alpha}\boldsymbol{z}_\alpha^{(2)}$
$\boldsymbol{B} = \sum \boldsymbol{x}_\alpha \boldsymbol{x}'_\alpha$	$\boldsymbol{A}_{22} = \sum \boldsymbol{z}_\alpha^{(2)}\boldsymbol{z}_\alpha^{(2)\prime}$
$1 = \sum x_{0\alpha}^2$	$a_{11} = \sum z_{1\alpha}^2$
$\dfrac{T^2}{N-1}$	$\dfrac{R^2}{1-R^2}$
p	$p-1$
N	n

16. (Sec. 5.2.2) Let

$$\frac{R^2}{1-R^2} = \frac{\sum u_\alpha \boldsymbol{x}'_\alpha \left(\sum \boldsymbol{x}_\alpha \boldsymbol{x}'_\alpha\right)^{-1} \sum u_\alpha \boldsymbol{x}_\alpha}{\sum u_\alpha^2 - \sum u_\alpha \boldsymbol{x}'_\alpha \left(\sum \boldsymbol{x}_\alpha \boldsymbol{x}'_\alpha\right)^{-1} \sum u_\alpha \boldsymbol{x}_\alpha},$$

where $u_1, \cdots, u_N$ are N numbers and $\boldsymbol{x}_1, \cdots, \boldsymbol{x}_N$ are independent, each with the distribution $N(\boldsymbol{0}, \boldsymbol{\Sigma})$. Prove that the distribution of $R^2/(1-R^2)$ is independent of $u_1, \cdots, u_N$. [*Hint:* There is an orthogonal $N \times N$ matrix $\boldsymbol{C}$ that carries $(u_1, \cdots, u_N)$ into a vector proportional to $(1/\sqrt{N}, \cdots, 1/\sqrt{N})$.]

17. (Sec. 5.2.2) Use Problems 15 and 16 to show that $[T^2/(N-1)][(N-p)/p]$ has the $F_{p,\,N-p}$-distribution (under the null hypothesis). [*Note:* This is the analysis that corresponds to Hotelling's geometric proof (1931).]

18. (Sec. 5.2.2) Let $T^2 = N\bar{\boldsymbol{x}}'\boldsymbol{S}^{-1}\bar{\boldsymbol{x}}$, where $\bar{\boldsymbol{x}}$ and $\boldsymbol{S}$ are the mean and covariance matrix of a sample of N from $N(\boldsymbol{\mu}, \boldsymbol{\Sigma})$. Show that T^2 is distributed the same as if $\boldsymbol{\mu}'$ is replaced by $\boldsymbol{\lambda}' = (\tau, 0, \cdots, 0)$, where $\tau^2 = \boldsymbol{\mu}'\boldsymbol{\Sigma}^{-1}\boldsymbol{\mu}$, and $\boldsymbol{\Sigma}$ is replaced by $\boldsymbol{I}$.

19. (Sec. 5.2.2) Let $u = [T^2/(N-1)]/[1 + T^2/(N-1)]$. Show that $u = \boldsymbol{\gamma}\boldsymbol{V}'(\boldsymbol{V}\boldsymbol{V}')^{-1}\boldsymbol{V}\boldsymbol{\gamma}'$, where $\boldsymbol{\gamma} = (1/\sqrt{N}, \cdots, 1/\sqrt{N})$ and

$$\boldsymbol{V} = \begin{pmatrix} \boldsymbol{v}_1 \\ \cdot \\ \cdot \\ \cdot \\ \boldsymbol{v}_p \end{pmatrix} = \begin{pmatrix} x_{11} & \cdots & x_{1N} \\ \cdot & & \cdot \\ \cdot & & \cdot \\ \cdot & & \cdot \\ x_{p1} & \cdots & x_{pN} \end{pmatrix}.$$

20. (Sec. 5.2.2) Let

$$v_1^* = v_1,$$

$$v_i^* = v_i - \frac{v_i v_1'}{v_1 v_1'} v_1 = v_i \left(I - \frac{1}{v_1 v_1'} v_1' v_1 \right), \qquad i \neq 1,$$

$$\gamma^* = \gamma - \frac{\gamma v_1'}{v_1 v_1'} v_1,$$

$$V^* = \begin{pmatrix} v_1^* \\ \cdot \\ \cdot \\ \cdot \\ v_p^* \end{pmatrix}.$$

Prove that $U = s + (1 - s)w$, where

$$s = \frac{(\gamma v_1^{*\prime})^2}{v_1^* v_1^{*\prime}} = \frac{(\gamma v_1')^2}{v_1 v_1'},$$

$$w = \frac{1}{\gamma^* \gamma^{*\prime}} \gamma^* \begin{pmatrix} v_2^* \\ \cdot \\ \cdot \\ \cdot \\ v_p^* \end{pmatrix}' \begin{pmatrix} v_2^* v_2^{*\prime} & \cdots & v_2^* v_p^{*\prime} \\ \cdot & & \cdot \\ \cdot & & \cdot \\ \cdot & & \cdot \\ v_p^* v_2^{*\prime} & \cdots & v_p^* v_p^{*\prime} \end{pmatrix}^{-1} \begin{pmatrix} v_2^* \\ \cdot \\ \cdot \\ \cdot \\ v_p^* \end{pmatrix} \gamma^{*\prime}.$$

Hint: $EV = V^*$, where

$$E = \begin{pmatrix} 1 & 0 \cdots 0 \\ -\frac{v_2 v_1'}{v_1 v_1'} & 1 \cdots 0 \\ \cdot & \cdot \quad \cdot \\ \cdot & \cdot \quad \cdot \\ \cdot & \cdot \quad \cdot \\ -\frac{v_p v_1'}{v_1 v_1'} & 0 \cdots 1 \end{pmatrix}.$$

21. (Sec. 5.2.2) Prove that w has the distribution of the square of a multiple correlation between one vector and $p - 1$ vectors in $(N - 1)$-space without subtracting means; that is, has density

$$\frac{\Gamma[\frac{1}{2}(N-1)]}{\Gamma[\frac{1}{2}(N-p)]\Gamma[\frac{1}{2}(p-1)]} w^{\frac{1}{2}(p-1)-1}(1-w)^{\frac{1}{2}(N-p)-1}.$$

[*Hint:* The transformation of Problem 20 is a projection of $v_2, \cdots, v_p, \gamma$ on the $(N - 1)$-space orthogonal to v_1.]

22. (Sec. 5.2.2) Verify that $r = s/(1 - s)$ multiplied by $(N - 1)/1$ has the noncentral F-distribution with 1 and $N - 1$ degrees of freedom and noncentrality parameter $N\tau^2$.

23. (Sec. 5.2.2) From Problems 18 through 22, verify Corollary 5.2.1.

CHAPTER 6

Classification of Observations

6.1. THE PROBLEM OF CLASSIFICATION

The problem of classification arises when an investigator makes a number of measurements on an individual and wishes to classify the individual into one of several categories on the basis of these measurements. The investigator cannot identify the individual with a category directly but must use these measurements. In many cases it can be assumed that there are a finite number of categories or populations from which the individual may have come and each population is characterized by a probability distribution of the measurements. Thus, an individual is considered as a random observation from this population. The question is: Given an individual with certain measurements, from which population did he arise?

The problem of classification may be considered as a problem of "statistical decision functions." We have a number of hypotheses: each hypothesis is that the distribution of the observation is a given one. We must accept one of these hypotheses and reject the others. If only two populations are admitted, we have an elementary problem of testing one hypothesis of a specified distribution against another.

In some instances, the categories are specified beforehand in the sense that the probability distributions of the measurements are completely known. In other cases, the form of each distribution may be known, but the parameters of the distribution must be estimated from a sample from that population.

Let us give an example of a problem of classification. Prospective students applying for admission into college are given a battery of tests; the vector of scores is a set of measurements $\boldsymbol{x}$. The prospective student may be a member of one population consisting of those students who will successfully complete college training or, rather, have potentialities for successfully completing training, or he may be a member of the other population, those who will not complete the college course successfully. The problem is to classify a student applying for admission on the basis of his scores on the entrance examination.

In this chapter we shall develop the theory of classification in general terms and then apply it to cases involving the normal distribution.

6.2. STANDARDS OF GOOD CLASSIFICATION

6.2.1. Preliminary Considerations

In constructing a procedure of classification, it is desired to minimize the probability of misclassification, or, more specifically, it is desired to minimize on the average the bad effects of misclassification. Now let us make this notion precise. For convenience we shall now consider the case of only two categories. Later we shall treat the more general case.

Suppose an individual is an observation from either population π_1 or population π_2. The classification of an observation depends on the vector of measurements $\boldsymbol{x}' = (x_1, \cdots, x_p)$ on that individual. We set up a rule that if an individual is characterized by certain sets of values of $x_1, \cdots, x_p$ he will be classified as from π_1; if he has other values he is classified as from π_2.

We can think of an observation as a point in a p-dimensional space. We divide this space into two regions. If the observation falls in R_1, we classify it as coming from population π_1, and if it falls in R_2 we classify it as coming from population π_2.

In following a given classification procedure, the statistician can make two kinds of errors in classification. If the individual is actually from π_1 the statistician can classify him as coming from population π_2; or if he is from π_2 the statistician may classify him as from π_1. We need to know the relative undesirability of these two kinds of misclassification. Let the "cost" of the first type of misclassification be $C(2|1)(>0)$ and let the cost of misclassifying an individual from π_2 as from π_1 be $C(1|2)(>0)$. These costs may be measured in any kind of units. As we shall see later, it is only the ratio of the two costs that is important. Although the statistician may not know these costs in each case, he will often have at least a rough idea of them.

Table 1, a 2 by 2 table, indicates the costs of correct and incorrect classification. Clearly, a good classification procedure is one which minimizes in some sense or other the cost of misclassification.

TABLE 1

		Statistician's Decision	
		π_1	π_2
Population	π_1	0	$C(2\|1)$
	π_2	$C(1\|2)$	0

6.2.2. Two Cases of Two Populations

We shall consider ways of defining "minimum cost" in two cases. In one case we shall suppose that we have a priori probabilities of the two populations. Let the probability that an observation comes from population π_1 be q_1 and from population π_2 be q_2. The probability properties of population π_1 are specified by a distribution function. For convenience we shall treat only the case that the distribution has a density, although the case of discrete probabilities lends itself to almost the same treatment. Let the density of population π_1 be $p_1(\boldsymbol{x})$ and that of π_2 be $p_2(\boldsymbol{x})$. If we have a region R_1 of classification as from π_1, the probability of correctly classifying an observation that actually is drawn from population π_1 is

$$P(1|1, R) = \int_{R_1} p_1(\boldsymbol{x})\, d\boldsymbol{x}, \tag{1}$$

where $d\boldsymbol{x} = dx_1 \cdots dx_p$, and the probability of misclassification of an observation from π_1 is

$$P(2|1, R) = \int_{R_2} p_1(\boldsymbol{x})\, d\boldsymbol{x}. \tag{2}$$

Similarly, the probability of correctly classifying an observation from π_2 is

$$P(2|2, R) = \int_{R_2} p_2(\boldsymbol{x})\, d\boldsymbol{x}, \tag{3}$$

and the probability of misclassifying such an observation is

$$P(1|2, R) = \int_{R_1} p_2(\boldsymbol{x})\, d\boldsymbol{x}. \tag{4}$$

Since the probability of drawing an observation from π_1 is q_1, the probability of drawing an observation from π_1 and correctly classifying it is $q_1P(1|1, R)$; that is, this is the probability of the situation in the upper left-hand corner of Table 1. Similarly, the probability of drawing an observation from π_1 and misclassifying it is $q_1P(2|1, R)$. The probability associated with the lower left-hand corner of Table 1 is $q_2P(1|2, R)$ and with the lower right-hand corner is $q_2P(2|2, R)$.

What is the average or expected loss from costs of misclassification? It is the sum of the products of costs of each misclassification multiplied by the probability of its occurrence; it is

$$C(2|1)P(2|1, R)q_1 + C(1|2)P(1|2, R)q_2. \tag{5}$$

It is this average loss that we wish to minimize. That is, we want to divide our space into regions R_1 and R_2 such that the expected loss is as small as possible. A procedure that minimizes (5) for a given q_1 and q_2 is called a Bayes procedure.

In the example of admission of students, the undesirability of misclassification is, in one instance, the expense of teaching a student who will not complete the course successfully and is, in the other instance, the undesirability of excluding from college a potentially good student.

The other case we shall treat is the case in which there are no known a priori probabilities. In this case the expected loss if the observation is from π_1 is

$$C(2|1)P(2|1, R) = r(1, R); \tag{6}$$

the expected loss if the observation is from π_2 is

$$C(1|2)P(1|2, R) = r(2, R). \tag{7}$$

We do not know whether the observation is from π_1 or from π_2, and we do not know probabilities of these two instances.

A procedure R is at least as good as a procedure R^* if $r(1, R) \leq r(1, R^*)$ and $r(2, R) \leq r(2, R^*)$; R is better than R^* if at least one of these inequalities is a strict inequality. Usually there is no procedure that is better than all other procedures or is at least as good as all other procedures. A procedure R is called *admissible* if there is no procedure better than R; we shall be interested in the entire class of admissible procedures. It will be shown that under certain conditions this class is the same as the class of Bayes procedures. A class of procedures is *complete* if for every procedure outside the class there is one in the class which is better; a class is called *essentially complete* if for every procedure outside the class there is one in the class which is at least as good. A *minimal complete class* (if it exists) is a complete class such that no proper subset is a complete class; a similar definition holds for a minimal essentially complete class. Under certain conditions we shall show that the admissible class is minimal complete. To simplify the discussion we shall consider procedures the same if they only differ on sets of probability zero. In fact, throughout the next section we shall make statements which are meant to hold "except for sets of probability zero" without saying so explicitly.

A principle that usually leads to a unique procedure is the minimax principle. A procedure is minimax if the maximum expected loss, $r(i, R)$, is a minimum. From a conservative point of view, this may be considered an optimum procedure. For a general discussion of the concepts in this section and the next see Wald (1950) and Blackwell and Girshick (1954).

6.3. PROCEDURES OF CLASSIFICATION INTO ONE OF TWO POPULATIONS WITH KNOWN PROBABILITY DISTRIBUTIONS

6.3.1. The Case when A Priori Probabilities Are Known

We now turn to the problem of choosing regions R_1 and R_2 so as to minimize (5) of Section 6.2. Since we have a priori probabilities, we can define joint probabilities of the population and the observed set of variables. The probability that an observation comes from π_1 and that each variate is less than the corresponding component in y is

$$\int_{-\infty}^{y_p} \cdots \int_{-\infty}^{y_1} q_1 p_1(x)\, dx_1 \cdots dx_p. \tag{1}$$

We can also define the conditional probability that an observation came from a certain population given the values of the observed variates. For instance, the conditional probability of coming from population π_1, given an observation x, is

$$\frac{q_1 p_1(x)}{q_1 p_1(x) + q_2 p_2(x)}. \tag{2}$$

Suppose for a moment that $C(1|2) = C(2|1) = 1$. Then the expected loss is

$$q_1 \int_{R_2} p_1(x)\, dx + q_2 \int_{R_1} p_2(x)\, dx. \tag{3}$$

This is also the probability of a misclassification; hence, we wish to minimize the probability of misclassification.

For a given observed point x we minimize the probability of a misclassification by assigning the population that has the higher conditional probability. If

$$\frac{q_1 p_1(x)}{q_1 p_1(x) + q_2 p_2(x)} \geq \frac{q_2 p_2(x)}{q_1 p_1(x) + q_2 p_2(x)}, \tag{4}$$

we choose population π_1. Otherwise we choose population π_2. Since we minimize the probability of misclassification at each point, we minimize it over the whole space. Thus the rule is

$$\begin{aligned} R_1&: \quad q_1 p_1(x) \geq q_2 p_2(x), \\ R_2&: \quad q_1 p_1(x) < q_2 p_2(x). \end{aligned} \tag{5}$$

If $q_1 p_1(x) = q_2 p_2(x)$, the point could be classified as either from π_1 or π_2; we have arbitrarily put it into R_1. If $q_1 p_1(x) + q_2 p_2(x) = 0$ for a given x, that point also may go into either region.

Now let us prove formally that (5) is the best procedure. For any procedure $R^* = (R_1^*, R_2^*)$, the probability of misclassification is

$$
q_1 \int_{R^*_2} p_1(x)dx + q_2 \int_{R^*_1} p_2(x)\,dx
= \int_{R^*_2} [q_1 p_1(x) - q_2 p_2(x)]\,dx + q_2 \int p_2(x)\,dx. \tag{6}
$$

On the right-hand side the second term is a given number; the first term is minimized if R_2^* includes the points x such that $q_1 p_1(x) - q_2 p_2(x) < 0$ and excludes the points for which $q_1 p_1(x) - q_2 p_2(x) > 0$. If we assume that

$$
\Pr\left\{\frac{p_1(x)}{p_2(x)} = \frac{q_2}{q_1}\,\middle|\,\pi_i\right\} = 0, \qquad i = 1, 2, \tag{7}
$$

then the Bayes procedure is unique except for sets of probability zero.

Now we notice that mathematically the problem was: Given nonnegative constants q_1 and q_2 and nonnegative functions $p_1(x)$ and $p_2(x)$, choose regions R_1 and R_2 so as to minimize (3). The solution is (5). If we wish to minimize (5) of Section 6.2, which can be written

$$
[C(2|1)q_1] \int_{R_2} p_1(x)\,dx + [C(1|2)q_2] \int_{R_1} p_2(x)\,dx, \tag{8}
$$

we choose R_1 and R_2 according to

$$
\begin{aligned}
R_1&: \quad [C(2|1)q_1]p_1(x) \geq [C(1|2)q_2]p_2(x),\\
R_2&: \quad [C(2|1)q_1]p_1(x) < [C(1|2)q_2]p_2(x),
\end{aligned} \tag{9}
$$

since $[C(2|1)q_1]$ and $[C(1|2)q_2]$ are nonnegative constants. Another way of writing (9) is

$$
\begin{aligned}
R_1&: \quad \frac{p_1(x)}{p_2(x)} \geq \frac{C(1|2)q_2}{C(2|1)q_1},\\
R_2&: \quad \frac{p_1(x)}{p_2(x)} < \frac{C(1|2)q_2}{C(2|1)q_1}.
\end{aligned} \tag{10}
$$

THEOREM 6.3.1. *If q_1 and q_2 are a priori probabilities of drawing an observation from population π_1 with density $p_1(x)$ and π_2 with density $p_2(x)$ respectively, and if the cost of misclassifying an observation from π_1 as from π_2 is $C(2|1)$ and an observation from π_2 as from π_1 is $C(1|2)$, then the regions of classification R_1 and R_2, defined by* (10), *minimize the expected cost. If*

$$
\Pr\left\{\frac{p_1(x)}{p_2(x)} = \frac{q_2 C(1|2)}{q_1 C(2|1)}\,\middle|\,\pi_i\right\} = 0, \qquad i = 1, 2, \tag{11}
$$

then the procedure is unique except for sets of probability zero.

6.3.2. The Case when No A Priori Probabilities Are Known

In many instances of classification the statistician cannot assign a priori probabilities to the two populations. In this case we shall look for the class of admissible procedures, that is, the set of procedures that cannot be improved upon.

First, let us prove that a Bayes procedure is admissible. Let $R = (R_1, R_2)$ be a Bayes procedure for a given q_1, q_2; is there a procedure $R^* = (R_1^*, R_2^*)$ such that $P(1|2, R^*) \leq P(1|2, R)$ and $P(2|1, R^*) \leq P(2|1, R)$ with at least one strict inequality? Since R is a Bayes procedure,

$$(12) \qquad q_1 P(2|1, R) + q_2 P(1|2, R) \leq q_1 P(2|1, R^*) + q_2 P(1|2, R^*).$$

This inequality can be written

$$(13) \qquad q_1[P(2|1, R) - P(2|1, R^*)] \leq q_2[P(1|2, R^*) - P(1|2, R)].$$

Suppose $q_1 > 0$. Then if $P(1|2, R^*) \leq P(1|2, R)$, the right-hand side of (13) is less than or equal to zero and therefore $P(2|1, R) \leq P(2|1, R^*)$. If $q_2 > 0$, then $P(2|1, R^*) \leq P(2|1, R)$ similarly implies $P(1|2, R) \leq P(1|2, R^*)$. Thus R^* is not better than R, and R is admissible. If $q_1 = 0$, then (13) implies $0 \leq P(1|2, R^*) - P(1|2, R)$. For a Bayes procedure, R_1 includes only points for which $p_2(\boldsymbol{x}) = 0$. Therefore, $P(1|2, R) = 0$ and if R^* is to be better $P(1|2, R^*) = 0$. If $\Pr\{p_2(\boldsymbol{x}) = 0|\pi_1\} = 0$, then $P(2|1, R) = \Pr\{p_2(\boldsymbol{x}) > 0|\pi_1\} = 1$. If $P(1|2, R^*) = 0$, then R_1^* contains only points for which $p_2(\boldsymbol{x}) = 0$. Then $P(2|1, R^*) = \Pr\{R_2^*|\pi_1\} = \Pr\{p_2(\boldsymbol{x}) > 0|\pi_1\} = 1$, and R^* is not better than R.

THEOREM 6.3.2. *If* $\Pr\{p_2(\boldsymbol{x}) = 0|\pi_1\} = 0$ *and* $\Pr\{p_1(\boldsymbol{x}) = 0|\pi_2\} = 0$, *then every Bayes procedure is admissible.*

Now let us prove the converse, namely, that every admissible procedure is a Bayes procedure. We assume†

$$(14) \qquad \Pr\left\{\frac{p_1(\boldsymbol{x})}{p_2(\boldsymbol{x})} = k \middle| \pi_i\right\} = 0, \quad i = 1, 2;\ 0 \leq k \leq \infty.$$

Then for any q_1 the Bayes procedure is unique. Moreover, the cdf of $p_1(\boldsymbol{x})/p_2(\boldsymbol{x})$ for π_1 and π_2 is continuous.

Let R be an admissible procedure. Then there exists a k such that

$$(15) \qquad \begin{aligned} P(2|1, R) &= \Pr\left\{\frac{p_1(\boldsymbol{x})}{p_2(\boldsymbol{x})} \leq k \middle| \pi_1\right\} \\ &= P(2|1, R^*), \end{aligned}$$

where R^* is the Bayes procedure corresponding to $q_2/q_1 = k$ [that is,

† $p_1(\boldsymbol{x})/p_2(\boldsymbol{x}) = \infty$ means $p_2(\boldsymbol{x}) = 0$.

$q_1 = 1/(1 + k)$]. Since R is admissible, $P(1|2, R) \leq P(1|2, R^*)$. However, since by Theorem 6.3.2 R^* is admissible, $P(1|2, R) \geq P(1|2, R^*)$; that is, $P(1|2, R) = P(1|2, R^*)$. Therefore, R is also a Bayes procedure; by the uniqueness of Bayes procedures R is the same as R^*.

THEOREM 6.3.3. *If* (14) *holds, then every admissible procedure is a Bayes procedure.*

The proof of Theorem 6.3.3 shows that the class of Bayes procedures is complete. For if R is any procedure outside the class, we construct a Bayes procedure R^* so that $P(2|1, R) = P(2|1, R^*)$. Then, since R^* is admissible, $P(1|2, R) \geq P(1|2, R^*)$. Furthermore, the class of Bayes procedures is minimal complete since it is identical with the class of admissible procedures.

THEOREM 6.3.4. *If* (14) *holds, the class of Bayes procedures is minimal complete.*

Finally, let us consider the minimax procedure. Let $P(i|j, q_1) = P(i|j, R)$, where R is the Bayes procedure corresponding to q_1. $P(i|j, q_1)$ is a continuous function of q_1. $P(2|1, q_1)$ varies from 1 to 0 as q_1 goes from 0 to 1; $P(1|2, q_1)$ varies from 0 to 1. Thus there is a value of q_1, say q_1^*, such that $P(2|1, q_1^*) = P(1|2, q_1^*)$. This is the minimax solution, for if there were another procedure R^* such that $\max\{P(2|1, R^*), P(1|2, R^*)\} \leq P(2|1, q_1^*) = P(1|2, q_1^*)$ this would contradict the fact that every Bayes solution is admissible.

6.4. CLASSIFICATION INTO ONE OF TWO KNOWN MULTIVARIATE NORMAL POPULATIONS

Now we shall use the general procedure outlined above in the case of two multivariate normal populations with equal covariance matrices, namely, $N(\boldsymbol{\mu}^{(1)}, \boldsymbol{\Sigma})$ and $N(\boldsymbol{\mu}^{(2)}, \boldsymbol{\Sigma})$, where $\boldsymbol{\mu}^{(i)\prime} = (\mu_1^{(i)}, \cdots, \mu_p^{(i)})$ is the vector of means of the ith population ($i = 1, 2$) and $\boldsymbol{\Sigma}$ is the matrix of variances and covariances of each population. [The approach was first used by Wald (1944).] Then the ith density is

$$p_i(\boldsymbol{x}) = \frac{1}{(2\pi)^{\frac{1}{2}p}|\boldsymbol{\Sigma}|^{\frac{1}{2}}} \exp\left[-\tfrac{1}{2}(\boldsymbol{x} - \boldsymbol{\mu}^{(i)})'\boldsymbol{\Sigma}^{-1}(\boldsymbol{x} - \boldsymbol{\mu}^{(i)})\right]. \tag{1}$$

The ratio of densities is

$$\begin{aligned} \frac{p_1(\boldsymbol{x})}{p_2(\boldsymbol{x})} &= \frac{\exp\left[-\tfrac{1}{2}(\boldsymbol{x} - \boldsymbol{\mu}^{(1)})'\boldsymbol{\Sigma}^{-1}(\boldsymbol{x} - \boldsymbol{\mu}^{(1)})\right]}{\exp\left[-\tfrac{1}{2}(\boldsymbol{x} - \boldsymbol{\mu}^{(2)})'\boldsymbol{\Sigma}^{-1}(\boldsymbol{x} - \boldsymbol{\mu}^{(2)})\right]} \\ &= \exp\left\{-\tfrac{1}{2}\left[(\boldsymbol{x} - \boldsymbol{\mu}^{(1)})'\boldsymbol{\Sigma}^{-1}(\boldsymbol{x} - \boldsymbol{\mu}^{(1)}) \right.\right. \\ &\qquad \left.\left. - (\boldsymbol{x} - \boldsymbol{\mu}^{(2)})'\boldsymbol{\Sigma}^{-1}(\boldsymbol{x} - \boldsymbol{\mu}^{(2)})\right]\right\}. \end{aligned} \tag{2}$$

The region of classification into π_1, R_1, is the set of $\boldsymbol{x}$'s for which (2) is $\geq k$ (for k suitably chosen). Since the logarithmic function is monotonic increasing, the inequality can be written in terms of the logarithm of (2) as

$$-\tfrac{1}{2}[(\boldsymbol{x} - \boldsymbol{\mu}^{(1)})'\boldsymbol{\Sigma}^{-1}(\boldsymbol{x} - \boldsymbol{\mu}^{(1)}) - (\boldsymbol{x} - \boldsymbol{\mu}^{(2)})'\boldsymbol{\Sigma}^{-1}(\boldsymbol{x} - \boldsymbol{\mu}^{(2)})] \geq \log k. \tag{3}$$

The left-hand side of (3) can be expanded as

$$\begin{aligned} -\tfrac{1}{2}[\boldsymbol{x}'\boldsymbol{\Sigma}^{-1}\boldsymbol{x} - \boldsymbol{x}'\boldsymbol{\Sigma}^{-1}\boldsymbol{\mu}^{(1)} - \boldsymbol{\mu}^{(1)\prime}\boldsymbol{\Sigma}^{-1}\boldsymbol{x} + \boldsymbol{\mu}^{(1)\prime}\boldsymbol{\Sigma}^{-1}\boldsymbol{\mu}^{(1)} \\ - \boldsymbol{x}'\boldsymbol{\Sigma}^{-1}\boldsymbol{x} + \boldsymbol{x}'\boldsymbol{\Sigma}^{-1}\boldsymbol{\mu}^{(2)} + \boldsymbol{\mu}^{(2)\prime}\boldsymbol{\Sigma}^{-1}\boldsymbol{x} - \boldsymbol{\mu}^{(2)\prime}\boldsymbol{\Sigma}^{-1}\boldsymbol{\mu}^{(2)}]. \end{aligned} \tag{4}$$

By rearrangement of the terms we obtain

$$\boldsymbol{x}'\boldsymbol{\Sigma}^{-1}(\boldsymbol{\mu}^{(1)} - \boldsymbol{\mu}^{(2)}) - \tfrac{1}{2}(\boldsymbol{\mu}^{(1)} + \boldsymbol{\mu}^{(2)})'\boldsymbol{\Sigma}^{-1}(\boldsymbol{\mu}^{(1)} - \boldsymbol{\mu}^{(2)}). \tag{5}$$

The first term is the well-known *discriminant function.* It is a linear function of the components of the observation vector.

The following theorem is now a direct consequence of Theorem 6.3.1.

THEOREM 6.4.1. *If π_i has the density* (1) ($i = 1, 2$), *the best regions of classification are given by*

$$\begin{aligned} R_1&: \ \boldsymbol{x}'\boldsymbol{\Sigma}^{-1}(\boldsymbol{\mu}^{(1)} - \boldsymbol{\mu}^{(2)}) - \tfrac{1}{2}(\boldsymbol{\mu}^{(1)} + \boldsymbol{\mu}^{(2)})'\boldsymbol{\Sigma}^{-1}(\boldsymbol{\mu}^{(1)} - \boldsymbol{\mu}^{(2)}) \geq \log k, \\ R_2&: \ \boldsymbol{x}'\boldsymbol{\Sigma}^{-1}(\boldsymbol{\mu}^{(1)} - \boldsymbol{\mu}^{(2)}) - \tfrac{1}{2}(\boldsymbol{\mu}^{(1)} + \boldsymbol{\mu}^{(2)})'\boldsymbol{\Sigma}^{-1}(\boldsymbol{\mu}^{(1)} - \boldsymbol{\mu}^{(2)}) < \log k. \end{aligned} \tag{6}$$

If a priori probabilities q_1 and q_2 are known, then k is given by

$$k = \frac{q_2 C(1|2)}{q_1 C(2|1)}. \tag{7}$$

In the particular case of the two populations being equally likely and the costs being equal, $k = 1$ and $\log k = 0$. Then the region of classification into π_1 is

$$R_1: \ \boldsymbol{x}'\boldsymbol{\Sigma}^{-1}(\boldsymbol{\mu}^{(1)} - \boldsymbol{\mu}^{(2)}) \geq \tfrac{1}{2}(\boldsymbol{\mu}^{(1)} + \boldsymbol{\mu}^{(2)})'\boldsymbol{\Sigma}^{-1}(\boldsymbol{\mu}^{(1)} - \boldsymbol{\mu}^{(2)}). \tag{8}$$

If we do not have a priori probabilities we may select $\log k = c$, say, on the basis of making the expected losses due to misclassification equal. Let X be a random observation. Then we wish to find the distribution of

$$U = \boldsymbol{X}'\boldsymbol{\Sigma}^{-1}(\boldsymbol{\mu}^{(1)} - \boldsymbol{\mu}^{(2)}) - \tfrac{1}{2}(\boldsymbol{\mu}^{(1)} + \boldsymbol{\mu}^{(2)})'\boldsymbol{\Sigma}^{-1}(\boldsymbol{\mu}^{(1)} - \boldsymbol{\mu}^{(2)}), \tag{9}$$

on the assumption that $\boldsymbol{X}$ is distributed according to $N(\boldsymbol{\mu}^{(1)}, \boldsymbol{\Sigma})$ and then on the assumption that $\boldsymbol{X}$ is distributed according to $N(\boldsymbol{\mu}^{(2)}, \boldsymbol{\Sigma})$. When $\boldsymbol{X}$ is distributed according to $N(\boldsymbol{\mu}^{(1)}, \boldsymbol{\Sigma})$, U is normally distributed with mean

$$\begin{aligned} \mathscr{E}_1 U &= \boldsymbol{\mu}^{(1)\prime}\boldsymbol{\Sigma}^{-1}(\boldsymbol{\mu}^{(1)} - \boldsymbol{\mu}^{(2)}) - \tfrac{1}{2}(\boldsymbol{\mu}^{(1)} + \boldsymbol{\mu}^{(2)})'\boldsymbol{\Sigma}^{-1}(\boldsymbol{\mu}^{(1)} - \boldsymbol{\mu}^{(2)}) \\ &= \tfrac{1}{2}(\boldsymbol{\mu}^{(1)} - \boldsymbol{\mu}^{(2)})'\boldsymbol{\Sigma}^{-1}(\boldsymbol{\mu}^{(1)} - \boldsymbol{\mu}^{(2)}) \end{aligned} \tag{10}$$

and variance

$$\text{(11)}\quad \begin{aligned} \operatorname{Var}_1(U) &= \mathscr{E}_1(\boldsymbol{\mu}^{(1)} - \boldsymbol{\mu}^{(2)})'\boldsymbol{\Sigma}^{-1}(X - \boldsymbol{\mu}^{(1)})(X - \boldsymbol{\mu}^{(1)})'\boldsymbol{\Sigma}^{-1}(\boldsymbol{\mu}^{(1)} - \boldsymbol{\mu}^{(2)}) \\ &= (\boldsymbol{\mu}^{(1)} - \boldsymbol{\mu}^{(2)})'\boldsymbol{\Sigma}^{-1}(\boldsymbol{\mu}^{(1)} - \boldsymbol{\mu}^{(2)}). \end{aligned}$$

Let the "distance" between $N(\boldsymbol{\mu}^{(1)}, \boldsymbol{\Sigma})$ and $N(\boldsymbol{\mu}^{(2)}, \boldsymbol{\Sigma})$ be

$$\text{(12)}\qquad (\boldsymbol{\mu}^{(1)} - \boldsymbol{\mu}^{(2)})'\boldsymbol{\Sigma}^{-1}(\boldsymbol{\mu}^{(1)} - \boldsymbol{\mu}^{(2)}) = \alpha,$$

say. Then U is distributed according to $N(\frac{1}{2}\alpha, \alpha)$ if X is distributed according to $N(\boldsymbol{\mu}^{(1)}, \boldsymbol{\Sigma})$. If X is distributed according to $N(\boldsymbol{\mu}^{(2)}, \boldsymbol{\Sigma})$, then

$$\text{(13)}\quad \begin{aligned} \mathscr{E}_2 U &= \boldsymbol{\mu}^{(2)\prime}\boldsymbol{\Sigma}^{-1}(\boldsymbol{\mu}^{(1)} - \boldsymbol{\mu}^{(2)}) - \tfrac{1}{2}(\boldsymbol{\mu}^{(1)} + \boldsymbol{\mu}^{(2)})'\boldsymbol{\Sigma}^{-1}(\boldsymbol{\mu}^{(1)} - \boldsymbol{\mu}^{(2)}) \\ &= \tfrac{1}{2}(\boldsymbol{\mu}^{(2)} - \boldsymbol{\mu}^{(1)})'\boldsymbol{\Sigma}^{-1}(\boldsymbol{\mu}^{(1)} - \boldsymbol{\mu}^{(2)}) \\ &= -\tfrac{1}{2}\alpha. \end{aligned}$$

The variance is the same as when X is distributed according to $N(\boldsymbol{\mu}^{(1)}, \boldsymbol{\Sigma})$ because it depends only on the second-order moments of X. Thus U is distributed according to $N(-\frac{1}{2}\alpha, \alpha)$.

The probability of misclassification if the observation is from π_1 is

$$\text{(14)}\quad \begin{aligned} P(2|1) &= \int_{-\infty}^{c} \frac{1}{\sqrt{2\pi\alpha}} e^{-\frac{1}{2}(z - \frac{1}{2}\alpha)^2/\alpha}\, dz \\ &= \int_{-\infty}^{(c-\frac{1}{2}\alpha)/\sqrt{\alpha}} \frac{1}{\sqrt{2\pi}} e^{-\frac{1}{2}y^2} dy, \end{aligned}$$

and the probability of misclassification if the observation is from π_2 is

$$\text{(15)}\quad P(1|2) = \int_{c}^{\infty} \frac{1}{\sqrt{2\pi\alpha}} e^{-\frac{1}{2}(z + \frac{1}{2}\alpha)^2/\alpha}\, dz = \int_{(c+\frac{1}{2}\alpha)/\sqrt{\alpha}}^{\infty} \frac{1}{\sqrt{2\pi}} e^{-\frac{1}{2}y^2} dy.$$

Figure 1 indicates the two probabilities as the shaded portion in the tails.

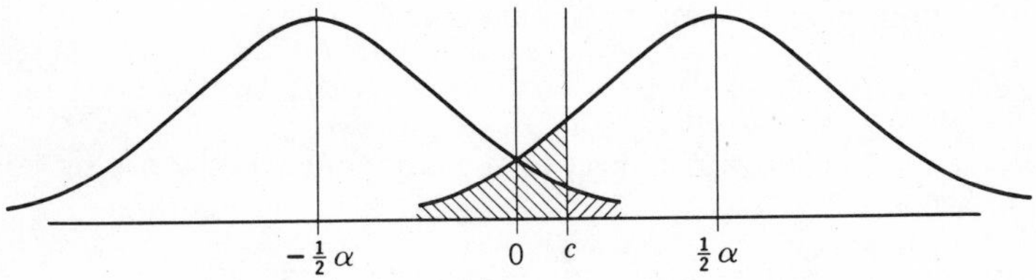

FIGURE 1

For the minimax solution we choose c so that

$$\text{(16)}\quad C(1|2)\int_{(c+\frac{1}{2}\alpha)/\sqrt{\alpha}}^{\infty} \frac{1}{\sqrt{2\pi}} e^{-\frac{1}{2}y^2}\, dy = C(2|1)\int_{-\infty}^{(c-\frac{1}{2}\alpha)/\sqrt{\alpha}} \frac{1}{\sqrt{2\pi}} e^{-\frac{1}{2}y^2} dy.$$

THEOREM 6.4.2. *If the π_i have densities* (1) $(i = 1, 2)$, *the minimax regions of classification are given by* (6) *where* $c = \log k$ *is chosen by the condition* (16) *with* $C(i|j)$ *the two costs of misclassification.*

It should be noted that if the costs of misclassification are equal, $c = 0$ and the probability of misclassification is

$$\int_{\sqrt{\alpha}/2}^{\infty} \frac{1}{\sqrt{2\pi}} e^{-\frac{1}{2}y^2} dy. \tag{17}$$

In case the costs of misclassification are unequal, c could be determined to sufficient accuracy by a trial-and-error method with the normal tables.

Both terms in (5) involve the vector

$$\boldsymbol{\delta} = \boldsymbol{\Sigma}^{-1}(\boldsymbol{\mu}^{(1)} - \boldsymbol{\mu}^{(2)}). \tag{18}$$

This is obtained as the solution of

$$\boldsymbol{\Sigma\delta} = (\boldsymbol{\mu}^{(1)} - \boldsymbol{\mu}^{(2)}) \tag{19}$$

by an efficient computing method such as the abbreviated Doolittle method.

It is interesting to note that $\boldsymbol{x}'\boldsymbol{\delta}$ is the linear function that maximizes

$$\frac{[\mathscr{E}_1(\boldsymbol{X}'\boldsymbol{d}) - \mathscr{E}_2(\boldsymbol{X}'\boldsymbol{d})]^2}{\operatorname{Var}(\boldsymbol{X}'\boldsymbol{d})} \tag{20}$$

for all choices of $\boldsymbol{d}$. The numerator of (20) is

$$[\boldsymbol{\mu}^{(1)\prime}\boldsymbol{d} - \boldsymbol{\mu}^{(2)\prime}\boldsymbol{d}]^2 = \boldsymbol{d}'[(\boldsymbol{\mu}^{(1)} - \boldsymbol{\mu}^{(2)})(\boldsymbol{\mu}^{(1)} - \boldsymbol{\mu}^{(2)})']\boldsymbol{d}; \tag{21}$$

the denominator is

$$\boldsymbol{d}'\mathscr{E}(\boldsymbol{X} - \mathscr{E}\boldsymbol{X})(\boldsymbol{X} - \mathscr{E}\boldsymbol{X})'\boldsymbol{d} = \boldsymbol{d}'\boldsymbol{\Sigma d}. \tag{22}$$

We wish to maximize (21) with respect to $\boldsymbol{d}$, holding (22) constant. If λ is a Lagrange multiplier we ask for the maximum of

$$\boldsymbol{d}'[(\boldsymbol{\mu}^{(1)} - \boldsymbol{\mu}^{(2)})(\boldsymbol{\mu}^{(1)} - \boldsymbol{\mu}^{(2)})']\boldsymbol{d} - \lambda(\boldsymbol{d}'\boldsymbol{\Sigma d} - 1). \tag{23}$$

The derivatives of (23) with respect to the components of $\boldsymbol{d}$ are set equal to zero to obtain

$$2[(\boldsymbol{\mu}^{(1)} - \boldsymbol{\mu}^{(2)})(\boldsymbol{\mu}^{(1)} - \boldsymbol{\mu}^{(2)})']\boldsymbol{d} = 2\lambda\boldsymbol{\Sigma d}. \tag{24}$$

Since $(\boldsymbol{\mu}^{(1)} - \boldsymbol{\mu}^{(2)})'\boldsymbol{d}$ is a scalar, say ν, we can write (24) as

$$\boldsymbol{\mu}^{(1)} - \boldsymbol{\mu}^{(2)} = \frac{\lambda}{\nu}\boldsymbol{\Sigma d}. \tag{25}$$

Thus the solution is proportional to $\boldsymbol{\delta}$.

We may finally note that if we have a sample of N from either π_1 or π_2, we use the mean of the sample and classify it as from $N[\boldsymbol{\mu}^{(1)}, (1/N)\boldsymbol{\Sigma}]$ or $N[\boldsymbol{\mu}^{(2)}, (1/N)\boldsymbol{\Sigma}]$.

6.5. CLASSIFICATION INTO ONE OF TWO MULTIVARIATE NORMAL POPULATIONS WHEN THE PARAMETERS ARE ESTIMATED

6.5.1. The Criterion of Classification

Thus far we have assumed that the two populations are known exactly. In most applications of this theory the populations are not known, but must be inferred from samples, one from each population. We shall now treat the case in which we have a sample from each of two normal populations and we wish to use that information in classifying another observation as coming from one of the two populations.

Suppose that we have a sample $\boldsymbol{x}_1^{(1)}, \cdots, \boldsymbol{x}_{N_1}^{(1)}$ from $N(\boldsymbol{\mu}^{(1)}, \boldsymbol{\Sigma})$ and a sample $\boldsymbol{x}_1^{(2)}, \cdots, \boldsymbol{x}_{N_2}^{(2)}$ from $N(\boldsymbol{\mu}^{(2)}, \boldsymbol{\Sigma})$. On the basis of this information we wish to classify the observation $\boldsymbol{x}$ as coming from π_1 or π_2. Clearly, our best estimate of $\boldsymbol{\mu}^{(1)}$ is $\bar{\boldsymbol{x}}^{(1)} = \sum_1^{N_1} \boldsymbol{x}_\alpha^{(1)}/N_1$, of $\boldsymbol{\mu}^{(2)}$ is $\bar{\boldsymbol{x}}^{(2)} = \sum_1^{N_2} \boldsymbol{x}_\alpha^{(2)}/N_2$, and of $\boldsymbol{\Sigma}$ is $\boldsymbol{S}$ defined by

$$(1)\quad (N_1 + N_2 - 2)\boldsymbol{S} = \sum_{\alpha=1}^{N_1} (\boldsymbol{x}_\alpha^{(1)} - \bar{\boldsymbol{x}}^{(1)})(\boldsymbol{x}_\alpha^{(1)} - \bar{\boldsymbol{x}}^{(1)})' + \sum_{\alpha=1}^{N_2} (\boldsymbol{x}_\alpha^{(2)} - \bar{\boldsymbol{x}}^{(2)})(\boldsymbol{x}_\alpha^{(2)} - \bar{\boldsymbol{x}}^{(2)})'.$$

We substitute these estimates for the parameters in (5) of Section 6.4 to obtain

$$(2)\qquad \boldsymbol{x}'\boldsymbol{S}^{-1}(\bar{\boldsymbol{x}}^{(1)} - \bar{\boldsymbol{x}}^{(2)}) - \tfrac{1}{2}(\bar{\boldsymbol{x}}^{(1)} + \bar{\boldsymbol{x}}^{(2)})'\boldsymbol{S}^{-1}(\bar{\boldsymbol{x}}^{(1)} - \bar{\boldsymbol{x}}^{(2)}).$$

The first term of (2) is the discriminant function based on two samples [suggested by Fisher (1936)]. It is the linear function that has greatest "variance between samples" relative to the "variance within samples." We propose that (2) be used as the criterion of classification in the same way that (5) of Section 6.4 is used.

When the populations are known, we can argue that the classification criterion is the best in the sense that its use minimizes the expected loss in the case of known a priori probabilities and generates the class of admissible procedures when a priori probabilities are not known. We cannot justify the use of (2) in the same way. However, it seems intuitively reasonable that (2) should give good results. Another criterion is indicated in Section 6.5.5.

Suppose we have a sample $\boldsymbol{x}_1, \cdots, \boldsymbol{x}_N$ from either π_1 or π_2, and we wish to classify the sample as a whole. Then we define $\boldsymbol{S}$ by

$$(3)\quad (N_1 + N_2 + N - 3)\boldsymbol{S} = \sum_{\alpha=1}^{N_1} (\boldsymbol{x}_\alpha^{(1)} - \bar{\boldsymbol{x}}^{(1)})(\boldsymbol{x}_\alpha^{(1)} - \bar{\boldsymbol{x}}^{(1)})' + \sum_{\alpha=1}^{N_2} (\boldsymbol{x}_\alpha^{(2)} - \bar{\boldsymbol{x}}^{(2)})(\boldsymbol{x}_\alpha^{(2)} - \bar{\boldsymbol{x}}^{(2)})' + \sum_{\alpha=1}^{N} (\boldsymbol{x}_\alpha - \bar{\boldsymbol{x}})(\boldsymbol{x}_\alpha - \bar{\boldsymbol{x}})',$$

where

$$(4)\qquad \bar{\boldsymbol{x}} = \frac{1}{N}\sum_{\alpha=1}^{N} \boldsymbol{x}_\alpha.$$

Then the criterion is

$$(5)\qquad [\bar{\boldsymbol{x}} - \tfrac{1}{2}(\bar{\boldsymbol{x}}^{(1)} + \bar{\boldsymbol{x}}^{(2)})]'\boldsymbol{S}^{-1}(\bar{\boldsymbol{x}}^{(1)} - \bar{\boldsymbol{x}}^{(2)}).$$

It can be shown that the larger N is, the smaller are the probabilities of misclassification.

6.5.2. On the Distribution of the Criterion

Let

$$(6)\qquad \begin{aligned} V &= \boldsymbol{X}'\boldsymbol{S}^{-1}(\bar{\boldsymbol{X}}^{(1)} - \bar{\boldsymbol{X}}^{(2)}) - \tfrac{1}{2}(\bar{\boldsymbol{X}}^{(1)} + \bar{\boldsymbol{X}}^{(2)})'\boldsymbol{S}^{-1}(\bar{\boldsymbol{X}}^{(1)} - \bar{\boldsymbol{X}}^{(2)}) \\ &= [\boldsymbol{X} - \tfrac{1}{2}(\bar{\boldsymbol{X}}^{(1)} + \bar{\boldsymbol{X}}^{(2)})]'\boldsymbol{S}^{-1}(\bar{\boldsymbol{X}}^{(1)} - \bar{\boldsymbol{X}}^{(2)}) \end{aligned}$$

for random $\boldsymbol{X}$, $\bar{\boldsymbol{X}}^{(1)}$, $\bar{\boldsymbol{X}}^{(2)}$, and $\boldsymbol{S}$.

The distribution of V is extremely complicated. It depends on the sample sizes and the unknown α. Let

$$(7)\qquad \boldsymbol{Z} = \boldsymbol{X} - \tfrac{1}{2}(\bar{\boldsymbol{X}}^{(1)} + \bar{\boldsymbol{X}}^{(2)}).$$

$$(8)\qquad \boldsymbol{Y} = \bar{\boldsymbol{X}}^{(1)} - \bar{\boldsymbol{X}}^{(2)}.$$

Then

$$(9)\qquad V = \boldsymbol{Z}'\boldsymbol{S}^{-1}\boldsymbol{Y}.$$

The expected value of $\boldsymbol{Y}$ is $\boldsymbol{\mu}^{(1)} - \boldsymbol{\mu}^{(2)}$ and the covariance matrix is $[(1/N_1) + (1/N_2)]\boldsymbol{\Sigma}$. $\boldsymbol{Z}$ is distributed normally with mean value

$$(10)\qquad \mathscr{E}_1\boldsymbol{Z} = \tfrac{1}{2}(\boldsymbol{\mu}^{(1)} - \boldsymbol{\mu}^{(2)})$$

if $\boldsymbol{X}$ is from π_1, and

$$(11)\qquad \mathscr{E}_2\boldsymbol{Z} = \tfrac{1}{2}(\boldsymbol{\mu}^{(2)} - \boldsymbol{\mu}^{(1)})$$

if $\boldsymbol{X}$ is from π_2. In either case the covariance matrix is $[1 + 1/(4N_1) + 1/(4N_2)]\boldsymbol{\Sigma}$. The covariance between $\boldsymbol{Z}$ and $\boldsymbol{Y}$ is

$$(12)\qquad -\left(\frac{1}{2N_1} - \frac{1}{2N_2}\right)\boldsymbol{\Sigma}.$$

If $N_1 = N_2$, this covariance is $\mathbf{0}$. In this case it is easily seen that the distribution of V for X from π_1 is the same as that of $-V$ for X from π_2. Thus, if $V \geq 0$ is the region of classification as π_1, then the probability of misclassifying X when it is from π_1 is equal to the probability of misclassifying it when it is from π_2.

The distribution of V is considered by Anderson (1951a), Sitgreaves (1952), and Wald (1944).

6.5.3. The Asymptotic Distribution of the Criterion

In the case of large samples from $N(\boldsymbol{\mu}^{(1)}, \boldsymbol{\Sigma})$ and $N(\boldsymbol{\mu}^{(2)}, \boldsymbol{\Sigma})$ we can apply limiting distribution theory. Since $\bar{\boldsymbol{X}}^{(1)}$ is the mean of a sample of N_1 independent observations from $N(\boldsymbol{\mu}^{(1)}, \boldsymbol{\Sigma})$, we know that

$$\operatorname*{plim}_{N_1 \to \infty} \bar{\boldsymbol{X}}^{(1)} = \boldsymbol{\mu}^{(1)}. \tag{13}$$

The explicit definition of (13) is as follows: Given arbitrary positive δ and ε, we can find N large enough so that for $N_1 \geq N$

$$\Pr\{|\bar{X}_i^{(1)} - \mu_i^{(1)}| < \delta, i = 1, \cdots, p\} > 1 - \varepsilon. \tag{14}$$

(See Problem 12 of Chapter 3.) This can be proved by using the Tchebycheff inequality. Similarly,

$$\operatorname*{plim}_{N_2 \to \infty} \bar{\boldsymbol{X}}^{(2)} = \boldsymbol{\mu}^{(2)}, \tag{15}$$

$$\operatorname{plim} \boldsymbol{S} = \boldsymbol{\Sigma}, \tag{16}$$

as $N_1 \to \infty$, $N_2 \to \infty$ or as both $N_1, N_2 \to \infty$. From (16) we obtain

$$\operatorname{plim} \boldsymbol{S}^{-1} = \boldsymbol{\Sigma}^{-1}, \tag{17}$$

since the probability limits of sums, differences, products, and quotients of random variables are the sums, differences, products, and quotients of the probability limits as long as the probability limit of each denominator is different from zero [Cramér (1946), p. 254]. Furthermore,

$$\operatorname*{plim}_{N_1, N_2 \to \infty} \boldsymbol{S}^{-1}(\bar{\boldsymbol{X}}^{(1)} - \bar{\boldsymbol{X}}^{(2)}) = \boldsymbol{\Sigma}^{-1}(\boldsymbol{\mu}^{(1)} - \boldsymbol{\mu}^{(2)}), \tag{18}$$

$$\operatorname*{plim}_{N_1, N_2 \to \infty} (\bar{\boldsymbol{X}}^{(1)} + \bar{\boldsymbol{X}}^{(2)})'\boldsymbol{S}^{-1}(\bar{\boldsymbol{X}}^{(1)} - \bar{\boldsymbol{X}}^{(2)}) = (\boldsymbol{\mu}^{(1)} + \boldsymbol{\mu}^{(2)})'\boldsymbol{\Sigma}^{-1}(\boldsymbol{\mu}^{(1)} - \boldsymbol{\mu}^{(2)}). \tag{19}$$

It follows then that the limiting distribution of V is the distribution of U. For sufficiently large samples from π_1 and π_2 we can use the criterion as if we knew the populations exactly and we make only a small error. [The result was first given by Wald (1944).]

THEOREM 6.5.1. *Let V be given by (6) with $\bar{X}^{(1)}$ the mean of a sample of N_1 from $N(\boldsymbol{\mu}^{(1)}, \boldsymbol{\Sigma})$, $\bar{X}^{(2)}$ the mean of a sample of N_2 from $N(\boldsymbol{\mu}^{(2)}, \boldsymbol{\Sigma})$, and S the estimate of $\boldsymbol{\Sigma}$ based on the pooled sample. The limiting distribution of V as $N_1 \to \infty$ and $N_2 \to \infty$ is $N(\frac{1}{2}\alpha, \alpha)$ if X is distributed according to $N(\boldsymbol{\mu}^{(1)}, \boldsymbol{\Sigma})$ and is $N(-\frac{1}{2}\alpha, \alpha)$ if X is distributed according to $N(\boldsymbol{\mu}^{(2)}, \boldsymbol{\Sigma})$.*

6.5.4. Another Derivation of the Criterion

A convenient mnemonic derivation of the criterion is the use of regression of a dummy variate [given by Fisher (1936)]. Let

$$y_\alpha^{(1)} = \frac{N_2}{N_1 + N_2} \quad (\alpha = 1, \cdots, N_1), \qquad y_\alpha^{(2)} = \frac{-N_1}{N_1 + N_2} \quad (\alpha = 1, \cdots, N_2).$$

Then formally find the regression on the variates $\boldsymbol{x}_\alpha^{(i)}$ by choosing $\boldsymbol{b}$ to minimize

$$\sum_{i=1}^{2} \sum_{\alpha=1}^{N_i} [y_\alpha^{(i)} - \boldsymbol{b}'(\boldsymbol{x}_\alpha^{(i)} - \bar{\boldsymbol{x}})]^2, \tag{20}$$

where

$$\bar{\boldsymbol{x}} = (N_1 \bar{\boldsymbol{x}}^{(1)} + N_2 \bar{\boldsymbol{x}}^{(2)})/(N_1 + N_2). \tag{21}$$

The "normal equations" are

$$\begin{aligned} \sum_{i=1}^{2} \sum_{\alpha=1}^{N_i} (\boldsymbol{x}_\alpha^{(i)} - \bar{\boldsymbol{x}})(\boldsymbol{x}_\alpha^{(i)} - \bar{\boldsymbol{x}})'\boldsymbol{b} &= \sum_{i=1}^{2} \sum_{\alpha=1}^{N_i} y_\alpha^{(i)}(\boldsymbol{x}_\alpha^{(i)} - \bar{\boldsymbol{x}}) \\ &= \frac{N_1 N_2}{N_1 + N_2} [(\bar{\boldsymbol{x}}^{(1)} - \bar{\boldsymbol{x}}) - (\bar{\boldsymbol{x}}^{(2)} - \bar{\boldsymbol{x}})] \\ &= \frac{N_1 N_2}{N_1 + N_2} (\bar{\boldsymbol{x}}^{(1)} - \bar{\boldsymbol{x}}^{(2)}). \end{aligned} \tag{22}$$

The matrix multiplying $\boldsymbol{b}$ can be written as

$$\begin{aligned} \sum_{i=1}^{2} \sum_{\alpha=1}^{N_i} (\boldsymbol{x}_\alpha^{(i)} - \bar{\boldsymbol{x}})(\boldsymbol{x}_\alpha^{(i)} - \bar{\boldsymbol{x}})' &= \sum_{i=1}^{2} \sum_{\alpha=1}^{N_i} (\boldsymbol{x}_\alpha^{(i)} - \bar{\boldsymbol{x}}^{(i)})(\boldsymbol{x}_\alpha^{(i)} - \bar{\boldsymbol{x}}^{(i)})' \\ &\quad + N_1(\bar{\boldsymbol{x}}^{(1)} - \bar{\boldsymbol{x}})(\bar{\boldsymbol{x}}^{(1)} - \bar{\boldsymbol{x}})' + N_2(\bar{\boldsymbol{x}}^{(2)} - \bar{\boldsymbol{x}})(\bar{\boldsymbol{x}}^{(2)} - \bar{\boldsymbol{x}})' \\ &= \sum_{i=1}^{2} \sum_{\alpha=1}^{N_i} (\boldsymbol{x}_\alpha^{(i)} - \bar{\boldsymbol{x}}^{(i)})(\boldsymbol{x}_\alpha^{(i)} - \bar{\boldsymbol{x}}^{(i)})' + \frac{N_1 N_2}{N_1 + N_2} (\bar{\boldsymbol{x}}^{(1)} - \bar{\boldsymbol{x}}^{(2)})(\bar{\boldsymbol{x}}^{(1)} - \bar{\boldsymbol{x}}^{(2)})'. \end{aligned} \tag{23}$$

Thus (22) can be written as

$$\boldsymbol{C}\boldsymbol{b} = (\bar{\boldsymbol{x}}^{(1)} - \bar{\boldsymbol{x}}^{(2)}) \left[\frac{N_1 N_2}{N_1 + N_2} - \frac{N_1 N_2}{N_1 + N_2} (\bar{\boldsymbol{x}}^{(1)} - \bar{\boldsymbol{x}}^{(2)})'\boldsymbol{b}\right], \tag{24}$$

where

$$C = \sum_{i=1}^{2} \sum_{\alpha=1}^{N_i} (\boldsymbol{x}_\alpha^{(i)} - \bar{\boldsymbol{x}}^{(i)})(\boldsymbol{x}_\alpha^{(i)} - \bar{\boldsymbol{x}}^{(i)})'. \tag{25}$$

Since $(\bar{\boldsymbol{x}}^{(1)} - \bar{\boldsymbol{x}}^{(2)})'\boldsymbol{b}$ is a scalar, we see that the solution $\boldsymbol{b}$ of (24) is proportional to $\boldsymbol{S}^{-1}(\bar{\boldsymbol{x}}^{(1)} - \bar{\boldsymbol{x}}^{(2)})$.

6.5.5. The Likelihood Ratio Criterion

Another criterion which can be used in classification is the likelihood ratio criterion. Consider testing the composite null hypothesis that $\boldsymbol{x}, \boldsymbol{x}_1^{(1)}, \cdots, \boldsymbol{x}_{N_1}^{(1)}$ are drawn from $N(\boldsymbol{\mu}^{(1)}, \boldsymbol{\Sigma})$ and $\boldsymbol{x}_1^{(2)}, \cdots, \boldsymbol{x}_{N_2}^{(2)}$ are drawn from $N(\boldsymbol{\mu}^{(2)}, \boldsymbol{\Sigma})$ against the composite alternative hypothesis that $\boldsymbol{x}_1^{(1)}, \cdots, \boldsymbol{x}_{N_1}^{(1)}$ are drawn from $N(\boldsymbol{\mu}_1^{(2)}, \boldsymbol{\Sigma})$ and $\boldsymbol{x}, \boldsymbol{x}_1^{(2)}, \cdots, \boldsymbol{x}_{N_2}^{(2)}$ are drawn from $N(\boldsymbol{\mu}^{(2)}, \boldsymbol{\Sigma})$, with $\boldsymbol{\mu}^{(1)}$, $\boldsymbol{\mu}^{(2)}$, and $\boldsymbol{\Sigma}$ unspecified. Under the first hypothesis the maximum likelihood estimates of $\boldsymbol{\mu}^{(1)}$, $\boldsymbol{\mu}^{(2)}$, and $\boldsymbol{\Sigma}$ are

$$\begin{aligned} \hat{\boldsymbol{\mu}}_1^{(1)} &= (N_1\bar{\boldsymbol{x}}^{(1)} + \boldsymbol{x})/(N_1 + 1), \\ \hat{\boldsymbol{\mu}}_1^{(2)} &= \bar{\boldsymbol{x}}^{(2)}, \end{aligned} \tag{26}$$

$$\begin{aligned} \hat{\boldsymbol{\Sigma}}_1 = \frac{1}{N_1 + N_2 + 1} \Bigg[& \sum_{\alpha=1}^{N_1} (\boldsymbol{x}_\alpha^{(1)} - \hat{\boldsymbol{\mu}}_1^{(1)})(\boldsymbol{x}_\alpha^{(1)} - \hat{\boldsymbol{\mu}}_1^{(1)})' \\ & + (\boldsymbol{x} - \hat{\boldsymbol{\mu}}_1^{(1)})(\boldsymbol{x} - \hat{\boldsymbol{\mu}}_1^{(1)})' + \sum_{\alpha=1}^{N_2} (\boldsymbol{x}_\alpha^{(2)} - \hat{\boldsymbol{\mu}}_1^{(2)})(\boldsymbol{x}_\alpha^{(2)} - \hat{\boldsymbol{\mu}}_1^{(2)})' \Bigg]. \end{aligned}$$

Since

$$\begin{aligned} \sum_{\alpha=1}^{N_1} & (\boldsymbol{x}_\alpha^{(1)} - \hat{\boldsymbol{\mu}}_1^{(1)})(\boldsymbol{x}_\alpha^{(1)} - \hat{\boldsymbol{\mu}}_1^{(1)})' + (\boldsymbol{x} - \hat{\boldsymbol{\mu}}_1^{(1)})(\boldsymbol{x} - \hat{\boldsymbol{\mu}}_1^{(1)})' \\ &= \sum_{\alpha=1}^{N_1} (\boldsymbol{x}_\alpha^{(1)} - \bar{\boldsymbol{x}}^{(1)})(\boldsymbol{x}_\alpha^{(1)} - \bar{\boldsymbol{x}}^{(1)})' + N_1(\bar{\boldsymbol{x}}^{(1)} - \hat{\boldsymbol{\mu}}_1^{(1)})(\bar{\boldsymbol{x}}^{(1)} - \hat{\boldsymbol{\mu}}_1^{(1)})' \\ &\qquad + (\boldsymbol{x} - \hat{\boldsymbol{\mu}}_1^{(1)})(\boldsymbol{x} - \hat{\boldsymbol{\mu}}_1^{(1)})' \\ &= \sum_{\alpha=1}^{N_1} (\boldsymbol{x}_\alpha^{(1)} - \bar{\boldsymbol{x}}^{(1)})(\boldsymbol{x}_\alpha^{(1)} - \bar{\boldsymbol{x}}^{(1)})' + \frac{N_1}{N_1 + 1} (\boldsymbol{x} - \bar{\boldsymbol{x}}^{(1)})(\boldsymbol{x} - \bar{\boldsymbol{x}}^{(1)})', \end{aligned} \tag{27}$$

we can write $\hat{\boldsymbol{\Sigma}}_1$ as

$$\hat{\boldsymbol{\Sigma}}_1 = \frac{1}{N_1 + N_2 + 1} \left[C + \frac{N_1}{N_1 + 1} (\boldsymbol{x} - \bar{\boldsymbol{x}}^{(1)})(\boldsymbol{x} - \bar{\boldsymbol{x}}^{(1)})' \right], \tag{28}$$

where C is given by (25). Under the assumptions of the alternative hypothesis we find (by considerations of symmetry) that the maximum likelihood estimates of the parameters are

$$\begin{aligned} \hat{\boldsymbol{\mu}}_2^{(1)} &= \bar{\boldsymbol{x}}^{(1)}, \\ \hat{\boldsymbol{\mu}}_2^{(2)} &= (N_2\bar{\boldsymbol{x}}^{(2)} + \boldsymbol{x})/(N_2 + 1), \\ \hat{\boldsymbol{\Sigma}}_2 &= \frac{1}{N_1 + N_2 + 1} \left[C + \frac{N_2}{N_2 + 1} (\boldsymbol{x} - \bar{\boldsymbol{x}}^{(2)})(\boldsymbol{x} - \bar{\boldsymbol{x}}^{(2)})' \right]. \end{aligned} \tag{29}$$

The likelihood ratio criterion is, therefore, the $(N_1 + N_2 + 1)/2$th power of

$$\frac{|\hat{\mathbf{\Sigma}}_2|}{|\hat{\mathbf{\Sigma}}_1|} = \frac{\left| C + \frac{N_2}{N_2 + 1} (\mathbf{x} - \bar{\mathbf{x}}^{(2)})(\mathbf{x} - \bar{\mathbf{x}}^{(2)})' \right|}{\left| C + \frac{N_1}{N_1 + 1} (\mathbf{x} - \bar{\mathbf{x}}^{(1)})(\mathbf{x} - \bar{\mathbf{x}}^{(1)})' \right|}. \tag{30}$$

This ratio can also be written

$$\frac{1 + \frac{N_2}{N_2 + 1} (\mathbf{x} - \bar{\mathbf{x}}^{(2)})' C^{-1} (\mathbf{x} - \bar{\mathbf{x}}^{(2)})}{1 + \frac{N_1}{N_1 + 1} (\mathbf{x} - \bar{\mathbf{x}}^{(1)})' C^{-1} (\mathbf{x} - \bar{\mathbf{x}}^{(1)})}. \tag{31}$$

The region of classification into π_1 consists of those points for which the ratio (31) is greater than a given number.

6.6. CLASSIFICATION INTO ONE OF SEVERAL POPULATIONS

Let us now consider the problem of classifying an observation into one of several populations. We shall extend the consideration of the previous sections to the cases of more than two populations. Let $\pi_1, \cdots, \pi_m$ be m populations with density functions $p_1(\mathbf{x}), \cdots, p_m(\mathbf{x})$ respectively. We wish to divide the space of observations into m mutually exclusive and exhaustive regions $R_1, \cdots, R_m$. If an observation falls into R_i we shall say that it comes from π_i. Let the cost of misclassifying an observation from π_i as coming from π_j be $C(j|i)$. The probability of this misclassification is

$$P(j|i, R) = \int_{R_j} p_i(\mathbf{x})\, d\mathbf{x}. \tag{1}$$

Suppose we have a priori probabilities of the populations, $q_1, \cdots, q_m$. Then the expected loss is

$$\sum_{i=1}^{m} q_i \left\{ \sum_{\substack{j=1 \\ j \neq i}}^{m} C(j|i) P(j|i, R) \right\}. \tag{2}$$

We should like to choose $R_1, \cdots, R_m$ to make this a minimum.

Since we have a priori probabilities for the populations, we can define the conditional probability of an observation coming from a population given the values of the components of the vector x. The conditional probability of the observation coming from π_i is

$$\frac{q_i p_i(x)}{\sum_{k=1}^{m} q_k p_k(x)}. \tag{3}$$

If we classify the observation as from π_j, the expected loss is

$$\sum_{\substack{i=1 \\ i \neq j}}^{m} \frac{q_i p_i(x)}{\sum_{k=1}^{m} q_k p_k(x)} C(j|i). \tag{4}$$

We minimize the expected loss at this point if we choose j so as to minimize (4); that is, we consider

$$\sum_{\substack{i=1 \\ i \neq j}}^{m} q_i p_i(x) C(j|i) \tag{5}$$

for all j and select that j that gives the minimum. (If two different indices give the minimum, it is irrelevant which index is selected.) This procedure assigns the point x to one of the R_j. Following this procedure for each x, we define our regions $R_1, \cdots, R_m$. The classification procedure, then, is to classify an observation as coming from π_j if it falls in R_j.

THEOREM 6.6.1. *If q_i is the a priori probability of drawing an observation from population π_i with density $p_i(x)$ $(i = 1, \cdots, m)$ and if the cost of misclassifying an observation from π_i as from π_j is $C(j|i)$, then the regions of classification, $R_1, \cdots, R_m$, that minimize the expected cost are defined by assigning x to R_k if*

$$\sum_{\substack{i=1 \\ i \neq k}}^{m} q_i p_i(x) C(k|i) < \sum_{\substack{i=1 \\ i \neq j}}^{m} q_i p_i(x) C(j|i) \qquad (j = 1, \cdots, m, j \neq k). \tag{6}$$

[*If* (6) *holds for all j $(j \neq k)$ except for h indices and the inequality is replaced by equality for those indices, then this point can be assigned to any of the $h + 1$ π's.*] *If the probability of equality between right-hand and left-hand sides of* (6) *is zero for each k and j under π_i (each i), then the minimizing procedure is unique except for sets of probability zero.*

We now verify this result. Let

$$h_j(x) = \sum_{\substack{i=1 \\ i \neq j}}^{m} q_i p_i(x) C(j|i). \tag{7}$$

Then the expected loss of a procedure R is

$$\sum_{j=1}^{m} \int_{R_j} h_j(x)\, dx = \int h(x)\, dx, \tag{8}$$

where $h(\mathbf{x}) = h_j(\mathbf{x})$ for $\mathbf{x}$ in R_j. For the Bayes procedure described in the theorem $h(\mathbf{x})$ is $h^*(\mathbf{x}) = \min_i h_i(\mathbf{x})$. Thus the difference between the expected loss for any procedure R and for R^* is

$$\int [h(\mathbf{x}) - h^*(\mathbf{x})]d\mathbf{x} = \sum_j \int_{R_j} [h_j(\mathbf{x}) - \min_i h_i(\mathbf{x})]\, d\mathbf{x} \geq 0. \tag{9}$$

Equality can hold only if $h_j(\mathbf{x}) = \min_i h_i(\mathbf{x})$ for $\mathbf{x}$ in R_j except for sets of probability zero.

Let us see how this method applies when $C(j|i) = 1$ for all i and j $(i \neq j)$. Then in R_k

$$\sum_{\substack{i=1\\ i\neq k}}^{m} q_i p_i(\mathbf{x}) < \sum_{\substack{i=1\\ i\neq j}}^{m} q_i p_i(\mathbf{x}) \qquad (j \neq k). \tag{10}$$

Subtracting $\sum_{\substack{i=1\\ i\neq k,j}}^{m} q_i p_i(\mathbf{x})$ from both sides of (10), we obtain

$$q_j p_j(\mathbf{x}) < q_k p_k(\mathbf{x}) \qquad (j \neq k). \tag{11}$$

In this case the point $\mathbf{x}$ is in R_k if k is the index for which $q_i p_i(\mathbf{x})$ is a maximum; that is, π_k is the most probable population.

Now suppose that we do not have a priori probabilities. Then we cannot define an unconditional expected loss for a classification procedure. However, we can define an expected loss on the condition that the observation comes from a given population. The conditional expected loss if the observation is from π_i is

$$\sum_{\substack{j=1\\ j\neq i}}^{m} C(j|i)P(j|i, R) = r(i, R). \tag{12}$$

A procedure R is at least as good as R^* if $r(i, R) \leq r(i, R^*)$, $i = 1, \cdots, m$; R is better if at least one inequality is strict. R is *admissible* if there is no procedure R^* that is better. A class of procedures is *complete* if for every procedure R outside the class there is a procedure R^* in the class that is better.

Now let us show that a Bayes procedure is admissible. Let R be a Bayes procedure; let R^* be another procedure. Since R is Bayes

$$\sum_{i=1}^{m} q_i r(i, R) \leq \sum_{i=1}^{m} q_i r(i, R^*). \tag{13}$$

Suppose $r(i, R^*) \leq r(i, R)$, $i = 2, \cdots, m$, and $q_1 > 0$. Then

$$q_1[r(1, R) - r(1, R^*)] \leq \sum_{i=2}^{m} q_i[r(i, R^*) - r(i, R)] \leq 0, \tag{14}$$

and $r(1, R) \leq r(1, R^*)$. Similarly, if $q_i > 0$, and $r(j, R^*) \leq r(j, R)$ for $j \neq i$ then $r(i, R) \leq r(i, R^*)$. Thus R^* cannot be better than R; R is admissible.

THEOREM 6.6.2. *If* $q_i > 0$ $(i = 1, \cdots, m)$, *then a Bayes procedure is admissible.*

We shall now assume that $C(i|j) = 1$, $i \neq j$, and $\Pr\{p_i(\boldsymbol{x}) = 0|\pi_j\} = 0$. The latter condition implies that all $p_i(\boldsymbol{x})$ are positive on the same set (except for a set of measure 0). Suppose $q_i = 0$ for $i = 1, \cdots, t$, and $q_i > 0$ for $i = t + 1, \cdots, m$. Then for the Bayes solution R_i $(i = 1, \cdots, t)$ is empty (except for a set of probability 0) as seen from (11) (that is, $p_m(\boldsymbol{x}) = 0$ for $\boldsymbol{x}$ in R_i). It follows that $r(i, R) = \sum_{j \neq i} P(j|i, R) = 1 - P(i|i, R) = 1$ for $i = 1, \cdots, t$. Then $(R_{t+1}, \cdots, R_m)$ is a Bayes solution for the problem involving $p_{t+1}(\boldsymbol{x}), \cdots, p_m(\boldsymbol{x})$ and $q_{t+1}, \cdots, q_m$. It follows from Theorem 6.6.2 that no procedure R^* for which $P(i|i, R^*) = 0$ $(i = 1, \cdots, t)$ can be better than the Bayes procedure. Now consider a procedure R^* such that R_1^* includes a set of positive probability so that $P(1|1, R^*) > 0$. For R^* to be better than R,

$$P(i|i, R) = \int_{R_i} p_i(\boldsymbol{x}) d\boldsymbol{x} \tag{15}$$

$$\leq P(i|i, R^*) = \int_{R_i^*} p_i(\boldsymbol{x})\, d\boldsymbol{x}, \qquad i = 2, \cdots, m.$$

In such a case a procedure R^{**} where R_i^{**} is empty, $i = 1, \cdots, t$, $R_i^{**} = R_i^*$, $i = t + 1, \cdots, m - 1$, and $R_m^{**} = R_m^* \cup R_1^* \cup \cdots \cup R_t^*$ would give risks such that

$$\begin{aligned} P(i|i, R^{**}) &= 0, & i &= 1, \cdots, t, \\ P(i|i, R^{**}) &= P(i|i, R^*) \geq P(i|i, R), & i &= t + 1, \cdots, m - 1, \\ P(m|m, R^{**}) &> P(m|m, R^*) \geq P(m|m, R). \end{aligned} \tag{16}$$

Then $(R_{t+1}^{**}, \cdots, R_m^{**})$ would be better than $(R_{t+1}, \cdots, R_m)$ for the $(m - t)$-decision problem which contradicts the preceding discussion.

THEOREM 6.6.3. *If* $C(i|j) = 1$, $i \neq j$, *and* $\Pr\{p_i(\boldsymbol{x}) = 0|\pi_j\} = 0$, *then a Bayes procedure is admissible.*

Now we want to show that admissible procedures are Bayes procedures. We shall restrict our attention to the case of $m = 3$. We shall assume that

$$\Pr\left\{\frac{p_i(\boldsymbol{x})}{p_j(\boldsymbol{x})} = k \,\middle|\, \pi_h\right\} = 0, \qquad i \neq j,\ 0 \leq k < \infty. \tag{17}$$

This implies that the cdf of $p_i(x)/p_j(x)$ for any π_h is continuous and that the joint cdf of two ratios is continuous (see Problem 45 of Chapter 2).

Let $\alpha_i(R) = 1 - P(i|i, R)$ be the probability of making a wrong decision when using procedure R and sampling from π_i. When R is a Bayes procedure, $\alpha_i(R)$ is a function of q_1, q_2, q_3, say $\alpha_i(q_1, q_2, q_3)$. It is a continuous function of q_1, q_2, q_3; for instance,

$$\alpha_1(q_1, q_2, q_3) = 1 - \Pr\left\{\frac{p_2(x)}{p_1(x)} \le \frac{q_1}{q_2}, \frac{p_3(x)}{p_1(x)} \le \frac{q_1}{q_3} \,\middle|\, \pi_1\right\}, \tag{18}$$

and the cdf of $p_2(x)/p_1(x)$ and $p_3(x)/p_1(x)$ is continuous. It is convenient to think of (q_1, q_2, q_3) as the barycentric coordinates of a point. The

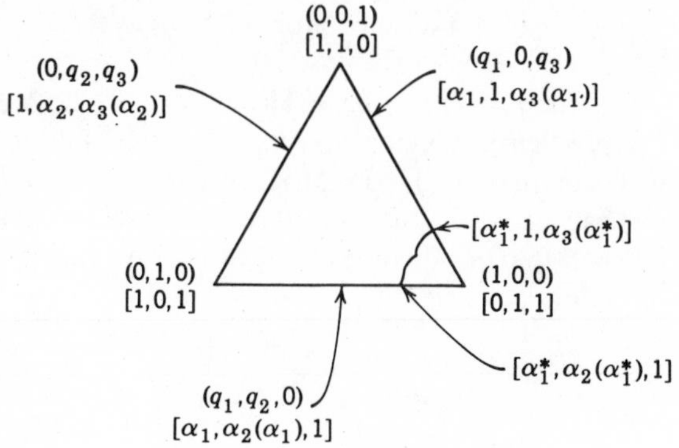

FIGURE 2

boundaries of the space of triplets and the values of the functions on the boundaries are indicated in Figure 2.

Now let R^* be an admissible procedure, and let $\alpha_i(R^*) = \alpha_i^*$. We shall show that R^* is a Bayes procedure. Consider the totality of Bayes procedures for which $\alpha_1(q_1, q_2, q_3) = \alpha_1^*$. When $q_3 = 0$, we have in effect a two-decision problem and then $\alpha_2 = \alpha_2(\alpha_1^*)$ which is the smallest α_2 given $\alpha_1 = \alpha_1^*$ (by the results in the case of two decisions); thus $\alpha_2(\alpha_1^*) \le \alpha_2^*$; and $\alpha_3 = 1$. Similarly, when $q_2 = 0$, $\alpha_3 = \alpha_3(\alpha_1^*) \le \alpha_3^*$, and $\alpha_2 = 1$. The set of points (q_1, q_2, q_3) for which $\alpha_1(q_1, q_2, q_3) = \alpha_1^*$ is a continuous curve† from the point with $[\alpha_1^*, \alpha_2(\alpha_1^*), 1]$ to the point with $[\alpha_1^*, 1, \alpha_3(\alpha_1^*)]$. Since α_2 varies continuously from $\alpha_2(\alpha_1^*)$ to 1, there is a point where $\alpha_2 = \alpha_2^*$. Thus there is a Bayes procedure $\bar{R}$ such that $\alpha_1(\bar{R}) = \alpha_1^*$ and $\alpha_2(\bar{R}) = \alpha_2^*$. Since $\bar{R}$ is admissible (by Theorem 6.6.3), $\alpha_3(\bar{R}) \le \alpha_3^*$. But since R^* is admissible $\alpha_3(\bar{R}) = \alpha_3^*$. By the uniqueness of Bayes solutions, $R^* = \bar{R}$.

† Along each ray $q_2 = k(1 - q_1)$, $q_3 = (1 - k)(1 - q_1)$, $0 < k < 1$, α_1 decreases continuously and monotonically from 1 to 0. Let $q_1 = q_1(k)$ be the value of q_1 so $\alpha_1 = \alpha_1^*$; then $q_1(k)$ is a continuous function of k [by the continuity of $\alpha_1(q_1, q_2, q_3)$ and the monotonicity of α_1 as a function of q_1 given k].

THEOREM 6.6.4. *If* (17) *holds, then every admissible procedure is a Bayes procedure.*

The proof of the above theorem shows that the class of Bayes procedures is complete. For given any procedure R^*, there is a Bayes procedure $\bar{R}$ that is at least as good (implying essential completeness). But if the $\bar{R}$ and R^* are equally good they are the same (to sets of probability 0).

Furthermore, since the class of Bayes procedures is identical with the admissible class, it is minimal complete.

THEOREM 6.6.5. *If* (17) *holds, the class of Bayes procedures is minimal complete.*

We can also consider the minimax solution. There is a Bayes solution for which $\alpha_1 = \alpha_2 = \alpha_3$, for the totality of points for which $\alpha_1 = \alpha_2$ is connected and includes a point for which $\alpha_3 = 1$ and for which $\alpha_3 = 0$. By continuity there is a point for which $\alpha_1 = \alpha_2 = \alpha_3$. Since this procedure is admissible, there is no other procedure which has a smaller maximum probability of error (that is, each risk being smaller). Thus this gives the minimax procedure.

For the general theory of statistical decision procedures, the reader is referred to Wald (1950) and Blackwell and Girshick (1954). Von Mises (1945) found the solution of the minimax problem by a different method.

6.7. CLASSIFICATION INTO ONE OF SEVERAL MULTIVARIATE NORMAL POPULATIONS

We shall now apply the theory of Section 6.6 to the case in which each population has a normal distribution [see von Mises (1945)]. We assume that the means are different and the covariance matrices are alike. Let $N(\boldsymbol{\mu}^{(i)}, \boldsymbol{\Sigma})$ be the distribution of π_i. The density is given by (1) of Section 6.4. At the outset the parameters are assumed known. For general costs with known a priori probabilities we can form the m functions (5) of Section 6.6 and define the region R_j as consisting of points $\boldsymbol{x}$ such that the jth function is minimum.

In the remainder of our discussion we shall assume that the costs of misclassification are equal. Then we use the functions

$$(1) \qquad u_{jk}(\boldsymbol{x}) = \log \frac{p_j(\boldsymbol{x})}{p_k(\boldsymbol{x})} = [\boldsymbol{x} - \tfrac{1}{2}(\boldsymbol{\mu}^{(j)} + \boldsymbol{\mu}^{(k)})]'\boldsymbol{\Sigma}^{-1}(\boldsymbol{\mu}^{(j)} - \boldsymbol{\mu}^{(k)}).$$

If a priori probabilities are known, the region R_j is defined by those $\boldsymbol{x}$ satisfying

$$(2) \qquad R_j\colon\ u_{jk}(\boldsymbol{x}) > \log \frac{q_k}{q_j}, \qquad k = 1, \cdots, m;\ k \neq j.$$

THEOREM 6.7.1. *If q_i is the a priori probability of drawing an observation from $\pi_i = N(\boldsymbol{\mu}^{(i)}, \boldsymbol{\Sigma})$ $(i = 1, \cdots, m)$ and if the costs of misclassification are equal, then the regions of classification, $R_1, \cdots, R_m$, that minimize the expected cost are defined by* (2), *where $u_{jk}(\boldsymbol{x})$ is given by* (1).

It should be noted that each $u_{jk}(\boldsymbol{x})$ is the classification function related to the jth and kth populations, and $u_{jk}(\boldsymbol{x}) = -u_{kj}(\boldsymbol{x})$. Since these are linear functions, the region R_i is bounded by hyperplanes. If the means span an $(m-1)$-dimensional hyperplane (for example, if the vectors $\boldsymbol{\mu}^{(i)}$ are linearly independent and $p \geq m-1$), then R_i is bounded by $m-1$ hyperplanes.

In the case of no a priori probabilities known, the region R_j is defined by inequalities

$$u_{jk}(\boldsymbol{x}) \geq c_j - c_k, \qquad k = 1, \cdots, m, k \neq j. \tag{3}$$

The constants c_k can be taken nonnegative. These sets of regions form the class of admissible procedures. For the minimax procedure these constants are determined so all $P(i|i, R)$ are equal.

We now show how to evaluate the probabilities of correct classification. If $\boldsymbol{X}$ is a random observation, we consider the random variables

$$U_{ji} = [\boldsymbol{X} - \tfrac{1}{2}(\boldsymbol{\mu}^{(i)} + \boldsymbol{\mu}^{(j)})]'\boldsymbol{\Sigma}^{-1}(\boldsymbol{\mu}^{(j)} - \boldsymbol{\mu}^{(i)}). \tag{4}$$

Here $U_{ji} = -U_{ij}$. Thus we use $m(m-1)/2$ classification functions if the means span an $(m-1)$-dimensional hyperplane. If $\boldsymbol{X}$ is from π_j, then U_{ji} is distributed according to $N(\frac{1}{2}\alpha_{jii}, \alpha_{jii})$ where

$$\alpha_{jii} = (\boldsymbol{\mu}^{(j)} - \boldsymbol{\mu}^{(i)})'\boldsymbol{\Sigma}^{-1}(\boldsymbol{\mu}^{(j)} - \boldsymbol{\mu}^{(i)}). \tag{5}$$

The covariance of U_{ji} and U_{jk} is

$$\alpha_{jki} = (\boldsymbol{\mu}^{(j)} - \boldsymbol{\mu}^{(k)})'\boldsymbol{\Sigma}^{-1}(\boldsymbol{\mu}^{(j)} - \boldsymbol{\mu}^{(i)}). \tag{6}$$

To determine the constants c_j we consider the integrals

$$P(j|j, R) = \int_{c_j - c_m}^{\infty} \cdots \int_{c_j - c_1}^{\infty} f_j \, du_{j1} \cdots du_{j,j-1}\, du_{j,j+1} \cdots du_{jm}, \tag{7}$$

where f_j is the density of U_{ji} $(i = 1, 2, \cdots, m)$ $(i \neq j)$.

THEOREM 6.7.2. *If π_i is $N(\boldsymbol{\mu}^{(i)}, \boldsymbol{\Sigma})$ and the costs of misclassification are equal, the regions of classification, $R_1, \cdots, R_m$, that minimize the maximum conditional expected loss are defined by* (3), *where $u_{jk}(\boldsymbol{x})$ is given by* (1). *The constants c_j are determined so that the integrals* (7) *are equal.*

As an example consider the case of $m = 3$. There is no loss of generality in taking $p = 2$, for the density for higher p can be projected on the two-dimensional plane determined by the means of the three populations if they are not collinear (that is, we can transform the vector $\boldsymbol{x}$ into u_{12}, u_{13} and $p - 2$ other coordinates, where these last $p - 2$ components are distributed independently of u_{12} and u_{13} and with zero means). The regions R_j are determined by three half-lines as shown in Figure 3. If this procedure is minimax, we cannot move the line between R_1 and R_2 nearer $(\mu_1^{(1)}, \mu_2^{(1)})$, the line between R_2 and R_3 nearer $(\mu_1^{(2)}, \mu_2^{(2)})$, and the line between R_3 and R_1 nearer $(\mu_1^{(3)}, \mu_2^{(3)})$ and still retain the equality $P(1|1, R) = P(2|2, R) = P(3|3, R)$ without leaving a triangle that is not included in any region. Thus, since the regions must exhaust the space, the lines must meet

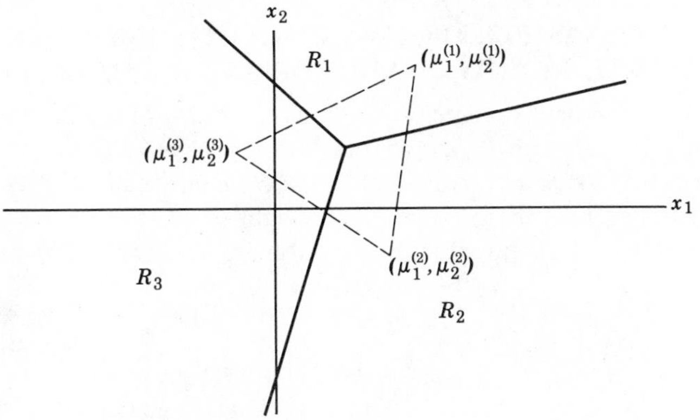

FIGURE 3

in a point, and the equality of probabilities determines $c_i - c_j$ uniquely.

To do this in a specific case in which we have numerical values for the components of the vectors $\boldsymbol{\mu}^{(1)}$, $\boldsymbol{\mu}^{(2)}$, $\boldsymbol{\mu}^{(3)}$, and the matrix $\boldsymbol{\Sigma}$, we would consider the three ($\leq p + 1$) joint distributions, each of $2U_{ij}$ ($j \neq i$). We could try the values of $c_i = 0$, and using tables [Pearson (1931)] of the bivariate normal distribution, compute $P(i|i, R)$. By a trial-and-error method we could obtain c_i to approximate the above condition.

The preceding theory has been given on the assumption that the parameters are known. If they are not known and if a sample from each population is available, the estimates of the parameters can be substituted in the definition of $u_{ij}(\boldsymbol{x})$. Let the observations be $\boldsymbol{x}_1^{(i)}, \cdots, \boldsymbol{x}_{N_i}^{(i)}$ from $N(\boldsymbol{\mu}^{(i)}, \boldsymbol{\Sigma})$, $i = 1, \cdots, m$. We estimate $\boldsymbol{\mu}^{(i)}$ by

$$\bar{\boldsymbol{x}}^{(i)} = \frac{1}{N_i} \sum_{\alpha=1}^{N_i} \boldsymbol{x}_\alpha^{(i)} \tag{8}$$

and $\boldsymbol{\Sigma}$ by S defined by

$$\left(\sum_{i=1}^{m} N_i - m\right) S = \sum_{i=1}^{m} \sum_{\alpha=1}^{N_i} (\boldsymbol{x}_\alpha^{(i)} - \bar{\boldsymbol{x}}^{(i)})(\boldsymbol{x}_\alpha^{(i)} - \bar{\boldsymbol{x}}^{(i)})'. \tag{9}$$

Then, the analogue of $u_{ij}(\boldsymbol{x})$ is

$$(10) \qquad v_{ij}(\boldsymbol{x}) = [\boldsymbol{x} - \tfrac{1}{2}(\bar{\boldsymbol{x}}^{(i)} + \bar{\boldsymbol{x}}^{(j)})]'S^{-1}(\bar{\boldsymbol{x}}^{(i)} - \bar{\boldsymbol{x}}^{(j)}).$$

If the variables above are random, the distributions are different from those of U_{ij}. However, as $N_i \to \infty$, the joint distributions approach those of U_{ij}. Hence, for sufficiently large samples one can use the theory given above.

6.8. AN EXAMPLE OF CLASSIFICATION INTO ONE OF SEVERAL MULTIVARIATE NORMAL POPULATIONS

Rao (1948a) considers three populations consisting of the Brahmin caste (π_1), the Artisan caste (π_2), and the Korwa caste (π_3) of India. The measurements for each individual of a caste are stature (x_1), sitting height (x_2), nasal depth (x_3), and nasal height (x_4). The means of these variables in the three populations are given in Table 2. The matrix of

TABLE 2

	Brahmin (π_1)	Artisan (π_2)	Korwa (π_3)
Stature (x_1)	164.51	160.53	158.17
Sitting height (x_2)	86.43	81.47	81.16
Nasal depth (x_3)	25.49	23.84	21.44
Nasal height (x_4)	51.24	48.62	46.72

correlations for all the populations is

$$(1) \qquad \begin{matrix} 1.0000 & 0.5849 & 0.1774 & 0.1974 \\ 0.5849 & 1.0000 & 0.2094 & 0.2170 \\ 0.1774 & 0.2094 & 1.0000 & 0.2910 \\ 0.1974 & 0.2170 & 0.2910 & 1.0000 \end{matrix}$$

The standard deviations are $\sigma_1 = 5.74$, $\sigma_2 = 3.20$, $\sigma_3 = 1.75$, $\sigma_4 = 3.50$. We assume that each population is normal. Our problem is to divide the space of the four variables x_1, x_2, x_3, x_4 into three regions of classification. We assume that the costs of misclassification are equal. We shall find (i) a set of regions under the assumption that drawing a new observation from each population is equally likely ($q_1 = q_2 = q_3 = \frac{1}{3}$), and (ii) a set of regions such that the largest probability of misclassification is minimized (the minimax solution).

We first compute the coefficients of $\boldsymbol{\Sigma}^{-1}(\boldsymbol{\mu}^{(1)} - \boldsymbol{\mu}^{(2)})$ and $\boldsymbol{\Sigma}^{-1}(\boldsymbol{\mu}^{(1)} - \boldsymbol{\mu}^{(3)})$. Then $\boldsymbol{\Sigma}^{-1}(\boldsymbol{\mu}^{(2)} - \boldsymbol{\mu}^{(3)}) = \boldsymbol{\Sigma}^{-1}(\boldsymbol{\mu}^{(1)} - \boldsymbol{\mu}^{(3)}) - \boldsymbol{\Sigma}^{-1}(\boldsymbol{\mu}^{(1)} - \boldsymbol{\mu}^{(2)})$. Then we calculate $\frac{1}{2}(\boldsymbol{\mu}^{(i)} + \boldsymbol{\mu}^{(j)})'\boldsymbol{\Sigma}^{-1}(\boldsymbol{\mu}^{(i)} - \boldsymbol{\mu}^{(j)})$. We obtain the discriminant functions*

$$
\begin{aligned}
u_{12}(\boldsymbol{x}) &= -0.0708x_1 + 0.4990x_2 + 0.3373x_3 + 0.0887x_4 - 43.13, \\
u_{13}(\boldsymbol{x}) &= 0.0003x_1 + 0.3550x_2 + 1.1063x_3 + 0.1375x_4 - 62.49, \qquad (2)\\
u_{23}(\boldsymbol{x}) &= 0.0711x_1 - 0.1440x_2 + 0.7690x_3 + 0.0488x_4 - 19.36.
\end{aligned}
$$

The other three functions are $u_{21}(\boldsymbol{x}) = -u_{12}(\boldsymbol{x})$, $u_{31}(\boldsymbol{x}) = -u_{13}(\boldsymbol{x})$, and $u_{32}(\boldsymbol{x}) = -u_{23}(\boldsymbol{x})$. If there are a priori probabilities and they are equal, the best set of regions of classification are $R_1\colon u_{12}(\boldsymbol{x}) \geq 0, u_{13}(\boldsymbol{x}) \geq 0$; $R_2\colon u_{21}(\boldsymbol{x}) \geq 0, u_{23}(\boldsymbol{x}) \geq 0$; and $R_3\colon u_{31}(\boldsymbol{x}) \geq 0, u_{32}(\boldsymbol{x}) \geq 0$. For example, if we obtain an individual with measurements $\boldsymbol{x}$ such that $u_{12}(\boldsymbol{x}) \geq 0$ and $u_{13}(\boldsymbol{x}) \geq 0$, we classify him as a Brahmin.

To find the probabilities of misclassification when an individual is drawn from population π_g we need the means, variances, and covariances of the proper pairs of u's. They are given in Table 3.†

TABLE 3

Population of $\boldsymbol{x}$	u	Means	Standard Deviation	Correlation
π_1	u_{12}	1.491	1.727	0.8658
	u_{13}	3.487	2.641	
π_2	u_{21}	1.491	1.727	−0.3894
	u_{23}	1.031	1.436	
π_3	u_{31}	3.487	2.641	0.7983
	u_{32}	1.031	1.436	

The probabilities of misclassification are then obtained by use of the tables for the bivariate normal distribution. These probabilities are 0.21 for π_1, 0.42 for π_2, and 0.25 for π_3. For example, if measurements are made on a Brahmin, the probability that he is classified as an Artisan or Korwa is 0.21.

The minimax solution is obtained by finding the constants c_1, c_2, and c_3 for (3) of Section 6.7 so that the probabilities of misclassification are equal. The regions of classification are

$$
\begin{aligned}
R_1'\colon\ & u_{12}(\boldsymbol{x}) \geq 0.54,\ u_{13}(\boldsymbol{x}) \geq 0.29; \\
R_2'\colon\ & u_{21}(\boldsymbol{x}) \geq -0.54,\ u_{23}(\boldsymbol{x}) \geq -0.25; \qquad (3)\\
R_3'\colon\ & u_{31}(\boldsymbol{x}) \geq -0.29,\ u_{32}(\boldsymbol{x}) \geq 0.25.
\end{aligned}
$$

* Due to an error in computations, Rao's discriminant functions are incorrect. I am indebted to Mr. Peter Frank for assistance in the computations.

† Some numerical errors in T. W. Anderson (1951*a*) are corrected in Table 3 and (3).

The common probability of misclassification (to two decimal places) is 0.30. Thus, the maximum probability of misclassification has been reduced from 0.42 to 0.30.

REFERENCES

Section 6.2. Berkson (1947); Blackwell and Girshick (1954); Burt (1944); Wald (1950).

Section 6.3. Blackwell and Girshick (1954); Hoel and Peterson (1949); Neyman and Pearson (1936); Wald (1950).

Section 6.4. Wald (1944).

Section 6.5. T. W. Anderson (1951a); Barnard (1935); Bartlett (1939a), (1939b); Beall (1945); Cramér (1946); Cochran and Bliss (1948); Fisher (1936), (1938), (1940); Horst and Smith (1950); Kendall (1946), 341–345; Kossack (1945); Sitgreaves (1952); Wald (1944).

Section 6.6. Blackwell and Girshick (1954); von Mises (1945); Wald (1950).

Section 6.7. von Mises (1945); K. Pearson (1931).

Section 6.8. Hamilton (1950); C. R. Rao (1948a).

Chapter 6. T. W. Anderson (1951a); Armitage (1950); Bartlett (1951a); Brown (1947), (1951); Lorge (1940); Lubin (1950); Mahalanobis (1936); Martin (1935); Nanda (1949); Narain (1949); Quenouille (1947); Penrose (1947); C. R. Rao (1946a), (1947), (1948a), (1948b), (1949b), (1951a), (1952), pp. 273–350, (1953), (1954a), (1954b); C. R. Rao and Slater (1949); S. N. Roy (1952a); C. A. B. Smith (1947); H. F. Smith (1936); Wallace and Travers (1938); Welch (1939); Wherry (1940).

PROBLEMS

1. (Sec. 6.5) Find the criterion (of Section 6.5.1) for classifying Iris as Iris setosa or Iris versicolor on the basis of data given in Section 5.3.3.

2. Let $u(\boldsymbol{x})$ be the classification criterion given by (2) in Section 6.5.1. Show that the T^2-criterion for testing $N(\boldsymbol{\mu}^{(1)}, \boldsymbol{\Sigma}) = N(\boldsymbol{\mu}^{(2)}, \boldsymbol{\Sigma})$ is proportional to $u(\bar{\boldsymbol{x}}^{(1)})$ and $u(\bar{\boldsymbol{x}}^{(2)})$.

3. (Sec. 6.3) Let π_i be $N(\boldsymbol{\mu}, \boldsymbol{\Sigma}_i)$, $i = 1, 2$. Find the form of the admissible classification procedures.

4. (Sec. 6.3) Let π_i be $N(\boldsymbol{\mu}^{(i)}, \boldsymbol{\Sigma}_i)$, $i = 1, 2$. Find the form of the admissible classification procedures.

5. (Sec. 6.7) Let π_i be $N(\boldsymbol{\mu}^{(i)}, \boldsymbol{\Sigma})$, $i = 1, \cdots, m$. If the $\boldsymbol{\mu}^{(i)}$ are on a line (that is, $\boldsymbol{\mu}^{(i)} = \boldsymbol{\mu} + \nu_i \boldsymbol{\beta}$), show that for admissible procedures the R_i are defined by parallel planes. Thus show that only one discriminant function $u_{jk}(\boldsymbol{x})$ need be used.

6. (Sec. 6.7) In Section 8.8 data are given on samples from four populations of skulls. Consider the first two measurements and the first three samples. Construct the classification functions $u_{ij}(\boldsymbol{x})$. Find the procedure for $q_i = N_i/(N_1 + N_2 + N_3)$. Find the minimax procedure.

7. (Sec. 6.5) Let $\boldsymbol{x}_\alpha^{*(i)} = \boldsymbol{B}\boldsymbol{x}_\alpha^{(i)} + \boldsymbol{c}(i = 1, 2;\ \alpha = 1, \cdots, N_i)$, where $\boldsymbol{B}$ is nonsingular, and $\boldsymbol{x}^* = \boldsymbol{B}\boldsymbol{x} + \boldsymbol{c}$. Show that the classification statistic (2) of Section 6.5 is invariant under such transformations.

8. (Sec. 6.4) Let $P(2|1)$ and $P(1|2)$ be defined by (14) and (15) of Section 6.4. Prove

that if α is increased, c can be modified so that at least one of $P(2|1)$ and $P(1|2)$ can be decreased while the other does not increase. *Hint:* Prove

$$\frac{\partial P(1|2)}{\partial \alpha}\frac{\partial P(2|1)}{\partial c} - \frac{\partial P(1|2)}{\partial c}\frac{\partial P(2|1)}{\partial \alpha} < 0.$$

9. (Sec. 6.4) Let $\boldsymbol{x}' = (\boldsymbol{x}^{(1)\prime}, \boldsymbol{x}^{(2)\prime})$. Using Problem 10 of Chapter 5 and Problem 8 of this chapter, prove that the class of classification procedures based on $\boldsymbol{x}$ is uniformly as good as the class of procedures based on $\boldsymbol{x}^{(1)}$.

10. (Sec. 6.3) Prove that every complete class of procedures includes the class of admissible procedures.

11. (Sec. 6.3) Prove that if the class of admissible procedures is complete it is minimal complete.

12. (Sec. 6.6) Prove that a (closed) curve can be drawn from $(q_1, q_2, 0)$ with $[\alpha_1^*, \alpha_2(\alpha_1^*), 1]$ to $(q_1, 0, q_3)$ with $[\alpha_1^*, 1, \alpha_3(\alpha_1^*)]$ so that $\alpha_1 = \alpha_1^*$ at every point on the curve.

13. (Sec. 6.5) Consider $\boldsymbol{d}'\boldsymbol{x}^{(i)}$. Prove that the ratio

$$\frac{[\boldsymbol{d}'\bar{\boldsymbol{x}}^{(1)} - \boldsymbol{d}'\bar{\boldsymbol{x}}^{(2)}]^2}{\sum_{\alpha=1}^{N_1} (\boldsymbol{d}'\boldsymbol{x}_\alpha^{(1)} - \boldsymbol{d}'\bar{\boldsymbol{x}}^{(1)})^2 + \sum_{\alpha=1}^{N_2} (\boldsymbol{d}'\boldsymbol{x}_\alpha^{(2)} - \boldsymbol{d}'\bar{\boldsymbol{x}}^{(2)})^2}$$

is maximized by $\boldsymbol{d} = \boldsymbol{S}^{-1}(\bar{\boldsymbol{x}}^{(1)} - \bar{\boldsymbol{x}}^{(2)})$.

14. (Sec. 6.3) The Neyman-Pearson fundamental lemma states that of all tests at a given significance level of the null hypothesis that $\boldsymbol{x}$ is drawn from $p_1(\boldsymbol{x})$ against alternatives that it is drawn from $p_2(\boldsymbol{x})$ the most powerful test has the critical region $p_1(\boldsymbol{x})/p_2(\boldsymbol{x}) < k$. Show that the discussion in Section 6.3 proves this result.

15. (Sec. 6.3) When $p(\boldsymbol{x}) = n(\boldsymbol{x}|\boldsymbol{\mu}, \boldsymbol{\Sigma})$ find the best test of $\boldsymbol{\mu} = \boldsymbol{0}$ against $\boldsymbol{\mu} = \boldsymbol{\mu}^*$ at significance level ε. Show that this test is uniformly most powerful against all alternatives $\boldsymbol{\mu} = c\boldsymbol{\mu}^*$, $c > 0$. Prove that there is no uniformly most powerful test against $\boldsymbol{\mu} = \boldsymbol{\mu}^{(1)}$ and $\boldsymbol{\mu} = \boldsymbol{\mu}^{(2)}$ unless $\boldsymbol{\mu}^{(1)} = c\boldsymbol{\mu}^{(2)}$ for some $c > 0$.

16. (Sec. 6.6) Prove that $q_1(k)$ is continuous, as asserted in the footnote to the proof of Theorem 6.6.4.

CHAPTER 7

The Distribution of the Sample Covariance Matrix and the Sample Generalized Variance

7.1. INTRODUCTION

The sample covariance matrix is $S = [1/(N-1)]\sum_\alpha(x_\alpha - \bar{x})(x_\alpha - \bar{x})'$ and is an estimate of the population covariance matrix $\mathbf{\Sigma}$. In Section 4.2 we found the density of $A = (N-1)S$ in the case of a 2×2 matrix. In Section 7.2 this result will be generalized to the case of a matrix A of any order. When $\mathbf{\Sigma} = I$, this distribution is in a sense a generalization of the χ^2-distribution. The distribution of A (or S), often called the Wishart distribution, is fundamental to multivariate statistical analysis. In Sections 7.3 and 7.4 we discuss some properties of the Wishart distribution.

The generalized variance of the sample is defined as $|S|$ in Section 7.5; it is a kind of measure of the scatter of the sample. Its distribution is characterized.

7.2. THE WISHART DISTRIBUTION

In this section we shall obtain the distribution of $A = \sum_{\alpha=1}^N (X_\alpha - \bar{X})(X_\alpha - \bar{X})'$, where the X_α are independent, each with the distribution $N(\boldsymbol{\mu}, \mathbf{\Sigma})$. As was shown in Section 3.3, A is distributed as $A = \sum_{\alpha=1}^n Z_\alpha Z_\alpha'$, where $n = N - 1$ and the Z_α are independent, each with the distribution $N(\mathbf{0}, \mathbf{\Sigma})$. We shall show that the density of A for A positive definite is

$$(1) \qquad \frac{|A|^{\frac{1}{2}(n-p-1)} \exp\left(-\frac{1}{2} \operatorname{tr} \mathbf{\Sigma}^{-1} A\right)}{2^{\frac{1}{2}np} \pi^{p(p-1)/4} |\mathbf{\Sigma}|^{\frac{1}{2}n} \prod_{i=1}^{p} \Gamma[\frac{1}{2}(n+1-i)]}.$$

We shall first derive (1) for $\mathbf{\Sigma} = I$. We use here repeatedly the following special case of Theorem 4.3.2. If the scalars U_α are independent and U_α has distribution $N(\mathbf{\Gamma} w_\alpha, \phi)$, then $\sum_\alpha U_\alpha^2 - \sum_\alpha U_\alpha w_\alpha'(\sum_\alpha w_\alpha w_\alpha')^{-1}\sum_\alpha U_\alpha w_\alpha$ is distributed as $\sum_{\alpha=1}^{n-q} V_\alpha^2$, where q is the number of components of w_α

and the V_α are independent, each with the distribution $N(0, \phi)$, and independently of $\sum U_\alpha \boldsymbol{w}'_\alpha(\sum \boldsymbol{w}_\alpha \boldsymbol{w}'_\alpha)^{-1}$. In particular, if $\phi = 1$, then $\sum U_\alpha^2 - \sum U_\alpha \boldsymbol{w}'_\alpha(\sum \boldsymbol{w}_\alpha \boldsymbol{w}'_\alpha)^{-1}\sum U_\alpha \boldsymbol{w}_\alpha$ has the χ^2-distribution with $n - q$ degrees of freedom, the density of which is

$$\text{(2)} \qquad \frac{1}{2^{\frac{1}{2}(n-q)}\Gamma[\frac{1}{2}(n-q)]} t^{\frac{1}{2}(n-q)-1}e^{-\frac{1}{2}t}.$$

We have further that $\sum U_\alpha \boldsymbol{w}_\alpha$ is normally distributed; if $\boldsymbol{\Gamma} = \boldsymbol{0}$, then $\mathscr{E}\sum U_\alpha \boldsymbol{w}_\alpha = \boldsymbol{0}$ and the covariance matrix is

$$\text{(3)} \qquad \begin{aligned} \mathscr{E}\sum U_\alpha \boldsymbol{w}_\alpha \sum U_\beta \boldsymbol{w}'_\beta &= \sum_{\alpha,\beta} \boldsymbol{w}_\alpha \boldsymbol{w}'_\beta \delta_{\alpha\beta} \\ &= \sum_\alpha \boldsymbol{w}_\alpha \boldsymbol{w}'_\alpha. \end{aligned}$$

Let $\boldsymbol{a}_{(i)} = (a_{i,i+1}, a_{i,i+2}, \cdots, a_{ip})$ and $\boldsymbol{A}_{ii} = (a_{jk}), j, k = i, \cdots, p$. Then

$$\text{(4)} \qquad \boldsymbol{A}_{ii} = \begin{pmatrix} a_{ii} & \boldsymbol{a}_{(i)} \\ \boldsymbol{a}'_{(i)} & \boldsymbol{A}_{i+1,i+1} \end{pmatrix}.$$

Let $a_{ii\cdot i+1,\cdots,p} = a_{ii} - \boldsymbol{a}_{(i)}\boldsymbol{A}^{-1}_{i+1,i+1}\boldsymbol{a}'_{(i)}$. The set $(Z_{i1}, \cdots, Z_{in})$ is distributed independently of $(Z_{j1}, \cdots, Z_{jn}), j \neq i$ (because $\boldsymbol{\Sigma} = \boldsymbol{I}$), and therefore, conditional on $Z_{j\alpha} = z_{j\alpha}$ ($j \neq i$; $\alpha = 1, \cdots, n$), the elements of $(Z_{i1}, \cdots, Z_{in})$ are distributed independently, each according to $N(0, 1)$, which is of the form $N(\boldsymbol{\Gamma}\boldsymbol{w}_\alpha, \phi)$ with $\boldsymbol{\Gamma} = \boldsymbol{0}$ and $\phi = 1$. Let $\boldsymbol{Z}_\alpha^{(j)\prime} = (Z_{j\alpha}, Z_{j+1,\alpha}, \cdots, Z_{p\alpha})$. Then $\boldsymbol{a}_{(i)} = \sum_\alpha Z_{i\alpha}\boldsymbol{Z}_\alpha^{(i+1)\prime}$ and $\boldsymbol{A}_{ii} = \sum_\alpha \boldsymbol{Z}_\alpha^{(i)}\boldsymbol{Z}_\alpha^{(i)\prime}$. From the special case of Theorem 4.3.2 we find that conditional on $\boldsymbol{Z}_\alpha^{(i+1)} = \boldsymbol{z}_\alpha^{(i+1)}$, $a_{ii\cdot i+1,\cdots,p}$ has a χ^2-distribution with $n - (p - i)$ degrees of freedom and independent of $\boldsymbol{a}'_{(i)}$ which is conditionally distributed according to $N(\boldsymbol{0}, \boldsymbol{A}_{i+1,i+1})$.

It will be observed that the conditional distribution depends on $\boldsymbol{z}_\alpha^{(i+1)}$ only through $\boldsymbol{A}_{i+1,i+1}$; that is, the density is of the form $f_i(a_{ii\cdot i+1,\cdots,p}, \boldsymbol{a}_{(i)}|\boldsymbol{A}_{i+1,i+1})$. Thus the joint density of $a_{11\cdot 2,\cdots,p}$, $\boldsymbol{a}_{(1)}$, $a_{22\cdot 3,\cdots,p}$, $\boldsymbol{a}_{(2)}, \cdots, a_{p-1,p-1\cdot p}, \boldsymbol{a}_{(p-1)}, a_{pp}$ is

$$\text{(5)} \quad \begin{aligned} &f_1(a_{11\cdot 2,\cdots,p}, \boldsymbol{a}_{(1)}|\boldsymbol{A}_{22}) \cdots f_{p-1}(a_{p-1,p-1\cdot p}, \boldsymbol{a}_{(p-1)}|a_{pp})f_p(a_{pp}) \\ &= \frac{a_{pp}^{\frac{1}{2}n-1}e^{-\frac{1}{2}a_{pp}}}{2^{\frac{1}{2}n}\Gamma(\frac{1}{2}n)} \prod_{i=1}^{p-1}\left\{\frac{a_{ii\cdot i+1,\cdots,p}^{\frac{1}{2}[n-(p-i)]-1}e^{-\frac{1}{2}a_{ii\cdot i+1,\cdots,p}}}{2^{\frac{1}{2}[n-(p-i)]}\Gamma\{\frac{1}{2}[n-(p-i)]\}} \frac{e^{-\frac{1}{2}\boldsymbol{a}_{(i)}\boldsymbol{A}^{-1}_{i+1,i+1}\boldsymbol{a}'_{(i)}}}{(2\pi)^{\frac{1}{2}(p-i)}|\boldsymbol{A}_{i+1,i+1}|^{\frac{1}{2}}}\right\}. \end{aligned}$$

We find the density of $a_{ii}, \boldsymbol{a}_{(1)}, \cdots, a_{pp}$ by substituting in (5) $a_{ii\cdot i+1,\cdots,p}$

$= a_{ii} - \boldsymbol{a}_{(i)}\boldsymbol{A}_{i+1,\,i+1}^{-1}\boldsymbol{a}'_{(i)}$, multiplying by the Jacobian, which is 1, and simplifying the expression. The exponent of e in (5) is

$$(6)\quad -\tfrac{1}{2}\left[a_{pp} + \sum_{i=1}^{p-1} a_{ii\cdot i+1,\cdots,p} + \sum_{i=1}^{p-1} \boldsymbol{a}_{(i)}\boldsymbol{A}_{i+1,\,i+1}^{-1}\boldsymbol{a}'_{(i)}\right]$$

$$= -\tfrac{1}{2}\left[a_{pp} + \sum_{i=1}^{p-1}(a_{ii} - \boldsymbol{a}_{(i)}\boldsymbol{A}_{i+1,\,i+1}^{-1}\boldsymbol{a}'_{(i)}) + \sum_{i=1}^{p-1}\boldsymbol{a}_{(i)}\boldsymbol{A}_{i+1,\,i+1}^{-1}\boldsymbol{a}'_{(i)}\right]$$

$$= -\tfrac{1}{2}\sum_{i=1}^{p} a_{ii} = -\tfrac{1}{2}\operatorname{tr}\boldsymbol{A}.$$

Using the fact (Theorem 5 of Appendix 1) that

$$(7)\qquad a_{ii\cdot i+1\cdots p} = \frac{\begin{vmatrix} a_{ii} & \boldsymbol{a}_{(i)} \\ \boldsymbol{a}'_{(i)} & \boldsymbol{A}_{i+1,\,i+1}\end{vmatrix}}{|\boldsymbol{A}_{i+1,\,i+1}|} = \frac{|\boldsymbol{A}_{ii}|}{|\boldsymbol{A}_{i+1,\,i+1}|}$$

we find ($a_{pp} = \boldsymbol{A}_{pp}$)

$$(8)\qquad a_{pp}\prod_{i=1}^{p-1} a_{ii\cdot i+1,\cdots,p} = a_{pp}\prod_{i=1}^{p-1}\frac{|\boldsymbol{A}_{ii}|}{|\boldsymbol{A}_{i+1,\,i+1}|} = |\boldsymbol{A}_{11}| = |\boldsymbol{A}|.$$

Then

$$(9)\qquad a_{pp}^{\frac{1}{2}n-1}\prod_{i=1}^{p-1}\frac{a_{ii\cdot i+1,\cdots,p}^{\frac{1}{2}[n-(p-i)]-1}}{|\boldsymbol{A}_{i+1,\,i+1}|^{\frac{1}{2}}} = |\boldsymbol{A}|^{\frac{1}{2}(n-p-1)}a_{pp}^{\frac{1}{2}(p-1)}\prod_{i=1}^{p-1}\frac{|\boldsymbol{A}_{ii}|^{\frac{1}{2}(i-1)}}{|\boldsymbol{A}_{i+1,\,i+1}|^{\frac{1}{2}i}}$$
$$= |\boldsymbol{A}|^{\frac{1}{2}(n-p-1)}.$$

The power of π in the denominator of (5) is $\frac{1}{2}[(p-1)+(p-2)+\cdots+1] = \frac{1}{2}[(p-1)p/2]$. Since $\Gamma(\frac{1}{2}n)\prod_{i=1}^{p-1}\Gamma\{\frac{1}{2}[n-(p-i)]\} = \prod_{i=1}^{p}\Gamma[\frac{1}{2}(n-i+1)]$, we find that the density of $a_{11}, a_{12}, \cdots, a_{pp}$ is

$$(10)\qquad \frac{|\boldsymbol{A}|^{\frac{1}{2}(n-p-1)}e^{-\frac{1}{2}\operatorname{tr}\boldsymbol{A}}}{2^{\frac{1}{2}np}\pi^{p(p-1)/4}\prod_{i=1}^{p}\Gamma[\frac{1}{2}(n+1-i)]}.$$

This is (1) with $\boldsymbol{\Sigma} = \boldsymbol{I}$.

Now let us derive (1) for arbitrary $\boldsymbol{\Sigma}$. Let $\boldsymbol{A} = \sum_{\alpha=1}^{n}\boldsymbol{Z}_\alpha\boldsymbol{Z}'_\alpha$, where the $\boldsymbol{Z}_\alpha$ are independent, each with the distribution $N(\boldsymbol{0}, \boldsymbol{\Sigma})$, and let $\boldsymbol{A}^* = \sum_{\alpha=1}^{n}\boldsymbol{Z}_\alpha^*\boldsymbol{Z}_\alpha^{*\prime}$, where the $\boldsymbol{Z}_\alpha^*$ are independent, each with distribution $N(\boldsymbol{0}, \boldsymbol{I})$. Then $\boldsymbol{A}^*$ has the density (10). Let $\boldsymbol{C}$ be a triangular matrix ($c_{ij} = 0$, $i > j$) such that $\boldsymbol{C\Sigma C}' = \boldsymbol{I}$ (Theorem 4 of Appendix 1). The distribution of $\boldsymbol{CAC}' = \sum_{\alpha=1}^{n}(\boldsymbol{CZ}_\alpha)(\boldsymbol{CZ}_\alpha)'$ is the same as that of $\boldsymbol{A}^*$ since the distribution of $\boldsymbol{CZ}_\alpha$ is that of $\boldsymbol{Z}_\alpha^*$. Therefore we obtain the density of $\boldsymbol{A}$ by substituting $\boldsymbol{A}^* = \boldsymbol{CAC}'$ in (10) and multiplying by the Jacobian.

LEMMA 7.2.1. *Let the symmetric matrix* $\boldsymbol{A}^*$ *be transformed into the symmetric matrix* $\boldsymbol{A}$ *by* $\boldsymbol{A}^* = \boldsymbol{CAC}'$, *where* $\boldsymbol{C}$ *is a nonsingular triangular matrix* ($c_{ij} = 0$, $i > j$). *The Jacobian of the transformation is* mod $|\boldsymbol{C}|^{p+1}$.

Proof. The transformation is

$$a_{ij}^* = \sum_{k,l} c_{ik} a_{kl} c_{jl}. \tag{11}$$

The partial derivatives are

$$\frac{\partial a_{ij}^*}{\partial a_{kk}} = c_{ik} c_{jk}, \tag{12}$$

$$\frac{\partial a_{ij}^*}{\partial a_{kl}} = c_{ik} c_{jl} + c_{il} c_{jk}, \qquad l \neq k. \tag{13}$$

We write down the matrix of partial derivatives in the order $a_{11}, a_{12}, \cdots, a_{1p}, a_{22}, \cdots, a_{pp}$; the row position corresponds to a_{ij}^* and the column position to a_{ij}. The matrix is

$$\begin{bmatrix} c_{11}^2 & 2c_{11}c_{12} \cdots & 2c_{11}c_{1p} & c_{12}^2 & \cdots & c_{1p}^2 \\ 0 & c_{11}c_{22} \cdots & c_{11}c_{2p} & c_{12}c_{22} & \cdots & c_{1p}c_{2p} \\ \cdot & \cdot & \cdot & \cdot & & \cdot \\ \cdot & \cdot & \cdot & \cdot & & \cdot \\ \cdot & \cdot & \cdot & \cdot & & \cdot \\ 0 & 0 \cdots & c_{11}c_{pp} & 0 & \cdots & c_{1p}c_{pp} \\ 0 & 0 \cdots & 0 & c_{22}^2 & \cdots & c_{2p}^2 \\ \cdot & \cdot & \cdot & \cdot & & \cdot \\ \cdot & \cdot & \cdot & \cdot & & \cdot \\ \cdot & \cdot & \cdot & \cdot & & \cdot \\ 0 & 0 \cdots & 0 & 0 & \cdots & c_{pp}^2 \end{bmatrix}, \tag{14}$$

which is a triangular matrix. The determinant is the product of the diagonal elements, which is $\prod_{j=1}^p c_{jj}^{p+1} = |C|^{p+1}$. This proves the lemma.

We now replace A in (10) by CAC'. From $C\Sigma C' = I$ we have $1 = |C\Sigma C'| = |C||\Sigma||C'| = |\Sigma| \cdot |CC'|$; thus $|CC'| = 1/|\Sigma|$ and mod $|C| = 1/\sqrt{|\Sigma|}$. Furthermore, $\Sigma = C^{-1}(C')^{-1} = (C'C)^{-1}$. Thus tr $CAC' =$ tr $AC'C =$ tr $A\Sigma^{-1}$, and $|CAC'| = |C||A||C'| = |A||C'C| = |A|/|\Sigma|$. These substitutions into (10) and multiplication by the Jacobian give (1).

THEOREM 7.2.1. *Suppose the p-component vectors* $Z_1, \cdots, Z_n$ $(n \geq p)$ *are independent, each distributed according to* $N(\mathbf{0}, \Sigma)$. *Then the density of* $A = \sum_{\alpha=1}^n Z_\alpha Z_\alpha'$ *is*

$$\frac{|A|^{\frac{1}{2}(n-p-1)} e^{-\frac{1}{2}\mathrm{tr} A\Sigma^{-1}}}{2^{\frac{1}{2}np} \pi^{p(p-1)/4} |\Sigma|^{\frac{1}{2}n} \prod_{i=1}^p \Gamma[\frac{1}{2}(n+1-i)]} \tag{15}$$

for A positive definite and 0 *otherwise.*

COROLLARY 7.2.1. *Suppose the p-component vectors* $X_1, \cdots, X_N$ $(N > p)$ *are independent, each with the distribution* $N(\mu, \Sigma)$. *Then the density of* $A = \sum_{\alpha=1}^N (X_\alpha - \bar{X})(X_\alpha - \bar{X})'$ *is* (15) *for* $n = N - 1$.

The density (15) will be denoted by $w(A|\mathbf{\Sigma}, n)$ and the associated distribution will be termed $W(\mathbf{\Sigma}, n)$.

The first derivation of this distribution [Wishart (1928)] was by a geometric argument, which is closely related to the proof given here. Let $(z_{i1}, \cdots, z_{in}) = \mathbf{v}_i'$ be a vector in n-space. The diagonal elements of A are the squared lengths of these vectors, $a_{ii} = \mathbf{v}_i'\mathbf{v}_i$, and the off-diagonal elements are related to the lengths and the angles between the vectors since $r_{ij} = a_{ij}/\sqrt{a_{ii}a_{jj}}$ is the cosine of the angle between $\mathbf{v}_i$ and $\mathbf{v}_j$. The matrix A describes the lengths and configuration of the vectors.

The probability element of $\sqrt{\mathbf{v}_i'\mathbf{v}_i}, \mathbf{v}_i'\mathbf{v}_{i+1}, \cdots, \mathbf{v}_i'\mathbf{v}_p$ given $\mathbf{v}_{i+1}, \cdots, \mathbf{v}_p$ is approximately the probability that $\mathbf{v}_i$ lies in the region for which $\sqrt{a_{ii}} < \sqrt{\mathbf{v}_i'\mathbf{v}_i} < \sqrt{a_{ii}} + d\sqrt{a_{ii}}, a_{i,\,i+1} < \mathbf{v}_i'\mathbf{v}_{i+1} < a_{i,\,i+1} + da_{i,\,i+1}, \cdots, a_{ip} < \mathbf{v}_i'\mathbf{v}_p < a_{ip} + da_{ip}$. The first pair of inequalities defines a spherical shell of inner radius $\sqrt{a_{ii}}$; each of the other pairs of inequalities defines the region between two hyperplanes. In this region the density $(2\pi)^{-\frac{1}{2}n} \exp(-\frac{1}{2}\mathbf{v}_i'\mathbf{v}_i)$ is approximately constant. The intersection of the regions is a spherical shell in $n - (p - i)$ dimensions with a cross section of $p - i$ dimensions. The volume of the cross section is approximately $d\sqrt{a_{ii}}\, da_{i,\,i+1} \cdots da_{ip}/|\mathbf{A}_{i+1,\,i+1}|^{\frac{1}{2}}$. The radius squared of the spherical shell is $a_{ii\cdot i+1,\cdots,p} = a_{ii} - \mathbf{a}_{(i)}\mathbf{A}_{i+1,\,i+1}^{-1}\mathbf{a}_{(i)}'$ (see Section 4.4.1). The surface area (or volume) is the $[n - (p - i) - 1]$th power of the radius times the surface area of a sphere [in $n - (p - i)$ dimensions] of unit radius. The surface area of the sphere of unit radius is $C[n - (p - i)] = 2\pi^{\frac{1}{2}[n-(p-i)]}/\Gamma\{\frac{1}{2}[n - (p - i)]\}$. Thus the probability element of $\sqrt{\mathbf{v}_i'\mathbf{v}_i}, \mathbf{v}_i'\mathbf{v}_{i+1}, \cdots, \mathbf{v}_i'\mathbf{v}_p$ is

$$(16) \qquad \frac{e^{-\frac{1}{2}\mathbf{v}_i'\mathbf{v}_i}}{(2\pi)^{\frac{1}{2}n}} \frac{2\pi^{\frac{1}{2}[n-(p-i)]}}{\Gamma\{\frac{1}{2}[n - (p - i)]\}} a_{ii\cdot i+1,\cdots,p}^{\frac{1}{2}[n-(p-i)-1]} \frac{d\sqrt{a_{ii}}\, da_{i,\,i+1} \cdots da_{ip}}{|\mathbf{A}_{i+1,\,i+1}|^{\frac{1}{2}}}.$$

The probability element of $\mathbf{v}_i'\mathbf{v}_i, \mathbf{v}_i'\mathbf{v}_{i+1}, \cdots, \mathbf{v}_i'\mathbf{v}_p$ involves the substitution of $d\sqrt{a_{ii}} = da_{ii}/(2\sqrt{a_{ii}})$. This leads to the ith term in the product in (5).

The analysis that exactly parallels the geometric derivation by Wishart [and later by Mahalanobis, Bose, and Roy (1937)] has been given by Sverdrup (1947) [and by Fog (1948) for $p = 3$]. Another method has been used by Madow (1938), who draws on the distribution of correlation coefficients (for $\mathbf{\Sigma} = \mathbf{I}$) obtained by Hotelling by considering certain partial correlation coefficients. Hsu (1939b) has given an inductive proof, and Rasch (1948) has given a method involving the use of a functional equation. A different method is to obtain the characteristic function and invert it, as done by Ingham (1933) and Wishart and Bartlett (1933).

Cramér (1946) verifies that the Wishart distribution has the characteristic function of A.

To close this section we give the joint distribution of the sample variances and covariances. We have shown that $N\hat{\boldsymbol{\Sigma}}$ [for a sample of N from $N(\boldsymbol{\mu}, \boldsymbol{\Sigma})$] is distributed as $\sum_{\alpha=1}^{n} \mathbf{Z}_\alpha \mathbf{Z}_\alpha'$ where the $\mathbf{Z}_\alpha$ are independent, each with distribution $N(\mathbf{0}, \boldsymbol{\Sigma})$ and $n = N - 1$, $\mathscr{E} N\hat{\boldsymbol{\Sigma}} = n\boldsymbol{\Sigma}$, and $\mathscr{E}\boldsymbol{S} = \boldsymbol{\Sigma}$, where $\boldsymbol{S} = (N/n)\hat{\boldsymbol{\Sigma}}$.

THEOREM 7.2.2. *Suppose* $X_1, \cdots, X_N$ $(N \geq p + 1)$ *are distributed independently, each according to* $N(\boldsymbol{\mu}, \boldsymbol{\Sigma})$. *Then the distribution of* $\boldsymbol{S} = (1/n)\sum_\alpha(\boldsymbol{X}_\alpha - \bar{\boldsymbol{X}})(\boldsymbol{X}_\alpha - \bar{\boldsymbol{X}})'$ *is* $W[(1/n)\boldsymbol{\Sigma}, n]$ *where* $n = N - 1$ *and* $W[(1/n)\boldsymbol{\Sigma}, n]$ *is the Wishart distribution with covariance matrix* $(1/n)\boldsymbol{\Sigma}$ *and degrees of freedom* n.

Proof. Clearly, $\boldsymbol{S} = (1/n)A = \sum_{\alpha=1}^{n}[(1/\sqrt{n})\mathbf{Z}_\alpha][(1/\sqrt{n})\mathbf{Z}_\alpha]'$, where the $(1/\sqrt{n})\mathbf{Z}_\alpha$ are independent, each with the distribution $N[\mathbf{0}, (1/n)\boldsymbol{\Sigma}]$. Then Theorem 7.2.2 follows directly from Theorem 7.2.1.

If $n < p$, the matrix $A = \sum_{\alpha=1}^{n} \mathbf{Z}_\alpha \mathbf{Z}_\alpha'$ does not have a density. We shall, however, refer to the distribution as the Wishart distribution.

7.3. SOME PROPERTIES OF THE WISHART DISTRIBUTION

7.3.1. The Characteristic Function

The characteristic function of the Wishart distribution can be obtained most easily from the distribution of the observations. Suppose $\mathbf{Z}_1, \cdots, \mathbf{Z}_n$ are distributed independently, each with density

$$\frac{1}{(2\pi)^{\frac{1}{2}p}|\boldsymbol{\Sigma}|^{\frac{1}{2}}} \exp\left(-\tfrac{1}{2}\mathbf{z}'\boldsymbol{\Sigma}^{-1}\mathbf{z}\right). \tag{1}$$

Let

$$A = \sum_{\alpha=1}^{n} \mathbf{Z}_\alpha \mathbf{Z}_\alpha'. \tag{2}$$

Introduce the $p \times p$ matrix $\boldsymbol{\Theta} = (\theta_{ij})$ with $\theta_{ij} = \theta_{ji}$. The characteristic function of $A_{11}, A_{22}, \cdots, A_{pp}, 2A_{12}, 2A_{13}, \cdots, 2A_{p-1,p}$ is

$$\begin{aligned} \mathscr{E} \exp[i \operatorname{tr}(A\boldsymbol{\Theta})] &= \mathscr{E} \exp\left(i \operatorname{tr} \sum_{\alpha=1}^{n} \mathbf{Z}_\alpha \mathbf{Z}_\alpha' \boldsymbol{\Theta}\right) \\ &= \mathscr{E} \exp\left(i \operatorname{tr} \sum_{\alpha=1}^{n} \mathbf{Z}_\alpha' \boldsymbol{\Theta} \mathbf{Z}_\alpha\right) \\ &= \mathscr{E} \exp\left(i \sum_{\alpha=1}^{n} \mathbf{Z}_\alpha' \boldsymbol{\Theta} \mathbf{Z}_\alpha\right) \end{aligned} \tag{3}$$

by virtue of the fact that tr $\boldsymbol{EFG} = \sum e_{ij} f_{jk} g_{ki} = $ tr $\boldsymbol{FGE}$. It follows from Lemma 2.6.1 that

$$\mathscr{E} \exp \left(i \sum_{\alpha=1}^{n} \boldsymbol{Z}_\alpha' \boldsymbol{\Theta} \boldsymbol{Z}_\alpha\right) = \prod_{\alpha=1}^{n} \mathscr{E} \exp (i \boldsymbol{Z}_\alpha' \boldsymbol{\Theta} \boldsymbol{Z}_\alpha) = [\mathscr{E} \exp (i \boldsymbol{Z}' \boldsymbol{\Theta} \boldsymbol{Z})]^n, \tag{4}$$

where $\boldsymbol{Z}$ has the density (1). For $\boldsymbol{\Theta}$ real, there is a real nonsingular matrix $\boldsymbol{B}$ such that

$$\boldsymbol{B}' \boldsymbol{\Sigma}^{-1} \boldsymbol{B} = \boldsymbol{I}, \tag{5}$$

$$\boldsymbol{B}' \boldsymbol{\Theta} \boldsymbol{B} = \boldsymbol{D}, \tag{6}$$

where $\boldsymbol{D}$ is a real diagonal matrix (Theorem 3 of Appendix 1). If we let

$$\boldsymbol{z} = \boldsymbol{B} \boldsymbol{y}, \tag{7}$$

then

$$\begin{aligned} \mathscr{E} \exp (i \boldsymbol{Z}' \boldsymbol{\Theta} \boldsymbol{Z}) &= \mathscr{E} \exp (i \boldsymbol{Y}' \boldsymbol{D} \boldsymbol{Y}) \\ &= \mathscr{E} \prod_{j=1}^{p} \exp (i d_{jj} Y_j^2) \\ &= \prod_{j=1}^{p} \mathscr{E} \exp (i d_{jj} Y_j^2) \end{aligned} \tag{8}$$

by Lemma 2.6.2. The jth term in the product is $\mathscr{E} \exp (i d_{jj} Y_j^2)$ where Y_j has the distribution $N(0, 1)$; this is the characteristic function of the χ^2-distribution with one degree of freedom, namely $(1 - 2id_{jj})^{-\frac{1}{2}}$ (and can be proved easily by expanding $\exp (i d_{jj} y_j^2)$ in a power series and integrating term by term). Thus

$$\mathscr{E} \exp (i \boldsymbol{Z}' \boldsymbol{\Theta} \boldsymbol{Z}) = \prod_{j=1}^{p} (1 - 2id_{jj})^{-\frac{1}{2}} = |\boldsymbol{I} - 2i\boldsymbol{D}|^{-\frac{1}{2}} \tag{9}$$

since $\boldsymbol{I} - 2i\boldsymbol{D}$ is a diagonal matrix. From (5) and (6) we see that

$$\begin{aligned} |\boldsymbol{I} - 2i\boldsymbol{D}| &= |\boldsymbol{B}' \boldsymbol{\Sigma}^{-1} \boldsymbol{B} - 2i \boldsymbol{B}' \boldsymbol{\Theta} \boldsymbol{B}| \\ &= |\boldsymbol{B}' (\boldsymbol{\Sigma}^{-1} - 2i\boldsymbol{\Theta}) \boldsymbol{B}| \\ &= |\boldsymbol{B}'| \cdot |\boldsymbol{\Sigma}^{-1} - 2i\boldsymbol{\Theta}| \cdot |\boldsymbol{B}| \\ &= |\boldsymbol{B}|^2 \cdot |\boldsymbol{\Sigma}^{-1} - 2i\boldsymbol{\Theta}|, \end{aligned} \tag{10}$$

$|\boldsymbol{B}'| \cdot |\boldsymbol{\Sigma}^{-1}| |\boldsymbol{B}| = |\boldsymbol{I}| = 1$, and $|\boldsymbol{B}|^2 = 1/|\boldsymbol{\Sigma}^{-1}|$. Combining the above results, we obtain

$$\mathscr{E} \exp [i \operatorname{tr} (\boldsymbol{A} \boldsymbol{\Theta})] = \frac{|\boldsymbol{\Sigma}^{-1}|^{\frac{1}{2}n}}{|\boldsymbol{\Sigma}^{-1} - 2i\boldsymbol{\Theta}|^{\frac{1}{2}n}}. \tag{11}$$

It can be shown that the result is valid provided $(\mathcal{R}(\sigma^{jk} - 2i\theta_{jk}))$ is positive definite. In particular, it is true for all real $\mathbf{\Theta}$.

THEOREM 7.3.1. *If $\mathbf{Z}_1, \cdots, \mathbf{Z}_n$ are independent, each with distribution $N(\mathbf{0}, \mathbf{\Sigma})$, the characteristic function of $A_{11}, \cdots, A_{pp}, 2A_{12}, \cdots, 2A_{p-1,p}$, where $(A_{ij}) = A = \sum_{\alpha=1}^n \mathbf{Z}_\alpha \mathbf{Z}'_\alpha$, is given by* (11).

We can obtain the moments of the elements of A either from their characteristic function or from the original normal distribution. The expected value of A_{ij} is

$$\mathcal{E}A_{ij} = \mathcal{E}\sum_{\alpha=1}^n Z_{i\alpha}Z_{j\alpha} = \sum_{\alpha=1}^n \sigma_{ij} = n\sigma_{ij}. \tag{12}$$

To obtain the covariances we need

$$\begin{aligned}\mathcal{E}A_{ij}A_{kl} &= \mathcal{E}\sum_{\alpha,\beta=1}^n Z_{i\alpha}Z_{j\alpha}Z_{k\beta}Z_{l\beta} \\ &= \mathcal{E}\sum_{\alpha=1}^n Z_{i\alpha}Z_{j\alpha}Z_{k\alpha}Z_{l\alpha} + \mathcal{E}\sum_{\substack{\alpha,\beta=1\\ \alpha\neq\beta}}^n Z_{i\alpha}Z_{j\alpha}Z_{k\beta}Z_{l\beta} \\ &= n(\sigma_{ij}\sigma_{kl} + \sigma_{ik}\sigma_{jl} + \sigma_{il}\sigma_{jk}) + n(n-1)\sigma_{ij}\sigma_{kl} \\ &= n^2\sigma_{ij}\sigma_{kl} + n\sigma_{ik}\sigma_{jl} + n\sigma_{il}\sigma_{jk},\end{aligned} \tag{13}$$

where the fourth moments have been taken from Section 2.6.2. Thus the covariance between A_{ij} and A_{kl} is

$$\mathcal{E}(A_{ij} - n\sigma_{ij})(A_{kl} - n\sigma_{kl}) = n(\sigma_{ik}\sigma_{jl} + \sigma_{il}\sigma_{jk}). \tag{14}$$

If $i = k$ and $j = l$, we obtain the variance of A_{ij}

$$\mathcal{E}(A_{ij} - n\sigma_{ij})^2 = n(\sigma_{ij}^2 + \sigma_{ii}\sigma_{jj}). \tag{15}$$

7.3.2. The Sum of Wishart Matrices

Suppose the A_i $(i = 1, 2)$ are distributed independently according to $W(\mathbf{\Sigma}, n_i)$ respectively. Then A_1 is distributed as $\sum_{\alpha=1}^{n_1} \mathbf{Z}_\alpha \mathbf{Z}'_\alpha$, and A_2 is distributed as $\sum_{\alpha=n_1+1}^{n_1+n_2} \mathbf{Z}_\alpha \mathbf{Z}'_\alpha$, where the $\mathbf{Z}_\alpha$ are independent, each with distribution $N(\mathbf{0}, \mathbf{\Sigma})$. Then

$$A = A_1 + A_2 \tag{16}$$

is distributed as $\sum_{\alpha=1}^n \mathbf{Z}_\alpha \mathbf{Z}'_\alpha$, where $n = n_1 + n_2$. Thus A is distributed according to $W(\mathbf{\Sigma}, n)$. Obviously, the sum of q matrices distributed independently, each according to a Wishart distribution with covariance $\mathbf{\Sigma}$, has a Wishart distribution with covariance matrix $\mathbf{\Sigma}$ and number of degrees of freedom equal to the sum of the numbers of degrees of freedom of the component matrices.

THEOREM 7.3.2. *If the* A_i $(i = 1, \cdots, q)$ *are independently distributed according to* $W(\Sigma, n_i)$ *respectively, then*

$$A = \sum_{i=1}^{q} A_i \tag{17}$$

is distributed according to $W(\Sigma, \sum_{i=1}^{q} n_i)$.

7.3.3. A Certain Linear Transformation

We shall frequently make the transformation

$$A = CBC', \tag{18}$$

where C is a nonsingular $p \times p$ matrix. If A is distributed according to $W(\Sigma, n)$, then B is distributed according to $W(\Phi, n)$ where

$$\Phi = C^{-1}\Sigma C'^{-1}. \tag{19}$$

This is proved by the following argument:

$$A = \sum_{\alpha=1}^{n} Z_\alpha Z'_\alpha, \tag{20}$$

where the Z_α are independently distributed, each according to $N(0, \Sigma)$. Then

$$Y_\alpha = C^{-1} Z_\alpha \tag{21}$$

is distributed according to $N(0, \Phi)$. However,

$$B = \sum_{\alpha=1}^{n} Y_\alpha Y'_\alpha = C^{-1} \sum_{\alpha=1}^{n} Z_\alpha Z'_\alpha C'^{-1} = C^{-1} A C'^{-1} \tag{22}$$

is distributed according to $W(\Phi, n)$. $\left|\frac{\partial(A)}{\partial(B)}\right|$, the Jacobian of the transformation (18), is

$$\left|\frac{\partial(A)}{\partial(B)}\right| = \frac{w(B, \Phi, n)}{w(A, \Sigma, n)} = \frac{|B|^{\frac{1}{2}(n-p-1)}|\Sigma|^{\frac{1}{2}n}}{|A|^{\frac{1}{2}(n-p-1)}|\Phi|^{\frac{1}{2}n}} = \operatorname{mod} |C|^{p+1}. \tag{23}$$

7.3.4. Marginal Distributions

If A is distributed according to $W(\Sigma, n)$, the marginal distribution of any arbitrary set of the elements of A may be awkward to obtain. However, the marginal distribution of some sets of elements can be found easily. We give some of these in the following two theorems.

THEOREM 7.3.3. *Let A and $\mathbf{\Sigma}$ be partitioned into q and $p - q$ rows and columns*

$$A = \begin{pmatrix} A_{11} & A_{12} \\ A_{21} & A_{22} \end{pmatrix}, \qquad \mathbf{\Sigma} = \begin{pmatrix} \mathbf{\Sigma}_{11} & \mathbf{\Sigma}_{12} \\ \mathbf{\Sigma}_{21} & \mathbf{\Sigma}_{22} \end{pmatrix}. \tag{24}$$

If A is distributed according to $W(\mathbf{\Sigma}, n)$, then A_{11} is distributed according to $W(\mathbf{\Sigma}_{11}, n)$.

Proof. A is distributed as $\sum_{\alpha=1}^{n} \mathbf{Z}_\alpha \mathbf{Z}'_\alpha$, where the $\mathbf{Z}_\alpha$ are independent, each with the distribution $N(\mathbf{0}, \mathbf{\Sigma})$. Partition $\mathbf{Z}_\alpha$ into subvectors of q and $p - q$ components,

$$\mathbf{Z}_\alpha = \begin{pmatrix} \mathbf{Z}_\alpha^{(1)} \\ \mathbf{Z}_\alpha^{(2)} \end{pmatrix}. \tag{25}$$

Then the $\mathbf{Z}_\alpha^{(1)}$ are independent, each with the distribution $N(\mathbf{0}, \mathbf{\Sigma}_{11})$, and A_{11} is distributed as $\sum_{\alpha=1}^{n} \mathbf{Z}_\alpha^{(1)} \mathbf{Z}_\alpha^{(1)\prime}$, which has the distribution $W(\mathbf{\Sigma}_{11}, n)$.

THEOREM 7.3.4. *Let A and $\mathbf{\Sigma}$ be partitioned into $p_1, p_2, \cdots, p_q$ rows and columns $(p_1 + \cdots + p_q = p)$*

$$A = \begin{pmatrix} A_{11} & \cdots & A_{1q} \\ \cdot & & \cdot \\ \cdot & & \cdot \\ \cdot & & \cdot \\ A_{q1} & \cdots & A_{qq} \end{pmatrix}, \qquad \mathbf{\Sigma} = \begin{pmatrix} \mathbf{\Sigma}_{11} & \cdots & \mathbf{\Sigma}_{1q} \\ \cdot & & \cdot \\ \cdot & & \cdot \\ \cdot & & \cdot \\ \mathbf{\Sigma}_{q1} & \cdots & \mathbf{\Sigma}_{qq} \end{pmatrix}. \tag{26}$$

If $\mathbf{\Sigma}_{ij} = \mathbf{0}$ for $i \neq j$ and if A is distributed according to $W(\mathbf{\Sigma}, n)$, then $A_{11}, A_{22}, \cdots, A_{qq}$ are independently distributed and A_{jj} is distributed according to $W(\mathbf{\Sigma}_{jj}, n)$.

Proof. A is distributed as $\sum_{\alpha=1}^{n} \mathbf{Z}_\alpha \mathbf{Z}'_\alpha$, where the $\mathbf{Z}_\alpha$ are independently distributed, each according to $N(\mathbf{0}, \mathbf{\Sigma})$. Let $\mathbf{Z}_\alpha$ be partitioned

$$\mathbf{Z}_\alpha = \begin{pmatrix} \mathbf{Z}_\alpha^{(1)} \\ \cdot \\ \cdot \\ \cdot \\ \mathbf{Z}_\alpha^{(q)} \end{pmatrix} \tag{27}$$

as A and $\mathbf{\Sigma}$ have been partitioned. Since $\mathbf{\Sigma}_{ij} = \mathbf{0}$, $\mathbf{Z}_\alpha^{(i)}$ is independent of $\mathbf{Z}_\beta^{(j)}$. Then $A_{ii} = \sum_{\alpha=1}^{n} \mathbf{Z}_\alpha^{(i)} \mathbf{Z}_\alpha^{(i)\prime}$ is independent of $A_{jj} = \sum_{\alpha=1}^{n} \mathbf{Z}_\alpha^{(j)} \mathbf{Z}_\alpha^{(j)\prime}$. The rest of Theorem 7.3.4 follows from Theorem 7.3.3.

7.4. THE COCHRAN THEOREM

The Cochran theorem [Cochran (1934)] is useful in proving that certain "vector quadratic forms" are distributed as sums of "vector squares." It is a statistical statement of an algebraic theorem. First we shall give the algebraic statement concerning scalar variables as a lemma.

LEMMA 7.4.1. *If*

$$q_i = \sum_{\alpha, \beta=1}^{N} a^i_{\alpha\beta} y_\alpha y_\beta, \qquad i = 1, \cdots, m, \tag{1}$$

is of rank r_i *and*

$$\sum_{i=1}^{m} q_i = \sum_{\alpha=1}^{N} y_\alpha^2, \tag{2}$$

then a necessary and sufficient condition that there exist an orthogonal transformation of $\{y_\alpha\}$ *to* $\{z_\alpha\}$ *such that*

$$q_i = \sum_{\alpha=r_1+\cdots+r_{i-1}+1}^{r_1+\cdots+r_i} z_\alpha^2 \tag{3}$$

is that

$$r_1 + \cdots + r_m = N. \tag{4}$$

Proof. The necessity of the condition is obvious because the sum of the ranks cannot be less than N if (2) is to hold and it cannot be more than N if the transformation to $z_1, \cdots, z_{r_1+\cdots+r_m}$ is to be nonsingular.

Let us prove the sufficiency. From Corollary 7 of Appendix 1 we know there is a (nonsingular) matrix $\boldsymbol{D}$ such that

$$\boldsymbol{D}' \begin{pmatrix} \boldsymbol{I} & \boldsymbol{0} & \boldsymbol{0} \\ \boldsymbol{0} & -\boldsymbol{I} & \boldsymbol{0} \\ \boldsymbol{0} & \boldsymbol{0} & \boldsymbol{0} \end{pmatrix} \boldsymbol{D} = \boldsymbol{A}_1, \tag{5}$$

where $\boldsymbol{A}_1 = (a^1_{\alpha\beta})$. Let $d_{\alpha\beta} = b^{(1)}_{\alpha\beta}$, $\alpha = 1, \cdots, r_1$; $\beta = 1, \cdots, N$. Then $z_\alpha = \sum_{\beta=1}^{N} b^{(1)}_{\alpha\beta} y_\beta$ $(\alpha = 1, \cdots, r_1)$ form a set of r_1 linear functions of $\{y_\beta\}$ such that*

$$q_1 = \sum_{\alpha=1}^{r_1} c_\alpha z_\alpha^2, \tag{6}$$

where $c_\alpha = 1$ or -1. In general, there is a set of r_i linear functions

$$z_\alpha = \sum_{\beta=1}^{N} b^{(i)}_{\alpha\beta} y_\beta, \qquad \alpha = r_1 + \cdots + r_{i-1} + 1, \cdots, r_1 + \cdots + r_i, \tag{7}$$

such that

$$q_i = \sum_{\alpha=r_1+\cdots+r_{i-1}+1}^{r_1+\cdots+r_i} c_\alpha z_\alpha^2. \tag{8}$$

Thus

$$\sum_{i=1}^{m} q_i = \sum_{\alpha=1}^{N} c_\alpha z_\alpha^2. \tag{9}$$

* The sum of the orders of $\boldsymbol{I}$ and $-\boldsymbol{I}$ in (5) is r_1.

Since $\sum_{i=1}^m q_i$ is positive definite, all the $c_\alpha = 1$. Thus (8) is equivalent to (3). Writing (7) for all i as

$$z_\alpha = \sum_{\beta=1}^{N} b_{\alpha\beta} y_\beta, \qquad \alpha = 1, \cdots, N, \tag{10}$$

we derive from (2) and (9)

$$\sum_{\alpha=1}^{N} y_\alpha^2 = \sum_{\alpha=1}^{N} z_\alpha^2. \tag{11}$$

Thus the transformation (10) is orthogonal (that is, $\boldsymbol{I} = \boldsymbol{B}'\boldsymbol{I}\boldsymbol{B} = \boldsymbol{B}'\boldsymbol{B}$).

An alternative statement of the lemma is as follows: Let the rank of the N-order symmetric matrix $\boldsymbol{A}_i$ be r_i $(i = 1, \cdots, m)$, and suppose $\sum_{i=1}^m \boldsymbol{A}_i = \boldsymbol{I}$. A necessary and sufficient condition that there exist an orthogonal matrix

$$\boldsymbol{B} = \begin{pmatrix} \boldsymbol{B}_1 \\ \cdot \\ \cdot \\ \cdot \\ \boldsymbol{B}_m \end{pmatrix} \tag{12}$$

such that $\boldsymbol{A}_i = \boldsymbol{B}_i'\boldsymbol{B}_i$ is that $\sum_{i=1}^m r_i = N$.

We now state Cochran's theorem.

THEOREM 7.4.1. *Suppose* $\mathbf{Y}_\alpha$ *is distributed according to* $N(\mathbf{0}, \boldsymbol{\Sigma})$ *independently of* $\mathbf{Y}_\beta (\alpha \neq \beta)$. *Suppose the matrix* $(a_{\alpha\beta}^i) = \boldsymbol{A}_i$ *used in forming*

$$\boldsymbol{Q}_i = \sum_{\alpha, \beta=1}^{N} a_{\alpha\beta}^i \mathbf{Y}_\alpha \mathbf{Y}_\beta' \qquad (i = 1, \cdots, m) \tag{13}$$

is of rank r_i *and suppose*

$$\sum_{i=1}^{m} \boldsymbol{Q}_i = \sum_{\alpha=1}^{N} \mathbf{Y}_\alpha \mathbf{Y}_\alpha'. \tag{14}$$

Then a necessary and sufficient condition that $\boldsymbol{Q}_i$ $(i = 1, \cdots, m)$ *is distributed as*

$$\sum_{r_1 + \cdots + r_{i-1}+1}^{r_1 + \cdots + r_i} \mathbf{Z}_\alpha \mathbf{Z}_\alpha', \tag{15}$$

where $\mathbf{Z}_\alpha$ *is distributed according to* $N(\mathbf{0}, \boldsymbol{\Sigma})$ *independently of* $\mathbf{Z}_\beta$ $(\alpha \neq \beta)$, *and* $\boldsymbol{Q}_i$ *is distributed independently of* $\boldsymbol{Q}_j$ $(i \neq j)$ *is that*

$$r_1 + \cdots + r_m = N. \tag{16}$$

COROLLARY 7.4.1. *If* $r_i \geq p$, *then* $\boldsymbol{Q}_i$ *is distributed according to* $W(\boldsymbol{\Sigma}, r_i)$.

Proof. If (16) holds, there is an orthogonal matrix $\boldsymbol{B}$ given by (12) such that $\boldsymbol{A}_i = \boldsymbol{B}_i'\boldsymbol{B}_i$. Since $\boldsymbol{B}$ is orthogonal,

$$\mathbf{Z}_\alpha = \sum_{\beta=1}^{N} b_{\alpha\beta} \mathbf{Y}_\beta \tag{17}$$

are independent, each with distribution $N(\mathbf{0}, \mathbf{\Sigma})$ (see Theorem 3.3.1). We see that

$$(18) \qquad Q_i = \sum_{\alpha, \beta=1}^{N} a_{\alpha\beta}^i Y_\alpha Y_\beta' = \sum_{\alpha, \beta} \sum_{\gamma} b_{\gamma\alpha} b_{\gamma\beta} Y_\alpha Y_\beta'$$

$$= \sum_{\gamma=r_1+\cdots r_{i-1}+1}^{r_1+\cdots+r_i} Z_\gamma Z_\gamma'.$$

The necessity of (16) follows from the argument in Lemma 7.4.1.

This theorem is useful in generalizing results from the univariate analysis of variance (see Chapter 8). As an example of the use of this theorem, let us prove that the mean of a sample of N times its transpose and a multiple of the sample covariance matrix are independently distributed with a singular and a nonsingular Wishart distribution respectively. Let $Y_1, \cdots, Y_N$ be independently distributed according to $N(\mathbf{0}, \mathbf{\Sigma})$. We shall use the matrices $(a_{\alpha\beta}^{(1)}) = (1/N)$ and $(a_{\alpha\beta}^{(2)}) = [\delta_{\alpha\beta} - (1/N)]$. Then

$$(19) \qquad Q_1 = \sum_{\alpha, \beta=1}^{N} \frac{1}{N} Y_\alpha Y_\beta' = N\bar{Y}\bar{Y}',$$

$$(20) \qquad Q_2 = \sum_{\alpha, \beta=1}^{N} \left(\delta_{\alpha\beta} - \frac{1}{N}\right) Y_\alpha Y_\beta'$$

$$= \sum_{\alpha=1}^{N} Y_\alpha Y_\alpha' - N\bar{Y}\bar{Y}'$$

$$= \sum_{\alpha=1}^{N} (Y_\alpha - \bar{Y})(Y_\alpha - \bar{Y})'.$$

Obviously, (14) is satisfied.

The first matrix is of rank 1; the second matrix is of rank $N - 1$ (since the rank of the sum of two matrices is less than or equal to the sum of the ranks of the matrices and the rank of the second matrix is less than N).

The conditions of the theorem are satisfied; therefore Q_1 is distributed as ZZ', where Z is distributed according to $N(\mathbf{0}, \mathbf{\Sigma})$, and Q_2 is distributed independently according to $W(\mathbf{\Sigma}, N - 1)$.

7.5. THE GENERALIZED VARIANCE

7.5.1. Definition of the Generalized Variance

One multivariate analogue of the variance σ^2 of a univariate distribution is the covariance matrix $\mathbf{\Sigma}$. Another multivariate analogue is the scalar $|\mathbf{\Sigma}|$, which is called the *generalized variance* of the multivariate distribution

[Wilks (1932a); see also Frisch (1929)]. Similarly, the generalized variance of the sample of vectors, $\boldsymbol{x}_1, \cdots, \boldsymbol{x}_N$ is

$$(1) \qquad |S| = \left| \frac{1}{N-1} \sum_{\alpha=1}^{N} (\boldsymbol{x}_\alpha - \bar{\boldsymbol{x}})(\boldsymbol{x}_\alpha - \bar{\boldsymbol{x}})' \right|.$$

In some sense each of these is a measure of spread. We consider them here because the sample generalized variance will recur in many likelihood ratio criteria for testing hypotheses.

A geometric interpretation of the sample generalized variance is in terms of p points in N-space. Let $\boldsymbol{z}_\alpha = \boldsymbol{x}_\alpha - \bar{\boldsymbol{x}}$, and let

$$(2) \qquad \begin{pmatrix} \boldsymbol{y}_1' \\ \cdot \\ \cdot \\ \cdot \\ \boldsymbol{y}_p' \end{pmatrix} = (\boldsymbol{z}_1, \cdots, \boldsymbol{z}_N).$$

The N components of the vector $\boldsymbol{y}_i$ consist of the ith components of $\boldsymbol{z}_1, \cdots, \boldsymbol{z}_N$. Then $a_{ii} = \boldsymbol{y}_i'\boldsymbol{y}_i$ is the square of the length of the ith vector and $a_{ij} = \boldsymbol{y}_i'\boldsymbol{y}_j$ is the product of the cosine of the angle between $\boldsymbol{y}_i$ and $\boldsymbol{y}_j$ and the lengths of $\boldsymbol{y}_i$ and $\boldsymbol{y}_j$.

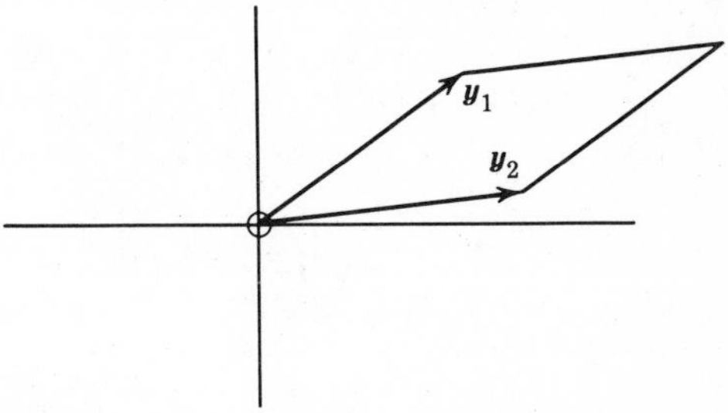

FIGURE 1

We now consider a geometric figure, Figure 1, based on these p vectors. If $p = 2$ we have a parallelogram with $\boldsymbol{y}_1$ and $\boldsymbol{y}_2$ as its principal edges; if $p = 3$ we have a parallelotope with $\boldsymbol{y}_1$, $\boldsymbol{y}_2$, and $\boldsymbol{y}_3$ as principal edges. For general p, the parallelotope is a figure in the p-dimensional hyperplane spanned by $\boldsymbol{y}_1, \cdots, \boldsymbol{y}_p$. It is cut out by pairs of parallel $(p-1)$-dimensional hyperplanes, one hyperplane of a pair being spanned by $p-1$ vectors and the other going through the end point of the other vector. The determinant $|A| = |(N-1)S| = (N-1)^p|S|$ is the square of the volume of this parallelotope.

THEOREM 7.5.1. *Let $Y = (\boldsymbol{y}_1, \cdots, \boldsymbol{y}_p)$, where $\boldsymbol{y}_i$ is an N-dimensional vector. Then the square of the p-dimensional volume of the parallelotope with $\boldsymbol{y}_1, \cdots, \boldsymbol{y}_p$ as principal edges is $|Y'Y| = |A| = (N-1)^p|S|$.*

Proof. The lemma is true for $p = 1$ since $|Y'Y| = \boldsymbol{y}_1'\boldsymbol{y}_1$, the length squared of $\boldsymbol{y}_1$. We assume the lemma true for $p = k - 1$ and prove it true for $p = k$. First we note that if two k-dimensional parallelotopes have bases consisting of $(k-1)$-dimensional parallelotopes of equal $(k-1)$-dimensional volumes and equal altitudes, their k-dimensional volumes are equal [since the k-dimensional volume is the integral of the $(k-1)$-dimensional volumes]. Since this is true for a rectangular parallelotope, the k-dimensional volume is the product of the altitude and the $(k-1)$-dimensional volume of the base. A linear combination of $y_1, \cdots, y_{k-1}$, say $c_1 y_1 + \cdots + c_{k-1} y_{k-1}$ is a vector in the base parallelotope and the minimum length of $\boldsymbol{v} = y_k - (c_1 y_1 + \cdots + c_{k-1} y_{k-1})$ is the altitude of the parallelotope with $y_1, \cdots, y_k$ as principal edges. The length of $\boldsymbol{v}$ is minimized by choosing $c_1, \cdots, c_{k-1}$ so that $0 = y_j' \boldsymbol{v} = y_j' y_k - (c_1 y_j' y_1 + \cdots + c_{k-1} y_j' y_{k-1})$, $j = 1, \cdots, k-1$. Let $\boldsymbol{Y}_{k-1} = (y_1 \cdots y_{k-1})$, $\boldsymbol{Y}_k = (y_1 \cdots y_k)$,

$$C = \begin{pmatrix} 1 & 0 \cdots 0 & -c_1 \\ 0 & 1 \cdots 0 & -c_2 \\ \cdot & \cdot \quad \cdot & \cdot \\ \cdot & \cdot \quad \cdot & \cdot \\ \cdot & \cdot \quad \cdot & \cdot \\ 0 & 0 \cdots 1 & -c_{k-1} \\ 0 & 0 \cdots 0 & 1 \end{pmatrix}. \tag{3}$$

Then $|C| = 1$ and $(\boldsymbol{Y}_{k-1}\ \boldsymbol{v}) = \boldsymbol{Y}_k C$. We have

$$\begin{aligned} |\boldsymbol{Y}_k'\boldsymbol{Y}_k| &= |C'|\,|\boldsymbol{Y}_k'\boldsymbol{Y}_k|\,|C| = |C'\boldsymbol{Y}_k'\boldsymbol{Y}_k C| \\ &= \left| \begin{pmatrix} \boldsymbol{Y}_{k-1}' \\ \boldsymbol{v}' \end{pmatrix} (\boldsymbol{Y}_{k-1}\ \boldsymbol{v}) \right| = \begin{vmatrix} \boldsymbol{Y}_{k-1}'\boldsymbol{Y}_{k-1} & \mathbf{0} \\ \mathbf{0} & \boldsymbol{v}'\boldsymbol{v} \end{vmatrix} \\ &= |\boldsymbol{Y}_{k-1}'\boldsymbol{Y}_{k-1}|\,\boldsymbol{v}'\boldsymbol{v}. \end{aligned} \tag{4}$$

Since we assumed $|\boldsymbol{Y}_{k-1}'\boldsymbol{Y}_{k-1}|$ was the square of the $(k-1)$-dimensional volume of the base parallelotope and $\boldsymbol{v}'\boldsymbol{v}$ is the square of the altitude, the product is the volume of the k-dimensional parallelotope. By induction the theorem is proved.

We shall see later that many multivariate statistics can be given an interpretation in terms of these volumes. These volumes are analogous to distances that arise in special cases when $p = 1$.

We now consider a geometric interpretation of $|A|$ in terms of N points in p-space. Let $z_1, \cdots, z_N$, as defined above, represent N points in p-space. When $p = 1$, $|A| = \sum_\alpha z_{1\alpha}^2$, which is the sum of squares of the distances the points are from the origin. In general $|A|$ is the sum of volumes of all parallelotopes formed by taking as principal edges p vectors from the set $z_1, \cdots, z_N$.

We see that

$$(5) \qquad |A| = \begin{vmatrix} \sum_\alpha z_{1\alpha}^2 & \cdots \sum_\alpha z_{1\alpha} z_{p-1,\alpha} & \sum_\beta z_{1\beta} z_{p\beta} \\ \vdots & \vdots & \vdots \\ \sum_\alpha z_{p-1,\alpha} z_{1\alpha} & \cdots \sum_\alpha z_{p-1,\alpha}^2 & \sum_\beta z_{p-1,\beta} z_{p\beta} \\ \sum_\alpha z_{p\alpha} z_{1\alpha} & \cdots \sum_\alpha z_{p\alpha} z_{p-1,\alpha} & \sum_\beta z_{p\beta}^2 \end{vmatrix}$$

$$= \sum_\beta \begin{vmatrix} \sum_\alpha z_{1\alpha}^2 & \cdots \sum_\alpha z_{1\alpha} z_{p-1,\alpha} & z_{1\beta} z_{p\beta} \\ \vdots & \vdots & \vdots \\ \sum_\alpha z_{p-1,\alpha} z_{1\alpha} & \cdots \sum_\alpha z_{p-1,\alpha}^2 & z_{p-1,\beta} z_{p\beta} \\ \sum_\alpha z_{p\alpha} z_{1\alpha} & \cdots \sum_\alpha z_{p\alpha} z_{p-1,\alpha} & z_{p\beta}^2 \end{vmatrix}$$

by the rule for expanding determinants [see (24) of Appendix 1]. In (5) the matrix A has been partitioned into $p - 1$ and 1 columns. Applying the rule successively to the columns, we find

$$(6) \qquad |A| = \sum_{\alpha_1, \cdots, \alpha_p = 1}^{N} |z_{i\alpha_j} z_{j\alpha_j}|.$$

By Theorem 7.5.1 the square of the volume of the parallelotope with $z_{\alpha_1}, \cdots, z_{\alpha_p}$ as principal edges is

$$(7) \qquad V_{\alpha_1, \cdots, \alpha_p}^2 = |\textstyle\sum_\beta z_{i\beta} z_{j\beta}|,$$

where the sum on β is over $(\alpha_1, \cdots, \alpha_p)$. If we now expand this determinant in the manner used for $|A|$ we obtain

$$(8) \qquad V_{\alpha_1, \alpha_2, \cdots, \alpha_p}^2 = \sum |z_{i\beta_j} z_{j\beta_j}|,$$

where the sum is for each β_j over the range $(\alpha_1, \cdots, \alpha_p)$. Summing (8) over all different sets $(\alpha_1, \cdots, \alpha_p)$ we obtain (6). ($|z_{i\beta_j} z_{j\beta_j}| = 0$ if two or more β_j are equal.) Thus $|A|$ is the sum of volumes squared of all parallelotopes formed by sets of p of the vectors z_α as principal edges. If we replace z_α by $x_\alpha - \bar{x}$, we can state the following theorem:

THEOREM 7.5.2. *Let $|S|$ be defined by* (1), *where $x_1, \cdots, x_N$ are the N vectors of a sample. Then $|S|$ is proportional to the sum of squares of the volumes of all parallelotopes formed by using as principal edges p vectors with p of $x_1, \cdots, x_N$ as one set of end points and $\bar{x}$ as the other, and the factor of proportionality is* $1/(N-1)^p$.

The population analogue of $|S|$ is $|\Sigma|$, which can also be given a geometric interpretation. From Section 3.3 we know that

$$\Pr\{X'\Sigma^{-1}X \leq \chi_p^2(\alpha)\} = 1 - \alpha \tag{9}$$

if X is distributed according to $N(\mathbf{0}, \Sigma)$; that is, the probability is $1 - \alpha$ that X fall inside the ellipsoid

$$x'\Sigma^{-1}x = \chi_p^2(\alpha). \tag{10}$$

The volume of this ellipsoid is $C(p)|\Sigma|^{\frac{1}{2}}[\chi_p^2(\alpha)]^{\frac{1}{2}p}/p$, where $C(p)$ is defined in Problem 3.

7.5.2. Distribution of the Sample Generalized Variance

The distribution of $|S|$ is the same as the distribution of $|A|/(N-1)^p$, where

$$A = \sum_{\alpha=1}^{n} Z_\alpha Z_\alpha' \tag{11}$$

and Z_α is distributed independently of Z_β $(\alpha \neq \beta)$ according to $N(\mathbf{0}, \Sigma)$ and $n = N - 1$. Let

$$W_\alpha = CZ_\alpha \tag{12}$$

with C chosen so $C\Sigma C' = I$. Then W_α is distributed independently of W_β $(\alpha \neq \beta)$ according to $N(\mathbf{0}, I)$ and

$$|B| = |A|/|\Sigma|, \tag{13}$$

where

$$B = \sum_{\alpha=1}^{n} W_\alpha W_\alpha' = \sum_{\alpha=1}^{n} CZ_\alpha Z_\alpha' C' = CAC'. \tag{14}$$

This follows because $|C| \cdot |\Sigma| \cdot |C| = 1$.

As in Section 7.2, let $B_{ii} = (b_{jk})$, $j, k = i, \cdots, p$, $b_{(i)} = (b_{i,i+1}, b_{i,i+2}, \cdots, b_{ip})$, and $b_{ii\cdot i+1, \cdots, p} = b_{ii} - b_{(i)}B_{i+1,i+1}^{-1}b_{(i)}'$. Then

$$|B| = b_{11\cdot 2, \cdots, p}\, b_{22\cdot 3, \cdots, p} \cdots b_{pp}. \tag{15}$$

As was shown in Section 7.2, the $b_{11\cdot 2, \cdots, p}, b_{22\cdot 3, \cdots, p}, \cdots, b_{pp}$ are independently distributed, and $b_{ii\cdot i+1,\cdots,p}$ has a χ^2-distribution with $n-(p-i)$ degrees of freedom. Thus $|\boldsymbol{B}|$ is distributed as $\chi^2_n \cdot \chi^2_{n-1} \cdot \cdots \cdot \chi^2_{n-p+1}$.

THEOREM 7.5.3. *The distribution of the generalized variance $|\boldsymbol{S}|$ of a sample $\boldsymbol{X}_1, \cdots, \boldsymbol{X}_N$ from $N(\boldsymbol{\mu}, \boldsymbol{\Sigma})$ is the same as the distribution of $|\boldsymbol{\Sigma}|/(N-1)^p$ times the product of p independent factors, the distribution of the ith factor being the χ^2-distribution with $N-i$ degrees of freedom.*

We can give a geometric interpretation of this theorem. Let $\boldsymbol{Y}_i = (W_{i1}, \cdots, W_{in})$ be a vector in an n-dimensional space. The p vectors, $\boldsymbol{Y}_1, \cdots, \boldsymbol{Y}_p$ are independently distributed; each component of $\boldsymbol{Y}_i$ is distributed according to $N(0, 1)$. $|\boldsymbol{B}|$ is the square of the volume of the parallelotope spanned by $\boldsymbol{Y}_1, \cdots, \boldsymbol{Y}_p$. Let $\boldsymbol{U}_i$ be the vector from $\boldsymbol{Y}_i$ orthogonal to $\boldsymbol{Y}_{i+1}, \cdots, \boldsymbol{Y}_p$. Then the volume squared of the parallelotope spanned by $\boldsymbol{Y}_i, \boldsymbol{Y}_{i+1}, \cdots, \boldsymbol{Y}_p$ is the volume squared of the parallelotope spanned by $\boldsymbol{Y}_{i+1}, \cdots, \boldsymbol{Y}_p$ times $\boldsymbol{U}_i'\boldsymbol{U}_i = a_{ii\cdot i+1,\cdots,p}$, which is the length squared of $\boldsymbol{U}_i$ (see the proof of Theorem 7.5.1). It follows that $|\boldsymbol{B}| = \boldsymbol{U}_1'\boldsymbol{U}_1 \cdot \boldsymbol{U}_2'\boldsymbol{U}_2 \cdot \cdots \cdot \boldsymbol{U}_p'\boldsymbol{U}_p$. Now $\boldsymbol{U}_i$ is orthogonal to $\boldsymbol{U}_{i+1}, \cdots, \boldsymbol{U}_p$; therefore the conditional distribution of $\boldsymbol{U}_i$ is normal in $n-(p-i)$ dimensions and the $n-(p-i)$ coordinates are distributed according to $N(0, 1)$. Thus $\boldsymbol{U}_i'\boldsymbol{U}_i$ has a χ^2-distribution with $n-(p-i)$ degrees of freedom (independent of the other $\boldsymbol{U}$'s).

For $p = 1$ or 2, the exact distribution of $|\boldsymbol{S}|$ can be obtained, but for higher values of p the integrals involved cannot be expressed nicely. However, the moments of $|\boldsymbol{S}|$ can be obtained easily from the fact that $|\boldsymbol{S}|$ can be written

$$|\boldsymbol{S}| = |\boldsymbol{A}|/(N-1)^p \tag{16}$$

and in turn $|\boldsymbol{A}|$ can be written

$$\begin{aligned} |\boldsymbol{A}| &= |\boldsymbol{\Sigma}| \cdot |\boldsymbol{B}| \\ &= |\boldsymbol{\Sigma}| \cdot \chi^2_{N-1} \cdot \chi^2_{N-2} \cdot \cdots \cdot \chi^2_{N-p}. \end{aligned} \tag{17}$$

Since the hth moment of a χ^2-variable with m degrees of freedom is $2^h\Gamma(\tfrac{1}{2}m + h)/\Gamma(\tfrac{1}{2}m)$ and the moment of a product of independent variables is the product of the moments of the variables, the hth moment of $|\boldsymbol{A}|$ is

$$|\boldsymbol{\Sigma}|^h \prod_{i=1}^{p} \left\{2^h \frac{\Gamma[\tfrac{1}{2}(N-i)+h]}{\Gamma[\tfrac{1}{2}(N-i)]}\right\} = 2^{hp}|\boldsymbol{\Sigma}|^h \frac{\prod_{i=1}^{p} \Gamma[\tfrac{1}{2}(N-i)+h]}{\prod_{i=1}^{p} \Gamma[\tfrac{1}{2}(N-i)]}. \tag{18}$$

Thus

(19) $$\mathscr{E}|A| = |\boldsymbol{\Sigma}| \prod_{i=1}^{p} (N - i),$$

(20) $$\operatorname{Var}(|A|) = |\boldsymbol{\Sigma}|^2 \prod_{i=1}^{p} (N - i)\left[\prod_{j=1}^{p} \Gamma(N - j + 2) - \prod_{j=1}^{p} \Gamma(N - j)\right],$$

where Var ($|A|$) is the variance of $|A|$.

In the case of $p = 1$ and $p = 2$ we can give the distribution of $V = |A|/|\boldsymbol{\Sigma}|$. For $p = 1$, V is distributed as χ^2 with $N - 1$ degrees of freedom. For $p = 2$, we find from (18) the $k/2$ moment of V as

(21) $$\mathscr{E}V^{k/2} = 2^k \frac{\Gamma[\frac{1}{2}(N - 1 + k)]\Gamma[\frac{1}{2}(N - 2 + k)]}{\Gamma[\frac{1}{2}(N - 1)]\Gamma[\frac{1}{2}(N - 2)]}$$

since (18) holds for every positive number h (which fact follows from Theorem 7.5.3 and the integral representation of the gamma function). We use the "duplication formula" for gamma functions

(22) $$2^{2\alpha-1}\Gamma(\alpha)\Gamma(\alpha + \tfrac{1}{2}) = \sqrt{\pi}\,\Gamma(2\alpha)$$

in (21) to obtain as the $k/2$th moment of V

(23) $$\mathscr{E}V^{\frac{1}{2}k} = \frac{\Gamma(N - 2 + k)}{\Gamma(N - 2)}.$$

This is the kth moment of $\frac{1}{2}$ times a variable with the χ^2-distribution with $2N - 4$ degrees of freedom. We need the following theorem:

CARLEMAN'S THEOREM. *If* $\{\mu_i\}$ $(i = 1, 2, \cdots)$ *is a sequence of numbers such that*

(24) $$\sum_{k=1}^{\infty} \left(\frac{1}{\mu_{2k}}\right)^{1/(2k)}$$

is divergent, then at most one distribution has the moment sequence $\{\mu_i\}$.

The proof is given by Shohat and Tamarkin (1943), for example. Using the criterion for divergence,

(25) $$\left(\frac{1}{\mu_{2k}}\right)^{1/(2k)} > \frac{1}{N - 3 + 2k} > C\,\frac{1}{k}$$

for a suitable C (> 0) and k sufficiently large, we can verify that the condition of the above theorem is satisfied by the moments of $\sqrt{V}$. Thus $2\sqrt{V}$ is distributed as χ^2 with $2N - 4$ degrees of freedom.

Hoel (1937) has suggested as an approximate density for $V^{1/p}$

(26) $$\frac{c^{\frac{1}{2}p(N-p)}y^{\frac{1}{2}p(N-p)-1}e^{-cy}}{\Gamma[\frac{1}{2}p(N - p)]},$$

where

$$c = \frac{p}{2}\left(1 - \frac{(p-1)(p-2)}{2N}\right)^{1/p}. \tag{27}$$

7.5.3. The Asymptotic Distribution of the Sample Generalized Variance

Let $|B|/n^p = V_1(n) \cdot V_2(n) \cdot \ldots \cdot V_p(n)$, where the V's are independently distributed and $nV_i(n) = \chi^2_{n-p+i}$. Since χ^2_{n-p+i} is distributed as $\sum_{\alpha=1}^{n-p+i} W_\alpha^2$, where the W_α are independent, each with distribution $N(0, 1)$, the central limit theorem (applied to W_α^2) states that

$$\frac{nV_i(n) - (n-p+i)}{\sqrt{2(n-p+i)}} = \sqrt{n}\,\frac{V_i(n) - 1 + \dfrac{p-i}{n}}{\sqrt{2}\sqrt{1 - \dfrac{p-i}{n}}} \tag{28}$$

is asymptotically distributed according to $N(0, 1)$. Then $\sqrt{n}[V_i(n) - 1]$ is asymptotically distributed according to $N(0, 2)$. We now apply Theorem 4.2.5. We have

$$U(n) = \begin{pmatrix} V_1(n) \\ \cdot \\ \cdot \\ \cdot \\ V_p(n) \end{pmatrix}, \qquad \boldsymbol{b} = \begin{pmatrix} 1 \\ \cdot \\ \cdot \\ \cdot \\ 1 \end{pmatrix}, \tag{29}$$

$|B|/n^p = w = f(u_1, \cdots, u_p) = u_1 u_2 \cdots u_p$, $\mathbf{T} = 2I$, $\partial f/\partial u_i|_{u=b} = 1$, and $\boldsymbol{\phi}_b' \mathbf{T} \boldsymbol{\phi}_b = 2p$. Thus

$$\sqrt{n}\left(\frac{|B|}{n^p} - 1\right) \tag{30}$$

is asymptotically distributed according to $N(0, 2p)$.

THEOREM 7.5.4. *Let S be a $p \times p$ sample covariance matrix with n degrees of freedom. Then $\sqrt{n}(|S|/|\Sigma| - 1)$ is asymptotically normally distributed with mean* 0 *and variance* $2p$.

7.6. DISTRIBUTION OF THE SET OF CORRELATION COEFFICIENTS FOR A DIAGONAL POPULATION COVARIANCE MATRIX

In Section 4.2.1 we found the distribution of a single sample correlation when the corresponding population correlation was zero. Here we shall find the density of the set r_{ij}, $i \neq j$, $i, j = 1, \cdots, p$, when $\rho_{ij} = 0$, $i \neq j$.

We start with the distribution of A when $\mathbf{\Sigma}$ is diagonal; that is, $W((\sigma_{ii}\delta_{ij}), n)$. The density of A is

$$(1)\qquad \frac{|a_{ij}|^{\frac{1}{2}(n-p-1)} \exp\left(-\frac{1}{2}\sum_{i=1}^{p} \frac{a_{ii}}{\sigma_{ii}}\right)}{2^{\frac{1}{2}np}\pi^{\frac{1}{4}p(p-1)} \prod_{i=1}^{p} \sigma_{ii}^{\frac{1}{2}n} \prod_{i=1}^{p} \Gamma[\frac{1}{2}(n+1-i)]},$$

since

$$(2)\qquad |\mathbf{\Sigma}| = \begin{vmatrix} \sigma_{11} & 0 \cdots & 0 \\ 0 & \sigma_{22} \cdots & 0 \\ \cdot & \cdot & \cdot \\ \cdot & \cdot & \cdot \\ \cdot & \cdot & \cdot \\ 0 & 0 \cdots & \sigma_{pp} \end{vmatrix} = \prod_{i=1}^{p} \sigma_{ii}.$$

We make the transformation

$$(3)\qquad a_{ij} = \sqrt{a_{ii}}\sqrt{a_{jj}}\, r_{ij}, \qquad i \neq j,$$

$$(4)\qquad a_{ii} = a_{ii}.$$

The Jacobian is the product of the Jacobian of (4) and that of (3) for a_{ii} fixed. The Jacobian of (3) is the determinant of a $p(p-1)/2$-order diagonal matrix with diagonal elements $\sqrt{a_{ii}}\sqrt{a_{jj}}$. Since each particular subscript k, say, appears in the set r_{ij}, $i < j$, $p-1$ times, the Jacobian is

$$(5)\qquad J = \prod_{i=1}^{p} a_{ii}^{\frac{1}{2}(p-1)}.$$

If we substitute from (3) and (4) into $w(A|(\sigma_{ii}\delta_{ij}), n)$ and multiply by (5) we obtain as the joint density of $\{a_{ii}\}$ and $\{r_{ij}\}$

$$(6)\qquad \frac{|\sqrt{a_{ii}}\sqrt{a_{jj}}\, r_{ij}|^{\frac{1}{2}(n-p-1)} \exp\left(-\frac{1}{2}\sum_{i=1}^{p} \frac{a_{ii}}{\sigma_{ii}}\right)}{2^{\frac{1}{2}np}\pi^{\frac{1}{4}p(p-1)} \prod_{i=1}^{p} \sigma_{ii}^{\frac{1}{2}n} \prod_{i=1}^{p} \Gamma[\frac{1}{2}(n+1-i)]} \prod_{i=1}^{p} a_{ii}^{\frac{1}{2}(p-1)}$$

$$= \frac{|r_{ij}|^{\frac{1}{2}(n-p-1)}}{\pi^{\frac{1}{4}p(p-1)} \prod_{i=1}^{p} \Gamma[\frac{1}{2}(n+1-i)]} \prod_{i=1}^{p} \left\{ \frac{a_{ii}^{\frac{1}{2}n-1} \exp\left(-\frac{1}{2}\frac{a_{ii}}{\sigma_{ii}}\right)}{2^{\frac{1}{2}n}\sigma_{ii}^{\frac{1}{2}n}} \right\}$$

since

$$(7)\qquad |\sqrt{a_{ii}}\sqrt{a_{jj}}\, r_{ij}| = \left(\prod a_{ii}\right)|r_{ij}|,$$

where $r_{ii} = 1$. In the ith term of the product on the right-hand side of (6), let $a_{ii}/(2\sigma_{ii}) = u_i$; then the integral of this term is

$$\int_0^\infty \frac{a_{ii}^{\frac{1}{2}n-1} \exp\left(-\frac{1}{2}\frac{a_{ii}}{\sigma_{ii}}\right)}{2^{\frac{1}{2}n}\,\sigma_{ii}^{\frac{1}{2}n}}\,da_{ii} = \int_0^\infty u_i^{\frac{1}{2}n-1}e^{-u_i}du_i = \Gamma(\tfrac{1}{2}n) \tag{8}$$

by definition of the gamma function (or by the fact that a_{ii}/σ_{ii} has the χ^2-density with n degrees of freedom). Hence, the density of r_{ij} is

$$\frac{\Gamma^p(\frac{1}{2}n)|r_{ij}|^{\frac{1}{2}(n-p-1)}}{\prod_{i=1}^{p}\Gamma[\frac{1}{2}(n+1-i)]\pi^{\frac{1}{4}p(p-1)}}. \tag{9}$$

THEOREM 7.6.1. *If $\mathbf{X}_1, \cdots, \mathbf{X}_N$ are independent, each with distribution $N(\boldsymbol{\mu}, (\sigma_{ii}\delta_{ij}))$, then the density of the sample correlation coefficients is given by* (9) *where $n = N - 1$.*

REFERENCES

Section 7.2. Aitken (1949); Cramér (1946), 390–394; Fog (1948); A. T. James (1954); P. L. Hsu (1939b); Ingham (1933); Kendall (1946), 330–335; Madow (1938); Mahalanobis, Bose, and Roy (1937); Mauldon (1955); Ogawa (1953); Rasch (1948); Steyn (1951); Sverdrup (1947); Wilks (1943), 226–232; Wishart (1928), (1948a); Wishart and Bartlett (1932), (1933).

Section 7.3. Wishart and Bartlett (1933).

Section 7.4. Aitken (1950); Cochran (1934); Craig (1947); G. S. James (1952); Matusita (1949); Ogasawara and Takahashi (1951); Ogawa (1949), (1950).

Section 7.5. Bennett (1955); Frisch (1929); Hoel (1937); Kullback (1934); Lévy (1948); Nyquist, Rice, and Riordan (1954); C. R. Rao (1944); Shohat and Tamarkin (1943); Wilks (1932a).

PROBLEMS

1. (Sec. 7.2) A transformation from rectangular to polar coordinates is

$$\begin{aligned}
y_1 &= w \sin\theta_1,\\
y_2 &= w\cos\theta_1 \sin\theta_2,\\
y_3 &= w\cos\theta_1\cos\theta_2\sin\theta_3,\\
&\;\,\vdots\\
y_{n-1} &= w\cos\theta_1\cos\theta_2\cdots\cos\theta_{n-2}\sin\theta_{n-1},\\
y_n &= w\cos\theta_1\cos\theta_2\cdots\cos\theta_{n-2}\cos\theta_{n-1},
\end{aligned}$$

where $-\frac{1}{2}\pi < \theta_i \le \frac{1}{2}\pi$ $(i = 1, \cdots, n-2)$ and $-\pi < \theta_{n-1} \le \pi$.

(*a*) Prove $w^2 = \sum y_\alpha^2$. [*Hint:* Compute in turn $y_n^2 + y_{n-1}^2$, $(y_n^2 + y_{n-1}^2) + y_{n-2}^2$, and so forth.]

(*b*) Show that the Jacobian is $w^{n-1}\cos^{n-2}\theta_1\cos^{n-3}\theta_2\cdots\cos\theta_{n-2}$. *Hint:* Prove that

$$\left|\frac{\partial(y_1,\cdots,y_n)}{\partial(\theta_1,\cdots,\theta_{n-1},w)}\right|\cdot\begin{vmatrix}\cos\theta_1 & 0 & \cdots & 0 & 0\\ 0 & \cos\theta_2 & \cdots & 0 & 0\\ \vdots & \vdots & & \vdots & \vdots\\ 0 & 0 & \cdots & \cos\theta_{n-1} & 0\\ w\sin\theta_1 & w\sin\theta_2 & \cdots & w\sin\theta_{n-1} & 1\end{vmatrix}$$

$$=\begin{vmatrix} w & x & \cdots & x & x\\ 0 & w\cos\theta_1 & \cdots & x & x\\ \vdots & \vdots & & \vdots & \vdots\\ 0 & 0 & \cdots & w\cos\theta_1\cdots\cos\theta_{n-2} & x\\ 0 & 0 & & 0 & \cos\theta_1\cdots\cos\theta_{n-1}\end{vmatrix},$$

where x denotes elements whose explicit values are not needed.

2. (Sec. 7.2) Prove that

$$\int_{-\pi/2}^{\pi/2}\cos^{h-1}\theta\,d\theta=\frac{\Gamma(\frac{1}{2}h)\Gamma(\frac{1}{2})}{\Gamma[\frac{1}{2}(h+1)]}.$$

[*Hint:* Let $\cos^2\theta = u$, and use the definition of $B(p, q)$.]

3. (Sec. 7.2) Use Problems 1 and 2 to prove that the surface area of a sphere of unit radius in n dimensions is

$$C(n)=\frac{2\pi^{\frac{1}{2}n}}{\Gamma(\frac{1}{2}n)}.$$

4. (Sec. 7.2) Use Problems 1, 2, and 3 to prove that if the density of $\boldsymbol{y}' = (y_1, \cdots, y_n)$ is $f(\boldsymbol{y}'\boldsymbol{y})$, then the density of $u = \boldsymbol{y}'\boldsymbol{y}$ is $\frac{1}{2}C(n)f(u)u^{\frac{1}{2}n-1}$.

5. (Sec. 7.2) Use Problem 4 to show that if the y_α are independent, each with the distribution $N(0, 1)$, then $U = \boldsymbol{y}'\boldsymbol{y}$ has the χ^2-distribution with n degrees of freedom.

6. (Sec. 7.3) Find the characteristic function of $\boldsymbol{A}$ from $W(\boldsymbol{\Sigma}, n)$. [*Hint:* From $\int w(\boldsymbol{A}|\boldsymbol{\Sigma}, n)\,d\boldsymbol{A} = 1$, one derives

$$\int\frac{|\boldsymbol{A}|^{\frac{1}{2}(n-p-1)}\exp(-\frac{1}{2}\operatorname{tr}\boldsymbol{\Phi}^{-1}\boldsymbol{A})d\boldsymbol{A}}{2^{\frac{1}{2}pn}\pi^{p(p-1)/4}\prod\Gamma[\frac{1}{2}(n+1-i)]}=|\boldsymbol{\Phi}|^{\frac{1}{2}n}$$

as an identity in $\boldsymbol{\Phi}$.] Note that comparison of this result with that of Section 7.3.1 is a proof of the Wishart distribution.

7. (Sec. 7.6) Prove that if $\boldsymbol{\Sigma} = \boldsymbol{I}$ the joint density of $r_{ij\cdot p}$, $i, j = 1, \cdots, p-1$, and $r_{1p}, \cdots, r_{p-1,p}$ is

$$\frac{\Gamma^{p-1}[\frac{1}{2}(n-1)]|\boldsymbol{R}_{11\cdot p}|^{\frac{1}{2}(n-p-1)}}{\pi^{(p-1)(p-2)/4}\prod_{i=1}^{p-1}\Gamma[\frac{1}{2}(n-i)]}\prod_{i=1}^{p-1}\frac{\Gamma(\frac{1}{2}n)}{\pi^{\frac{1}{2}}\Gamma[\frac{1}{2}(n-1)]}(1-r_{ip}^2)^{\frac{1}{2}(n-3)}$$

where $\boldsymbol{R}_{11\cdot p} = (r_{ij\cdot p})$. [*Hint:* $r_{ij\cdot p} = (r_{ij} - r_{ip}r_{jp})/(\sqrt{1-r_{ip}^2}\sqrt{1-r_{jp}^2})$ and $|r_{ij}| = |\sqrt{1-r_{ip}^2}\,\sqrt{1-r_{jp}^2}\,r_{ij\cdot p}|$. Use (9).]

8. (Sec. 7.6) Prove that the joint density of $r_{12\cdot 3,\cdots,p}, r_{13\cdot 4,\cdots,p}, r_{23\cdot 4,\cdots,p},\cdots, r_{1p},\cdots, r_{p-1,p}$ is

$$\frac{\Gamma\{\frac{1}{2}[n-(p-2)]\}}{\pi^{\frac{1}{2}}\Gamma\{\frac{1}{2}[n-(p-1)]\}}(1-r_{12\cdot 3,\cdots,p}^2)^{\frac{1}{2}[n-(p+1)]}$$

$$\prod_{i=1}^{2}\frac{\Gamma\{\frac{1}{2}[n-(p-3)]\}}{\pi^{\frac{1}{2}}\Gamma\{\frac{1}{2}[n-(p-2)]\}}(1-r_{i3\cdot 4,\cdots,p}^2)^{\frac{1}{2}(n-p)}$$

$$\cdots\prod_{i=1}^{p-2}\frac{\Gamma[\frac{1}{2}(n-1)]}{\pi^{\frac{1}{2}}\Gamma[\frac{1}{2}(n-2)]}(1-r_{i,p-1\cdot p}^2)^{\frac{1}{2}(n-4)}\prod_{i=1}^{p-1}\frac{\Gamma(\frac{1}{2}n)}{\pi^{\frac{1}{2}}\Gamma[\frac{1}{2}(n-1)]}(1-r_{ip}^2)^{\frac{1}{2}(n-3)}$$

[*Hint:* Use the result of Problem 7 inductively.]

9. (Sec. 7.6) Prove (without the use of Problem 8) that if $\boldsymbol{\Sigma} = \boldsymbol{I}$, then $r_{1p},\cdots, r_{p-1,p}$ are independently distributed. [*Hint:* $r_{ip} = a_{ip}/(\sqrt{a_{ii}}\sqrt{a_{pp}})$. Prove that the pairs $(a_{1p}, a_{11}),\cdots,(a_{p-1,p}, a_{p-1,p-1})$ are independent when $(z_{1p},\cdots, z_{np})$ are fixed, and note from Section 4.2.1 that the marginal distribution of r_{ip}, conditional on $z_{\alpha p}$, does not depend on $z_{\alpha p}$.]

10. (Sec. 7.6) Prove (without the use of Problems 7 and 8) that if $\boldsymbol{\Sigma} = \boldsymbol{I}$, then the set $r_{1p},\cdots, r_{p-1,p}$ is independent of the set $r_{ij\cdot p}$, $i, j = 1,\cdots, p-1$. [*Hint:* From Section 4.3.2 a_{pp}, (a_{ip}) are independent of $(a_{ij\cdot p})$. Prove that a_{pp}, (a_{ip}) and $a_{ii}(i = 1,\cdots, p-1)$ are independent of $(r_{ij\cdot p})$ by proving that $a_{ii\cdot p}$ are independent of $(r_{ij\cdot p})$. See Problem 36 of Chapter 4.]

11. (Sec. 7.6) Prove the conclusion of Problem 8 by using Problems 9 and 10.

12. (Sec. 7.6) Reverse the steps in Problem 8 to derive (9) of Section 7.6.

13. (Sec. 7.2) Use (9) of Section 7.6 to derive the distribution of $\boldsymbol{A}$.

14. (Sec. 7.3.2) Prove Theorem 7.3.2 by use of characteristic functions.

15. (Sec. 7.3.1) Find the first two moments of the elements of $\boldsymbol{A}$ by differentiating the characteristic function (11).

16. (Sec. 7.4) Let $\boldsymbol{x}_\alpha$ be an observation from $N(\boldsymbol{\beta} z_\alpha, \boldsymbol{\Sigma})$, $\alpha = 1,\cdots, N$, where z_α is a scalar. Let $\boldsymbol{b} = \sum_\alpha z_\alpha \boldsymbol{x}_\alpha / \sum_\alpha z_\alpha^2$. Use Theorem 7.4.1 to show that $\sum_\alpha \boldsymbol{x}_\alpha \boldsymbol{x}_\alpha' - \boldsymbol{b}\boldsymbol{b}'\sum_\alpha z_\alpha^2$ and $\boldsymbol{b}\boldsymbol{b}'$ are independent.

17. (Sec. 7.5) Find $\mathscr{E}|\boldsymbol{A}|^h$ directly from $W(\boldsymbol{\Sigma}, n)$. [*Hint:* The fact that

$$\int w(\boldsymbol{A}|\boldsymbol{\Sigma}, n)\, d\boldsymbol{A} \equiv 1$$

shows

$$\int \frac{|\boldsymbol{A}|^{\frac{1}{2}(n-p-1)}\exp(-\frac{1}{2}\operatorname{tr}\boldsymbol{\Sigma}^{-1}\boldsymbol{A})}{\pi^{p(p-1)/4}}\, d\boldsymbol{A} = 2^{\frac{1}{2}np}|\boldsymbol{\Sigma}|^{\frac{1}{2}n}\prod_{i=1}^{p}\Gamma[\tfrac{1}{2}(n+1-i)]$$

as an identity in n.]

18. (Sec. 7.5) Consider the confidence region for $\boldsymbol{\mu}$ given by

$$N(\bar{\boldsymbol{x}}-\boldsymbol{\mu}^*)'\boldsymbol{S}^{-1}(\bar{\boldsymbol{x}}-\boldsymbol{\mu}^*) \le \frac{(N-1)p}{N-p}F_{p,\,N-p}(\varepsilon),$$

where $\bar{\boldsymbol{x}}$ and $\boldsymbol{S}$ are based on a sample of N from $N(\boldsymbol{\mu}, \boldsymbol{\Sigma})$. Find the expected value of the volume of the confidence region.

19. Let the density of $\boldsymbol{Y}$ be $f(\boldsymbol{y}) = K$ for $\boldsymbol{y}'\boldsymbol{y} \le p + 2$ and 0 elsewhere. Prove that $K = \Gamma(\frac{1}{2}p+1)/[(p+2)\pi]^{\frac{1}{2}p}$ and show that $\mathscr{E}\boldsymbol{Y} = \boldsymbol{0}$ and $\mathscr{E}\boldsymbol{Y}\boldsymbol{Y}' = \boldsymbol{I}$.

CHAPTER 8

Testing the General Linear Hypothesis; Analysis of Variance

8.1. INTRODUCTION

In this chapter we generalize the univariate least squares theory (that is, "regression analysis") and analysis of variance to vector variates. The algebra of the multivariate case is essentially the same as that of the univariate case. This leads to distribution theory that is analogous to that of the univariate case and to test criteria that are analogous to F-tests. In fact, given a univariate test, we shall be able to write down immediately a corresponding multivariate test. Since analysis of variance based on the model of fixed effects can be obtained from least squares theory, we obtain directly a multivariate analysis of variance theory. However, in the multivariate case there is more latitude in the choice of tests of significance.

In univariate least squares we consider scalar dependent variates $x_1, \cdots, x_N$ drawn from populations with expected values $\boldsymbol{\beta} \boldsymbol{z}_1, \cdots, \boldsymbol{\beta} \boldsymbol{z}_N$ respectively, where $\boldsymbol{\beta}$ is a row vector of q components and each of the $\boldsymbol{z}_\alpha$ is a column vector of q known components. Under the assumption that the variance in each population is the same, the least squares estimates of the components of $\boldsymbol{\beta}$ are

$$(1) \qquad \boldsymbol{b} = \Big(\sum_{\alpha=1}^{N} x_\alpha \boldsymbol{z}_\alpha'\Big)\Big(\sum_{\alpha=1}^{N} \boldsymbol{z}_\alpha \boldsymbol{z}_\alpha'\Big)^{-1}.$$

If the populations are normal, the vector is the maximum likelihood estimate of $\boldsymbol{\beta}$. The unbiased estimate of the common variance σ^2 is

$$(2) \qquad s^2 = \sum_{\alpha=1}^{N} (x_\alpha - \boldsymbol{b}\boldsymbol{z}_\alpha)^2/(N-q),$$

and under the assumption of normality, the maximum likelihood estimate of σ^2 is $\hat{\sigma}^2 = (N-q)s^2/N$.

In the multivariate case $\boldsymbol{x}_\alpha$ is a vector, $\boldsymbol{\beta}$ is replaced by a matrix $\mathbf{B}$ and σ^2 replaced by a covariance matrix $\boldsymbol{\Sigma}$. The estimates of $\mathbf{B}$ and $\boldsymbol{\Sigma}$, given in Section 8.2, are matric analogues of (1) and (2).

To test a hypothesis concerning $\boldsymbol{\beta}$, say the hypothesis $\boldsymbol{\beta} = \mathbf{0}$, we use an F-test. A criterion equivalent to the F-ratio is

$$\text{(3)} \qquad \frac{1}{[q/(N-q)]F + 1} = \hat{\sigma}^2/\hat{\sigma}_0^2,$$

where $\hat{\sigma}_0^2$ is the maximum likelihood estimate of σ^2 under the null hypothesis. We shall find that the likelihood ratio criterion for the corresponding multivariate hypothesis, say $\mathbf{B} = \mathbf{0}$, is the above with the variances replaced by generalized variances. This will give a general method for setting up a multivariate analogue to any F-test. We shall also consider other analogues of the F-test.

8.2. ESTIMATES OF PARAMETERS IN MULTIVARIATE LINEAR REGRESSION

8.2.1. The Maximum Likelihood Estimates

Suppose $\boldsymbol{x}_1, \cdots, \boldsymbol{x}_N$ are a set of N observations, $\boldsymbol{x}_\alpha$ being drawn from $N(\mathbf{B}\boldsymbol{z}_\alpha, \boldsymbol{\Sigma})$. Ordinarily the vectors $\boldsymbol{z}_\alpha$ (with q components) are known vectors and the $p \times p$ matrix $\boldsymbol{\Sigma}$ and the $p \times q$ matrix $\mathbf{B}$ are unknown. We assume $N \geq p + q$ and the rank of

$$\text{(1)} \qquad Z = (\boldsymbol{z}_1, \cdots, \boldsymbol{z}_N)$$

is q. We shall estimate $\boldsymbol{\Sigma}$ and $\mathbf{B}$ by the method of maximum likelihood. The likelihood function is

$$\text{(2)} \quad L = (2\pi)^{-\frac{1}{2}Np}|\boldsymbol{\Sigma}^{*-1}|^{\frac{1}{2}N} \exp\left[-\tfrac{1}{2}\sum_{\alpha=1}^{N}(\boldsymbol{x}_\alpha - \mathbf{B}^*\boldsymbol{z}_\alpha)'\boldsymbol{\Sigma}^{*-1}(\boldsymbol{x}_\alpha - \mathbf{B}^*\boldsymbol{z}_\alpha)\right].$$

In (2), the elements of $\boldsymbol{\Sigma}^*$ and $\mathbf{B}^*$ are indeterminates. The method of maximum likelihood specifies the estimates of $\boldsymbol{\Sigma}$ and $\mathbf{B}$ based on the given sample $\boldsymbol{x}_1, \boldsymbol{z}_1, \cdots, \boldsymbol{x}_N, \boldsymbol{z}_N$ as the $\boldsymbol{\Sigma}^*$ and $\mathbf{B}^*$ that maximize (2). It is convenient to write L in terms of the coordinates; we let

$$\text{(3)} \qquad \boldsymbol{x}_\alpha = \begin{pmatrix} x_{1\alpha} \\ \cdot \\ \cdot \\ \cdot \\ x_{p\alpha} \end{pmatrix}, \qquad \boldsymbol{z}_\alpha = \begin{pmatrix} z_{1\alpha} \\ \cdot \\ \cdot \\ \cdot \\ z_{q\alpha} \end{pmatrix},$$

$$\text{(4)} \qquad \mathbf{B}^* = (\beta_{iu}^*); \qquad i = 1, \cdots, p; \; u = 1, \cdots, q,$$

$$\text{(5)} \qquad \boldsymbol{\Sigma}^* = (\sigma_{ij}^*) = (\sigma^{*ij})^{-1}.$$

The logarithm of the likelihood function is

$$\text{(6)}\quad \log L = -\tfrac{1}{2}Np\log(2\pi) + \tfrac{1}{2}N\log|\sigma^{*ij}| - \tfrac{1}{2}\sum_{i,j=1}^{p}\sum_{\alpha=1}^{N}\Big(x_{i\alpha} - \sum_{g=1}^{q}\beta^*_{ig}z_{g\alpha}\Big)\sigma^{*ij}\Big(x_{j\alpha} - \sum_{h=1}^{q}\beta^*_{jh}z_{h\alpha}\Big).$$

We note that at the maximum of $\log L$, $\mathbf{\Sigma}^*$ must be nonsingular; hence by Lemma 3.2.3, the maximum of $\log L$ with respect to $\mathbf{\Sigma}^*$ and $\boldsymbol{\beta}^*$ is equal to the maximum of $\log L$ with respect to $\mathbf{\Sigma}^{*-1}$ and $\boldsymbol{\beta}^*$ and the maximizing value of $\mathbf{\Sigma}^*$ is the inverse of the maximizing value of $\mathbf{\Sigma}^{*-1}$. We can verify that for $\log L$ to attain its maximum, the derivatives of $\log L$ with respect to β^*_{ig} must vanish. To find the maximum of $\log L$ we find it convenient to use the following lemma:

LEMMA 8.2.1. *If* $f(u_1, \cdots, u_m) = \sum_{i,j=1}^{m} d_{ij}u_iu_j$, *where* $d_{ij} = d_{ji}$, *then*

$$\text{(7)}\qquad \frac{\partial f}{\partial v} = 2\sum_{i,j} d_{ij}\frac{\partial u_i}{\partial v}u_j,$$

and

$$\text{(8)}\qquad \frac{\partial}{\partial v}\sum_{\alpha} f(u_{1\alpha}, \cdots, u_{m\alpha}) = 2\sum_{\alpha}\sum_{i,j} d_{ij}\frac{\partial u_{i\alpha}}{\partial v}u_{j\alpha}.$$

Proof.

$$\text{(9)}\qquad \begin{aligned}\frac{\partial}{\partial v}\sum_{i,j=1}^{m} d_{ij}u_iu_j &= \sum d_{ij}\frac{\partial u_j}{\partial v}u_i + \sum d_{ij}\frac{\partial u_i}{\partial v}u_j\\ &= \sum(d_{ij} + d_{ji})\frac{\partial u_i}{\partial v}u_j\\ &= 2\sum d_{ij}\frac{\partial u_i}{\partial v}u_j.\end{aligned}$$

Then (7) and (8) follow directly.

The derivative of $\log L$ with respect to β^*_{kf} is

$$\text{(10)}\qquad \begin{aligned}\frac{\partial \log L}{\partial \beta^*_{kf}} &= \sum_{j=1}^{p}\sum_{\alpha=1}^{N}\sigma^{*kj}z_{f\alpha}\Big(x_{j\alpha} - \sum_{h=1}^{q}\beta^*_{jh}z_{h\alpha}\Big)\\ &= \sum_{j=1}^{p}\sigma^{*kj}\Big(c_{jf} - \sum_{h}\beta^*_{jh}a_{hf}\Big),\end{aligned}$$

where

$$\text{(11)}\qquad c_{jf} = \sum_{\alpha} x_{j\alpha}z_{f\alpha},$$

$$\text{(12)}\qquad a_{hf} = \sum_{\alpha} z_{h\alpha}z_{f\alpha}.$$

(10) results from the lemma by taking $v = \beta^*_{kf}$ and $u_{i\alpha} = x_{i\alpha} - \sum_h \beta^*_{ih}z_{h\alpha}$.

Since (σ^{*kj}) is nonsingular at a maximum of $\log L$, setting (10) equal to zero yields

$$c_{kf} - \sum_h \beta^*_{kh} a_{hf} = 0. \tag{13}$$

These equations can be summarized in matrix form as $C - \boldsymbol{\beta}^* A = \mathbf{0}$, where $C = (c_{kf})$ and $A = (a_{hf})$. The solution of this equation (for A nonsingular),

$$\hat{\boldsymbol{\beta}} = CA^{-1}, \tag{14}$$

is unique and defines the maximum likelihood estimate of $\boldsymbol{\beta}$. It will be seen that each row of $\hat{\boldsymbol{\beta}}$ is of the form of $\hat{\boldsymbol{\beta}} = \boldsymbol{b}$ given in (1) of Section 8.1.

By Lemma 3.2.2, we find that $\log L$ is maximized with respect to σ^{*ij} when

$$\boldsymbol{\Sigma}^{*-1} = N[\sum_\alpha (\boldsymbol{x}_\alpha - \boldsymbol{\beta}^* \boldsymbol{z}_\alpha)(\boldsymbol{x}_\alpha - \boldsymbol{\beta}^* \boldsymbol{z}_\alpha)']^{-1}. \tag{15}$$

Thus, the maximum likelihood estimate of $\boldsymbol{\Sigma}$ is

$$\hat{\boldsymbol{\Sigma}} = \frac{1}{N} \sum_\alpha (\boldsymbol{x}_\alpha - \hat{\boldsymbol{\beta}} \boldsymbol{z}_\alpha)(\boldsymbol{x}_\alpha - \hat{\boldsymbol{\beta}} \boldsymbol{z}_\alpha)'. \tag{16}$$

This is the multivariate analogue of $\hat{\sigma}^2 = (N - q)s^2/N$ defined by (2) of Section 8.1.

THEOREM 8.2.1. *If $\boldsymbol{x}_\alpha$ is an observation from $N(\boldsymbol{\beta} \boldsymbol{z}_\alpha, \boldsymbol{\Sigma})$, $\alpha = 1, \cdots, N$, with $(\boldsymbol{z}_1, \cdots, \boldsymbol{z}_N)$ of rank q, the maximum likelihood estimate of $\boldsymbol{\beta}$ is given by* (14), *where $C = \sum_\alpha \boldsymbol{x}_\alpha \boldsymbol{z}_\alpha'$ and $A = \sum_\alpha \boldsymbol{z}_\alpha \boldsymbol{z}_\alpha'$. The maximum likelihood estimate of $\boldsymbol{\Sigma}$ is given by* (16).

A useful algebraic result follows from (16):

$$\begin{aligned} N\hat{\boldsymbol{\Sigma}} &= \sum_\alpha \boldsymbol{x}_\alpha \boldsymbol{x}_\alpha' - \hat{\boldsymbol{\beta}} \sum_\alpha \boldsymbol{z}_\alpha \boldsymbol{x}_\alpha' - \sum_\alpha \boldsymbol{x}_\alpha \boldsymbol{z}_\alpha' \hat{\boldsymbol{\beta}}' + \hat{\boldsymbol{\beta}} \sum_\alpha \boldsymbol{z}_\alpha \boldsymbol{z}_\alpha' \hat{\boldsymbol{\beta}}' \\ &= \sum_\alpha \boldsymbol{x}_\alpha \boldsymbol{x}_\alpha' - \hat{\boldsymbol{\beta}} A \hat{\boldsymbol{\beta}}', \end{aligned} \tag{17}$$

by virtue of (14).

Now let us consider a geometric interpretation of the estimation procedure. Let the ith row of $(\boldsymbol{x}_1, \cdots, \boldsymbol{x}_N)$ be $\boldsymbol{x}_i^*$ (with N components) and the ith row of $(\boldsymbol{z}_1, \cdots, \boldsymbol{z}_N)$ be $\boldsymbol{z}_i^*$ (with N components). Then $\sum_j \hat{\beta}_{ij} \boldsymbol{z}_j^*$, being a linear combination of the vectors $\boldsymbol{z}_1^*, \cdots, \boldsymbol{z}_q^*$, is a vector in the q-space spanned by $\boldsymbol{z}_1^*, \cdots, \boldsymbol{z}_q^*$, and is in fact of all such vectors the one nearest to $\boldsymbol{x}_i^*$; hence, it is the projection of $\boldsymbol{x}_i^*$ on the q-space. Thus $\boldsymbol{x}_i^* - \sum_j \hat{\beta}_{ij} \boldsymbol{z}_j^*$ is the vector orthogonal to the q-space going from the projection of $\boldsymbol{x}_i^*$ on the q-space to $\boldsymbol{x}_i^*$. Translate this vector so that one end point of this vector is at the origin. Then the set of p vectors, $\boldsymbol{x}_1^* - \sum_j \hat{\beta}_{1j} \boldsymbol{z}_j^*, \cdots, \boldsymbol{x}_p^* - \sum_j \hat{\beta}_{pj} \boldsymbol{z}_j^*$, is a set of vectors emanating from the origin.

$N\hat{\sigma}_{ii} = (x_i^* - \sum_j \hat{\beta}_{ij} z_j^*)(x_i^* - \sum_j \hat{\beta}_{ij} z_j^*)'$ is the square of the length of the ith such vector, and $N\hat{\sigma}_{ij} = (x_i^* - \sum_h \hat{\beta}_{ih} z_h^*)(x_j^* - \sum_g \hat{\beta}_{jg} z_g^*)'$ is the product of the length of the ith vector, the length of the jth vector, and the cosine of the angle between them.

8.2.2. Distribution of $\hat{\boldsymbol\beta}$ and $\hat{\boldsymbol\Sigma}$

Now let us find the joint distribution of $\hat{\beta}_{iu}(i = 1, \cdots, p; u = 1, \cdots, q)$. Clearly, the joint distribution is normal since the $\hat{\beta}_{iu}$ are linear combinations of the $X_{i\alpha}$. From (14) we see that

$$\begin{aligned}
\text{(18)}\qquad \mathscr{E}\hat{\boldsymbol\beta} &= \mathscr{E} \sum_{\alpha=1}^{N} \mathbf{X}_\alpha \mathbf{z}_\alpha' A^{-1} \\
&= \sum_{\alpha=1}^{N} \boldsymbol\beta \mathbf{z}_\alpha \mathbf{z}_\alpha' A^{-1} = \boldsymbol\beta A A^{-1} \\
&= \boldsymbol\beta.
\end{aligned}$$

The covariance between $\hat{\boldsymbol\beta}_i$ and $\hat{\boldsymbol\beta}_j$, two rows of $\hat{\boldsymbol\beta}$, is

$$\begin{aligned}
\text{(19)}\qquad & \mathscr{E}(\hat{\boldsymbol\beta}_i - \boldsymbol\beta_i)'(\hat{\boldsymbol\beta}_j - \boldsymbol\beta_j) \\
&= A^{-1}\mathscr{E} \sum_{\alpha=1}^{N} (X_{i\alpha} - \mathscr{E}X_{i\alpha})\mathbf{z}_\alpha \sum_{\gamma=1}^{N} (X_{j\gamma} - \mathscr{E}X_{j\gamma})\mathbf{z}_\gamma' A^{-1} \\
&= A^{-1} \sum_{\alpha, \gamma=1}^{N} \mathscr{E}(X_{i\alpha} - \mathscr{E}X_{i\alpha})(X_{j\gamma} - \mathscr{E}X_{j\gamma})\mathbf{z}_\alpha \mathbf{z}_\gamma' A^{-1} \\
&= A^{-1} \sum_{\alpha, \gamma=1}^{N} \delta_{\alpha\gamma}\sigma_{ij}\mathbf{z}_\alpha \mathbf{z}_\gamma' A^{-1} \\
&= A^{-1} \sum_{\alpha=1}^{N} \sigma_{ij}\mathbf{z}_\alpha \mathbf{z}_\alpha' A^{-1} \\
&= \sigma_{ij} A^{-1} A A^{-1} \\
&= \sigma_{ij} A^{-1}.
\end{aligned}$$

To summarize, the row vector of pq components $(\hat{\boldsymbol\beta}_1, \cdots, \hat{\boldsymbol\beta}_p)$ is normally distributed with mean $(\boldsymbol\beta_1, \cdots, \boldsymbol\beta_p)$ and covariance matrix

$$\text{(20)}\qquad \begin{pmatrix}
\sigma_{11}A^{-1} & \sigma_{12}A^{-1} \cdots & \sigma_{1p}A^{-1} \\
\sigma_{21}A^{-1} & \sigma_{22}A^{-1} \cdots & \sigma_{2p}A^{-1} \\
\cdot & \cdot & \cdot \\
\cdot & \cdot & \cdot \\
\cdot & \cdot & \cdot \\
\sigma_{p1}A^{-1} & \sigma_{p2}A^{-1} \cdots & \sigma_{pp}A^{-1}
\end{pmatrix}.$$

The matrix (20) is the Kronecker (or direct) product of the matrices $\mathbf{\Sigma}$ and A^{-1}.

From Theorem 4.3.2 it follows that $N\hat{\mathbf{\Sigma}} = \sum_{\alpha=1}^{N} \boldsymbol{x}_\alpha \boldsymbol{x}_\alpha' - \hat{\boldsymbol{\beta}} A \hat{\boldsymbol{\beta}}'$ is distributed according to $W(\mathbf{\Sigma}, N - q)$. From this we see that an unbiased estimate of $\mathbf{\Sigma}$ is $S = [N/(N - q)]\hat{\mathbf{\Sigma}}$.

THEOREM 8.2.2. *The maximum likelihood estimate $\hat{\boldsymbol{\beta}}$ based on a set of N observations, the* αth *from $N(\boldsymbol{\beta} \boldsymbol{z}_\alpha, \mathbf{\Sigma})$, is normally distributed with mean $\boldsymbol{\beta}$, and the covariance matrix of the* ith *and* jth *rows of $\hat{\boldsymbol{\beta}}$ is $\sigma_{ij} A^{-1}$, where $A = \sum_\alpha \boldsymbol{z}_\alpha \boldsymbol{z}_\alpha'$. The maximum likelihood estimate $\hat{\mathbf{\Sigma}}$ multiplied by N is independently distributed according to $W(\mathbf{\Sigma}, N - q)$, where q is the number of components of $\boldsymbol{z}_\alpha$.*

In the exponent of the density of $\boldsymbol{x}_1, \cdots, \boldsymbol{x}_N$ we have

$$\sum_\alpha (\boldsymbol{x}_\alpha - \boldsymbol{\beta} \boldsymbol{z}_\alpha)' \mathbf{\Sigma}^{-1} (\boldsymbol{x}_\alpha - \boldsymbol{\beta} \boldsymbol{z}_\alpha) = \operatorname{tr} \mathbf{\Sigma}^{-1} \sum_\alpha (\boldsymbol{x}_\alpha - \boldsymbol{\beta} \boldsymbol{z}_\alpha)(\boldsymbol{x}_\alpha - \boldsymbol{\beta} \boldsymbol{z}_\alpha)'.$$

We can write

$$\begin{aligned} (21) \quad \sum_\alpha (\boldsymbol{x}_\alpha - \boldsymbol{\beta} \boldsymbol{z}_\alpha)(\boldsymbol{x}_\alpha - \boldsymbol{\beta} \boldsymbol{z}_\alpha)' &= \sum_\alpha [(\boldsymbol{x}_\alpha - \hat{\boldsymbol{\beta}} \boldsymbol{z}_\alpha) + (\hat{\boldsymbol{\beta}} - \boldsymbol{\beta}) \boldsymbol{z}_\alpha][(\boldsymbol{x}_\alpha - \hat{\boldsymbol{\beta}} \boldsymbol{z}_\alpha) + (\hat{\boldsymbol{\beta}} - \boldsymbol{\beta}) \boldsymbol{z}_\alpha]' \\ &= \sum_\alpha (\boldsymbol{x}_\alpha - \hat{\boldsymbol{\beta}} \boldsymbol{z}_\alpha)(\boldsymbol{x}_\alpha - \hat{\boldsymbol{\beta}} \boldsymbol{z}_\alpha)' + (\hat{\boldsymbol{\beta}} - \boldsymbol{\beta}) A (\hat{\boldsymbol{\beta}} - \boldsymbol{\beta})'. \end{aligned}$$

The density then can be written

$$(22) \qquad \frac{1}{(2\pi)^{\frac{1}{2}pN} |\mathbf{\Sigma}|^{\frac{1}{2}N}} \exp\left(-\tfrac{1}{2} \operatorname{tr} \{\mathbf{\Sigma}^{-1}[(\hat{\boldsymbol{\beta}} - \boldsymbol{\beta}) A (\hat{\boldsymbol{\beta}} - \boldsymbol{\beta})' + N\hat{\mathbf{\Sigma}}]\}\right).$$

This proves the following:

COROLLARY 8.2.1. *$\hat{\boldsymbol{\beta}}$ and $\hat{\mathbf{\Sigma}}$ form a sufficient set of statistics for $\boldsymbol{\beta}$ and $\mathbf{\Sigma}$.*

THEOREM 8.2.3. *Let $\boldsymbol{X}_\alpha$ be distributed according to $N(\boldsymbol{\beta} \boldsymbol{z}_\alpha, \mathbf{\Sigma})$, $\alpha = 1, \cdots, N$, independently of $\boldsymbol{X}_\beta$ $(\beta \neq \alpha)$.*

(a) Then if $\boldsymbol{w}_\alpha = \boldsymbol{H} \boldsymbol{z}_\alpha$ and $\mathbf{\Gamma} = \boldsymbol{\beta} H^{-1}$, then $\boldsymbol{X}_\alpha$ is distributed according to $N(\mathbf{\Gamma} \boldsymbol{w}_\alpha, \mathbf{\Sigma})$, independently of $\boldsymbol{X}_\beta$ $(\beta \neq \alpha)$.

(b) The maximum likelihood estimate of $\mathbf{\Gamma}$ based on observations $\boldsymbol{x}_\alpha$ on $\boldsymbol{X}_\alpha (\alpha = 1, \cdots, N)$ is $\hat{\mathbf{\Gamma}} = \hat{\boldsymbol{\beta}} H^{-1}$, where $\hat{\boldsymbol{\beta}}$ is the maximum likelihood estimate of $\boldsymbol{\beta}$.

(c) $\hat{\mathbf{\Gamma}}(\sum_\alpha \boldsymbol{w}_\alpha \boldsymbol{w}_\alpha') \hat{\mathbf{\Gamma}}' = \hat{\boldsymbol{\beta}} A \hat{\boldsymbol{\beta}}'$, where $A = \sum_\alpha \boldsymbol{z}_\alpha \boldsymbol{z}_\alpha'$, and the maximum likelihood estimate of $N\mathbf{\Sigma}$ is $N\hat{\mathbf{\Sigma}} = \sum_\alpha \boldsymbol{x}_\alpha \boldsymbol{x}_\alpha' - \hat{\mathbf{\Gamma}}(\sum_\alpha \boldsymbol{w}_\alpha \boldsymbol{w}_\alpha') \hat{\mathbf{\Gamma}}' = \sum_\alpha \boldsymbol{x}_\alpha \boldsymbol{x}_\alpha' - \hat{\boldsymbol{\beta}} A \hat{\boldsymbol{\beta}}'$.

(d) $\hat{\mathbf{\Gamma}}$ and $\hat{\mathbf{\Sigma}}$ are independently distributed.

(e) $\hat{\mathbf{\Gamma}}$ is normally distributed with mean $\mathbf{\Gamma}$ and the covariance matrix of the ith *and* jth *rows of $\hat{\mathbf{\Gamma}}$ is $\sigma_{ij}(\boldsymbol{H} \boldsymbol{A} \boldsymbol{H}')^{-1} = \sigma_{ij} \boldsymbol{H}'^{-1} \boldsymbol{A}^{-1} \boldsymbol{H}^{-1}$.*

The proof is left to the reader.

8.2.3. Computation of $\hat{\boldsymbol{\beta}}$ and $\hat{\boldsymbol{\Sigma}}$

For each k the equations (13), namely,

$$\sum_{h=1}^{q} a_{hf}\hat{\beta}_{kh} = c_{kf}, \tag{23}$$

are a set of q linear equations in q unknowns, namely, $\hat{\beta}_{k1}, \cdots, \hat{\beta}_{kq}$. An efficient method of solving such a set of equations numerically is the abbreviated Doolittle method. In using this method on p sets of such equations, the forward solution is the same for each set (with the exception of one column) and hence does not involve p times as much computation as the solution of one set.

In giving the computational method it will be convenient to replace the equation $\hat{\boldsymbol{\beta}}A = C$ by $A\hat{\boldsymbol{\beta}}' = C'$ or $A\bar{B} = \bar{C}$, where $\bar{B} = \hat{\boldsymbol{\beta}}'$ and $\bar{C} = C'$. In components this is

$$\sum_{j=1}^{q} a_{ij}\bar{b}_{jk} = \bar{c}_{ik}, \qquad i = 1, \cdots, q;\ k = 1, \cdots, p. \tag{24}$$

In both the abbreviated Doolittle method and the method of pivotal condensation, the forward solution consists of operations which replace (24) by a set of linear combinations of the component equations resulting in a new set

$$\sum_{h=f}^{q} a_{fh}^{**}\bar{b}_{hk} = \bar{c}_{fk}^{**}, \tag{25}$$

or

$$A^{**}\bar{B} = \bar{C}^{**}, \tag{26}$$

where

$$A^{**} = \begin{pmatrix} 1 & a_{12}^{**} & \cdots & a_{1q}^{**} \\ 0 & 1 & \cdots & a_{2q}^{**} \\ \cdot & \cdot & & \cdot \\ \cdot & \cdot & & \cdot \\ \cdot & \cdot & & \cdot \\ 0 & 0 & \cdots & 1 \end{pmatrix}. \tag{27}$$

The backward solution consists in solving (25) rewritten as

$$\bar{b}_{fk} = \bar{c}_{fk}^{**} - \sum_{h=f+1}^{q} a_{fh}^{**}\bar{b}_{hk}, \tag{28}$$

successively for $\bar{b}_{qk}, \bar{b}_{q-1,k}, \cdots, \bar{b}_{1k}(k = 1, \cdots, p)$. This is relatively easy because of the form of A^{**}.

The method of pivotal condensation is the straightforward method of successive elimination of variables. Before the jth step, the equations for the last $q - j + 1$ $\bar{b}$'s (for a given k) are

$$\sum_{h=j}^{q} \tilde{a}_{gh}^{(j-1)} \bar{b}_{hk} = \tilde{c}_{gk}^{(j-1)}, \qquad g = j, \cdots, q \tag{29}$$

($\tilde{a}_{ij}^{(0)} = a_{ij}$). The jth step of the computation is

$$\tilde{a}_{jh}^{(j)} = \frac{\tilde{a}_{jh}^{(j-1)}}{\tilde{a}_{jj}^{(j-1)}}, \qquad h = j, \cdots, q, \tag{30}$$

$$\tilde{c}_{jk}^{(j)} = \frac{\tilde{c}_{jk}^{(j-1)}}{a_{jj}^{(j-1)}},$$

($\tilde{a}_{jj}^{(j)} = 1$) and

$$\tilde{a}_{gh}^{(j)} = \tilde{a}_{gh}^{(j-1)} - \tilde{a}_{gj}^{(j-1)} \tilde{a}_{jh}^{(j)}, \qquad g = j+1, \cdots, q, \; h = j, \cdots, q, \tag{31}$$

$$\tilde{c}_{gk}^{(j)} = \tilde{c}_{gk}^{(j-1)} - \tilde{a}_{gj}^{(j-1)} \tilde{c}_{jk}^{(j)}.$$

The equations (30) and (31) indicate that at the jth step we divide equation j by its leading coefficient to obtain

$$\bar{b}_{jk} + \sum_{h=j+1}^{q} \tilde{a}_{jh}^{(j)} \bar{b}_{hk} = \tilde{c}_{jk}^{(j)}. \tag{32}$$

In (31) we subtract $\tilde{a}_{gj}^{(j-1)}$ times this result from each of the succeeding equations. Then the last $q - j$ equations involve only the last $q - j$ $\bar{b}$'s (since $a_{gj}^{(j)} = 0$, $g > j$), and they are of the form (29) with $j - 1$ replaced by j. If we assemble the equations (32) for $j = 1, \cdots, p$ we obtain (26) with A^{**} of the form of (27) (that is, $a_{jh}^{**} = \tilde{a}_{jh}^{(j)}$, $h = j, \cdots, q$).

In the abbreviated Doolittle procedure one performs the same computations, but in a different order. Let $a_{jh}^{*} = \tilde{a}_{jh}^{(j-1)}$, $h \geq j$. Then

$$A^{*} = \begin{pmatrix} a_{11}^{*} & a_{12}^{*} & \cdots & a_{1q}^{*} \\ 0 & a_{22}^{*} & \cdots & a_{2q}^{*} \\ \cdot & \cdot & & \cdot \\ \cdot & \cdot & & \cdot \\ \cdot & \cdot & & \cdot \\ 0 & 0 & \cdots & a_{qq}^{*} \end{pmatrix} \tag{33}$$

and

$$a_{ij}^{**} = a_{ij}^{*} / a_{ii}^{*}, \qquad j = i, i+1, \cdots, q. \tag{34}$$

The computation of the elements of A^{*} in the abbreviated Doolittle method is by (note $a_{jg}^{*} a_{jh}^{**} = a_{jg}^{**} a_{jh}^{*}$)

$$a_{gh}^{*} = a_{gh} - \sum_{j=1}^{g-1} a_{jg}^{*} a_{jh}^{**}, \qquad h = g, g+1, \cdots, q. \tag{35}$$

These computations can be done easily on a desk calculator with a minimum of recording. The same operations are carried out on $\bar{C}$:

$$\bar{c}_{ik}^{*} = \bar{c}_{ik} - \sum_{h=1}^{i-1} a_{hi}^{*} \bar{c}_{hk}^{**}, \tag{36}$$

$$\bar{c}_{ik}^{**} = \bar{c}_{ik}^{*} / a_{ii}^{*}. \tag{37}$$

The computation of the elements in the gth row involves only the elements in the first $g - 1$ rows of A^* and A^{**}. It is shown in Appendix 1 that the computations for the abbreviated Doolittle method are the same as for the pivotal condensation method.

The operations in either case amount to multiplication of A and $\bar{C}$ on the left by a triangular matrix

$$F = \begin{pmatrix} 1 & 0 & \cdots & 0 \\ f_{21} & 1 & \cdots & 0 \\ \cdot & \cdot & & \cdot \\ \cdot & \cdot & & \cdot \\ \cdot & \cdot & & \cdot \\ f_{q1} & f_{q2} & \cdots & 1 \end{pmatrix}; \tag{38}$$

that is,

$$FA = A^*. \tag{39}$$

This is shown in Appendix 1. Then

$$FAF' = FA^{*\prime} = \begin{pmatrix} 1 & 0 & \cdots & 0 \\ f_{21} & 1 & \cdots & 0 \\ \cdot & \cdot & & \cdot \\ \cdot & \cdot & & \cdot \\ \cdot & \cdot & & \cdot \\ f_{q1} & f_{q2} & \cdots & 1 \end{pmatrix} \begin{pmatrix} a_{11}^* & 0 & \cdots & 0 \\ a_{12}^* & a_{22}^* & \cdots & 0 \\ \cdot & \cdot & & \cdot \\ \cdot & \cdot & & \cdot \\ \cdot & \cdot & & \cdot \\ a_{1q}^* & a_{2q}^* & \cdots & a_{qq}^* \end{pmatrix} \tag{40}$$

is triangular with diagonal elements a_{ii}^*, but since FAF' is symmetric it must be diagonal; that is,

$$FAF' = \begin{pmatrix} a_{11}^* & 0 & \cdots & 0 \\ 0 & a_{22}^* & \cdots & 0 \\ \cdot & \cdot & & \cdot \\ \cdot & \cdot & & \cdot \\ \cdot & \cdot & & \cdot \\ 0 & 0 & \cdots & a_{qq}^* \end{pmatrix} = D, \tag{41}$$

say, a diagonal matrix. Furthermore $A^{**} = D^{-1}A^*$, $\bar{C}^* = F\bar{C}$, and $\bar{C}^{**} = D^{-1}\bar{C}^*$. From (41) we obtain

$$FAF'D^{-1} = I, \tag{42}$$

$$A^{-1} = F'D^{-1}F. \tag{43}$$

Thus,

$$\begin{aligned} \hat{\boldsymbol{\beta}} A \hat{\boldsymbol{\beta}}' &= \bar{B}' A \bar{B} \\ &= \bar{B}' A A^{-1} A \bar{B} \\ &= \bar{C}' A^{-1} \bar{C} \\ &= \bar{C}' F' D^{-1} F \bar{C} \\ &= \bar{C}^{*\prime} \bar{C}^{**} \\ &= \bar{C}^{**\prime} \bar{C}^{*}. \end{aligned} \tag{44}$$

This gives us a way to compute $\hat{\boldsymbol{\beta}} A \hat{\boldsymbol{\beta}}'$ and hence to compute $\hat{\boldsymbol{\Sigma}}$ for $N\hat{\boldsymbol{\Sigma}} = \sum \boldsymbol{x}_\alpha \boldsymbol{x}_\alpha' - \hat{\boldsymbol{\beta}} A \hat{\boldsymbol{\beta}}'$. Alternatively, $\hat{\boldsymbol{\beta}} A \hat{\boldsymbol{\beta}}' = \hat{\boldsymbol{\beta}} C'$.

It may be observed that the forward solution can be checked by carrying along a sum column on the right $\sum_{h=1}^{q} a_{fh} + \sum_{k=1}^{p} \bar{c}_{fk}$; the operation carried out on this sum should equal $\sum_{h=1}^{q} a_{fh}^{**} + \sum_{k=1}^{p} \bar{c}_{fk}^{**}$. The solution $\bar{B}$ can be checked by comparing $A\bar{B}$ with $\bar{C}$.

The calculation of A^{-1} can be carried along with the calculation of $\bar{B}$, for A^{-1} satisfies $A(A^{-1}) = I$ which is of the same form as $A\bar{B} = \bar{C}$. The operations carried out on $\bar{C}$ are also carried out on I. At the end of the forward solution one has the resulting equation $A^{**}(A^{-1}) = D^{-1}F$; each column of this matrix equation constitutes a set of q simultaneous equations which are solved for the elements of that column of A^{-1}.

We have defined $a_{ij} = \sum_\alpha z_{i\alpha} z_{j\alpha}$ and $c_{ki} = \sum_\alpha x_{k\alpha} z_{i\alpha} = \bar{c}_{ik}$. If $z_{1\alpha} \equiv 1$, then $a_{11} = N$, $a_{1j} = N\bar{z}_j$, and $c_{k1} = \bar{c}_{1k} = N\bar{x}_k$. In this case the first step of the pivotal condensation yields

$$\begin{aligned} \tilde{a}_{gh}^{(1)} &= a_{gh}^{(0)} - \frac{a_{g1}^{(0)} a_{1h}^{(0)}}{a_{11}^{(0)}} \\ &= \sum_\alpha z_{g\alpha} z_{h\alpha} - N\bar{z}_g \bar{z}_h, \\ \tilde{c}_{gk}^{(1)} &= \sum_\alpha x_{k\alpha} z_{g\alpha} - N\bar{x}_k \bar{z}_g, \qquad g, h = 2, \cdots, q. \end{aligned} \tag{45}$$

8.3. LIKELIHOOD RATIO CRITERIA FOR TESTING LINEAR HYPOTHESES ABOUT REGRESSION COEFFICIENTS

Suppose we partition

$$\boldsymbol{\beta} = (\boldsymbol{\beta}_1 \; \boldsymbol{\beta}_2) \tag{1}$$

such that $\boldsymbol{\beta}_1$ has q_1 columns and $\boldsymbol{\beta}_2$ has q_2 columns. We shall derive the likelihood ratio criterion for testing the hypothesis

$$H: \; \boldsymbol{\beta}_1 = \boldsymbol{\beta}_1^*, \tag{2}$$

where $\boldsymbol{\beta}_1^*$ is a given matrix. The maximum of the likelihood function L for the sample $x_1, \cdots, x_N$ is

$$\max_{\boldsymbol{\beta}, \boldsymbol{\Sigma}} L = (2\pi)^{-\frac{1}{2}pN} |\hat{\boldsymbol{\Sigma}}_\Omega|^{-\frac{1}{2}N} e^{-\frac{1}{2}pN}, \tag{3}$$

where $\hat{\boldsymbol{\Sigma}}_\Omega$ is given by (16) or (17) of Section 8.2.

To find the maximum of the likelihood function for the parameters restricted to ω defined by (2) we let

$$y_\alpha = x_\alpha - \boldsymbol{\beta}_1^* z_\alpha^{(1)}, \tag{4}$$

where

$$z_\alpha = \begin{pmatrix} z_\alpha^{(1)} \\ z_\alpha^{(2)} \end{pmatrix} \tag{5}$$

is partitioned in a manner corresponding to the partition of $\boldsymbol{\beta}$. Then y_α can be considered as an observation from $N(\boldsymbol{\beta}_2 z_\alpha^{(2)}, \boldsymbol{\Sigma})$. The estimate of $\boldsymbol{\beta}_2$ is obtained by the procedure of Section 8.2 as

$$\begin{aligned} \hat{\boldsymbol{\beta}}_{2\omega} &= \sum_{\alpha=1}^N y_\alpha z_\alpha^{(2)\prime} A_{22}^{-1} = \sum_{\alpha=1}^N (x_\alpha - \boldsymbol{\beta}_1^* z_\alpha^{(1)}) z_\alpha^{(2)\prime} A_{22}^{-1} \\ &= (C_2 - \boldsymbol{\beta}_1^* A_{12}) A_{22}^{-1} \end{aligned} \tag{6}$$

with C and A being partitioned in the manner corresponding to the partitioning of $\boldsymbol{\beta}$ and z_α,

$$C = (C_1 \; C_2), \tag{7}$$

$$A = \begin{pmatrix} A_{11} & A_{12} \\ A_{21} & A_{22} \end{pmatrix}. \tag{8}$$

The estimate of $\boldsymbol{\Sigma}$ is given by

$$\begin{aligned} N\hat{\boldsymbol{\Sigma}}_\omega &= \sum_{\alpha=1}^N (y_\alpha - \hat{\boldsymbol{\beta}}_{2\omega} z_\alpha^{(2)})(y_\alpha - \hat{\boldsymbol{\beta}}_{2\omega} z_\alpha^{(2)})' \\ &= \sum_{\alpha=1}^N y_\alpha y_\alpha' - \hat{\boldsymbol{\beta}}_{2\omega} A_{22} \hat{\boldsymbol{\beta}}_{2\omega}' \\ &= \sum_{\alpha=1}^N (x_\alpha - \boldsymbol{\beta}_1^* z_\alpha^{(1)})(x_\alpha - \boldsymbol{\beta}_1^* z_\alpha^{(1)})' - \hat{\boldsymbol{\beta}}_{2\omega} A_{22} \hat{\boldsymbol{\beta}}_{2\omega}'. \end{aligned} \tag{9}$$

Thus the maximum of the likelihood function over ω is

$$\max_{\boldsymbol{\beta}_2, \boldsymbol{\Sigma}} L = (2\pi)^{-\frac{1}{2}pN} |\hat{\boldsymbol{\Sigma}}_\omega|^{-\frac{1}{2}N} e^{-\frac{1}{2}pN}. \tag{10}$$

The likelihood ratio criterion for testing H is (10) divided by (3), namely,

$$\lambda = \frac{|\hat{\boldsymbol{\Sigma}}_\Omega|^{\frac{1}{2}N}}{|\hat{\boldsymbol{\Sigma}}_\omega|^{\frac{1}{2}N}}. \tag{11}$$

In testing H, one rejects the hypothesis if $\lambda < \lambda_0$, where λ_0 is a suitably chosen number.

An insight into the algebra developed here can be given in terms of a geometric interpretation. It will be convenient to use the following lemma:

LEMMA 8.3.1.

$$\hat{\boldsymbol{\beta}}_{2\omega} - \hat{\boldsymbol{\beta}}_{2\Omega} = (\hat{\boldsymbol{\beta}}_{1\Omega} - \boldsymbol{\beta}_1^*)A_{12}A_{22}^{-1}, \tag{12}$$

Proof. From (14) of Section 8.2 we have that

$$C_2 = \hat{\boldsymbol{\beta}}_{\Omega}\begin{pmatrix} A_{12} \\ A_{22} \end{pmatrix} = \hat{\boldsymbol{\beta}}_{1\Omega}A_{12} + \hat{\boldsymbol{\beta}}_{2\Omega}A_{22}. \tag{13}$$

Thus $\hat{\boldsymbol{\beta}}_{2\Omega} = C_2A_{22}^{-1} - \hat{\boldsymbol{\beta}}_{1\Omega}A_{12}A_{22}^{-1}$. The lemma follows by comparison with (6).

We can now write

$$\begin{aligned} X - \boldsymbol{\beta}Z &= (X - \hat{\boldsymbol{\beta}}_{\Omega}Z) + (\hat{\boldsymbol{\beta}}_{2\Omega} - \boldsymbol{\beta}_2)Z_2 + (\hat{\boldsymbol{\beta}}_{1\Omega} - \boldsymbol{\beta}_1^*)Z_1 \\ &= (X - \hat{\boldsymbol{\beta}}_{\Omega}Z) + (\hat{\boldsymbol{\beta}}_{2\omega} - \boldsymbol{\beta}_2)Z_2 \\ &\qquad - (\hat{\boldsymbol{\beta}}_{2\omega} - \hat{\boldsymbol{\beta}}_{2\Omega})Z_2 + (\hat{\boldsymbol{\beta}}_{1\Omega} - \boldsymbol{\beta}_1^*)Z_1 \\ &= (X - \hat{\boldsymbol{\beta}}_{\Omega}Z) + (\hat{\boldsymbol{\beta}}_{2\omega} - \boldsymbol{\beta}_2)Z_2 \\ &\qquad + (\hat{\boldsymbol{\beta}}_{1\Omega} - \boldsymbol{\beta}_1^*)(Z_1 - A_{12}A_{22}^{-1}Z_2) \end{aligned} \tag{14}$$

as an identity; here $X = (x_1, \cdots, x_N)$, $Z_1 = (z_1^{(1)}, \cdots, z_N^{(1)})$, and $Z_2 = (z_1^{(2)}, \cdots, z_N^{(2)})$. The rows of $Z = \begin{pmatrix} Z_1 \\ Z_2 \end{pmatrix}$ span a q-dimensional space. Each row of $\boldsymbol{\beta}Z$ is a vector in the q-space, and hence each row of $X - \boldsymbol{\beta}Z$ is a vector from a vector in the q-space to the corresponding row vector of X. Each row vector of $X - \boldsymbol{\beta}Z$ is expressed above as the sum of three row vectors. The first matrix on the right of (14) has as its ith row a vector orthogonal to the q-space and leading to the ith row vector of X (as shown in the preceding section). The row vectors of $(\hat{\boldsymbol{\beta}}_{2\omega} - \boldsymbol{\beta}_2)Z_2$ are vectors in the q_2-space spanned by the rows of Z_2 (since they are linear combinations of the rows of Z_2). The row vectors of $(\hat{\boldsymbol{\beta}}_{1\Omega} - \boldsymbol{\beta}_1^*)(Z_1 - A_{12}A_{22}^{-1}Z_2)$ are vectors in the q_1-space of $Z_1 - A_{12}A_{22}^{-1}Z_2$ and this space is in the q-space of Z, but orthogonal to the q_2-space of Z_2 [since $(Z_1 - A_{12}A_{22}^{-1}Z_2)Z_2' = \mathbf{0}$]. Thus each row of $X - \boldsymbol{\beta}Z$ is indicated in Figure 1 as the sum of three orthogonal vectors; one vector is in the space orthogonal to Z, one is in the space of Z_2, and one is in the subspace of Z which is orthogonal to Z_2.

From the orthogonality relations we have

$$(15)\quad \begin{aligned}(X-\boldsymbol{\beta}Z)(X-\boldsymbol{\beta}Z)' &= (X-\hat{\boldsymbol{\beta}}_\Omega Z)(X-\hat{\boldsymbol{\beta}}_\Omega Z)' + (\hat{\boldsymbol{\beta}}_{2\omega}-\boldsymbol{\beta}_2)Z_2Z_2'(\hat{\boldsymbol{\beta}}_{2\omega}-\boldsymbol{\beta}_2)' \\ &\quad + (\hat{\boldsymbol{\beta}}_{1\Omega}-\boldsymbol{\beta}_1^*)(Z_1-A_{12}A_{22}^{-1}Z_2)(Z_1-A_{12}A_{22}^{-1}Z_2)'(\hat{\boldsymbol{\beta}}_{1\Omega}-\boldsymbol{\beta}_1^*)' \\ &= N\hat{\boldsymbol{\Sigma}}_\Omega + (\hat{\boldsymbol{\beta}}_{2\omega}-\boldsymbol{\beta}_2)A_{22}(\hat{\boldsymbol{\beta}}_{2\omega}-\boldsymbol{\beta}_2)' \\ &\quad + (\hat{\boldsymbol{\beta}}_{1\Omega}-\boldsymbol{\beta}_1^*)(A_{11}-A_{12}A_{22}^{-1}A_{21})(\hat{\boldsymbol{\beta}}_{1\Omega}-\boldsymbol{\beta}_1^*)'.\end{aligned}$$

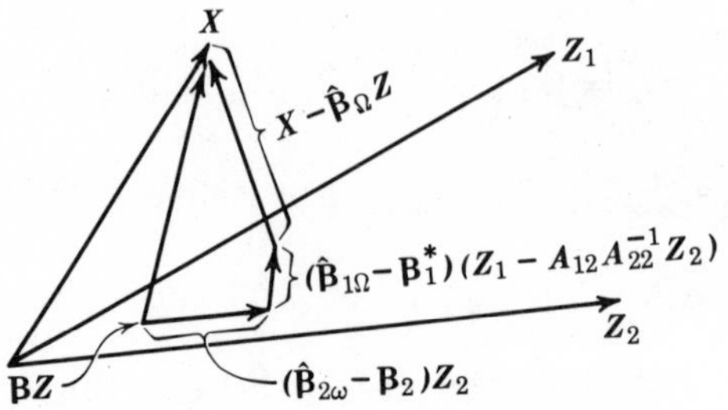

FIGURE 1

If we subtract $(\hat{\boldsymbol{\beta}}_{2\omega}-\boldsymbol{\beta}_2)Z_2$ from both sides of (14) we have

$$(16)\quad X-\boldsymbol{\beta}_1^*Z_1-\hat{\boldsymbol{\beta}}_{2\omega}Z_2 = (X-\hat{\boldsymbol{\beta}}_\Omega Z) + (\hat{\boldsymbol{\beta}}_{1\Omega}-\boldsymbol{\beta}_1^*)(Z_1-A_{12}A_{22}^{-1}Z_2).$$

From this we obtain

$$(17)\quad \begin{aligned}N\hat{\boldsymbol{\Sigma}}_\omega &= (X-\boldsymbol{\beta}_1^*Z_1-\hat{\boldsymbol{\beta}}_{2\omega}Z_2)(X-\boldsymbol{\beta}_1^*Z_1-\hat{\boldsymbol{\beta}}_{2\omega}Z_2)' \\ &= (X-\hat{\boldsymbol{\beta}}_\Omega Z)(X-\hat{\boldsymbol{\beta}}_\Omega Z)' \\ &\quad + (\hat{\boldsymbol{\beta}}_{1\Omega}-\boldsymbol{\beta}_1^*)(Z_1-A_{12}A_{22}^{-1}Z_2)(Z_1-A_{12}A_{22}^{-1}Z_2)'(\hat{\boldsymbol{\beta}}_{1\Omega}-\boldsymbol{\beta}_1^*)' \\ &= N\hat{\boldsymbol{\Sigma}}_\Omega + (\hat{\boldsymbol{\beta}}_{1\Omega}-\boldsymbol{\beta}_1^*)(A_{11}-A_{12}A_{22}^{-1}A_{21})(\hat{\boldsymbol{\beta}}_{1\Omega}-\boldsymbol{\beta}_1^*)'.\end{aligned}$$

The determinant $|\hat{\boldsymbol{\Sigma}}_\Omega| = (1/N^p)|(X-\hat{\boldsymbol{\beta}}_\Omega Z)(X-\hat{\boldsymbol{\beta}}_\Omega Z)'|$ is proportional to the volume squared of the parallelotope spanned by the row vectors of $X-\hat{\boldsymbol{\beta}}_\Omega Z$ (translated to the origin). The determinant $|\hat{\boldsymbol{\Sigma}}_\omega| = (1/N^p)$ $|(X-\boldsymbol{\beta}_1^*Z_1-\hat{\boldsymbol{\beta}}_{2\omega}Z_2)(X-\boldsymbol{\beta}_1^*Z_1-\hat{\boldsymbol{\beta}}_{2\omega}Z_2)'|$ is proportional to the volume squared of the parallelotope spanned by the row vectors of $X-\boldsymbol{\beta}_1^*Z_1-\hat{\boldsymbol{\beta}}_{2\omega}Z_2$ (translated to the origin); each of these vectors is the part of the vector of $X-\boldsymbol{\beta}_1^*Z_1$ that is orthogonal to Z_2. Thus the test based on the likelihood ratio criterion depends on the ratio of volumes of parallelotopes. One parallelotope involves vectors orthogonal to Z and the other involves vectors orthogonal to Z_2.

From (15) and (22) of Section 8.2 we see that the density of $\boldsymbol{x}_1, \cdots, \boldsymbol{x}_N$ can be written as

$$(18)\quad \frac{1}{(2\pi)^{\frac{1}{2}pN}|\boldsymbol{\Sigma}|^{\frac{1}{2}N}} \exp\left(-\tfrac{1}{2}\operatorname{tr}\{\boldsymbol{\Sigma}^{-1}[N\hat{\boldsymbol{\Sigma}} + (\hat{\boldsymbol{\beta}}_{2\omega} - \boldsymbol{\beta}_2)A_{22}(\hat{\boldsymbol{\beta}}_{2\omega} - \boldsymbol{\beta}_2)' + (\hat{\boldsymbol{\beta}}_{1\Omega} - \boldsymbol{\beta}_1^*)(A_{11} - A_{12}A_{22}^{-1}A_{21})(\hat{\boldsymbol{\beta}}_{1\Omega} - \boldsymbol{\beta}_1^*)']\}\right).$$

Thus $\hat{\boldsymbol{\Sigma}}$, $\hat{\boldsymbol{\beta}}_{1\Omega}$, and $\hat{\boldsymbol{\beta}}_{2\omega}$ form a sufficient set of statistics for $\boldsymbol{\Sigma}$, $\boldsymbol{\beta}_1$, and $\boldsymbol{\beta}_2$.

Wilks (1932a) first gave the likelihood ratio criterion for testing the equality of mean vectors from several populations (Section 8.8). Wilks (1934) and Bartlett (1934) extended its use to regression coefficients.

8.4. THE MOMENTS OF THE LIKELIHOOD RATIO CRITERION WHEN THE HYPOTHESIS IS TRUE

The likelihood ratio criterion is the $N/2$ power of

$$(1)\quad U = \lambda^{2/N} = \frac{|\hat{\boldsymbol{\Sigma}}_\Omega|}{|\hat{\boldsymbol{\Sigma}}_\omega|} = \frac{|N\hat{\boldsymbol{\Sigma}}_\Omega|}{|N\hat{\boldsymbol{\Sigma}}_\Omega + (\hat{\boldsymbol{\beta}}_{1\Omega} - \boldsymbol{\beta}_1^*)A_{11\cdot 2}(\hat{\boldsymbol{\beta}}_{1\Omega} - \boldsymbol{\beta}_1^*)'|},$$

where $A_{11\cdot 2} = A_{11} - A_{12}A_{22}^{-1}A_{21}$. We shall find the moments of U when $\boldsymbol{\beta}_1 = \boldsymbol{\beta}_1^*$. It has been shown in Section 8.2 that $N\hat{\boldsymbol{\Sigma}}_\Omega$ is distributed according to $W(\boldsymbol{\Sigma}, n)$, where $n = N - q$, and the elements of $\hat{\boldsymbol{\beta}}_\Omega - \boldsymbol{\beta}$ have a joint normal distribution independent of $N\hat{\boldsymbol{\Sigma}}_\Omega$ and the covariance of the ith and jth rows is $\sigma_{ij}\boldsymbol{A}^{-1}$.

LEMMA 8.4.1. *$(\hat{\boldsymbol{\beta}}_{1\Omega} - \boldsymbol{\beta}_1^*)A_{11\cdot 2}(\hat{\boldsymbol{\beta}}_{1\Omega} - \boldsymbol{\beta}_1^*)'$ is distributed as $\sum_{\nu=1}^{q_1} \boldsymbol{Y}_\nu \boldsymbol{Y}_\nu'$, where the $\boldsymbol{Y}_\nu$ are independently distributed, each according to $N(\boldsymbol{0}, \boldsymbol{\Sigma})$.*

Proof. The covariance between the ith and jth rows of $\hat{\boldsymbol{\beta}}_{1\Omega}$ is σ_{ij} times the submatrix of $\boldsymbol{A}^{-1}$ consisting of the first q_1 rows and columns; that is, it is $\sigma_{ij}\boldsymbol{A}_{11\cdot 2}^{-1}$ (see Problem 18 of Chapter 2). Let $\boldsymbol{E}$ be any matrix satisfying $\boldsymbol{E}\boldsymbol{A}_{11\cdot 2}\boldsymbol{E}' = \boldsymbol{I}$ and let

$$(2)\quad \hat{\boldsymbol{\beta}}_{1\Omega} - \boldsymbol{\beta}_1^* = \boldsymbol{Y}\boldsymbol{E} = (\boldsymbol{Y}_1, \cdots, \boldsymbol{Y}_{q_1})\boldsymbol{E}.$$

Then

$$(3)\quad (\hat{\boldsymbol{\beta}}_{1\Omega} - \boldsymbol{\beta}_1^*)A_{11\cdot 2}(\hat{\boldsymbol{\beta}}_{1\Omega} - \boldsymbol{\beta}_1^*)' = \boldsymbol{Y}\boldsymbol{Y}' = \sum_{\nu=1}^{q_1} \boldsymbol{Y}_\nu \boldsymbol{Y}_\nu'.$$

Clearly $\mathscr{E}\boldsymbol{Y} = \mathscr{E}(\hat{\boldsymbol{\beta}}_{1\Omega} - \boldsymbol{\beta}_1^*)\boldsymbol{E}^{-1} = \boldsymbol{0}$ since $\mathscr{E}\hat{\boldsymbol{\beta}}_{1\Omega} = \boldsymbol{\beta}_1^*$. Let the ith row of $\hat{\boldsymbol{\beta}}_{1\Omega}$ be $\hat{\boldsymbol{\beta}}_{i1}$, and the ith row of $\boldsymbol{Y}$ be $\tilde{\boldsymbol{Y}}_i$. Then

$$(4)\quad \begin{aligned} \mathscr{E}\tilde{\boldsymbol{Y}}_i'\tilde{\boldsymbol{Y}}_j &= \mathscr{E}(\boldsymbol{E}^{-1})'(\hat{\boldsymbol{\beta}}_{i1} - \boldsymbol{\beta}_{i1}^*)'(\hat{\boldsymbol{\beta}}_{j1} - \boldsymbol{\beta}_{j1}^*)\boldsymbol{E}^{-1} \\ &= \sigma_{ij}(\boldsymbol{E}^{-1})'\boldsymbol{A}_{11\cdot 2}^{-1}\boldsymbol{E}^{-1} \\ &= \sigma_{ij}(\boldsymbol{E}\boldsymbol{A}_{11\cdot 2}\boldsymbol{E}')^{-1} \\ &= \sigma_{ij}\boldsymbol{I}. \end{aligned}$$

This proves Lemma 8.4.1.

Let $G = N\hat{\mathbf{\Sigma}}_\Omega$. Then the criterion is

$$U = \frac{|G|}{|G + \sum Y_\nu Y_\nu'|}. \tag{5}$$

Let $F = G + \sum Y_i Y_i'$. The hth moment of U is

$$\begin{aligned}\mathscr{E}(U^h) &= \mathscr{E}(|G|^h |F|^{-h}) \\ &= \int \cdots \int K(\mathbf{\Sigma}, n)|G|^h |F|^{-h} |G|^{\frac{1}{2}(n-p-1)} \exp(-\tfrac{1}{2}\operatorname{tr} \mathbf{\Sigma}^{-1} G) \\ &\qquad \cdot (2\pi)^{-\frac{1}{2}q_1 p} |\mathbf{\Sigma}|^{-\frac{1}{2}q_1} \exp(-\tfrac{1}{2}\textstyle\sum_i y_i' \mathbf{\Sigma}^{-1} y_i)\, dG dY,\end{aligned} \tag{6}$$

where

$$K^{-1}(\mathbf{\Sigma}, n) = 2^{\frac{1}{2}np} \pi^{p(p-1)/4} |\mathbf{\Sigma}|^{\frac{1}{2}n} \prod_{i=1}^{p} \Gamma[\tfrac{1}{2}(n + 1 - i)] \tag{7}$$

and $dG = dg_{11} \cdots dg_{pp}$, $dY = dy_{11} \cdots dy_{pq_1}$. The moment can be written

$$\begin{aligned}\mathscr{E}U^h = \frac{K(\mathbf{\Sigma}, n)}{K(\mathbf{\Sigma}, n + 2h)} \int \cdots \int |F|^{-h} \\ \cdot K(\mathbf{\Sigma}, n + 2h)|G|^{\frac{1}{2}(n+2h-p-1)} \exp(-\tfrac{1}{2}\operatorname{tr} \mathbf{\Sigma}^{-1} G) \\ \cdot (2\pi)^{-\frac{1}{2}q_1 p} |\mathbf{\Sigma}|^{-\frac{1}{2}q_1} \exp(-\tfrac{1}{2}\textstyle\sum y_i' \mathbf{\Sigma}^{-1} y_i)\, dG\, dY.\end{aligned} \tag{8}$$

The part of the integrand involving G can be thought of as the density of $W(\mathbf{\Sigma}, n + 2h)$. Since $F = G + \sum y_i y_i'$, the integral can be thought of as the expected value of $|F|^{-h}$ when F has the distribution $W(\mathbf{\Sigma}, n + 2h + q_1)$. Thus (8) can be written

$$\begin{aligned}\mathscr{E}U^h &= \frac{K(\mathbf{\Sigma}, n)}{K(\mathbf{\Sigma}, n + 2h)} \int \cdots \int |F|^{-h} w(F|\mathbf{\Sigma}, n + 2h + q_1)\, dF \\ &= \frac{K(\mathbf{\Sigma}, n)K(\mathbf{\Sigma}, n + 2h + q_1)}{K(\mathbf{\Sigma}, n + 2h)K(\mathbf{\Sigma}, n + q_1)}\end{aligned} \tag{9}$$

by (18) of Section 7.5. This is

$$\begin{aligned}\mathscr{E}U^h &= \prod_{i=1}^{p} \left\{\frac{\Gamma[\frac{1}{2}(N - q_1 - q_2 + 1 - i) + h]\Gamma[\frac{1}{2}(N - q_2 + 1 - i)]}{\Gamma[\frac{1}{2}(N - q_1 - q_2 + 1 - i)]\Gamma[\frac{1}{2}(N - q_2 + 1 - i) + h]}\right\} \\ &= \prod_{i=1}^{p} \left\{\frac{\Gamma[\frac{1}{2}(n + 1 - i) + h]\Gamma[\frac{1}{2}(n + q_1 + 1 - i)]}{\Gamma[\frac{1}{2}(n + 1 - i)]\Gamma[\frac{1}{2}(n + q_1 + 1 - i) + h]}\right\}.\end{aligned} \tag{10}$$

We shall denote the random variable U when H is true by $U_{p, q_1, n}$. This is an analogue to $\dfrac{1}{1 + (q_1/n)F_{q_1, n}}$ for $p = 1$.

THEOREM 8.4.1. *The* hth *moment of* $\lambda^{2/N}$ *defined in Section* 8.3 *is given by* (10) *when the hypothesis is true.*

Since U lies between 0 and 1, the moments determine the distribution uniquely.

It is easy to show that p and q_1 can be interchanged in the above expression when $N - q_2$ is held fixed; that is, that

$$(11)\quad \mathscr{E}U^h = \prod_{i=1}^{q_1} \left\{ \frac{\Gamma[\frac{1}{2}(N - q_2 + 1 - i)]\Gamma[\frac{1}{2}(N - q_2 - p + 1 - i) + h]}{\Gamma[\frac{1}{2}(N - q_2 + 1 - i) + h]\Gamma[\frac{1}{2}(N - q_2 - p + 1 - i)]} \right\}.$$

Suppose $q_1 < p$. Then (10) can be written as

$$(12)\quad \prod_{i=1}^{q_1} \frac{\Gamma[\frac{1}{2}(N - q_2 + 1 - i)]}{\Gamma[\frac{1}{2}(N - q_2 + 1 - i) + h]} \prod_{i=q_1+1}^{p} \frac{\Gamma[\frac{1}{2}(N - q_2 + 1 - i)]}{\Gamma[\frac{1}{2}(N - q_2 + 1 - i) + h]}$$

$$\cdot \prod_{i=1}^{p-q_1} \frac{\Gamma[\frac{1}{2}(N - q + 1 - i) + h]}{\Gamma[\frac{1}{2}(N - q + 1 - i)]} \prod_{i=p-q_1+1}^{p} \frac{\Gamma[\frac{1}{2}(N - q + 1 - i) + h]}{\Gamma[\frac{1}{2}(N - q + 1 - i)]}$$

$$= \prod_{i=1}^{q_1} \frac{\Gamma[\frac{1}{2}(N - q_2 + 1 - i)]}{\Gamma[\frac{1}{2}(N - q_2 + 1 - i) + h]} \left\{ \prod_{i=1}^{p-q_1} \frac{\Gamma[\frac{1}{2}(N - q + 1 - i)]}{\Gamma[\frac{1}{2}(N - q + 1 - i) + h]} \right.$$

$$\left. \prod_{i=1}^{p-q_1} \frac{\Gamma[\frac{1}{2}(N - q + 1 - i) + h]}{\Gamma[\frac{1}{2}(N - q + 1 - i)]} \right\} \prod_{i=1}^{q_1} \frac{\Gamma[\frac{1}{2}(N - q_2 - p + 1 - i) + h]}{\Gamma[\frac{1}{2}(N - q_2 - p + 1 - i)]}.$$

Since the quantity within braces is 1, this expression is equal to (11). Of course, if $q_1 > p$, we can reverse the demonstration. This demonstrates the following theorem:

THEOREM 8.4.2. *When the hypothesis is true, the distribution of* $U_{p,q_1,N-q_1-q_2}$ *is the same as that of* $U_{q_1,p,N-p-q_2}$ *(that is, that of* $U_{p,q_1,n}$ *is that of* $U_{q_1,p,n+q_1-p}$*).*

8.5. SOME DISTRIBUTIONS OF THE CRITERIA

8.5.1. Introduction

To set significance points for tests involving the likelihood ratio criterion, we desire to use the distribution of $U_{p,q_1,n}$. In this section we show that the distribution is that of a product of independent variates. In special cases this leads to explicit expressions for the distributions. In other cases the distributions can only be indicated as integrals.

In view of Theorem 8.4.2, we assume here that $p \leq q_1$. We shall denote q_1 by m.

8.5.2. *U* as a Product of Independent Variates

We use the fact that

$$(1)\qquad \frac{\Gamma[\frac{1}{2}(a+b)]\Gamma(\frac{1}{2}a+h)}{\Gamma[\frac{1}{2}(a+b)+h]\Gamma(\frac{1}{2}a)} = \int_0^1 \frac{\Gamma[\frac{1}{2}(a+b)]}{\Gamma(\frac{1}{2}a)\Gamma(\frac{1}{2}b)} x^{\frac{1}{2}a+h-1}(1-x)^{\frac{1}{2}b-1}\,dx.$$

Let

$$(2)\qquad \beta(x;\tfrac{1}{2}a,\tfrac{1}{2}b) = \frac{\Gamma[\frac{1}{2}(a+b)]}{\Gamma(\frac{1}{2}a)\Gamma(\frac{1}{2}b)} x^{\frac{1}{2}a-1}(1-x)^{\frac{1}{2}b-1}$$

in the unit interval and zero elsewhere. If X has the density $\beta(x;\frac{1}{2}a,\frac{1}{2}b)$, then the hth moment of X is (1). From (10) of the preceding section we see that

$$(3)\qquad \mathscr{E}U^h = \prod_{i=1}^{p}\left\{\frac{\Gamma[\frac{1}{2}(n+1-i)+h]\Gamma[\frac{1}{2}(n+m+1-i)]}{\Gamma[\frac{1}{2}(n+1-i)]\Gamma[\frac{1}{2}(n+m+1-i)+h]}\right\}$$
$$= \prod_{i=1}^{p}\{\mathscr{E}X_i^h\},$$

where X_i has the density $\beta[x;\frac{1}{2}(n+1-i),\frac{1}{2}m]$, and $X_1,\cdots,X_p$ are independent. Thus $\mathscr{E}U^h = \prod\{\mathscr{E}X_i^h\} = \mathscr{E}\{\prod(X_i^h)\} = \mathscr{E}\{(\prod X_i)^h\}$. Since the distribution of U is of finite range and is therefore determined uniquely by its moments, we have the following theorem:

THEOREM 8.5.1. *The distribution of* $U_{p,m,n}$ *is the distribution of* $\prod_{i=1}^{p}\{X_i\}$, *where the* X_i *are independent and* X_i *has density* $\beta[x;\frac{1}{2}(n+1-i),\frac{1}{2}m]$.

The cdf of U can be found by integrating the joint density of $X_1,\cdots,X_p$ over the range

$$(4)\qquad \prod_{i=1}^{p} x_i \le u.$$

Suppose p is even, that is, $p = 2r$. We use the fact that

$$(5)\qquad \Gamma(\alpha+\tfrac{1}{2})\Gamma(\alpha+1) = \frac{\sqrt{\pi}\,\Gamma(2\alpha+1)}{2^{2\alpha}}.$$

Then the hth moment of U is

$$(6)\qquad \mathscr{E}U^h = \prod_{j=1}^{r}\left\{\frac{\Gamma[\frac{1}{2}(m+n+2)-j]}{\Gamma[\frac{1}{2}(m+n+2)-j+h]}\frac{\Gamma[\frac{1}{2}(m+n+1)-j]}{\Gamma[\frac{1}{2}(m+n+1)-j+h]}\cdot\frac{\Gamma[\frac{1}{2}(n+2)-j+h]\Gamma[\frac{1}{2}(n+1)-j+h]}{\Gamma[\frac{1}{2}(n+2)-j]\Gamma[\frac{1}{2}(n+1)-j]}\right\}$$
$$= \prod_{j=1}^{r}\left\{\frac{\Gamma(m+n+1-2j)\Gamma(n+1-2j+2h)}{\Gamma(m+n+1-2j+2h)\Gamma(n+1-2j)}\right\}.$$

It is clear from (1) that (6) is

$$\prod_{j=1}^{r}\left\{\int_0^1 \frac{\Gamma(m+n+1-2j)}{\Gamma(n+1-2j)\Gamma(m)} y^{(n+1-2j)+2h-1}(1-y)^{m-1}dy\right\} \tag{7}$$

$$= \prod_{j=1}^{r} \mathscr{E} Y_j^{2h}$$

$$= \mathscr{E}\left(\prod_{j=1}^{r} Y_j^2\right)^h,$$

where the Y_j are independent and Y_j has density $\beta(y; n+1-2j, m)$.

Suppose p is odd; that is, $p = 2s + 1$. Then

$$\mathscr{E} U^h = \mathscr{E}\left(\prod_{i=1}^{s} Z_i^2 Z_{s+1}\right)^h, \tag{8}$$

where the Z_i are independent and Z_i has density $\beta(z; n+1-2i, m)$ for $i = 1, \cdots, s$ and Z_{s+1} is distributed with density $\beta[z; (n+1-p)/2, m/2]$.

THEOREM 8.5.2. *$U_{2r,m,n}$ is distributed as $\prod_{i=1}^{r} Y_i^2$, where the Y_i are independent and Y_i has density $\beta(y; n+1-2i, m)$; $U_{2s+1,m,n}$ is distributed as $\prod_{i=1}^{s} Z_i^2 Z_{s+1}$, where the Z_i $(i = 1, \cdots, s)$ are independent and Z_i has density $\beta(z; n+1-2i, m)$ and Z_{s+1} is independently distributed with density $\beta[z; \frac{1}{2}(n+1-p), \frac{1}{2}m]$.*

8.5.3. Some Special Distributions

CASE 1: $p = 1$. From the preceding section we see that the density of $U_{1,m,n}$ is

$$\frac{\Gamma[\frac{1}{2}(n+m)]}{\Gamma(\frac{1}{2}n)\Gamma(\frac{1}{2}m)} u^{\frac{1}{2}n-1}(1-u)^{\frac{1}{2}m-1}. \tag{9}$$

Another way of writing $U_{1,m,n}$ is

$$U_{1,m,n} = \frac{1}{1 + \sum_{i=1}^{m} Y_i^2/g_{11}} = \frac{1}{1 + (m/n)F_{m,n}}, \tag{10}$$

where g_{11} is the one element of $G = N\hat{\boldsymbol{\Sigma}}_\Omega$ and $F_{m,n}$ is an F-statistic. Thus

$$\frac{1 - U_{1,m,n}}{U_{1,m,n}} \cdot \frac{n}{m} = F_{m,n}. \tag{11}$$

The F-distribution is well known.

THEOREM 8.5.3. *The distribution of $\dfrac{1 - U_{1,m,n}}{U_{1,m,n}} \cdot \dfrac{n}{m}$ is the F-distribution with m and n degrees of freedom; the distribution of $\dfrac{1 - U_{p,1,n}}{U_{p,1,n}} \cdot \dfrac{n+1-p}{p}$ is the F-distribution with p and $n+1-p$ degrees of freedom.*

CASE 2: $p = 2$. From Theorem 8.5.2 we see that the density of $\sqrt{U_{2,m,n}}$ is

$$\frac{\Gamma(n+m-1)}{\Gamma(n-1)\Gamma(m)} x^{n-2}(1-x)^{m-1}, \tag{12}$$

and thus the density of $U_{2,m,n}$ is

$$\frac{\Gamma(n+m-1)}{2\Gamma(n-1)\Gamma(m)} U^{\frac{1}{2}(n-3)}(1-\sqrt{U})^{m-1}. \tag{13}$$

From (12) it follows that

$$\frac{1-\sqrt{U_{2,m,n}}}{\sqrt{U_{2,m,n}}} \cdot \frac{n-1}{m} = F_{2m,\,2(n-1)}. \tag{14}$$

THEOREM 8.5.4. *The distribution of* $\dfrac{1-\sqrt{U_{2,m,n}}}{\sqrt{U_{2,m,n}}} \cdot \dfrac{n-1}{m}$ *is the F-distribution with* $2m$ *and* $2(n-1)$ *degrees of freedom; the distribution of* $\dfrac{1-\sqrt{U_{p,2,n}}}{\sqrt{U_{p,2,n}}} \cdot \dfrac{n+1-p}{p}$ *is the F-distribution with* $2p$ *and* $2(n+1-p)$ *degrees of freedom.*

CASE 3: $p = 3$. Here $U = Z_1^2 Z_2$, where the density of Z_1 and Z_2 is

$$f(z_1, z_2) = \frac{\Gamma(n+m-1)}{\Gamma(n-1)\Gamma(m)} \frac{\Gamma[\frac{1}{2}(n+m-2)]}{\Gamma[\frac{1}{2}(n-2)]\Gamma(\frac{1}{2}m)} z_1^{n-2}(1-z_1)^{m-1} z_2^{\frac{1}{2}(n-4)}(1-z_2)^{\frac{1}{2}(m-2)}. \tag{15}$$

We wish to find the probability that $U \le u\ (\le 1)$. Thus we integrate over the shaded region in Figure 2. The integral over I is

$$\int_0^u \int_0^1 f(z_1, z_2)\, dz_1\, dz_2 = I_u[\tfrac{1}{2}(n-2), \tfrac{1}{2}m]. \tag{16}$$

where

$$I_u(a, b) = B^{-1}(a, b) \int_0^u x^{a-1}(1-x)^{b-1}\, dx$$

has been tabulated by K. Pearson (1932). The integral over II is

$$\int_u^1 \int_0^{\sqrt{u/z_2}} f(z_1, z_2) dz_1 dz_2. \tag{17}$$

To evaluate this we expand $f(z_1)$ in powers of z_1. Thus the area II is

$$(18)\quad C\int_u^1\left\{\int_0^{\sqrt{u/z_2}}\sum_{i=0}^{m-1}\frac{(m-1)!}{i!(m-1-i)!}z_1^{n-2+i}(-1)^i\,dz_1\right\}\cdot z_2^{\frac{1}{2}(n-4)}(1-z_2)^{\frac{1}{2}(m-2)}dz_2$$

$$= C\int_u^1\left\{\sum_{i=0}^{m-1}\frac{(m-1)!}{i!(m-1-i)!}\cdot\frac{(-1)^i}{n-1+i}\left(\frac{u}{z_2}\right)^{\frac{1}{2}(n-1+i)}\right\}z_2^{\frac{1}{2}(n-4)}(1-z_2)^{\frac{1}{2}(m-2)}dz_2$$

$$= C\sum_{i=0}^{m-1}\frac{(m-1)!(-1)^i u^{\frac{1}{2}(n-1+i)}}{i!(m-1-i)!(n-1+i)}\int_u^1 z_2^{-\frac{1}{2}(i+3)}(1-z_2)^{\frac{1}{2}(m-2)}dz_2,$$

where

$$(19)\qquad C=\frac{\Gamma(n+m-1)\Gamma[\frac{1}{2}(n+m-2)]}{\Gamma(n-1)\Gamma(m)\Gamma[\frac{1}{2}(n-2)]\Gamma(\frac{1}{2}m)}.$$

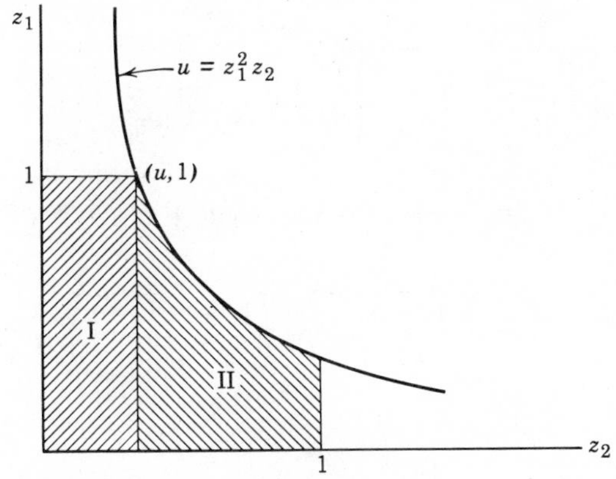

FIGURE 2

If m is even, the term $(1-z_2)^{\frac{1}{2}(m-2)}$ can be expanded in powers of z_2 and the integration effected immediately. If m is odd, we can expand $(1-z_2)^{\frac{1}{2}(m-3)}$ in powers of z_2. Then the integrand is a sum of powers of z_2 times $\sqrt{1-z_2}$.

As an example, let us consider the case $m = 3$ (the case of $m = 4$ being essentially treated later). Then the integral for II is

$$(20)\quad C\left\{\frac{u^{\frac{1}{2}(n-1)}}{n-1}\int_u^1 z_2^{-\frac{3}{2}}\sqrt{1-z_2}\,dz_2-\frac{2u^{\frac{1}{2}n}}{n}\int_u^1 z_2^{-2}\sqrt{1-z_2}\,dz_2+\frac{u^{\frac{1}{2}(n+1)}}{n+1}\int_u^1 z_2^{-\frac{5}{2}}\sqrt{1-z_2}\,dz_2\right\},$$

where

$$C = \frac{\Gamma(n+2)\Gamma[\frac{1}{2}(n+1)]}{\Gamma(n-1)\Gamma[\frac{1}{2}(n-2)]\Gamma(3)\Gamma(\frac{3}{2})}. \tag{21}$$

The first term is $C\dfrac{u^{\frac{1}{2}(n-1)}}{n-1}$ times

$$\int_u^1 \frac{\sqrt{1-z_2}}{z^{\frac{3}{2}}}\,dz = -2\sqrt{\frac{1-z}{z}}\Big|_u^1 - \int_u^1 \frac{dz}{\sqrt{z-z^2}} \tag{22}$$

$$= 2\sqrt{\frac{1-u}{u}} - \frac{\pi}{2} + \text{arc sin}\,(2u-1).$$

The second term is $-C\dfrac{2u^{\frac{1}{2}n}}{n}$ times

$$\int_u^1 \frac{\sqrt{1-z}}{z^2}\,dz = \frac{\sqrt{1-z}}{z}\Big|_u^1 - \tfrac{1}{2}\int_u^1 \frac{dz}{z\sqrt{1-z}} \tag{23}$$

$$= \frac{\sqrt{1-u}}{u} + \log\,(\sec\theta + \tan\theta)\Big|_{\theta=0}^{\theta=\text{arc cos}\sqrt{u}}$$

$$= \frac{\sqrt{1-u}}{u} - \log\left(\frac{1+\sqrt{1-u}}{\sqrt{u}}\right).$$

The third term is $C\dfrac{u^{\frac{1}{2}(n+1)}}{n+1}$ times

$$\int_u^1 \frac{\sqrt{1-z}}{z^{\frac{5}{2}}}\,dz = -\frac{2}{3}\frac{\sqrt{1-z}}{z\sqrt{z}}\Big|_u^1 - \frac{1}{3}\int_u^1 \frac{dz}{z\sqrt{z-z^2}} \tag{24}$$

$$= \frac{2}{3}\frac{\sqrt{1-u}}{u\sqrt{u}} + \frac{2}{3}\frac{\sqrt{z-z^2}}{z}\Big|_u^1$$

$$= \frac{2}{3}\left(\frac{\sqrt{1-u}}{u\sqrt{u}} - \frac{\sqrt{u-u^2}}{u}\right)$$

$$= \frac{2}{3}\left(\frac{1-u}{u}\right)^{\frac{3}{2}}.$$

Now we can write the cumulative distribution of $U_{3,3,n}$ as

$$(25)\quad \Pr\{U \le u\} = I_u(\tfrac{1}{2}n - 1, \tfrac{3}{2}) + \frac{\Gamma(n+2)\Gamma(\frac{1}{2}[n+1])}{\Gamma(n-1)\Gamma(\frac{1}{2}n-1)\sqrt{\pi}} \left\{ \frac{2u^{\frac{1}{2}n-1}\sqrt{1-u}}{n(n-1)} + \frac{u^{\frac{1}{2}(n-1)}}{n-1}[\arcsin(2u-1) - \tfrac{1}{2}\pi] + \frac{2u^{\frac{1}{2}n}}{n}\log\left(\frac{1+\sqrt{1-u}}{\sqrt{u}}\right) + \frac{2u^{\frac{1}{2}n-1}(1-u)^{\frac{3}{2}}}{3(n+1)} \right\}.$$

Table 1 gives the values of u for which $\Pr\{U \le u\} = 0.05$ and 0.01. The numbers are not accurately computed (perhaps to only two significant figures). For larger values of n the method of finding probabilities given in Section 8.6 is adequate.

TABLE 1. u FOR WHICH $\Pr\{U_{3,3,n} \le u\} = 0.05$ AND 0.01

n	5%	1%
3	8.59×10^{-5}	4.09×10^{-6}
4	3.30×10^{-3}	6.6×10^{-4}
5	0.0183	0.00626
6	0.0447	
7	0.0794	
8	0.105	
9	0.131	
10	0.162	0.0979

It will be shown in the next section that for $m = 4$

$$(26)\quad \Pr\{U_{3,4,n} \le u\} = \frac{(n+2)(n+1)n^2(n-1)(n-2)}{48} u^{\frac{1}{2}(n-2)} \left\{ \frac{1}{n-2} - \frac{8\sqrt{u}}{n-1} + \frac{12}{n^2}u - \frac{6}{n}u\log u + \frac{8}{n+1}u^{\frac{3}{2}} - \frac{1}{n+2}u^2 \right\}.$$

CASE 4: $p = 4$. Here $U = Z_1^2 Z_2^2$ where the density of Z_1 and Z_2 is

$$(27)\quad f(z_1, z_2) = \frac{\Gamma(n+m-1)\Gamma(n+m-3)}{\Gamma(n-1)\Gamma(m)\Gamma(n-3)\Gamma(m)} z_1^{n-2}(1-z_1)^{m-1} z_2^{n-4}(1-z_2)^{m-1}.$$

To find the probability that $U \le u$ we integrate over the shaded portion of Figure 3. The integral over I is $I_{\sqrt{u}}(n-1, m)$. The integral over II is

$$\int_{\sqrt{u}}^{1}\int_{0}^{\sqrt{u}/z_1} f(z_1, z_2)\,dz_2\,dz_1 \tag{28}$$

$$= C\int_{\sqrt{u}}^{1}\left\{\sum_{i=0}^{m-1}\frac{(m-1)!(-1)^i}{i!(m-1-i)!}\int_0^{\sqrt{u}/z_1} z_2^{n-4+i}\,dz_2\right\} z_1^{n-2}(1-z_1)^{m-1}\,dz_1$$

$$= C\int_{\sqrt{u}}^{1}\left\{\sum_{i=0}^{m-1}\frac{(m-1)!(-1)^i}{i!(m-1-i)!(n-3+i)}\left(\frac{\sqrt{u}}{z_1}\right)^{n-3+i}\right\} z_1^{n-2}(1-z_1)^{m-1}dz_1$$

$$= C\sum_{i=0}^{m-1}\frac{(m-1)!(-1)^i u^{\frac{1}{2}(n-3+i)}}{i!(m-1-i)!(n-3+i)}\int_{\sqrt{u}}^{1} z_1^{1-i}(1-z_1)^{m-1}\,dz_1$$

$$= C\sum_{i,j=0}^{m-1}\frac{(m-1)!(-1)^{i+j}(m-1)!u^{\frac{1}{2}(n-3+i)}}{i!(m-1-i)!j!(m-1-j)!(n-3+i)}\int_{\sqrt{u}}^{1} z_1^{1-i+j}\,dz_1,$$

where

$$C = \frac{\Gamma(n+m-1)\Gamma(n+m-3)}{\Gamma(n-1)\Gamma(n-3)[\Gamma(m)]^2}. \tag{29}$$

The integration indicated in (28) can be carried out specifically for any given m. The distribution function is the sum of $I_{\sqrt{u}}(n-1, m)$ and (28).

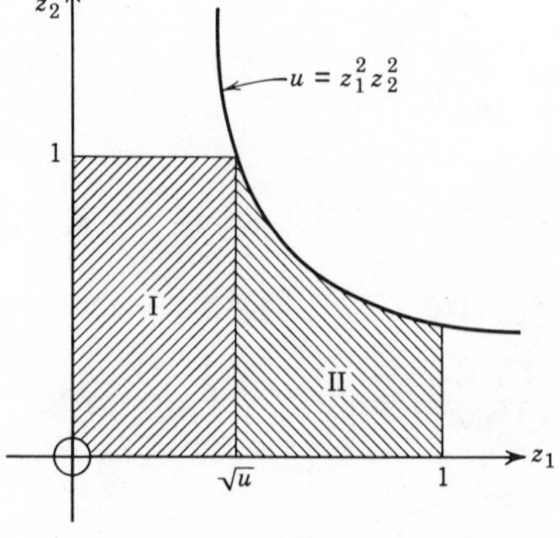

FIGURE 3

Let us consider the special case of $m = 3$. The integration on z_2 gives

$$(30)\quad C\left\{\frac{u^{\frac{1}{2}(n-3)}}{n-3}\int_{\sqrt{u}}^{1} z_1(1-z_1)^2\,dz_1 - 2\frac{u^{\frac{1}{2}(n-2)}}{n-2}\int_{\sqrt{u}}^{1}(1-z_1)^2\,dz_1 + \frac{u^{\frac{1}{2}(n-1)}}{n-1}\int_{\sqrt{u}}^{1}\frac{(1-z_1)^2}{z_1}\,dz_1\right\}$$

$$= C\left\{\frac{u^{\frac{1}{2}(n-3)}}{n-3}\left[\int_{\sqrt{u}}^{1}(1-z_1)^2\,dz_1 - \int_{\sqrt{u}}^{1}(1-z_1)^3\,dz_1\right] - 2\frac{u^{\frac{1}{2}(n-2)}}{n-2}\frac{(1-\sqrt{u})^3}{3} + \frac{u^{\frac{1}{2}(n-1)}}{n-1}\left[\int_{\sqrt{u}}^{1}\frac{dz_1}{z_1} - 2\int_{\sqrt{u}}^{1}dz_1 + \int_{\sqrt{u}}^{1}z_1\,dz_1\right]\right\}$$

$$= C\left\{\frac{u^{\frac{1}{2}(n-3)}}{n-3}\left[\frac{(1-\sqrt{u})^3}{3} - \frac{(1-\sqrt{u})^4}{4}\right] - \frac{2u^{\frac{1}{2}(n-2)}(1-\sqrt{u})^3}{3(n-2)} + \frac{u^{\frac{1}{2}(n-1)}}{n-1}\left[-\log\sqrt{u} - 2(1-\sqrt{u}) + \frac{1-u}{2}\right]\right\},$$

where

$$(31)\qquad C = \frac{\Gamma(n+2)\Gamma(n)}{4\Gamma(n-1)\Gamma(n-3)}.$$

Thus

$$(32)\quad \Pr\{U \le u\} = I_{\sqrt{u}}(n-1, 3) + \frac{\Gamma(n+2)\Gamma(n)}{4\Gamma(n-1)\Gamma(n-3)} \cdot\left\{\frac{u^{\frac{1}{2}(n-3)}}{n-3}\left[\frac{(1-\sqrt{u})^3}{3} - \frac{(1-\sqrt{u})^4}{4}\right] - \frac{2u^{\frac{1}{2}(n-2)}(1-\sqrt{u})^3}{3(n-2)} + \frac{u^{\frac{1}{2}(n-1)}}{2(n-1)}[-\log u + 4\sqrt{u} - u - 3]\right\}.$$

This can be written

$$(33)\quad \frac{(n+1)n(n-1)^2(n-2)(n-3)}{48}\,u^{\frac{1}{2}(n-3)}\left\{\frac{1}{n-3} - \frac{8\sqrt{u}}{n-2} + \frac{12}{(n-1)^2}u - \frac{6}{n-1}u\log u + \frac{8}{n}u^{\frac{3}{2}} - \frac{1}{n+1}u^2\right\}.$$

We shall also treat the case $m = 4$. The integration on z_2 in (28) gives

$$
\begin{aligned}
(34)\quad & C\left\{\frac{u^{\frac{1}{2}(n-3)}}{n-3}\int_{\sqrt{u}}^{1} z_1(1-z_1)^3\,dz_1 - \frac{3u^{\frac{1}{2}(n-2)}}{n-2}\int_{\sqrt{u}}^{1}(1-z_1)^3\,dz_1\right. \\
&\qquad \left. + \frac{3u^{\frac{1}{2}(n-1)}}{n-1}\int_{\sqrt{u}}^{1}\frac{(1-z_1)^3}{z_1}\,dz_1 - \frac{u^{\frac{1}{2}n}}{n}\int_{\sqrt{u}}^{1}\frac{(1-z_1)^3}{z_1^2}\,dz_1\right\} \\
&= C\left\{\frac{u^{\frac{1}{2}(n-3)}}{n-3}\left[\int_{\sqrt{u}}^{1}(1-z_1)^3\,dz_1 - \int_{\sqrt{u}}^{1}(1-z_1)^4\,dz_1\right]\right. \\
&\qquad - \frac{3u^{\frac{1}{2}(n-2)}}{n-2}\frac{(1-\sqrt{u})^4}{4} + \frac{3u^{\frac{1}{2}(n-1)}}{n-1}\left[\int_{\sqrt{u}}^{1}\frac{dz_1}{z_1} - 3\int_{\sqrt{u}}^{1}dz_1\right. \\
&\qquad \left. + 3\int_{\sqrt{u}}^{1} z_1\,dz_1 - \int_{\sqrt{u}}^{1} z_1^2\,dz_1\right] - \frac{u^{\frac{1}{2}n}}{n}\left[\int_{\sqrt{u}}^{1}\frac{dz_1}{z_1^2} - 3\int_{\sqrt{u}}^{1}\frac{dz_1}{z_1}\right. \\
&\qquad \left.\left. + 3\int_{\sqrt{u}}^{1}dz_1 - \int_{\sqrt{u}}^{1} z_1\,dz_1\right]\right\} \\
&= C\left\{\frac{u^{\frac{1}{2}(n-3)}}{n-3}\left[\frac{(1-\sqrt{u})^4}{4} - \frac{(1-\sqrt{u})^5}{5}\right] - \frac{3u^{\frac{1}{2}(n-2)}(1-\sqrt{u})^4}{4(n-2)}\right. \\
&\qquad + \frac{3u^{\frac{1}{2}(n-1)}}{n-1}\left[-\log\sqrt{u} - 3(1-\sqrt{u}) + \frac{3}{2}(1-u) - \frac{(1-u^{\frac{3}{2}})}{3}\right] \\
&\qquad \left. - \frac{u^{\frac{1}{2}n}}{n}\left[\frac{1}{\sqrt{u}} - 1 + 3\log\sqrt{u} + 3(1-\sqrt{u}) - \frac{(1-u)}{2}\right]\right\},
\end{aligned}
$$

where

$$
(35)\qquad C = \frac{\Gamma(n+3)\Gamma(n+1)}{36\Gamma(n-1)\Gamma(n-3)}.
$$

The distribution function is the sum of (34) and $I_{\sqrt{u}}(n-1, 4)$.

Wald and Brookner (1941) give a method for finding the distribution of U if either p or q_1 or both are even. These distributions can also be obtained by integrating the densities given in Section 8.5.2 (the effect of asking for p or q_1 to be even is that then U is distributed as $\prod_1^r Y_i^2$ and the density of $Y_1, \cdots, Y_r$ can be expanded in powers).

Wilks (1935a) gives explicitly the distributions of U for $p = 1$, $p = 2$, $p = 3$ with $m = 3$; $p = 3$ with $m = 4$; and $p = 4$ with $m = 4$. Wilks' formula for $p = 3$ with $m = 4$ disagrees with ours and appears to be incorrect.

8.6. AN ASYMPTOTIC EXPANSION OF THE DISTRIBUTION OF THE LIKELIHOOD RATIO CRITERION

8.6.1. General Theory of Asymptotic Expansions

In this section we develop a large sample distribution theory for the criterion studied in this chapter. First we develop a general asymptotic expansion of the distribution of a random variable whose moments are certain functions of gamma functions [Box (1949)]. Then we apply it to the case of the likelihood ratio criterion for the linear hypothesis.

We consider a random variable $W (0 \leq W \leq 1)$ with hth moment*

$$\mathscr{E} W^h = K \left(\frac{\prod_{j=1}^{b} y_j^{y_j}}{\prod_{k=1}^{a} x_k^{x_k}} \right)^h \frac{\prod_{k=1}^{a} \Gamma[x_k(1+h) + \xi_k]}{\prod_{j=1}^{b} \Gamma[y_j(1+h) + \eta_j]}, \qquad h = 0, 1, \cdots, \tag{1}$$

where K is a constant (such that $\mathscr{E} W^0 = 1$) and

$$\sum_{k=1}^{a} x_k = \sum_{j=1}^{b} y_j. \tag{2}$$

It will be observed that the hth moment of $\lambda = U_{p, q_1, n}^{\frac{1}{2}N}$ is of this form where $x_k = \frac{1}{2}N = y_j$, $\xi_k = \frac{1}{2}(-q + 1 - k)$, $\eta_j = \frac{1}{2}(-q_2 + 1 - j)$, $a = b = p$. We treat a more general case here because applications later in this book require it.

If we let

$$M = -2 \log W \tag{3}$$

the characteristic function of ρM $(0 \leq \rho < 1)$ is

$$\begin{aligned} \phi(t) &= \mathscr{E} e^{it\rho M} \\ &= \mathscr{E} W^{-2it\rho} \\ &= K \left(\frac{\prod y_j^{y_j}}{\prod x_k^{x_k}} \right)^{-2it\rho} \frac{\prod \Gamma[x_k(1 - 2it\rho) + \xi_k]}{\prod \Gamma[y_j(1 - 2it\rho) + \eta_j]}. \end{aligned} \tag{4}$$

Here ρ is arbitrary; later it will depend on N. If $a = b$, $x_k = y_k$, $\xi_k \leq \eta_k$, (1) is the hth moment of the product of powers of variables with beta distributions, and then (1) holds for all h for which the gamma functions exist. In this case (4) is valid for all real t. We shall assume here that (4) holds for all real t, and in each case where we apply the result we shall verify this assumption.

* In all cases where we apply this result the parameters x_k, ξ_k, y_j, and η_j will be such that there is a distribution with such moments.

Let

$$\Phi(t) = \log \phi(t) = g(t) - g(0), \tag{5}$$

where

$$g(t) = 2it\rho(\sum x_k \log x_k - \sum y_j \log y_j) + \sum_k \log \Gamma[\rho x_k(1 - 2it) + \beta_k + \xi_k] - \sum_j \log \Gamma[\rho y_j(1 - 2it) + \varepsilon_j + \eta_j],$$

where $\beta_k = (1 - \rho)x_k$ and $\varepsilon_j = (1 - \rho)y_j$. The form $g(t) - g(0)$ makes $\Phi(0) = 0$ which agrees with the fact that K is such that $\phi(0) = 1$. We make use of an expansion formula for the gamma function [Barnes (1899), p. 64] which is asymptotic in x for bounded h:

$$\log \Gamma(x + h) = \log \sqrt{2\pi} + (x + h - \tfrac{1}{2}) \log x - x - \sum_{r=1}^{m} (-1)^r \frac{B_{r+1}(h)}{r(r + 1)x^r} + R_{m+1}(x), \tag{6}$$

where† $R_{m+1}(x) = O(x^{-(m+1)})$ and $B_r(h)$ is the Bernoulli polynomial of degree r and order unity defined by‡

$$\frac{\tau e^{h\tau}}{e^\tau - 1} = \sum_{r=0}^{\infty} \frac{\tau^r}{r!} B_r(h). \tag{7}$$

The first three polynomials are $[B_0(h) = 1]$

$$\begin{aligned} B_1(h) &= h - \tfrac{1}{2}, \\ B_2(h) &= h^2 - h + \tfrac{1}{6}, \\ B_3(h) &= h^3 - \tfrac{3}{2}h^2 + \tfrac{1}{2}h. \end{aligned} \tag{8}$$

Taking $x = \rho x_k(1 - 2it), \rho y_j(1 - 2it)$ and $h = \beta_k + \xi_k, \varepsilon_j + \eta_j$ in turn, we obtain

$$\Phi(t) = Q - g(0) - \frac{f}{2} \log (1 - 2it) + \sum_{r=1}^{m} \omega_r(1 - 2it)^{-r} + \sum O(x_k^{-(m+1)}) + \sum O(y_j^{-(m+1)}), \tag{9}$$

† $R_{m+1}(x) = O(x^{-(m+1)})$ means $|x^{m+1}R_{m+1}(x)|$ is bounded as $|x| \to \infty$.

‡ This definition differs slightly from that of Whittaker and Watson [(1943), p. 126] who expand $\tau(e^{h\tau} - 1)/(e^\tau - 1)$. If $B_r^*(h)$ is this second type of polynomial, $B_1(h) = B_1^*(h) - \frac{1}{2}$, $B_{2r}(h) = B_{2r}^*(h) + (-1)^{r+1}B_r$, where B_r is the rth Bernoulli number, and $B_{2r+1}(h) = B_{2r+1}^*(h)$.

where

(10) $$f = -2\{\sum \xi_i - \sum \eta_j - \tfrac{1}{2}(a - b)\},$$

(11) $$\omega_r = \frac{(-1)^{r+1}}{r(r+1)} \left\{ \sum_k \frac{B_{r+1}(\beta_k + \xi_k)}{(\rho x_k)^r} - \sum_j \frac{B_{r+1}(\varepsilon_j + \eta_j)}{(\rho y_j)^r} \right\},$$

(12) $$Q = \tfrac{1}{2}(a - b) \log 2\pi - \frac{f}{2} \log \rho + \sum_k (x_k + \xi_k - \tfrac{1}{2}) \log x_k - \sum_j (y_j + \eta_j - \tfrac{1}{2}) \log y_j.$$

One resulting form for $\phi(t)$ (which we shall not use here) is

(13) $$\phi(t) = e^{\Phi(t)} = e^{Q - g(0)}(1 - 2it)^{-\frac{1}{2}f} \sum_{v=0}^{m} a_v (1 - 2it)^{-v} + R^*_{m+1},$$

where $\sum_{v=0}^{m} a_v z^{-m}$ are the first $m + 1$ terms in the expansion of exp $(-\sum_{r=0}^{m} \omega_r z^{-r})$ and R^*_{m+1} is a remainder term. Alternatively

(14) $$\Phi(t) = -\frac{f}{2} \log (1 - 2it) + \sum_{r=1}^{m} \omega_r [(1 - 2it)^{-r} - 1] + R'_{m+1},$$

where

(15) $$R'_{m+1} = \sum O(x_k^{-(m+1)}) + \sum O(y_j^{-(m+1)}).$$

In (14) we have expanded $g(0)$ in the same way we had expanded $g(t)$ and have collected similar terms.

Then

(16) $$\begin{aligned} \phi(t) &= e^{\Phi(t)} \\ &= (1 - 2it)^{-\frac{1}{2}f} \exp \Big[\sum_{r=1}^{m} \omega_r (1 - 2it)^{-r} - \sum_{r=1}^{m} \omega_r + R'_{m+1} \Big] \\ &= (1 - 2it)^{-\frac{1}{2}f} \left\{ \prod_{r=1}^{m} \Big[1 + \omega_r (1 - 2it)^{-r} + \frac{1}{2!} \omega_r^2 (1 - 2it)^{-2r} \cdots \Big] \prod_{r=1}^{m} \Big(1 - \omega_r + \frac{1}{2!} \omega_r^2 - \cdots \Big) + R''_{m+1} \right\} \\ &= (1 - 2it)^{-\frac{1}{2}f} [1 + T_1(t) + T_2(t) + \cdots + T_m(t) + R'''_{m+1}], \end{aligned}$$

where $T_r(t)$ is the term in the expansion with terms $\omega_1^{s_1} \cdots \omega_r^{s_r}$, $\sum i s_i = r$; for example,

(17) $$T_1(t) = \omega_1 [(1 - 2it)^{-1} - 1],$$

(18) $T_2(t) = \omega_2[(1-2it)^{-2} - 1] + \frac{1}{2}\omega_1^2[(1-2it)^{-2} - 2(1-2it)^{-1} + 1].$

In most applications, we will have $x_k = c_k\theta$ and $y_j = d_j\theta$, where c_k and d_j will be constant and θ will vary (that is, will grow with the sample size). In this case if ρ is chosen so $(1-\rho)x_k$ and $(1-\rho)y_j$ have limits, then R'''_{m+1} is $O(\theta^{-(m+1)})$. We collect in (16) all terms $\omega_1^{s_1}\cdots\omega_r^{s_r}$, $\sum is_i = r$ because these terms are $O(\theta^{-r})$.

It will be observed that $T_r(t)$ is a polynomial of degree r in $(1-2it)^{-1}$ and each term of $(1-2it)^{-\frac{1}{2}f}T_r(t)$ is a constant times $(1-2it)^{-\frac{1}{2}v}$ for an integral v. We know that $(1-2it)^{-\frac{1}{2}v}$ is the characteristic function of the χ^2-density with v degrees of freedom; that is,

$$g_v(z) = \frac{1}{2^{\frac{1}{2}v}\Gamma(\frac{1}{2}v)} z^{\frac{1}{2}v-1}e^{-\frac{1}{2}z} = \int_{-\infty}^{\infty} \frac{1}{2\pi}(1-2it)^{-\frac{1}{2}v}e^{-itz}\,dt. \tag{19}$$

Let

$$\begin{aligned} S_r(z) &= \int_{-\infty}^{\infty}\frac{1}{2\pi}(1-2it)^{-\frac{1}{2}f}T_r(t)e^{-itz}\,dt,\\ R^{\text{iv}}_{m+1} &= \int_{-\infty}^{\infty}\frac{1}{2\pi}(1-2it)^{-\frac{1}{2}f}R'''_{m+1}e^{-itz}\,dt. \end{aligned} \tag{20}$$

Then the density of ρM is

$$\begin{aligned} \int_{-\infty}^{\infty}\frac{1}{2\pi}\phi(t)e^{-itz}\,dt &= \sum_{r=0}^{m} S_r(z) + R^{\text{iv}}_{m+1}\\ &= g_f(z) + \omega_1[g_{f+2}(z) - g_f(z)]\\ &\quad + \left\{\omega_2[g_{f+4}(z) - g_f(z)] + \frac{\omega_1^2}{2}[g_{f+4}(z) - 2g_{f+2}(z) + g_f(z)]\right\}\\ &\quad + \cdots + S_m(z) + R^{\text{iv}}_{m+1}. \end{aligned} \tag{21}$$

Let

$$\begin{aligned} U_r(z_0) &= \int_0^{z_0} S_r(z)\,dz,\\ R^{\text{v}}_{m+1} &= \int_0^{z_0} R^{\text{iv}}_{m+1}\,dz. \end{aligned} \tag{22}$$

The cdf of M_0 is written in terms of the cdf of ρM_0, which is the integral of the density, namely,

$$\begin{aligned}
(23)\quad \Pr\{M \le M_0\} &= \Pr\{\rho M \le \rho M_0\} \\
&= \sum_{r=0}^{m} U_r(\rho M_0) + R_{m+1}^{\text{v}} \\
&= \Pr\{\chi_f^2 \le \rho M_0\} + \omega_1(\Pr\{\chi_{f+2}^2 \le \rho M_0\} - \Pr\{\chi_f^2 \le \rho M_0\}) \\
&\quad + [\omega_2(\Pr\{\chi_{f+4}^2 \le \rho M_0\} - \Pr\{\chi_f^2 \le \rho M_0\}) \\
&\quad + \frac{\omega_1^2}{2}(\Pr\{\chi_{f+4}^2 \le \rho M_0\} - 2\Pr\{\chi_{f+2}^2 \le \rho M_0\} \\
&\quad + \Pr\{\chi_f^2 \le \rho M_0\})] + \cdots + U_m(\rho M_0) + R_{m+1}^{\text{v}}.
\end{aligned}$$

The remainder R_{m+1}^{v} is $O(\theta^{-(m+1)})$; this last statement can be verified by following the remainder terms along. (In fact, to make the proof rigorous one needs to verify that each remainder is of the proper order in a uniform sense.)

In many cases it is desirable to choose ρ so that $\omega_1 = 0$. In such a case using only the first term of (23) gives an error of order θ^{-2}.

Further details of the expansion can be found in Box's paper (1949).

THEOREM 8.6.1. *Suppose that* $\mathscr{E}W^h$ *is given by* (1) *for all purely imaginary* h, *with* (2) *holding. Then the* cdf *of* $-2\rho \log W$ *is given by* (23). *The error,* R_{m+1}^{v}, *is* $O(\theta^{-(m+1)})$ *if* $x_k \ge c_k\theta$, $y_j \ge d_j\theta$ $(c_k > 0,\ d_j > 0)$, *and* $(1-\rho)x_k$, $(1-\rho)y_j$ *have limits, where* ρ *may depend on* θ.

Box also considers approximating the distribution of $-2\rho \log W$ by an F-distribution. He finds that the error in this approximation can be made to be of order θ^{-3}. We shall not extend our discussion to the F-approximation.

8.6.2. Asymptotic Distributions of Criterion

We now apply Theorem 8.6.1 to the distribution of $-2 \log \lambda$, the likelihood ratio criterion developed in Section 8.3. We let $W = \lambda$. The hth moment of λ is

$$(24)\qquad \mathscr{E}\lambda^h = K \frac{\prod_{k=1}^{p} \Gamma[\frac{1}{2}(N - q + 1 - k + Nh)]}{\prod_{j=1}^{p} \Gamma[\frac{1}{2}(N - q_2 + 1 - j + Nh)]},$$

and this holds for all h for which the gamma functions exist including purely imaginary h. We let $a = b = p$,

$$(25)\qquad \begin{aligned}
x_k &= \tfrac{1}{2}N, & \xi_k &= \tfrac{1}{2}(-q + 1 - k), & \beta_k &= \tfrac{1}{2}(1-\rho)N, \\
y_j &= \tfrac{1}{2}N, & \eta_j &= \tfrac{1}{2}(-q_2 + 1 - j), & \varepsilon_j &= \tfrac{1}{2}(1-\rho)N.
\end{aligned}$$

We observe that

$$(26)\quad 2\omega_1 = \sum_{k=1}^{p}\left\{\frac{\{\frac{1}{2}[(1-\rho)N-q+1-k]\}^2 - \frac{1}{2}[(1-\rho)N-q+1-k]}{\frac{1}{2}\rho N}\right.$$

$$\left. - \frac{\{\frac{1}{2}[(1-\rho)N-q_2+1-k]\}^2 - \frac{1}{2}[(1-\rho)N-q_2+1-k]}{\frac{1}{2}\rho N}\right\}$$

$$= \frac{2}{\rho N}\sum_k\left[\frac{-2[(1-\rho)N-q_2+1-k]q_1+q_1^2}{4}+\frac{q_1}{2}\right]$$

$$= \frac{pq_1}{2\rho N}[-2(1-\rho)N+2q_2-2+(p+1)+q_1+2].$$

To make this zero, we require that

$$(27)\quad \rho = \frac{N-q_2-\frac{1}{2}(p+q_1+1)}{N}.$$

Then

$$(28)\quad \Pr\left\{-2\frac{m}{N}\log\lambda \le z\right\} = \Pr\{-m\log U_{p,q_1,N-q} \le z\}$$

$$= \Pr\{\chi^2_{pq_1} \le z\} + \frac{\gamma_2}{m^2}(\Pr\{\chi^2_{pq_1+4} \le z\} - \Pr\{\chi^2_{pq_1} \le z\})$$

$$+ \frac{1}{m^4}[\gamma_4(\Pr\{\chi^2_{pq_1+8} \le z\} - \Pr\{\chi^2_{pq_1} \le z\})$$

$$- \gamma_2^2(\Pr\{\chi^2_{pq_1+4} \le z\} - \Pr\{\chi^2_{pq_1} \le z\})] + R_5^{\mathrm{V}},$$

where

$$(29)\quad m = \rho N = N - q_2 - \tfrac{1}{2}(p+q_1+1),$$

$$(30)\quad \gamma_2 = \frac{pq_1(p^2+q_1^2-5)}{48},$$

$$(31)\quad \gamma_4 = \frac{\gamma_2^2}{2} + \frac{pq_1}{1920}[3p^4+3q_1^4+10p^2q_1^2-50(p^2+q_1^2)+159].$$

Since $\lambda = U_{p,q_1,n}^{\frac{1}{2}N}$, where $n = N - q$, (28) gives $\Pr\{-m\log U_{p,q_1,n} \le z\}$.

THEOREM 8.6.2. *The cdf of* $-m\log U_{p,q_1,n}$ *is given by* (28) *with* $m = n - \frac{1}{2}(p - q_1 + 1)$, *and* γ_2 *and* γ_4 *given by* (30) *and* (31) *respectively. The remainder term is* $O(N^{-6})$.

If the first term of (28) is used, the error is of the order N^{-2}; if the second, N^{-4}; and if the third,* N^{-6}. The second term is always negative

* Box has shown that the term of order N^{-5} is 0 and gives the coefficients to be used in the term of order N^{-6}.

and is numerically maximum for $z = \sqrt{(pq_1 + 2)(pq_1)}$ ($= pq_1 + 1$, approximately). For $p \geq 3$, $q_1 \geq 3$, $\gamma_2/m^2 \leq [(p^2 + q_1^2)/m]^2/96$ and the contribution of the second term lies between $-0.005\,[(p^2 + q_1^2)/m]^2$ and 0. For $p \geq 3$, $q_1 \geq 3$, $\gamma_4 \leq \gamma_2^2$ and the contribution of the third term is numerically less than $(\gamma_2/m^2)^2$. A rough rule that may be followed is that use of the first term is accurate to three decimal places if $p^2 + q_1^2 \leq m/3$.

As an example of the calculation, consider the case of $p = 3$, $q_1 = 6$, $N - q_2 = 24$, and $z = 26.0$ (the 10% significance point of χ^2_{18}, $18 = pq_1$). In this case $\gamma_2/m^2 = 0.048$ and the second term is -0.007; $\gamma_4/m^4 = 0.0015$ and the third term is -0.0001. Thus the probability of $-19 \log U_{3,6,18} \leq 26.0$ is 0.093 to three decimal places.

Table 2 is presented to give an idea of the accuracy of some of these approximations; the approximations use only the first term of the expansion. Tabulated is $U_{p,q_1,N-q}$ at the 5% point. [Box (1949) gives more examples.]

TABLE 2

p	q_1	$N-q$	Exact	Using $-2 \log \lambda$	Using $-m \log U$
1	1	50	0.925	0.935	0.924
	2		0.883	0.892	0.883
	3		0.849	0.856	0.849
2	2		0.819	0.827	0.819
	3		0.766	0.778	0.767
1	1	25	0.850		0.849
2	3		0.564		0.564
1	1	15	0.752		0.752
2	3		0.352		0.350
4	4	24	0.232		0.236

In Section 8.5 Table 1 gave significance points at the 5% significance level for $p = 3$, $q_1 = 3$, $n = 3$ to 10. Table 3 gives significance points for $-m \log U$ for $n = 8$ to ∞.

TABLE 3. SIGNIFICANCE POINTS FOR $-m \log U_{3,3,n}$

n	5%	1%
8	17.4	22.3
9	17.3	22.2
10	17.2	22.1
11	17.2	22.0
12–14	17.1	21.9
15–21	17.0	21.8
22–28	17.0	21.7
29–∞	16.9	21.7

The value of (28) for $z = 16.92$ (5% point for χ_9^2) is 0.9440 for $n = 10$, 0.9487 for $n = 21$, and 0.94959 for $n = 37$. Thus the first term is sufficient for $n \geq 21$ to give the 5% significance level to within 0.001.

Wald and Brookner (1941) obtained an asymptotic expansion for $-2 \log \lambda$. Bartlett (1938) suggested χ^2-significance points for $-m \log U$ and Rao (1948b) modified the expansion of Wald and Brookner to obtain (28).

8.7. TESTS OF HYPOTHESES ABOUT MATRICES OF REGRESSION COEFFICIENTS AND CONFIDENCE REGIONS

8.7.1. Testing Hypotheses

Suppose we are given a set of vector observations $x_1, \cdots, x_N$ with accompanying fixed vectors $z_1, \cdots, z_N$, where x_α is an observation from $N(\mathbf{B}z_\alpha, \mathbf{\Sigma})$. We let $\mathbf{B} = (\mathbf{B}_1\ \mathbf{B}_2)$ and $z'_\alpha = (z_\alpha^{(1)\prime}, z_\alpha^{(2)\prime})$, where $\mathbf{B}_1$ and $z_\alpha^{(1)\prime}$ have $q_1 (= q - q_2)$ columns. The null hypothesis is

$$H: \ \mathbf{B}_1 = \mathbf{B}_1^*, \tag{1}$$

where $\mathbf{B}_1^*$ is a specified matrix. Suppose the desired significance level is α. The test procedure is to compute

$$U = \frac{|N\hat{\mathbf{\Sigma}}_\Omega|}{|N\hat{\mathbf{\Sigma}}_\omega|} \tag{2}$$

and compare this number with $U_{p, q_1, n}(\alpha)$, the α significance point of the $U_{p, q_1, n}$-distribution. For certain cases this can be obtained from the distribution theory given in Section 7.5.2; otherwise this can be obtained from the asymptotic theory if sufficient terms are used. Alternatively, $\Pr\{U_{p, q_1, n} \leq U\}$ can be computed; if this is less than α, the null hypothesis is rejected.

Now we consider the computational problem of calculating (2). If we let $y_\alpha = x_\alpha - \mathbf{B}_1^* z_\alpha^{(1)}$, then y_α can be considered as an observation from $N(\mathbf{\Delta} z_\alpha, \mathbf{\Sigma})$, where $\mathbf{\Delta} = (\mathbf{\Delta}_1\ \mathbf{\Delta}_2) = (\mathbf{B}_1 - \mathbf{B}_1^*\ \mathbf{B}_2)$. Then the null hypothesis is H: $\mathbf{\Delta}_1 = \mathbf{0}$. Then

$$\sum y_\alpha y'_\alpha = \sum x_\alpha x'_\alpha - \mathbf{B}_1^* C'_1 - C_1 \mathbf{B}_1^{*\prime} + \mathbf{B}_1^* A_{11} \mathbf{B}_1^{*\prime}, \tag{3}$$

$$\sum y_\alpha z'_\alpha = C - \mathbf{B}_1^*(A_{11}\ A_{12}). \tag{4}$$

Thus the problem of testing the hypothesis $\mathbf{B}_1 = \mathbf{B}_1^*$ is equivalent to testing the hypothesis $\mathbf{\Delta}_1 = \mathbf{0}$ where $\mathscr{E} y_\alpha = \mathbf{\Delta} z_\alpha$. Hence let us suppose the problem is testing the hypothesis $\mathbf{B}_1 = \mathbf{0}$. Then $N\hat{\mathbf{\Sigma}}_\omega = \sum x_\alpha x'_\alpha - \hat{\mathbf{B}}_{2\omega} A_{22} \hat{\mathbf{B}}'_{2\omega}$ and $N\hat{\mathbf{\Sigma}}_\Omega = \sum x_\alpha x'_\alpha - \hat{\mathbf{B}}_\Omega A \hat{\mathbf{B}}'_\Omega$. We have shown in Section 8.2.2 how to compute $\hat{\mathbf{B}}_\Omega A \hat{\mathbf{B}}'_\Omega$, and hence $N\hat{\mathbf{\Sigma}}_\Omega$, by the Doolittle solution.

Clearly, $\hat{\boldsymbol{\beta}}_{2\omega}A_{22}\hat{\boldsymbol{\beta}}_{2\omega}'$ can be computed in a similar manner. If the Doolittle method is laid out

$$\begin{pmatrix} A_{22} & A_{21} \\ A_{12} & A_{11} \end{pmatrix}\begin{pmatrix} \hat{\boldsymbol{\beta}}_{2\Omega}' \\ \hat{\boldsymbol{\beta}}_{1\Omega}' \end{pmatrix} = \begin{pmatrix} C_2' \\ C_1' \end{pmatrix}, \tag{5}$$

the first q_2 rows and columns of A^* and of A^{**} are the same as the result of applying the forward solution to the left-hand side of

$$A_{22}\hat{\boldsymbol{\beta}}_{2\omega}' = C_2' \tag{6}$$

and the first q_2 rows of $\bar{C}^*$ and $\bar{C}^{**}$ are the same as the result of applying the forward solution to the right-hand side of (6). Thus $\hat{\boldsymbol{\beta}}_{2\omega}A_{22}\hat{\boldsymbol{\beta}}_{2\omega}' = \bar{C}_2^*\bar{C}_2^{**\prime}$, where $\bar{C}^* = \begin{pmatrix} \bar{C}_2^* \\ \bar{C}_1^* \end{pmatrix}$ and $\bar{C}^{**} = \begin{pmatrix} \bar{C}_2^{**} \\ \bar{C}_1^{**} \end{pmatrix}$.

It might be noticed that the abbreviated Doolittle method implies a method for computing a determinant. In Section 8.2.3 it was shown that the result of the forward solution is $FA = A^*$. Thus $|F| \cdot |A| = |A^*|$. Since the determinant of a triangular matrix is the product of its diagonal elements, $|F| = 1$ and $|A| = |A^*| = \prod_{i=1}^{q} a_{ii}^*$. This result holds for any positive definite matrix in place of A (with a suitable modification of F) and hence can be used to compute $|N\hat{\boldsymbol{\Sigma}}_\Omega|$ and $|N\hat{\boldsymbol{\Sigma}}_\omega|$.

8.7.2. Confidence Regions

We have considered tests of hypotheses $\boldsymbol{\beta}_1 = \boldsymbol{\beta}_1^*$ where $\boldsymbol{\beta}_1^*$ is specified. In the usual way we can deduce from the family of tests a confidence region for $\boldsymbol{\beta}_1$. From the theory given before, we know that the probability is $1 - \alpha$ of drawing a sample so that

$$\frac{|N\hat{\boldsymbol{\Sigma}}_\Omega|}{|N\hat{\boldsymbol{\Sigma}}_\Omega + (\hat{\boldsymbol{\beta}}_{1\Omega} - \boldsymbol{\beta}_1)(A_{11} - A_{12}A_{22}^{-1}A_{21})(\hat{\boldsymbol{\beta}}_{1\Omega} - \boldsymbol{\beta}_1)'|} \geq U_{p, q_1, n}(\alpha). \tag{7}$$

Thus if we make the confidence region statement that $\boldsymbol{\beta}_1$ satisfies

$$\frac{|N\hat{\boldsymbol{\Sigma}}_\Omega|}{|N\hat{\boldsymbol{\Sigma}}_\Omega + (\hat{\boldsymbol{\beta}}_{1\Omega} - \bar{\boldsymbol{\beta}}_1)(A_{11} - A_{12}A_{22}^{-1}A_{21})(\hat{\boldsymbol{\beta}}_{1\Omega} - \bar{\boldsymbol{\beta}}_1)'|} \geq U_{p, q_1, n}(\alpha), \tag{8}$$

where (8) is interpreted as an inequality on $\boldsymbol{\beta}_1 = \bar{\boldsymbol{\beta}}_1$, the probability is $1 - \alpha$ of drawing a sample such that the statement is true.

THEOREM 8.7.1. *The region* (8) *in the* $\boldsymbol{\beta}_1$ *space is a confidence region for* $\boldsymbol{\beta}_1$ *with confidence coefficient* $1 - \alpha$.

8.8. TESTING EQUALITY OF MEANS OF SEVERAL NORMAL DISTRIBUTIONS WITH COMMON COVARIANCE MATRIX

In univariate analysis it is well known that many hypotheses can be put in the form of hypotheses concerning regression coefficients. The same is true for the corresponding multivariate cases. As an example we consider testing the hypothesis that the means of, say, q normal distributions with a common covariance matrix are equal.

Let $\boldsymbol{y}_\alpha^{(i)}$ be an observation from $N(\boldsymbol{\mu}^{(i)}, \boldsymbol{\Sigma})$, $\alpha = 1, \cdots, N_i$; $i = 1, \cdots, q$. The null hypothesis is

$$H: \quad \boldsymbol{\mu}^{(1)} = \cdots = \boldsymbol{\mu}^{(q)}. \tag{1}$$

To put the problem in the form considered earlier in this chapter, let

$$\boldsymbol{X} = (\boldsymbol{x}_1\ \boldsymbol{x}_2 \cdots \boldsymbol{x}_{N_1}\ \boldsymbol{x}_{N_1+1} \cdots \boldsymbol{x}_N) = (\boldsymbol{y}_1^{(1)}\ \boldsymbol{y}_2^{(1)} \cdots \boldsymbol{y}_{N_1}^{(1)}\ \boldsymbol{y}_1^{(2)} \cdots \boldsymbol{y}_{N_q}^{(q)}) \tag{2}$$

with $N = N_1 + \cdots + N_q$. Let

$$\begin{aligned} \boldsymbol{Z} &= (\boldsymbol{z}_1\ \boldsymbol{z}_2 \cdots \boldsymbol{z}_{N_1} \boldsymbol{z}_{N_1+1} \cdots \boldsymbol{z}_N) \\ &= \begin{pmatrix} 1 & 1 \cdots 1 & 0 \cdots 0 \\ 0 & 0 \cdots 0 & 1 \cdots 0 \\ 0 & 0 \cdots 0 & 0 \cdots 0 \\ \cdot & \cdot \quad\ \ \cdot & \cdot \quad\ \ \cdot \\ \cdot & \cdot \quad\ \ \cdot & \cdot \quad\ \ \cdot \\ \cdot & \cdot \quad\ \ \cdot & \cdot \quad\ \ \cdot \\ 0 & 0 \cdots 0 & 0 \cdots 0 \\ 1 & 1 \cdots 1 & 1 \cdots 1 \end{pmatrix}; \end{aligned} \tag{3}$$

that is, $z_{i\alpha} = 1$ if $N_1 + \cdots + N_{i-1} < \alpha \leq N_1 + \cdots + N_i$, and $z_{i\alpha} = 0$ otherwise, for $i = 1, \cdots, q - 1$, and $z_{q\alpha} = 1$ (all α). Let $\boldsymbol{\beta} = (\boldsymbol{\beta}_1\ \boldsymbol{\beta}_2)$ where

$$\begin{aligned} \boldsymbol{\beta}_1 &= (\boldsymbol{\mu}_1 - \boldsymbol{\mu}_q, \cdots, \boldsymbol{\mu}_{q-1} - \boldsymbol{\mu}_q), \\ \boldsymbol{\beta}_2 &= \boldsymbol{\mu}_q. \end{aligned} \tag{4}$$

Then $\boldsymbol{x}_\alpha$ is an observation from $N(\boldsymbol{\beta} \boldsymbol{z}_\alpha, \boldsymbol{\Sigma})$ and the null hypothesis is $\boldsymbol{\beta}_1 = \boldsymbol{0}$. Thus we can use the above theory for finding the criterion for testing the hypothesis.

We have

$$\boldsymbol{A} = \sum \boldsymbol{z}_\alpha \boldsymbol{z}_\alpha' = \begin{pmatrix} N_1 & 0 & \cdots & 0 & N_1 \\ 0 & N_2 & \cdots & 0 & N_2 \\ \cdot & \cdot & & \cdot & \cdot \\ \cdot & \cdot & & \cdot & \cdot \\ \cdot & \cdot & & \cdot & \cdot \\ 0 & 0 & \cdots & N_{q-1} & N_{q-1} \\ N_1 & N_2 & \cdots & N_{q-1} & N. \end{pmatrix}, \tag{5}$$

$$\boldsymbol{C} = \sum \boldsymbol{x}_\alpha \boldsymbol{z}_\alpha' = (\textstyle\sum_\alpha \boldsymbol{y}_\alpha^{(1)}\ \sum_\alpha \boldsymbol{y}_\alpha^{(2)} \cdots \sum_\alpha \boldsymbol{y}_\alpha^{(q-1)}\quad \sum_{i,\alpha} \boldsymbol{y}_\alpha^{(i)}). \tag{6}$$

Here $A_{22} = N$ and $C_2 = \sum_{i,\alpha} y_\alpha^{(i)}$. Thus $\hat{\boldsymbol\beta}_{2\omega} = \sum_{i,\alpha} y_\alpha^{(i)} \cdot (1/N) = \bar{\boldsymbol y}$, say, and

$$N\hat{\boldsymbol\Sigma}_\omega = \sum \boldsymbol x_\alpha \boldsymbol x_\alpha' - \bar{\boldsymbol y} N \bar{\boldsymbol y}' \tag{7}$$
$$= \sum_{i,\alpha} \boldsymbol y_\alpha^{(i)} \boldsymbol y_\alpha^{(i)\prime} - N\bar{\boldsymbol y}\bar{\boldsymbol y}'$$
$$= \sum_{i,\alpha} (\boldsymbol y_\alpha^{(i)} - \bar{\boldsymbol y})(\boldsymbol y_\alpha^{(i)} - \bar{\boldsymbol y})'.$$

For $\hat{\boldsymbol\Sigma}_\Omega$, we use the formula $N\hat{\boldsymbol\Sigma}_\Omega = \sum \boldsymbol x_\alpha \boldsymbol x_\alpha' - \hat{\boldsymbol\beta}_\Omega A \hat{\boldsymbol\beta}_\Omega' = \sum \boldsymbol x_\alpha \boldsymbol x_\alpha' - CA^{-1}C'$. Let

$$D = \begin{pmatrix} 1 & 0 \cdots & 0 & 0 \\ 0 & 1 \cdots & 0 & 0 \\ \cdot & \cdot & \cdot & \cdot \\ \cdot & \cdot & \cdot & \cdot \\ \cdot & \cdot & \cdot & \cdot \\ 0 & 0 \cdots & 1 & 0 \\ -1 & -1 \cdots & -1 & 1 \end{pmatrix}; \tag{8}$$

then

$$D^{-1} = \begin{pmatrix} 1 & 0 \cdots & 0 & 0 \\ 0 & 1 \cdots & 0 & 0 \\ \cdot & \cdot & \cdot & \cdot \\ \cdot & \cdot & \cdot & \cdot \\ \cdot & \cdot & \cdot & \cdot \\ 0 & 0 \cdots & 1 & 0 \\ 1 & 1 \cdots & 1 & 1 \end{pmatrix}. \tag{9}$$

Thus

$$CA^{-1}C' = CD'D^{-1\prime}A^{-1}D^{-1}DC' \tag{10}$$
$$= CD'(DAD')^{-1}DC'$$
$$= \left(\sum_\alpha \boldsymbol y_\alpha^{(1)} \cdots \sum_\alpha \boldsymbol y_\alpha^{(q)}\right) \begin{pmatrix} N_1 & 0 & \cdots & 0 \\ 0 & N_2 & \cdots & 0 \\ \cdot & \cdot & & \cdot \\ \cdot & \cdot & & \cdot \\ \cdot & \cdot & & \cdot \\ 0 & 0 & \cdots & N_q \end{pmatrix}^{-1} \begin{pmatrix} \sum_\alpha \boldsymbol y_\alpha^{(1)\prime} \\ \cdot \\ \cdot \\ \cdot \\ \sum_\alpha \boldsymbol y_\alpha^{(q)\prime} \end{pmatrix}$$
$$= \sum_i \left(\sum_\alpha \boldsymbol y_\alpha^{(i)} \frac{1}{N_i} \sum_\gamma \boldsymbol y_\gamma^{(i)\prime}\right)$$
$$= \sum_i N_i \bar{\boldsymbol y}^{(i)} \bar{\boldsymbol y}^{(i)\prime},$$

where $\bar{\mathbf{y}}^{(i)} = \frac{1}{N_i} \sum_\alpha \mathbf{y}_\alpha^{(i)}$. Thus

$$N\hat{\mathbf{\Sigma}}_\Omega = \sum_{i,\alpha} \mathbf{y}_\alpha^{(i)} \mathbf{y}_\alpha^{(i)\prime} - \sum_i N_i \bar{\mathbf{y}}^{(i)} \bar{\mathbf{y}}^{(i)\prime} = \sum_{i,\alpha} (\mathbf{y}_\alpha^{(i)} - \bar{\mathbf{y}}^{(i)})(\mathbf{y}_\alpha^{(i)} - \bar{\mathbf{y}}^{(i)})'. \tag{11}$$

It will be seen that $\hat{\mathbf{\Sigma}}_\omega$ is the estimate of $\mathbf{\Sigma}$ when $\boldsymbol{\mu}^{(1)} = \cdots = \boldsymbol{\mu}^{(q)}$ and $\hat{\mathbf{\Sigma}}_\Omega$ is the weighted average of the estimates of $\mathbf{\Sigma}$ based on the separate samples.

When the null hypothesis is true $|N\hat{\mathbf{\Sigma}}_\Omega|/|N\hat{\mathbf{\Sigma}}_\omega|$ is distributed as $U_{p,\,q-1,\,n}$, where $n = N - q$. Therefore, the rejection region at the α significance level is

$$\frac{|N\hat{\mathbf{\Sigma}}_\Omega|}{|N\hat{\mathbf{\Sigma}}_\omega|} < U_{p,\,q-1,\,n}(\alpha). \tag{12}$$

We observe that

$$N\hat{\mathbf{\Sigma}}_\omega - N\hat{\mathbf{\Sigma}}_\Omega = \sum_{i,\alpha} \mathbf{y}_\alpha^{(i)} \mathbf{y}_\alpha^{(i)\prime} - N\bar{\mathbf{y}}\bar{\mathbf{y}}' - \Big(\sum_{i,\alpha} \mathbf{y}_\alpha^{(i)} \mathbf{y}_\alpha^{(i)\prime} - \sum_i N_i \bar{\mathbf{y}}^{(i)} \bar{\mathbf{y}}^{(i)\prime}\Big) = \sum N_i (\bar{\mathbf{y}}^{(i)} - \bar{\mathbf{y}})(\bar{\mathbf{y}}^{(i)} - \bar{\mathbf{y}})'. \tag{13}$$

It will be seen that when $p = 1$, this test reduces to the usual F-test

$$\frac{\sum N_i (\bar{y}^{(i)} - \bar{y})^2}{\sum (y_\alpha^{(i)} - \bar{y}^{(i)})^2} \cdot \frac{n}{q-1} > F_{q-1,\,n}(\alpha). \tag{14}$$

We give an example of the analysis. The data are taken from Barnard's study of Egyptian skulls (1935). The 4 $(= q)$ populations are Late Predynastic $(i = 1)$, Sixth to Twelfth $(i = 2)$, Twelfth to Thirteenth $(i = 3)$, and Ptolemaic Dynasties $(i = 4)$. The 4 $(= p)$ measurements (that is, components of $\mathbf{y}_\alpha^{(i)}$) are maximum breadth, basialveolar length, nasal height, and basibregmatic height. The numbers of observations are $N_1 = 91$, $N_2 = 162$, $N_3 = 70$, $N_4 = 75$. The data are summarized as

$$(\bar{\mathbf{y}}^{(1)}\ \bar{\mathbf{y}}^{(2)}\ \bar{\mathbf{y}}^{(3)}\ \bar{\mathbf{y}}^{(4)}) = \begin{pmatrix} 133.582\,418 & 134.265\,432 & 134.371\,429 & 135.306\,667 \\ 98.307\,692 & 96.462\,963 & 95.857\,143 & 95.040\,000 \\ 50.835\,165 & 51.148\,148 & 50.100\,000 & 52.093\,333 \\ 133.000\,000 & 134.882\,716 & 133.642\,857 & 131.466\,667 \end{pmatrix}, \tag{15}$$

$$N\hat{\mathbf{\Sigma}}_\Omega = \begin{pmatrix} 9661.997\,470 & 445.573\,301 & 1130.623\,900 & 2148.584\,210 \\ 445.573\,301 & 9073.115\,027 & 1239.211\,990 & 2255.812\,722 \\ 1130.623\,900 & 1239.211\,990 & 3938.320\,351 & 1271.054\,662 \\ 2148.584\,210 & 2255.812\,722 & 1271.054\,662 & 8741.508\,829 \end{pmatrix}. \tag{16}$$

From these data we find

$$(17)\quad N\hat{\boldsymbol{\Sigma}}_\omega$$

$$= \begin{pmatrix} 9785.178\,098 & 214.197\,666 & 1217.929\,248 & 2019.820\,216 \\ 214.197\,666 & 9559.460\,890 & 1131.716\,372 & 2381.126\,040 \\ 1217.929\,248 & 1131.716\,372 & 4088.731\,856 & 1133.473\,898 \\ 2019.820\,216 & 2381.126\,040 & 1133.473\,898 & 9382.242\,720 \end{pmatrix}.$$

The ratio of determinants is

$$(18)\qquad U = \frac{|N\hat{\boldsymbol{\Sigma}}_\Omega|}{|N\hat{\boldsymbol{\Sigma}}_\omega|} = \frac{2.426\,905\,4 \times 10^5}{2.954\,477\,5 \times 10^5} = 0.821\,434\,4.$$

Here $N = 398$, $n = 394$, $p = 4$, and $q = 4$. Thus $m = 393$. Since n is very large, we may assume $-m \log U_{4,3,394}$ is distributed as χ^2 with 12 degrees of freedom (when the null hypothesis is true). Here $-m \log U = 77.30$. Since the 1% point of the χ^2_{12}-distribution is 26.2, the hypothesis of $\boldsymbol{\mu}^{(1)} = \boldsymbol{\mu}^{(2)} = \boldsymbol{\mu}^{(3)} = \boldsymbol{\mu}^{(4)}$ is rejected.*

8.9. GENERALIZED ANALYSIS OF VARIANCE

The univariate analysis of variance has a direct generalization for vector variables leading to an analysis of vector sums of squares (that is, sums such as $\sum \boldsymbol{x}_\alpha \boldsymbol{x}'_\alpha$). In fact, in the preceding section this generalization was considered for an analysis of variance problem involving a single classification.

As another example consider a two-way layout. Suppose that we are interested in the question whether the column effects are zero. We shall review the analysis for a scalar variable and then show the analysis for a vector variable. Let $Y_{ij} (i = 1, \cdots, r;\ j = 1, \cdots, c)$ be a set of rc random variables. We assume that

$$(1)\qquad \mathscr{E} Y_{ij} = \mu + \lambda_i + \nu_j, \qquad i = 1, \cdots, r;\ j = 1, \cdots c,$$

with the restrictions

$$(2)\qquad \sum_{i=1}^{r} \lambda_i = \sum_{j=1}^{c} \nu_j = 0,$$

that the variance of Y_{ij} is σ^2, and that the Y_{ij} are independently normally distributed. To test that column effects are zero is to test that

$$(3)\qquad \nu_j = 0, \qquad j = 1, \cdots, c.$$

* The above computations have been given by Bartlett (1947a).

It is well known that this problem can be treated as a problem of regression by the introduction of dummy fixed variates. Let

$$
\begin{aligned}
(4)\qquad z_{00,\,ij} &= 1, & \\
z_{k0,\,ij} &= 1, & k = i, \\
&= 0, & k \neq i, \\
z_{0k,\,ij} &= 1, & k = j, \\
&= 0, & k \neq j.
\end{aligned}
$$

Then (1) can be written

$$
(5)\qquad \mathscr{E} Y_{ij} = \mu z_{00,\,ij} + \sum_{k=1}^{r} \lambda_k z_{k0,\,ij} + \sum_{k=1}^{c} \nu_k z_{0k,\,ij}.
$$

The hypothesis is that the coefficients of $z_{0k,\,ij}$ are zero. Since the matrix of fixed variates here

$$
(6)\qquad \begin{pmatrix}
z_{00,\,11} & \cdots & z_{00,\,rc} \\
z_{10,\,11} & \cdots & z_{10,\,rc} \\
z_{20,\,11} & \cdots & z_{20,\,rc} \\
\cdot & & \cdot \\
\cdot & & \cdot \\
\cdot & & \cdot \\
z_{0c,\,11} & \cdots & z_{0c,\,rc}
\end{pmatrix}
$$

is singular (for example, row 00 is the sum of rows 10, 20, $\cdots$, $r0$), one must elaborate the regression theory. When one does, one finds that the test criterion indicated by the regression theory is the usual F-test of analysis of variance.

Let

$$
\begin{aligned}
(7)\qquad Y_{\cdot\cdot} &= \frac{1}{rc} \sum_{i,\,j} Y_{ij}, \\
Y_{i\cdot} &= \frac{1}{c} \sum_{j} Y_{ij}, \\
Y_{\cdot j} &= \frac{1}{r} \sum_{j} Y_{ij},
\end{aligned}
$$

and let

$$
\begin{aligned}
(8)\qquad a &= \sum_{i,\,j} (Y_{ij} - Y_{i\cdot} - Y_{\cdot j} + Y_{\cdot\cdot})^2 \\
&= \sum_{i,\,j} Y_{ij}^2 - c\sum_{i} Y_{i\cdot}^2 - r\sum_{j} Y_{\cdot j}^2 + rcY_{\cdot\cdot}^2, \\
b &= r\sum_{j} (Y_{\cdot j} - Y_{\cdot\cdot})^2 \\
&= r\sum_{j} Y_{\cdot j}^2 - rcY_{\cdot\cdot}^2.
\end{aligned}
$$

Then the F-statistic is given by

$$F = \frac{b}{a} \cdot \frac{(c-1)(r-1)}{c-1}. \tag{9}$$

Under the null hypothesis, this has the F-distribution with $c - 1$ and $(r - 1)(c - 1)$ degrees of freedom. The likelihood ratio criterion for the hypothesis is the $rc/2$ power of

$$\frac{a}{a+b} = \frac{1}{1 + (c-1)/[(r-1)(c-1)]F}. \tag{10}$$

Now let us turn to the multivariate analysis of variance. We have a set of p-dimensional random vectors $\boldsymbol{Y}_{ij}$ $(i = 1, \cdots, r;\ j = 1, \cdots, c)$ with expected values (1), where $\boldsymbol{\mu}$, $\boldsymbol{\lambda}$'s, and $\boldsymbol{\nu}$'s are vectors, and covariance matrix $\boldsymbol{\Sigma}$, and they are independently normally distributed. Then the same algebra may be used to reduce this problem to the regression problem. We define $\boldsymbol{Y}_{..}$, $\boldsymbol{Y}_{i.}$, $\boldsymbol{Y}_{.j}$ by (7) and

$$\begin{aligned} \boldsymbol{A} &= \sum_{i,j} (\boldsymbol{Y}_{ij} - \boldsymbol{Y}_{i.} - \boldsymbol{Y}_{.j} + \boldsymbol{Y}_{..})(\boldsymbol{Y}_{ij} - \boldsymbol{Y}_{i.} - \boldsymbol{Y}_{.j} + \boldsymbol{Y}_{..})' \\ &= \sum_{i,j} \boldsymbol{Y}_{ij}\boldsymbol{Y}_{ij}' - c\sum_{i} \boldsymbol{Y}_{i.}\boldsymbol{Y}_{i.}' - r\sum_{j} \boldsymbol{Y}_{.j}\boldsymbol{Y}_{.j}' + rc\boldsymbol{Y}_{..}\boldsymbol{Y}_{..}', \\ \boldsymbol{B} &= r\sum_{j} (\boldsymbol{Y}_{.j} - \boldsymbol{Y}_{..})(\boldsymbol{Y}_{.j} - \boldsymbol{Y}_{..})' \\ &= r\sum_{j} \boldsymbol{Y}_{.j}\boldsymbol{Y}_{.j}' - rc\boldsymbol{Y}_{..}\boldsymbol{Y}_{..}'. \end{aligned} \tag{11}$$

The statistic analogous to (10) is

$$\frac{|\boldsymbol{A}|}{|\boldsymbol{A} + \boldsymbol{B}|}. \tag{12}$$

Under the null hypothesis, this has the distribution of U for p, $n = (r - 1)(c - 1)$ and $q_1 = c - 1$ given in Section 8.5. In order for $\boldsymbol{A}$ to be nonsingular (with probability 1) we must require $p \leq (r - 1)(c - 1)$.

As an example we use data first published by Immer, Hayes, and Powers (1934), and later used by Fisher (1947a), by Yates and Cochran (1938), and by Tukey (1949). The first component of the observation vector is a barley yield in a given year; the second component is the same measurement made the following year. Column indices run over the varieties of barley and row indices over the locations. The data are given in Table 3. [for example, $\begin{pmatrix} 81 \\ 81 \end{pmatrix}$ in the upper left-hand corner indicates a yield of 81 in each year of variety M in location UF.] The numbers along the borders are sums.

TABLE 3

		Varieties					
		M	*S*	*V*	*T*	*P*	
Location	*UF*	81 81	105 82	120 80	110 87	98 84	514 414
	W	147 100	142 116	151 112	192 148	146 108	778 584
	M	82 103	77 105	78 117	131 140	90 130	458 595
	C	120 99	121 62	124 96	141 126	125 76	631 459
	GR	99 66	89 50	69 97	89 62	104 80	450 355
	D	87 68	77 67	79 67	102 92	96 94	441 388
		616 517	611 482	621 569	765 655	659 572	3272 2795

We consider the square of (147, 100) to be

$$\begin{pmatrix} 21609 & 14700 \\ 14700 & 10000 \end{pmatrix}.$$

Then

$$\sum_{i,j} \boldsymbol{Y}_{ij}\boldsymbol{Y}'_{ij} = \begin{pmatrix} 380,944 & 315,381 \\ 315,381 & 277,625 \end{pmatrix}, \tag{13}$$

$$\sum_j (6\boldsymbol{Y}_{.j})(6\boldsymbol{Y}_{.j})' = \begin{pmatrix} 2,157,924 & 1,844,346 \\ 1,844,346 & 1,579,583 \end{pmatrix}, \tag{14}$$

$$\sum_i (5\boldsymbol{Y}_{i.})(5\boldsymbol{Y}_{i.})' = \begin{pmatrix} 1,874,386 & 1,560,145 \\ 1,560,145 & 1,353,727 \end{pmatrix}, \tag{15}$$

$$(30\boldsymbol{Y}_{..})(30\boldsymbol{Y}_{..})' = \begin{pmatrix} 10,705,984 & 9,145,240 \\ 9,145,240 & 7,812,025 \end{pmatrix}. \tag{16}$$

Then the "error sum of squares" is

$$A = \begin{pmatrix} 3279 & 802 \\ 802 & 4017 \end{pmatrix}, \tag{17}$$

the "row sum of squares" is

$$(18)\qquad 5\sum_i (\boldsymbol{Y}_{i\cdot} - \boldsymbol{Y}_{\cdot\cdot})(\boldsymbol{Y}_{i\cdot} - \boldsymbol{Y}_{\cdot\cdot})' = \begin{pmatrix} 18{,}011 & 7{,}188 \\ 7{,}188 & 10{,}345 \end{pmatrix},$$

and the "column sum of squares" is

$$(19)\qquad \boldsymbol{B} = \begin{pmatrix} 2788 & 2550 \\ 2550 & 2863 \end{pmatrix}.$$

The test criterion is

$$(20)\qquad \frac{|\boldsymbol{A}|}{|\boldsymbol{A}+\boldsymbol{B}|} = \frac{\begin{vmatrix} 3279 & 802 \\ 802 & 4017 \end{vmatrix}}{\begin{vmatrix} 6067 & 3352 \\ 3352 & 6880 \end{vmatrix}} = 0.4107.$$

This result is to be compared with the significance point for $U_{2,\,4,\,20}$. Using the result of Section 8.5, we see that

$$\frac{1 - \sqrt{0.4107}}{\sqrt{0.4107}} \cdot \frac{19}{4} = 2.66$$

is to be compared with the significance point of $F_{8,\,38}$. This is significant at the 5% level. Our data show that there are differences between varieties.

Now let us see that each F-test in the univariate analysis of variance has an analogous U-test in the multivariate analysis of variance. In the linear hypothesis model for the univariate analysis of variance, one assumes that the random variables $Y_1, \cdots, Y_N$ have expected values that are linear combinations of unknown parameters

$$(21)\qquad \mathscr{E}\, Y_\alpha = \sum_g \beta_g z_{g\alpha},$$

where the β's are the parameters and the z's are the known coefficients. The variables $\{Y_\alpha\}$ are assumed to be normally and independently distributed with common variance σ^2. In this model there are a set of linear combinations, say $\sum_\alpha \gamma_{i\alpha} Y_\alpha$, where the γ's are known, such that

$$(22)\qquad a = \sum_i \Big(\sum_\alpha \gamma_{i\alpha} Y_\alpha\Big)^2 = \sum_{\alpha,\,\beta} d_{\alpha\beta} Y_\alpha Y_\beta$$

is distributed as $\sigma^2\chi^2$ with n degrees of freedom. There is another set of linear combinations, say $\sum_\alpha \phi_{g\alpha} Y_\alpha$, where the ϕ's are known, such that

$$(23)\qquad b = \sum_g \Big(\sum_\alpha \phi_{g\alpha} Y_\alpha\Big)^2 = \sum_{\alpha,\,\beta} c_{\alpha\beta} Y_\alpha Y_\beta$$

is distributed as $\sigma^2\chi^2$ with m degrees of freedom when the null hypothesis is true and as σ^2 times a noncentral χ^2 when the null hypothesis is not true; and in either case b is distributed independently of a. Then

$$\frac{b}{a}\cdot\frac{n}{m} = \frac{\sum c_{\alpha\beta} Y_\alpha Y_\beta}{\sum d_{\alpha\beta} Y_\alpha Y_\beta}\cdot\frac{n}{m} \tag{24}$$

has the F-distribution with m and n degrees of freedom respectively when the null hypothesis is true. The null hypothesis is that certain β's are zero.

In the multivariate analysis of variance, $\mathbf{Y}_1, \cdots, \mathbf{Y}_N$ are vector variables with p components. The expected value of $\mathbf{Y}_\alpha$ is given by (21) where $\boldsymbol{\beta}_g$ is a vector of p parameters. We assume that the $\{\mathbf{Y}_\alpha\}$ are normally and independently distributed with common covariance matrix $\boldsymbol{\Sigma}$. The linear combinations $\sum \gamma_{i\alpha}\mathbf{Y}_\alpha$ can be formed for the vectors. Then

$$\boldsymbol{A} = \sum_i \left(\sum_\alpha \gamma_{i\alpha}\mathbf{Y}_\alpha\right)\left(\sum_\alpha \gamma_{i\alpha}\mathbf{Y}_\alpha\right)' = \sum_{\alpha,\beta} d_{\alpha\beta}\mathbf{Y}_\alpha\mathbf{Y}_\beta' \tag{25}$$

has the distribution $W(\boldsymbol{\Sigma}, n)$. When the null hypothesis is true,

$$\boldsymbol{B} = \sum_g \left(\sum_\alpha \phi_{g\alpha}\mathbf{Y}_\alpha\right)\left(\sum_\alpha \phi_{g\alpha}\mathbf{Y}_\alpha\right)' = \sum_{\alpha,\beta} c_{\alpha\beta}\mathbf{Y}_\alpha\mathbf{Y}_\beta' \tag{26}$$

has the distribution $W(\boldsymbol{\Sigma}, m)$, and $\boldsymbol{B}$ is independent of $\boldsymbol{A}$. Then

$$\frac{|\boldsymbol{A}|}{|\boldsymbol{A}+\boldsymbol{B}|} = \frac{\left|\sum d_{\alpha\beta}\mathbf{Y}_\alpha\mathbf{Y}_\beta'\right|}{\left|\sum d_{\alpha\beta}\mathbf{Y}_\alpha\mathbf{Y}_\beta' + \sum c_{\alpha\beta}\mathbf{Y}_\alpha\mathbf{Y}_\beta'\right|} \tag{27}$$

has the $U_{p,m,n}$-distribution.

The argument for the distribution of a and b involves showing that $\mathscr{E}\sum_\alpha \gamma_{i\alpha} Y_\alpha = 0$ and $\mathscr{E}\sum_\alpha \phi_{g\alpha} Y_\alpha = 0$ when certain β's are equal to zero as specified by the null hypothesis (as identities in the unspecified β's). Clearly this argument holds for the vector case as well. Secondly, one argues, in the univariate case, that there is an orthogonal matrix $\boldsymbol{\Psi} = (\psi_{\alpha\beta})$ such that when the transformation $Y_\beta = \sum_\alpha \psi_{\beta\alpha} Z_\alpha$ is made

$$\begin{aligned} a &= \sum_{\alpha,\beta,\gamma,\delta} d_{\alpha\beta}\psi_{\alpha\gamma}\psi_{\beta\delta} Z_\gamma Z_\delta = \sum_{\alpha=1}^{n} Z_\alpha^2, \\ b &= \sum_{\alpha,\beta,\gamma,\delta} c_{\alpha\beta}\psi_{\alpha\gamma}\psi_{\beta\delta} Z_\gamma Z_\delta = \sum_{\alpha=n+1}^{n+m} Z_\alpha^2. \end{aligned} \tag{28}$$

Because the transformation is orthogonal the $\{Z_\alpha\}$ are independently and normally distributed with common variance σ^2. Since the Z_α, $\alpha = 1, \cdots, n$, must be linear combinations of $\sum_\alpha \gamma_{i\alpha} Y_\alpha$ and Z_α, $\alpha = n+1, \cdots, n+m$, must be linear combinations of $\sum_\alpha \phi_{g\alpha} Y_\alpha$, they must have means zero (under the null hypothesis). Thus a/σ^2 and b/σ^2 have the stated independent χ^2-distributions.

In the multivariate case the transformation $Y_\beta = \sum_\alpha \psi_{\beta\alpha} Z_\alpha$ is used, where Y_β and Z_α are vectors. Then

$$(29) \qquad A = \sum_{\alpha,\beta,\gamma,\delta} d_{\alpha\beta}\psi_{\alpha\gamma}\psi_{\beta\delta} Z_\gamma Z_\delta' = \sum_{\alpha=1}^{n} Z_\alpha Z_\alpha'$$

$$B = \sum_{\alpha,\beta,\gamma,\delta} c_{\alpha\beta}\psi_{\alpha\gamma}\psi_{\beta\delta} Z_\gamma Z_\delta' = \sum_{\alpha=n+1}^{n+m} Z_\alpha Z_\alpha'$$

because it follows from (28) that $\sum_{\alpha,\beta} d_{\alpha\beta}\psi_{\alpha\gamma}\psi_{\beta\delta} = 1$, $\gamma = \delta \leq n$, $= 0$, otherwise, and $\sum_{\alpha,\beta} c_{\alpha\beta}\psi_{\alpha\gamma}\psi_{\beta\delta} = 1, n + 1 \leq \gamma = \delta \leq n + m, = 0$, otherwise. Since Ψ is orthogonal, the $\{Z_\alpha\}$ are independently normally distributed with covariance matrix Σ. The same argument shows $\mathscr{E} Z_\alpha = 0$, $\alpha = 1, \cdots, n + m$, under the null hypothesis. Thus A and B are independently distributed according to $W(\Sigma, n)$ and $W(\Sigma, m)$ respectively.

8.10. OTHER CRITERIA FOR TESTING THE LINEAR HYPOTHESIS

Thus far the only test of the linear hypothesis we have considered is the likelihood ratio test. In this section we consider other test procedures.

Let $\hat{\Sigma}_\Omega$, $\hat{\beta}_{1\Omega}$, and $\hat{\beta}_{2\omega}$ be the estimates of the parameters in $N(\beta z, \Sigma)$, based on a sample of N observations. These are a sufficient set of statistics, and we shall base test procedures on them. As was shown in Section 8.3, if the hypothesis is $\beta_1 = \beta_1^*$, one can reformulate the hypothesis as $\beta_1 = 0$ (by replacing x_α by $x_\alpha - \beta_1^* z_\alpha^{(1)}$). Moreover,

$$(1) \qquad \begin{aligned} \beta z_\alpha &= \beta_1 z_\alpha^{(1)} + \beta_2 z_\alpha^{(2)} \\ &= \beta_1(z_\alpha^{(1)} - A_{12}A_{22}^{-1} z_\alpha^{(2)}) + (\beta_2 + \beta_1 A_{12} A_{22}^{-1}) z_\alpha^{(2)} \\ &= \beta_1 z_\alpha^{*(1)} + \beta_2^* z_\alpha^{(2)} \end{aligned}$$

where $\sum_\alpha z_\alpha^{*(1)} z_\alpha^{(2)\prime} = 0$ and $\sum_\alpha z_\alpha^{*(1)} z_\alpha^{*(1)\prime} = A_{11\cdot 2}$. Then $\hat{\beta}_1 = \hat{\beta}_{1\Omega}$ and $\hat{\beta}_2^* = \hat{\beta}_{2\omega}$.

We shall use the principle of invariance to reduce the set of tests to be considered. First, if we make the transformation $X_2^* = X_\alpha + \Gamma z_\alpha^{(2)}$, we leave the null hypothesis invariant since $\mathscr{E} X_\alpha^* = \beta_1 z_\alpha^{*(1)} + (\beta_2^* + \Gamma) z_\alpha^{(2)}$ and $B_2^* + \Gamma$ is unspecified. The only invariants of the sufficient statistics are $\hat{\Sigma}$ and $\hat{\beta}_1$ (since for each $\hat{\beta}_2^*$, there is a Γ that transforms it to 0, namely $-\hat{\beta}_2^*$).

Second, the null hypothesis is invariant under the transformation $z_\alpha^{**(1)} = C z_\alpha^{*(1)}$ (C nonsingular); the transformation carries β_1 to $\beta_1 C^{-1}$. Under this transformation $\hat{\Sigma}$ and $\hat{\beta}_1 A_{11\cdot 2} \hat{\beta}_1'$ are invariant; we consider $A_{11\cdot 2}$ as information relevant to inference. But these are the only invariants. For consider a function of $\hat{\beta}_1$ and $A_{11\cdot 2}$, say $f(\hat{\beta}_1, A_{11\cdot 2})$. Then there is a C^* that carries this into $f(\hat{\beta}_1 C^{*-1}, I)$ and a further orthogonal transformation carries this into $f(T, I)$, where $t_{i\nu} = 0$, $i < \nu$, $t_{ii} \geq 0$. (If each row of T is considered a vector in q_1-space, the rotation of coordinate axes can be done so the first vector is along the first coordinate axis, the second vector is in the plane determined by the first two coordinate axes, and so forth.) But T is a function of $TT' = \hat{\mathbf{\beta}}_1 A_{11\cdot2}\hat{\mathbf{\beta}}_1'$; that is, the elements of T are uniquely determined by this equation and the preceding restrictions. Thus our tests will depend on $\hat{\mathbf{\Sigma}}$ and $\hat{\mathbf{\beta}}_1 A_{11\cdot2}\hat{\mathbf{\beta}}_1'$. Let $N\hat{\mathbf{\Sigma}} = \boldsymbol{G}$ and $\hat{\mathbf{\beta}}_1 A_{11\cdot2}\hat{\mathbf{\beta}}_1' = \boldsymbol{H}$.

Third, the null hypothesis is invariant when $\boldsymbol{x}_\alpha$ is replaced by $\boldsymbol{K}\boldsymbol{x}_\alpha$, for $\mathbf{\Sigma}$ and $\mathbf{\beta}_2^*$ are unspecified. This transforms $\boldsymbol{G}$ to $\boldsymbol{KGK'}$ and $\boldsymbol{H}$ to $\boldsymbol{KHK'}$. The only invariants of $\boldsymbol{G}$ and $\boldsymbol{H}$ under such transformations are the roots of

$$|\boldsymbol{H} - \theta \boldsymbol{G}| = 0. \tag{2}$$

It is clear the roots are invariant, for

$$\begin{aligned} 0 &= |\boldsymbol{KHK'} - \theta \boldsymbol{KGK'}| \\ &= |\boldsymbol{K}(\boldsymbol{H} - \theta \boldsymbol{G})\boldsymbol{K'}| \\ &= |\boldsymbol{K}| \cdot |\boldsymbol{H} - \theta \boldsymbol{G}| \cdot |\boldsymbol{K'}|. \end{aligned} \tag{3}$$

On the other hand, these are the only invariants, for given $\boldsymbol{G}$ and $\boldsymbol{H}$ there is a $\boldsymbol{K}$ such that $\boldsymbol{KGK'} = \boldsymbol{I}$ and

$$\boldsymbol{KHK'} = \mathbf{\Theta} = \begin{pmatrix} \theta_1 & 0 \cdots & 0 \\ 0 & \theta_2 \cdots & 0 \\ \cdot & \cdot & \cdot \\ \cdot & \cdot & \cdot \\ \cdot & \cdot & \cdot \\ 0 & 0 \cdots & \theta_p \end{pmatrix}, \tag{4}$$

where $\theta_1 \geq \cdots \geq \theta_p$ are the roots of (2) (see Theorem 3 of Appendix 1).

THEOREM 8.10.1. *Let $\boldsymbol{x}_\alpha$ be an observation from $N(\mathbf{\beta}_1 \boldsymbol{z}_\alpha^{*(1)} + \mathbf{\beta}_2^* \boldsymbol{z}_\alpha^{(2)}, \mathbf{\Sigma})$, where $\sum_\alpha \boldsymbol{z}_\alpha^{*(1)} \boldsymbol{z}_\alpha^{(2)\prime} = \boldsymbol{0}$ and $\sum_\alpha \boldsymbol{z}_\alpha^{*(1)} \boldsymbol{z}_\alpha^{*(1)\prime} = A_{11\cdot2}$. The only functions of the sufficient statistics and $A_{11\cdot2}$ invariant under the transformations $\boldsymbol{x}_\alpha^* = \boldsymbol{x}_\alpha + \mathbf{\Gamma} \boldsymbol{z}_\alpha^{(2)}$, $\boldsymbol{z}_\alpha^{**(1)} = \boldsymbol{C} \boldsymbol{z}_\alpha^{*(1)}$, and $\boldsymbol{x}_\alpha^* = \boldsymbol{K}\boldsymbol{x}_\alpha$ are the roots of (2), where $\boldsymbol{G} = N\hat{\mathbf{\Sigma}}$ and $\boldsymbol{H} = \hat{\mathbf{\beta}}_1 A_{11\cdot2}\hat{\mathbf{\beta}}_1'$.*

The likelihood ratio criterion is a function of

$$\begin{aligned} U &= \frac{|\boldsymbol{G}|}{|\boldsymbol{G} + \boldsymbol{H}|} = \frac{|\boldsymbol{KGK'}|}{|\boldsymbol{KGK'} + \boldsymbol{KHK'}|} = \frac{|\boldsymbol{I}|}{|\boldsymbol{I} + \mathbf{\Theta}|} \\ &= \prod_{i=1}^{p} (1 + \theta_i)^{-1}. \end{aligned} \tag{5}$$

It is clearly invariant under the transformations.

Intuitively it would appear that good tests should reject the null hypothesis when the roots in some sense are large, for if $\boldsymbol{\beta}_1$ is very different from $\mathbf{0}$, $\hat{\boldsymbol{\beta}}_1$ will tend to be large and so will $\boldsymbol{H}$. Some other criteria that have been suggested are (*a*) $\sum\theta_i$, (*b*) $\sum\theta_i/(1+\theta_i)$, (*c*) $\max\theta_i$, and (*d*) $\min\theta_i$. In each case we reject the null hypothesis if the criterion exceeds some specified number.

We can write the first two in simple functions of $\boldsymbol{H}$ and $\boldsymbol{G}$. Let $\boldsymbol{K}$ be the matrix such that $\boldsymbol{KGK'} = \boldsymbol{I}\,[\boldsymbol{G} = \boldsymbol{K}^{-1}(\boldsymbol{K}')^{-1}$ or $\boldsymbol{G}^{-1} = \boldsymbol{K'K})]$ and so (4) holds. Then

$$\sum_{i=1}^{p}\theta_i = \operatorname{tr}\boldsymbol{\Theta} = \operatorname{tr}\boldsymbol{KHK'} = \operatorname{tr}\boldsymbol{HK'K} = \operatorname{tr}\boldsymbol{HG}^{-1}. \tag{6}$$

This criterion was suggested by Lawley (1938) and Hotelling (1947). The second criterion can be written as

$$\begin{aligned}\sum_{i=1}^{p}\frac{\theta_i}{1+\theta_i} &= \operatorname{tr}\boldsymbol{\Theta}(\boldsymbol{I}+\boldsymbol{\Theta})^{-1}\\ &= \operatorname{tr}\boldsymbol{KHK'}(\boldsymbol{KGK'}+\boldsymbol{KHK'})^{-1}\\ &= \operatorname{tr}\boldsymbol{HK'}[\boldsymbol{K}(\boldsymbol{G}+\boldsymbol{H})\boldsymbol{K'}]^{-1}\boldsymbol{K}\\ &= \operatorname{tr}\boldsymbol{H}(\boldsymbol{G}+\boldsymbol{H})^{-1}.\end{aligned} \tag{7}$$

The third criterion has been proposed by Roy (1953).

In principle the probability that one of the criteria exceed a given number can be obtained from the distribution of the roots (Chapter 13); in practice this can be done fairly easily for $p = 2$, but with considerably more difficulty for $p > 2$. We shall develop an asymptotic theory. For $\operatorname{tr}\boldsymbol{HG}^{-1}$ an asymptotic expansion has been given* [Morrow (1948)].

Under the null hypothesis $\boldsymbol{G}$ is distributed as $\sum_{\alpha=1}^{n}\boldsymbol{Z}_\alpha\boldsymbol{Z}'_\alpha$ $(n = N - q)$ and $\boldsymbol{H}$ is distributed as $\sum_{\nu=1}^{q_1}\boldsymbol{Y}_\nu\boldsymbol{Y}'_\nu$, where the $\boldsymbol{Z}_\alpha$ and $\boldsymbol{Y}_\nu$ are independent, each with distribution $N(\mathbf{0}, \boldsymbol{\Sigma})$. Since the roots are invariant under the previously specified linear transformation, we can choose $\boldsymbol{K}$ so that $\boldsymbol{K\Sigma K'} = \boldsymbol{I}$ and let $\boldsymbol{G}^* = \boldsymbol{KGK'}[=\sum(\boldsymbol{KZ}_\alpha)(\boldsymbol{KZ}_\alpha)']$ and $\boldsymbol{H}^* = \boldsymbol{KHK'}$. This is equivalent to assuming at the outset that $\boldsymbol{\Sigma} = \boldsymbol{I}$.

Now

$$\operatorname*{plim}_{N\to\infty}\frac{1}{N}\boldsymbol{G} = \operatorname*{plim}_{n\to\infty}\frac{n}{n+q}\frac{1}{n}\sum_{\alpha=1}^{n}\boldsymbol{Z}_\alpha\boldsymbol{Z}'_\alpha = \boldsymbol{I}. \tag{8}$$

* Lawley (1938) purports to derive the exact distribution of $\operatorname{tr}\boldsymbol{HG}^{-1}$, but the result is in error.

This result follows applying the (weak) law of large numbers to each element of $(1/n)G$,

$$\operatorname*{plim}_{n\to\infty} \frac{1}{n}\sum_{\alpha=1}^{n} Z_{i\alpha}Z_{j\alpha} = \mathscr{E}Z_{i\alpha}Z_{j\alpha} = \delta_{ij}. \tag{9}$$

THEOREM 8.10.2. *Let $f(\boldsymbol{H})$ be a function whose discontinuities form a set of probability zero when $\boldsymbol{H}$ is distributed as $\sum_{\nu=1}^{q_1} \boldsymbol{Y}_\nu \boldsymbol{Y}_\nu'$ with the $\boldsymbol{Y}_\nu$ independent, each with distribution $N(\boldsymbol{0}, \boldsymbol{I})$. Then the limiting distribution of $f(N\boldsymbol{H}\boldsymbol{G}^{-1})$ is the distribution of $f(\boldsymbol{H})$.*

Proof. This is a straightforward application of a general theorem [for example, Theorem 2, Chernoff (1956)] to the effect that if the cdf of $\boldsymbol{X}_n$ converges to that of $\boldsymbol{X}$ (in every continuity point of the latter) and if $g(\boldsymbol{x})$ is a function whose discontinuities form a set of probability 0 according to the distribution of $\boldsymbol{X}$, then the cdf of $g(\boldsymbol{X}_n)$ converges to that of $g(\boldsymbol{X})$. In our case $\boldsymbol{X}_n$ consists of the components of $\boldsymbol{H}$ and $\boldsymbol{G}$, and $\boldsymbol{X}$ consists of the components of $\boldsymbol{H}$ and $\boldsymbol{I}$.

COROLLARY 8.10.1. *The limiting distribution of* $N\operatorname{tr}\boldsymbol{H}\boldsymbol{G}^{-1}$ *is the χ^2-distribution with pq_1 degrees of freedom.*

This follows from Theorem 8.10.2 because

$$\operatorname{tr}\boldsymbol{H} = \sum_{i=1}^{p} h_{ii} = \sum_{i=1}^{p}\sum_{\nu=1}^{q_1} Y_{i\nu}^2. \tag{10}$$

One can argue similarly that $N\operatorname{tr}\boldsymbol{H}(\boldsymbol{H}+\boldsymbol{G})^{-1}$ also has a limiting χ^2-distribution with pq_1 degrees of freedom. Theorem 8.10.2 also tells us that N times the largest root of (2), that is, the largest characteristic root of $N\boldsymbol{H}\boldsymbol{G}^{-1}$, has in the limit the distribution of the largest characteristic root of $\boldsymbol{H}$.

It is seen that there are many reasonable criteria to be used. Unfortunately, there is no theory to tell us what is a good class of tests (a complete class) nor how to choose among the proposed tests when we want power against certain kinds of alternatives. Narain (1950) has shown that the likelihood ratio test is unbiased. Roy (1953) has given some properties of the test based on the largest root.

8.11. THE CANONICAL FORM

We can reduce the problem of testing a submatrix of regression coefficients being $\boldsymbol{0}$ to a simpler form. As in Section 8.2, we consider $\boldsymbol{X}_1, \cdots, \boldsymbol{X}_N$ as independent random vectors with $\boldsymbol{X}_\alpha$ having the distribution $N(\boldsymbol{\beta}\boldsymbol{z}_\alpha, \boldsymbol{\Sigma})$, where the q-dimensional "fixed" vectors $\boldsymbol{z}_1, \cdots, \boldsymbol{z}_N$ are such that $\boldsymbol{Z} = (\boldsymbol{z}_1, \cdots, \boldsymbol{z}_N)$ is of rank $q(\leq N)$. We partition

$$\boldsymbol{\beta} = (\boldsymbol{\beta}_1\ \boldsymbol{\beta}_2), \tag{1}$$

$$Z = \begin{pmatrix} Z_1 \\ Z_2 \end{pmatrix}. \tag{2}$$

Since the hypothesis $\boldsymbol{\beta}_1 = \boldsymbol{\beta}_1^*$ can be transformed to $\boldsymbol{\beta}_1 = \mathbf{0}$ by subtraction of $\boldsymbol{\beta}_1^* \mathbf{z}_\alpha^{(1)}$ from X_α, we shall consider testing the hypothesis

$$H: \quad \boldsymbol{\beta}_1 = \mathbf{0}. \tag{3}$$

Let $X = (X_1, \cdots, X_N)$. Then we can write

$$\mathscr{E} X = \boldsymbol{\beta} Z. \tag{4}$$

Let

$$W_1 = Z_1 - Z_1 Z_2'(Z_2 Z_2')^{-1} Z_2 = Z_1 - A_{12} A_{22}^{-1} Z_2, \qquad W_2 = Z_2. \tag{5}$$

Then W_1 and W_2 are orthogonal to each other; that is,

$$\begin{aligned} W_1 W_2' &= Z_1 Z_2' - Z_1 Z_2'(Z_2 Z_2')^{-1} Z_2 Z_2' \\ &= Z_1 Z_2' - Z_1 Z_2' \\ &= 0. \end{aligned} \tag{6}$$

Then

$$\mathscr{E} X = \boldsymbol{\beta}_1 W_1 + \boldsymbol{\beta}_2^* W_2 \tag{7}$$

where

$$\boldsymbol{\beta}_2^* = \boldsymbol{\beta}_2 + \boldsymbol{\beta}_1 Z_1 Z_2'(Z_2 Z_2')^{-1}. \tag{8}$$

The hypothesis $\boldsymbol{\beta}_1 = \mathbf{0}$ is unchanged by the transformation.

Next consider a transformation

$$W_1 = D_1 V_1, \qquad W_2 = D_2 V_2, \tag{9}$$

where $D_1(q_1 \times q_1)$ and $D_2(q_2 \times q_2)$ are two nonsingular matrices such that $V_1(q_1 \times N)$ and $V_2(q_2 \times N)$ each consist of rows of an orthogonal matrix; that is

$$V_1 V_1' = I, \qquad V_2 V_2' = I. \tag{10}$$

It can be seen that V_2 is orthogonal to V_1, for

$$V_1 V_2' = D_1^{-1} W_1 W_2' (D_2^{-1})' = 0 \tag{11}$$

since $W_1 W_2' = \mathbf{0}$.

Let us choose $V_3[(N-q) \times N]$ so that

$$V = \begin{pmatrix} V_1 \\ V_2 \\ V_3 \end{pmatrix} \begin{matrix} q_1 \\ q_2 \\ N-q \end{matrix} \tag{12}$$

is an orthogonal (square) matrix of the Nth order. Make a transformation $U = XV'$. We have, from (7) and (9) and the orthogonality of V,

$$\begin{aligned}(13)\qquad \mathscr{E}U &= \mathscr{E}XV' \\ &= (\beta_1 D_1 V_1 + \beta_2^* D_2 V_2)\cdot(V_1' \quad V_2' \quad V_3') \\ &= (\beta_1 D_1 \quad \beta_2^* D_2 \quad 0).\end{aligned}$$

Let $U = (U_1 \; U_2 \; U_3)$ where U_1, U_2, U_3 are the first q_1 columns, the next q_2 columns, and the last $N - q$ columns of U respectively. We have

$$(14)\qquad (U_1 \; U_2 \; U_3) = X(V_1' \; V_2' \; V_3').$$

By Section 8.2 the first q variate vectors $(U_1 \; U_2)$ are the sample regression coefficients of X's on $\begin{pmatrix} V_1 \\ V_2 \end{pmatrix}$ with expected values

$$(15)\qquad \mathscr{E}(U_1 \; U_2) = (\beta_1 D_1 \quad \beta_2^* D_2).$$

The expected values of the elements of U_3 are 0, $\mathscr{E}U_3 = \mathbf{0}$. Since V is orthogonal the columns of U are independently normally distributed with covariance matrix Σ.

Let

$$(16)\qquad \beta_1 D_1 = (\mu_1, \cdots, \mu_{q_1}),$$

$$(17)\qquad \beta_2^* D_2 = (\mu_{q_1+1}, \cdots, \mu_q).$$

Then the joint density of $U = (U_1, \cdots, U_N)$ is

$$(18)\qquad (2\pi)^{-\frac{1}{2}pN}|\Sigma|^{-\frac{1}{2}N} \exp\Big[-\tfrac{1}{2}\sum_{\alpha=1}^{q}(u_\alpha - \mu_\alpha)'\Sigma^{-1}(u_\alpha - \mu_\alpha) - \tfrac{1}{2}\sum_{\alpha=q+1}^{N} u_\alpha'\Sigma^{-1}u_\alpha\Big]$$

The hypothesis (3) is equivalent to

$$(19)\qquad \beta_1 D_1 = (\mu_1, \cdots, \mu_{q_1}) = \mathbf{0}.$$

The density (18) is called the canonical form of the density. The criterion (1) of Section 8.4 is transformed to

$$(20)\qquad \frac{\left|\sum_{\alpha=q+1}^{N} U_\alpha U_\alpha'\right|}{\left|\sum_{\alpha=1}^{q_1} U_\alpha U_\alpha' + \sum_{\alpha=q+1}^{N} U_\alpha U_\alpha'\right|}.$$

REFERENCES

Section 8.2. Aitken (1948a); Bartlett (1938); Finney (1946); Wilks (1932a), (1934).
Section 8.3. Bartlett (1934); P. K. Bose (1947a); Wilks (1932a), (1934).

Section 8.4. Wilks (1932a).

Section 8.5. Hartley and Fitch (1951); Jambunathan (1954); Wald and Brookner (1941); Wilks (1935a).

Section 8.6. Barnes (1899); Bartlett (1938); Box (1949); C. R. Rao (1948b), (1951b); J. Roy (1951); Tukey and Wilks (1946); Wald and Brookner (1941); Whittaker and Watson (1943).

Section 8.8. Barnard (1935); Bartlett (1947a).

Section 8.9. Fisher (1947a); Immer, Hayes, and Powers (1934); C. R. Rao (1955a); Tukey (1949); Yates and Cochran (1938).

Section 8.10. Chernoff (1956); Hotelling (1947), (1951); Kullback (1956); Lawley (1938); Morrow (1948); Nanda (1951); Narain (1950); Pillai (1955); S. N. Roy (1950a), (1953).

Section 8.11. P. L. Hsu (1941d).

Chapter 8. R. L. Anderson and Bancroft (1952); T. W. Anderson (1955); Carter (1949); Dunnett and Sobel (1954); Fisher (1936); Gulliksen and Wilks (1950); P. L. Hsu (1940a); Kendall (1946), 338–341, 345–348; S. N. Roy (1946e), (1950b); Wishart (1948b); Woltz, Reid, and Colwell (1948).

PROBLEMS

1. (Sec. 8.2.2) Consider the following sample (for $N = 8$):

Weight of grain	40	17	9	15	6	12	5	9
Weight of straw	53	19	10	29	13	27	19	30
Amount of fertilizer	24	11	5	12	7	14	11	18

Let $z_{2\alpha} = 1$, and let $z_{1\alpha}$ be the amount of fertilizer on the αth plot. Estimate $\boldsymbol{\beta}$ for this sample. Test the hypothesis $\boldsymbol{\beta}_1 = \mathbf{0}$ at the 0·01 significance level.

2. (Sec. 8.3) In the following data [Woltz, Reid, and Colwell (1948), used by R. L. Anderson and Bancroft (1952)] the variables are x_1, rate of cigarette burn; x_2, the per cent nicotine; z_1, per cent of nitrogen; z_2, per cent of chlorine; z_3 of potassium; z_4 of phosphorus; z_5 of calcium; and z_6 of magnesium; and $z_7 = 1$, and $N = 25$.

$$\sum \boldsymbol{x}_\alpha = \begin{pmatrix} 42.20 \\ 54.03 \end{pmatrix}, \qquad \sum \boldsymbol{z}_\alpha = \begin{pmatrix} 53.92 \\ 62.02 \\ 56.00 \\ 12.25 \\ 89.79 \\ 24.10 \\ 25 \end{pmatrix}$$

$$\sum (\boldsymbol{x}_\alpha - \bar{\boldsymbol{x}})(\boldsymbol{x}_\alpha - \bar{\boldsymbol{x}})' = \begin{pmatrix} 0.6690 & 0.4527 \\ 0.4527 & 6.5921 \end{pmatrix}$$

$$\sum (\boldsymbol{z}_\alpha - \bar{\boldsymbol{z}})(\boldsymbol{z}_\alpha - \bar{\boldsymbol{z}})' = \begin{pmatrix} 1.8311 & -0.3589 & -0.0125 & -0.0244 & 1.6379 & 0.5057 & 0 \\ -0.3589 & 8.8102 & -0.3469 & 0.0352 & 0.7920 & 0.2173 & 0 \\ -0.0125 & -0.3469 & 1.5818 & -0.0415 & -1.4278 & -0.4753 & 0 \\ -0.0244 & 0.0352 & -0.0415 & 0.0258 & 0.0043 & 0.0154 & 0 \\ 1.6379 & 0.7920 & -1.4278 & 0.0043 & 3.7248 & 0.9120 & 0 \\ 0.5057 & 0.2173 & -0.4753 & 0.0154 & 0.9120 & 0.3828 & 0 \\ 0 & 0 & 0 & 0 & 0 & 0 & 0 \end{pmatrix}$$

$$\sum(z_\alpha - \bar{z})(x_\alpha - \bar{x})' = \begin{pmatrix} 0.2501 & 2.6691 \\ -1.5136 & -2.0617 \\ 0.5007 & -0.9503 \\ -0.0421 & -0.0187 \\ -0.1914 & 3.4020 \\ -0.1586 & 1.1663 \\ 0 & 0 \end{pmatrix}$$

(*a*) Estimate the regression of x_1 and x_2 on z_1, z_5, z_6, and z_7.
(*b*) Estimate the regression on all seven variables.
(*c*) Test the hypothesis that the regression on z_2, z_3, and z_4 is 0.

3. (Sec. 8.2) Show that Theorem 3.2.1 is a special case of Theorem 8.2.1. [*Hint:* Let $q = 1$, $z_\alpha = 1$, $\boldsymbol{\beta} = \boldsymbol{\mu}$.]

4. (Sec. 8.2) Prove Theorem 8.2.3.

5. (Sec. 8.2) Show that $\hat{\boldsymbol{\beta}}$ minimizes the generalized variance

$$\left|\sum_{\alpha=1}^{N} (x_\alpha - \boldsymbol{\beta} z_\alpha)(x_\alpha - \boldsymbol{\beta} z_\alpha)'\right|.$$

6. (Sec. 8.3) Let $q = 2$, $z_{1\alpha} = w_\alpha$ (scalar), $z_{2\alpha} = 1$. Show that the U-statistic for testing the hypothesis $\boldsymbol{\beta}_1 = \mathbf{0}$ is a monotonic function of a T^2-statistic and give the T^2-statistic in a simple form.

7. (Sec. 8.3) Let $z_{q\alpha} = 1$, let $q_2 = 1$, and let

$$A^* = \left[\sum_\alpha (z_{i\alpha} - \bar{z}_i)(z_{j\alpha} - \bar{z}_j)\right], \quad i, j = 1, \cdots, q_1 = q - 1.$$

Prove that

$$(\hat{\boldsymbol{\beta}}_{1\Omega} - \boldsymbol{\beta}_1)(A_{11} - A_{12}A_{22}^{-1}A_{21})(\hat{\boldsymbol{\beta}}_{1\Omega} - \boldsymbol{\beta}_1)' = (\hat{\boldsymbol{\beta}}_{1\Omega} - \boldsymbol{\beta}_1)A^*(\hat{\boldsymbol{\beta}}_{1\Omega} - \boldsymbol{\beta}_1)'.$$

8. (Sec. 8.3) Let $q_1 = q_2$. How do you test the hypothesis $\boldsymbol{\beta}_1 = \boldsymbol{\beta}_2$?

9. (Sec. 8.3) Prove

$$\begin{aligned} \hat{\boldsymbol{\beta}}_{1\Omega} &= \sum_\alpha x_\alpha (z_\alpha^{(1)} - A_{12}A_{22}^{-1}z_\alpha^{(2)})'\left[\sum (z_\alpha^{(1)} - A_{12}A_{22}^{-1}z_\alpha^{(2)})(z_\alpha^{(1)} - A_{12}A_{22}^{-1}z_\alpha^{(2)})'\right]^{-1} \\ &= (C_1 - C_2A_{22}^{-1}A_{21})(A_{11} - A_{12}A_{22}^{-1}A_{21})^{-1}. \end{aligned}$$

10. (Sec. 8.4) By comparing Theorem 8.2.2 and Problem 9, prove Lemma 8.4.1.

11. (Sec. 8.4) Prove Lemma 8.4.1 by showing that the density of $\hat{\boldsymbol{\beta}}_{1\Omega}$ and $\hat{\boldsymbol{\beta}}_{2\omega}$ is

$$K_1 \exp\left[-\tfrac{1}{2} \operatorname{tr} \boldsymbol{\Sigma}^{-1}(\hat{\boldsymbol{\beta}}_{1\Omega} - \boldsymbol{\beta}_1^*)A_{11\cdot2}(\hat{\boldsymbol{\beta}}_{1\Omega} - \boldsymbol{\beta}_1^*)'\right]$$
$$\cdot K_2 \exp\left[-\tfrac{1}{2} \operatorname{tr} \boldsymbol{\Sigma}^{-1}(\hat{\boldsymbol{\beta}}_{2\omega} - \boldsymbol{\beta}_2)A_{22}(\hat{\boldsymbol{\beta}}_{2\omega} - \boldsymbol{\beta}_2)'\right]$$

12. (Sec. 8.5) (*a*) Show that when p is even, the characteristic function of $Y = \log U_{p,m,n}$, say $\phi(t) = \mathscr{E}e^{itY}$, is the reciprocal of a polynomial. (*b*) Sketch a method of inverting the characteristic function of Y by the method of residues. (*c*) Show that the resulting density of U is a polynomial in $\sqrt{u}$ and $\log u$ with possibly a factor of $u^{-\frac{1}{2}}$.

13. (Sec. 8.6) Use the asymptotic expansion of the distribution to compute $\Pr\{-m \log U_{3,3,n} \le M^*\}$ for

(*a*) $n = 8$, $M^* = 14.7$,
(*b*) $n = 8$, $M^* = 21.7$,

(*c*) $n = 16,\ M^* = 14.7,$

(*d*) $n = 16,\ M^* = 21.7.$

(Either compute to the third decimal place or use the expansion to the m^{-4} term.)

14. (Sec. 8.6) In case $p = 3,\ q_1 = 4,$ and $n = N - q = 20$" by "$n = 20,$ and $N = 24$" point for $m \log U$ using (*a*) $-2 \log \lambda$ as χ^2 and (*b*) using $-m \log U$ as χ^2. Using more terms of this expansion, evaluate the true significance levels for your answers to (*a*) and (*b*).

15. (Sec. 8.9) Let Y_{ij} (a *p*-component vector) be distributed according to $N(\boldsymbol{\mu}_{ij}, \boldsymbol{\Sigma})$, where $\mathscr{E}Y_{ij} = \boldsymbol{\mu}_{ij} = \boldsymbol{\mu} + \boldsymbol{\lambda}_i + \boldsymbol{\nu}_j + \boldsymbol{\gamma}_{ij},\ \sum_i \boldsymbol{\lambda}_i = \mathbf{0} = \sum_j \boldsymbol{\nu}_j = \sum_i \boldsymbol{\gamma}_{ij} = \sum_j \boldsymbol{\gamma}_{ij}$; the $\boldsymbol{\gamma}_{ij}$ are the interactions if m observations are made on each Y_{ij} (say $y_{ij1}, \cdots, y_{ijm}$). How do you test the hypothesis $\boldsymbol{\lambda}_i = \mathbf{0}\ (i = 1, \cdots, r)$? How do you test the hypothesis $\boldsymbol{\gamma}_{ij} = \mathbf{0}\ (i = 1, \cdots, r;\ j = 1, \cdots, c)$?

16. (Sec. 8.9) Consider the Latin square. Let $Y_{ij} (i, j = 1, \cdots, r)$ be distributed according to $N(\boldsymbol{\mu}_{ij}, \boldsymbol{\Sigma})$, where $\mathscr{E}Y_{ij} = \boldsymbol{\mu}_{ij} = \boldsymbol{\gamma} + \boldsymbol{\lambda}_i + \boldsymbol{\nu}_j + \boldsymbol{\mu}_k$ and $k = j - i + 1$ (mod r) with $\sum \boldsymbol{\lambda}_i = \sum \boldsymbol{\nu}_i = \sum \boldsymbol{\mu}_i = \mathbf{0}$.

(*a*) Give the univariate analysis of variance table for main effects and error (including sums of squares, numbers of degrees of freedom, and so forth).

(*b*) Give the table for the vector case.

(*c*) Indicate in the vector case how to test the hypothesis $\boldsymbol{\lambda}_i = \mathbf{0},\ i = 1, \cdots, r$.

17. (Sec. 8.9) Let x_1 be the yield of a process and x_2 a quality measure. Let $z_1 = 1$, $z_2 = \pm 10°$ (temperature relative to average), $z_3 = \pm\ 0.75$ (relative measure of flow of one agent), and $z_4 = \pm\ 1.50$ (relative measure of flow of another agent). [See T. W. Anderson (1955) for details.] Three observations were made on x_1 and x_2 for each possible triplet of values of z_2, z_3, and z_4. The estimate of $\boldsymbol{\beta}$ is

$$\hat{\boldsymbol{\beta}} = \begin{pmatrix} 58.529 & -0.3829 & -5.050 & 2.308 \\ 98.675 & 0.1558 & 4.144 & -0.700 \end{pmatrix};$$

$s_1 = 3.090,\ s_2 = 1.619$, and $r = -0.6632$ can be used to compute S or $\hat{\boldsymbol{\Sigma}}$.

(*a*) Formulate an analysis of variance model for this situation.

(*b*) Find a confidence region for the effects of temperature (that is, β_{12}, β_{22}).

(*c*) Test the hypothesis that the two agents have no effect on the yield and quántity.

18. (Sec. 8.10) Interpret the transformations referred to in Theorem 8.10.1 in the original terms; that is, $H\colon \boldsymbol{\beta}_1 = \boldsymbol{\beta}_1^*$ and $z_\alpha^{(1)}$.

19. (Sec. 8.10) Find the cdf of tr HG^{-1} for $p = 2$. [*Hint:* Use the distribution of the roots given in Chapter 13.]

20. Let $x_\alpha^{(\nu)}\ (\alpha = 1, \cdots, N_\nu)$ be observations from $N(\boldsymbol{\mu}^{(\nu)}, \boldsymbol{\Sigma}),\ (\nu = 1, \cdots, q)$. What criterion may be used to test the hypothesis that

$$\boldsymbol{\mu}^{(\nu)} = \sum_{h=1}^{m} \boldsymbol{\gamma}_h c_{h\nu} + \boldsymbol{\mu},$$

where $c_{h\nu}$ are given numbers and $\boldsymbol{\gamma}_\nu$, $\boldsymbol{\mu}$ are unknown vectors? [*Note:* This hypothesis (that the means lie on an m-dimensional hyperplane with ratios of distances known) can be put in the form of the general linear hypothesis.]

21. Let x_α be an observation from $N(\boldsymbol{\beta} z_\alpha, \boldsymbol{\Sigma}),\ \alpha = 1, \cdots, N$. Suppose there is a known fixed vector $\boldsymbol{\gamma}$ such that $\boldsymbol{\beta}\boldsymbol{\gamma} = \mathbf{0}$. How do you estimate $\boldsymbol{\beta}$?

CHAPTER 9

Testing Independence of Sets of Variates

9.1. INTRODUCTION

In this section we divide a set of p variates with a joint normal distribution into q subsets and ask whether the q sets are mutually independent; this is equivalent to testing the hypothesis that each variable in one set is uncorrelated with each variable in the others. We find the likelihood ratio criterion for this hypothesis, the moments of the criterion under the null hypothesis, some particular distributions, and an asymptotic expansion of the distribution. It will be shown that, in the case of two subsets, this theory is closely related to that given in the preceding chapter.

9.2. THE LIKELIHOOD RATIO CRITERION FOR TESTING INDEPENDENCE OF SETS OF VARIATES

Let the p-component vector X be distributed according to $N(\boldsymbol{\mu}, \boldsymbol{\Sigma})$. We partition X into q subvectors with $p_1, p_2, \cdots, p_q$ components respectively; that is,

$$X = \begin{pmatrix} X^{(1)} \\ X^{(2)} \\ \cdot \\ \cdot \\ \cdot \\ X^{(q)} \end{pmatrix}. \tag{1}$$

The vector of means $\boldsymbol{\mu}$ and the covariance matrix $\boldsymbol{\Sigma}$ are partitioned similarly,

$$\boldsymbol{\mu} = \begin{pmatrix} \boldsymbol{\mu}^{(1)} \\ \boldsymbol{\mu}^{(2)} \\ \cdot \\ \cdot \\ \cdot \\ \boldsymbol{\mu}^{(q)} \end{pmatrix}, \tag{2}$$

$$\boldsymbol{\Sigma} = \begin{pmatrix} \boldsymbol{\Sigma}_{11} & \boldsymbol{\Sigma}_{12} & \cdots & \boldsymbol{\Sigma}_{1q} \\ \boldsymbol{\Sigma}_{21} & \boldsymbol{\Sigma}_{22} & \cdots & \boldsymbol{\Sigma}_{2q} \\ \cdot & \cdot & & \cdot \\ \cdot & \cdot & & \cdot \\ \cdot & \cdot & & \cdot \\ \boldsymbol{\Sigma}_{q1} & \boldsymbol{\Sigma}_{q2} & \cdots & \boldsymbol{\Sigma}_{qq} \end{pmatrix}. \tag{3}$$

The null hypothesis we wish to test is that the subvectors $X^{(1)}, \cdots, X^{(q)}$ are mutually independently distributed, that is, that the density of X factors into the densities of $X^{(1)}, \cdots, X^{(q)}$. It is

$$H: \quad n(\boldsymbol{x} | \boldsymbol{\mu}, \boldsymbol{\Sigma}) = \prod_{i=1}^{q} n(\boldsymbol{x}^{(i)} | \boldsymbol{\mu}^{(i)}, \boldsymbol{\Sigma}_{ii}). \tag{4}$$

If $X^{(1)}, \cdots, X^{(q)}$ are independent subvectors,

$$\mathscr{E}(X^{(i)} - \boldsymbol{\mu}^{(i)})(X^{(j)} - \boldsymbol{\mu}^{(j)})' = \boldsymbol{\Sigma}_{ij} = \mathbf{0}, \qquad i \neq j. \tag{5}$$

(See Section 2.4.) Conversely, if (5) holds, then (4) is true. Thus the null hypothesis is equivalently H: $\boldsymbol{\Sigma}_{ij} = \mathbf{0}$, $i \neq j$. This can be stated alternatively as the hypothesis that $\boldsymbol{\Sigma}$ is of the form

$$\boldsymbol{\Sigma}_0 = \begin{pmatrix} \boldsymbol{\Sigma}_{11} & \mathbf{0} & \cdots & \mathbf{0} \\ \mathbf{0} & \boldsymbol{\Sigma}_{22} & \cdots & \mathbf{0} \\ \cdot & \cdot & & \cdot \\ \cdot & \cdot & & \cdot \\ \cdot & \cdot & & \cdot \\ \mathbf{0} & \mathbf{0} & \cdots & \boldsymbol{\Sigma}_{qq} \end{pmatrix}. \tag{6}$$

Given a sample $\boldsymbol{x}_1, \cdots, \boldsymbol{x}_N$ of N observations on X, the likelihood ratio criterion is

$$\lambda = \frac{\max_{\boldsymbol{\mu}, \boldsymbol{\Sigma}_0} L(\boldsymbol{\mu}, \boldsymbol{\Sigma}_0)}{\max_{\boldsymbol{\mu}, \boldsymbol{\Sigma}} L(\boldsymbol{\mu}, \boldsymbol{\Sigma})}, \tag{7}$$

where

$$L(\boldsymbol{\mu}, \boldsymbol{\Sigma}) = \prod_{\alpha=1}^{N} \frac{1}{(2\pi)^{\frac{1}{2}p} |\boldsymbol{\Sigma}|^{\frac{1}{2}}} e^{-\frac{1}{2}(\boldsymbol{x}_\alpha - \boldsymbol{\mu})' \boldsymbol{\Sigma}^{-1} (\boldsymbol{x}_\alpha - \boldsymbol{\mu})} \tag{8}$$

and $L(\boldsymbol{\mu}, \boldsymbol{\Sigma}_0)$ is $L(\boldsymbol{\mu}, \boldsymbol{\Sigma})$ with $\boldsymbol{\Sigma}_{ij} = \mathbf{0}$, $i \neq j$, and the maximum is taken with respect to all vectors $\boldsymbol{\mu}$ and positive definite $\boldsymbol{\Sigma}$ and $\boldsymbol{\Sigma}_0$ (that is, $\boldsymbol{\Sigma}_{ii}$). As derived in Section 5.2, equation (6),

$$\max_{\boldsymbol{\mu}, \boldsymbol{\Sigma}} L(\boldsymbol{\mu}, \boldsymbol{\Sigma}) = \frac{1}{(2\pi)^{\frac{1}{2}pN} |\hat{\boldsymbol{\Sigma}}_\Omega|^{\frac{1}{2}N}} e^{-\frac{1}{2}pN}, \tag{9}$$

where

$$\hat{\boldsymbol{\Sigma}}_\Omega = \frac{1}{N} A = \frac{1}{N} \sum_{\alpha=1}^{N} (\boldsymbol{x}_\alpha - \bar{\boldsymbol{x}})(\boldsymbol{x}_\alpha - \bar{\boldsymbol{x}})'. \tag{10}$$

Under the null hypothesis

$$L(\boldsymbol{\mu}, \boldsymbol{\Sigma}_0) = \prod_{i=1}^{q} L_i(\boldsymbol{\mu}^{(i)}, \boldsymbol{\Sigma}_{ii}), \tag{11}$$

where

$$L_i(\boldsymbol{\mu}^{(i)}, \boldsymbol{\Sigma}_{ii}) = \prod_{\alpha=1}^{N} \frac{1}{(2\pi)^{\frac{1}{2}p_i}|\boldsymbol{\Sigma}_{ii}|^{\frac{1}{2}}} e^{-\frac{1}{2}(\boldsymbol{x}_\alpha^{(i)}-\boldsymbol{\mu}^{(i)})'\boldsymbol{\Sigma}_{ii}^{-1}(\boldsymbol{x}_\alpha^{(i)}-\boldsymbol{\mu}^{(i)})}. \tag{12}$$

Clearly

$$\begin{aligned} \max_{\boldsymbol{\mu}, \boldsymbol{\Sigma}_0} L(\boldsymbol{\mu}, \boldsymbol{\Sigma}_0) &= \prod_{i=1}^{q} \max_{\boldsymbol{\mu}^{(i)}, \boldsymbol{\Sigma}_{ii}} L_i(\boldsymbol{\mu}^{(i)}, \boldsymbol{\Sigma}_{ii}) \\ &= \prod_{i=1}^{q} \frac{1}{(2\pi)^{\frac{1}{2}p_iN}|\hat{\boldsymbol{\Sigma}}_{ii\omega}|^{\frac{1}{2}N}} e^{-\frac{1}{2}p_iN} \\ &= \frac{1}{(2\pi)^{\frac{1}{2}pN} \prod_{i=1}^{q} |\hat{\boldsymbol{\Sigma}}_{ii\omega}|^{\frac{1}{2}N}} e^{-\frac{1}{2}pN}, \end{aligned} \tag{13}$$

where

$$\hat{\boldsymbol{\Sigma}}_{ii\omega} = \frac{1}{N} \sum_{\alpha=1}^{N} (\boldsymbol{x}_\alpha^{(i)} - \bar{\boldsymbol{x}}^{(i)})(\boldsymbol{x}_\alpha^{(i)} - \bar{\boldsymbol{x}}^{(i)})'. \tag{14}$$

If we partition $\boldsymbol{A}$ and $\hat{\boldsymbol{\Sigma}}_\Omega$ as we have $\boldsymbol{\Sigma}$,

$$\boldsymbol{A} = \begin{pmatrix} \boldsymbol{A}_{11} & \boldsymbol{A}_{12} & \cdots & \boldsymbol{A}_{1q} \\ \boldsymbol{A}_{21} & \boldsymbol{A}_{22} & \cdots & \boldsymbol{A}_{2q} \\ \vdots & \vdots & & \vdots \\ \boldsymbol{A}_{q1} & \boldsymbol{A}_{q2} & \cdots & \boldsymbol{A}_{qq} \end{pmatrix}, \qquad \hat{\boldsymbol{\Sigma}}_\Omega = \begin{pmatrix} \hat{\boldsymbol{\Sigma}}_{11} & \hat{\boldsymbol{\Sigma}}_{12} & \cdots & \hat{\boldsymbol{\Sigma}}_{1q} \\ \hat{\boldsymbol{\Sigma}}_{21} & \hat{\boldsymbol{\Sigma}}_{22} & \cdots & \hat{\boldsymbol{\Sigma}}_{2q} \\ \vdots & \vdots & & \vdots \\ \hat{\boldsymbol{\Sigma}}_{q1} & \hat{\boldsymbol{\Sigma}}_{q2} & \cdots & \hat{\boldsymbol{\Sigma}}_{qq} \end{pmatrix}, \tag{15}$$

we see that $\hat{\boldsymbol{\Sigma}}_{ii\omega} = \hat{\boldsymbol{\Sigma}}_{ii} = \frac{1}{N}\boldsymbol{A}_{ii}$.

The likelihood ratio criterion is

$$\begin{aligned} \lambda = \frac{\max_{\boldsymbol{\mu}, \boldsymbol{\Sigma}_0} L(\boldsymbol{\mu}, \boldsymbol{\Sigma}_0)}{\max_{\boldsymbol{\mu}, \boldsymbol{\Sigma}} L(\boldsymbol{\mu}, \boldsymbol{\Sigma})} &= \frac{|\hat{\boldsymbol{\Sigma}}_\Omega|^{\frac{1}{2}N}}{\prod_{i=1}^{q} |\hat{\boldsymbol{\Sigma}}_{ii}|^{\frac{1}{2}N}} \\ &= \frac{|\boldsymbol{A}|^{\frac{1}{2}N}}{\prod_{i=1}^{q} |\boldsymbol{A}_{ii}|^{\frac{1}{2}N}}. \end{aligned} \tag{16}$$

The critical region of the likelihood ratio test is

$$\lambda \leq \lambda(\varepsilon), \tag{17}$$

where $\lambda(\varepsilon)$ is a number such that the probability of (17) is ε when $\boldsymbol{\Sigma} = \boldsymbol{\Sigma}_0$. (It remains to show that such a number can be found.) Let

$$V = \frac{|A|}{\prod_{i=1}^{q} |A_{ii}|}. \tag{18}$$

Then $\lambda = V^{\frac{1}{2}N}$ is a monotonic increasing function of V. The critical region (17) can be equivalently written as

$$V \leq V(\varepsilon). \tag{19}$$

THEOREM 9.2.1. *Let $\boldsymbol{x}_1, \cdots, \boldsymbol{x}_N$ be a sample of N observations drawn from $N(\boldsymbol{\mu}, \boldsymbol{\Sigma})$, where $\boldsymbol{x}_\alpha$, $\boldsymbol{\mu}$, and $\boldsymbol{\Sigma}$ are partitioned into $p_1, \cdots, p_q$ rows (and columns in the case of $\boldsymbol{\Sigma}$) as indicated in* (1), (2), *and* (3). *The likelihood ratio criterion that the q sets of components are mutually independent is given by* (16), *where $\boldsymbol{A}$ is defined by* (10) *and partitioned according to* (15). *The likelihood ratio test is given by* (17) *and equivalently by* (19), *where V is defined by* (18) *and $\lambda(\varepsilon)$ or $V(\varepsilon)$ is chosen to obtain the significance level ε.*

Since $r_{ij} = a_{ij}/\sqrt{a_{ii}a_{jj}}$, we have

$$|A| = |R| \prod_{i=1}^{p} a_{ii}, \tag{20}$$

where

$$R = (r_{ij}) = \begin{pmatrix} R_{11} & R_{12} & \cdots & R_{1q} \\ R_{21} & R_{22} & \cdots & R_{2q} \\ \cdot & \cdot & & \cdot \\ \cdot & \cdot & & \cdot \\ \cdot & \cdot & & \cdot \\ R_{q1} & R_{q2} & \cdots & R_{qq} \end{pmatrix} \tag{21}$$

and

$$|A_{ii}| = |R_{ii}| \prod_{j=p_1+\cdots+p_{i-1}+1}^{p_1+\cdots+p_i} a_{jj}. \tag{22}$$

Thus

$$V = \frac{|A|}{\prod |A_{ii}|} = \frac{|R|}{\prod |R_{ii}|}. \tag{23}$$

That is, V can be expressed entirely in terms of sample correlation coefficients.

We can interpret the criterion V in terms of generalized variance. Each set $(x_{i1}, \cdots, x_{iN})$ can be considered as a vector in N-space; the set $(x_{i1} - \bar{x}_i, \cdots, x_{iN} - \bar{x}_i) = \boldsymbol{z}_i$, say, is the projection on the plane orthogonal to the equiangular line. The determinant $|A|$ is the p-dimensional volume squared of the parallelotope with $\boldsymbol{z}_1, \cdots, \boldsymbol{z}_p$ as principal edges. The determinant $|A_{ii}|$ is the (p_i-dimensional) volume squared of the parallelotope having as principal edges the ith set of vectors. If each set of vectors is orthogonal to each other set (that is, $\boldsymbol{R}_{ij} = \boldsymbol{0}$, $i \neq j$), then the volume squared $|A|$ is the product of the volumes squared $|\boldsymbol{A}_{ii}|$. For example, if $p = 2$, $p_1 = p_2 = 1$, this statement is that the area of a parallelogram is the product of the lengths of the sides if the sides are at right angles. If the sets are almost orthogonal $|A|$ is almost $\prod |\boldsymbol{A}_{ii}|$ and V is almost 1.

The criterion has a certain invariance property. Let $\boldsymbol{C}_i$ be an arbitrary nonsingular matrix of order p_i and let

$$\boldsymbol{C} = \begin{pmatrix} \boldsymbol{C}_1 & \boldsymbol{0} \cdots & \boldsymbol{0} \\ \boldsymbol{0} & \boldsymbol{C}_2 \cdots & \boldsymbol{0} \\ \cdot & \cdot & \cdot \\ \cdot & \cdot & \cdot \\ \cdot & \cdot & \cdot \\ \boldsymbol{0} & \boldsymbol{0} \cdots & \boldsymbol{C}_q \end{pmatrix}. \tag{24}$$

Let $\boldsymbol{C}\boldsymbol{x}_\alpha = \boldsymbol{x}_\alpha^*$. Then the criterion for independence in terms of $\boldsymbol{x}_\alpha^*$ is identical to the criterion in terms of $\boldsymbol{x}_\alpha$. Let $\boldsymbol{A}^* = \sum_\alpha(\boldsymbol{x}_\alpha^* - \bar{\boldsymbol{x}}^*)(\boldsymbol{x}_\alpha^* - \bar{\boldsymbol{x}}^*)'$ be partitioned into submatrices $\boldsymbol{A}_{ij}^*$. Then

$$\begin{aligned} \boldsymbol{A}_{ij}^* &= \textstyle\sum_\alpha(\boldsymbol{x}_\alpha^{*(i)} - \bar{\boldsymbol{x}}^{*(i)})(\boldsymbol{x}_\alpha^{*(j)} - \bar{\boldsymbol{x}}^{*(j)})' \\ &= \boldsymbol{C}_i \textstyle\sum_\alpha(\boldsymbol{x}_\alpha^{(i)} - \bar{\boldsymbol{x}}^{(i)})(\boldsymbol{x}_\alpha^{(j)} - \bar{\boldsymbol{x}}^{(j)})'\boldsymbol{C}_j' \\ &= \boldsymbol{C}_i \boldsymbol{A}_{ij} \boldsymbol{C}_j' \end{aligned} \tag{25}$$

and $\boldsymbol{A}^* = \boldsymbol{C}\boldsymbol{A}\boldsymbol{C}'$. Thus

$$\begin{aligned} V^* &= \frac{|\boldsymbol{A}^*|}{\prod |\boldsymbol{A}_{ii}^*|} = \frac{|\boldsymbol{C}\boldsymbol{A}\boldsymbol{C}'|}{\prod |\boldsymbol{C}_i\boldsymbol{A}_{ii}\boldsymbol{C}_i'|} \\ &= \frac{|\boldsymbol{C}| \cdot |\boldsymbol{A}| \cdot |\boldsymbol{C}'|}{\prod |\boldsymbol{C}_i| \cdot |\boldsymbol{A}_{ii}| \cdot |\boldsymbol{C}_i'|} = \frac{|\boldsymbol{A}|}{\prod |\boldsymbol{A}_{ii}|} = V, \end{aligned} \tag{26}$$

for $|\boldsymbol{C}| = \prod |\boldsymbol{C}_i|$. Thus the test is invariant with respect to linear transformations within each set.

Narain (1950) has shown that the test based on V is strictly unbiased; that is, the probability of rejecting the null hypothesis is greater than the significance level if the hypothesis is not true [see also Daly (1940)].

9.3. MOMENTS OF THE LIKELIHOOD RATIO CRITERION UNDER THE NULL HYPOTHESIS

In order to find the significance points $V(\varepsilon)$ [used in (19) of Section 9.2] we would like to have the distribution of V when the null hypothesis is true. To determine the distribution of V, we find the moments when the null hypothesis is true.

The matrix $A = N\hat{\boldsymbol{\Sigma}}_{\Omega}$ is distributed according to $W(\boldsymbol{\Sigma}_0, n)$, where $n = N - 1$ and $\boldsymbol{\Sigma}_0$ is given in (6) of Section 9.2. The marginal distribution of the matrix $A_{ii} = N\hat{\boldsymbol{\Sigma}}_{ii\omega}$ is a Wishart distribution $W(\boldsymbol{\Sigma}_{ii}, n)$, and A_{ii} is distributed independently of A_{jj} $(i \neq j)$ since $X^{(i)}$ is independent of $X^{(j)}$ under the null hypothesis.

The hth moment of $|A|/|A_0| = V$ is

$$(1) \quad \mathscr{E}V^h = \int \cdots \int |A|^h \prod_{i=1}^{q} |A_{ii}|^{-h} K(\boldsymbol{\Sigma}_0, n) \cdot |A|^{\frac{1}{2}(n-p-1)} \exp\left(-\tfrac{1}{2} \operatorname{tr} \boldsymbol{\Sigma}_0^{-1} A\right) dA$$
$$= \frac{K(\boldsymbol{\Sigma}_0, n)}{K(\boldsymbol{\Sigma}_0, n+2h)} \int \cdots \int \prod_{i=1}^{q} |A_{ii}|^{-h} \{K(\boldsymbol{\Sigma}_0, n+2h) \cdot |A|^{\frac{1}{2}(n+2h-p-1)} \exp\left(-\tfrac{1}{2} \operatorname{tr} \boldsymbol{\Sigma}_0^{-1} A\right)\} \, dA,$$

where $K(\boldsymbol{\Sigma}, n)$ is given by (7) of Section 8.4, and the integration with respect to $dA = da_{11} \cdots da_{pp}$ is over the range A positive semidefinite. The term within braces is the density of $W(\boldsymbol{\Sigma}, n + 2h)$. The integral of this density with respect to the elements in A_{ij}, $i \neq j$, must give the marginal density of $A_{11}, \cdots, A_{qq}$, which is the product of $w(A_{ii}|\boldsymbol{\Sigma}_{ii}, n + 2h)$. (Note that this integration does not involve $|A_{ii}|$.) Thus (1) is

$$(2) \quad \mathscr{E}V^h = \frac{K(\boldsymbol{\Sigma}_0, n)}{K(\boldsymbol{\Sigma}_0, n+2h)} \int \cdots \int \prod |A_{ii}|^{-h} \cdot \prod \{K(\boldsymbol{\Sigma}_{ii}, n+2h) |A_{ii}|^{\frac{1}{2}(n+2h-p_i-1)} \exp\left(-\tfrac{1}{2} \operatorname{tr} \boldsymbol{\Sigma}_{ii}^{-1} A_{ii}\right) dA_{ii}\}$$
$$= \frac{K(\boldsymbol{\Sigma}_0, n)}{K(\boldsymbol{\Sigma}_0, n+2h)} \prod_{i=1}^{q} \int \cdots \int K(\boldsymbol{\Sigma}_{ii}, n+2h) \cdot |A_{ii}|^{\frac{1}{2}(n-p_i-1)} \exp\left(-\tfrac{1}{2} \operatorname{tr} \boldsymbol{\Sigma}_{ii}^{-1} A_{ii}\right) dA_{ii}$$
$$= \frac{K(\boldsymbol{\Sigma}_0, n)}{K(\boldsymbol{\Sigma}_0, n+2h)} \prod_{i=1}^{q} \frac{K(\boldsymbol{\Sigma}_{ii}, n+2h)}{K(\boldsymbol{\Sigma}_{ii}, n)}.$$

Since $|\boldsymbol{\Sigma}| = \prod_{i=1}^{q} |\boldsymbol{\Sigma}_{ii}|$, (2) reduces to

$$(3) \qquad \mathscr{E}V^h = \frac{\prod_{i=1}^{p} \Gamma[\frac{1}{2}(n+1-i)+h] \prod_{i=1}^{q} \left\{\prod_{j=1}^{p_i} \Gamma[\frac{1}{2}(n+1-j)]\right\}}{\prod_{i=1}^{p} \Gamma[\frac{1}{2}(n+1-i)] \prod_{i=1}^{q} \left\{\prod_{j=1}^{p_i} \Gamma[\frac{1}{2}(n+1-j)+h]\right\}}.$$

Since $0 \leq V \leq 1$, these moments determine the distribution of V uniquely. It will be observed that the moments do not depend on $\boldsymbol{\Sigma}_0$. Thus the distribution of V does not depend on nuisance parameters when the null hypothesis is true.

9.4. SOME DISTRIBUTIONS OF THE CRITERION

From the moments of V we can in some cases derive the exact distribution; a number of these are given in Section 9.4.2. Some of the derivations are based on expressing V as a product of independent random variables (Section 9.4.1); in principle the distribution of V can be described as an integral of the joint density of the independent variables. For purposes of obtaining significance points the asymptotic expansion studied in Section 9.5 is useful.

9.4.1. V as a Product of Independent Variates

The hth moment of V is

$$(1)\quad \mathscr{E}V^h = \prod_{i=p_1+1}^{p} \left\{\frac{\Gamma[\frac{1}{2}(n+1-i)+h]}{\Gamma[\frac{1}{2}(n+1-i)]}\right\} \prod_{i=2}^{q}\left\{\prod_{j=1}^{p_i}\frac{\Gamma[\frac{1}{2}(n+1-j)]}{\Gamma[\frac{1}{2}(n+1-j)+h]}\right\}$$
$$= \prod_{i=2}^{q}\left\{\prod_{j=1}^{p_i}\frac{\Gamma[\frac{1}{2}(n+1-\bar{p}_i-j)+h]\Gamma[\frac{1}{2}(n+1-j)]}{\Gamma[\frac{1}{2}(n+1-\bar{p}_i-j)]\Gamma[\frac{1}{2}(n+1-j)+h]}\right\},$$

where $\bar{p}_i = p_1 + \cdots + p_{i-1}$. This can be expressed as

$$(2)\quad \mathscr{E}V^h = \prod_{i=2}^{q}\left\{\prod_{j=1}^{p_i}\frac{\Gamma[\frac{1}{2}(n+1-j)]\Gamma[\frac{1}{2}(n+1-\bar{p}_i-j)+h]\Gamma(\frac{1}{2}\bar{p}_i)}{\Gamma[\frac{1}{2}(n+1-\bar{p}_i-j)]\Gamma(\frac{1}{2}\bar{p}_i)\Gamma[n+1-j)+h]}\right\}$$
$$= \prod_{i=2}^{q}\left\{\prod_{j=1}^{p_i} B^{-1}[\tfrac{1}{2}(n+1-\bar{p}_i-j),\tfrac{1}{2}\bar{p}_i]\int_0^1 x^{\frac{1}{2}(n+1-\bar{p}_i-j)+h-1}\cdot(1-x)^{\frac{1}{2}\bar{p}_i-1}\,dx\right\}.$$

Thus V is distributed as $\prod_{i=2}^{q}\{\prod_{j=1}^{p_i} X_{ij}\}$, where the X_{ij} are independent and X_{ij} has density $\beta[x;\frac{1}{2}(n+1-\bar{p}_i-j),\frac{1}{2}\bar{p}_i]$.

If the p_i are even, say $p_i = 2r_i$, $i > 1$, then by using the duplication formula $[\Gamma(\alpha+\frac{1}{2})\Gamma(\alpha+1) = \sqrt{\pi}\,\Gamma(2\alpha+1)2^{-2\alpha}]$ for the gamma function we can reduce the hth moment of V to

$$(3)\quad \mathscr{E}V^h = \prod_{i=2}^{q}\left\{\prod_{k=1}^{r_i}\frac{\Gamma(n+1-\bar{p}_i-2k+2h)\Gamma(n+1-2k)}{\Gamma(n+1-\bar{p}_i-2k)\Gamma(n+1-2k+2h)}\right\}$$
$$= \prod_{i=2}^{q}\left\{\prod_{k=1}^{r_i} B^{-1}(n+1-\bar{p}_i-2k,\bar{p}_i)\int_0^1 x^{n+1-\bar{p}_i-2k+2h-1}(1-x)^{\bar{p}_i-1}\,dx\right\}.$$

Thus V is distributed as $\prod_{i=2}^{q}\{\prod_{k=1}^{r_i} Y_{ik}^2\}$ where the Y_{ik} are independent, and Y_{ik} has density $\beta(y; n+1-\bar{p}_i-2k,\bar{p}_i)$.

9.4.2. Some Special Distributions

In some special cases we can give explicitly the distribution of $V = \lambda^{2/N}$.

CASE 1: $q = 2$. In this case the hth moment of V is

$$(4)\quad \mathscr{E}V^h = \prod_{i=p_2+1}^{p} \left\{\frac{\Gamma[\frac{1}{2}(n+1-i)+h]}{\Gamma[\frac{1}{2}(n+1-i)]}\right\} \prod_{j=1}^{p_1} \left\{\frac{\Gamma[\frac{1}{2}(n+1-j)]}{\Gamma[\frac{1}{2}(n+1-j)+h]}\right\}$$

$$= \prod_{i=1}^{p_1} \left\{\frac{\Gamma[\frac{1}{2}(n-p_2+1-i)+h]\Gamma[\frac{1}{2}(n+1-i)]}{\Gamma[\frac{1}{2}(n-p_2+1-i)]\Gamma[\frac{1}{2}(n+1-i)+h]}\right\}.$$

This is of the form (10) of Section 8.4 with the relations between parameters

Chapter 8	Chapter 9
$N - q_2$	$n = N - 1$
q_1	p_2
p	p_1

Thus the distribution of V for $q = 2$ is the distribution of $U_{p_1, p_2, N-1-p_2}$. A number of special cases have been treated in Section 8.5.

CASE 2: $q = p$. Here $p_i = 1$ and the test is a test of lack of correlation between every pair of variables. From Section 9.4.1 we see that V is distributed as $\prod_{i=1}^{p-1} X_i$, where the X_i are independent and X_i has density $\beta[x; \frac{1}{2}(n-i), \frac{1}{2}i]$.

Let us consider the case $p = 3$. If

$$K^{-1} = B[\tfrac{1}{2}(n-1), \tfrac{1}{2}]B[\tfrac{1}{2}(n-2), 1],$$

then

$$(5)\quad \Pr\{V \le v\} = \Pr\{X_1 X_2 \le v\}$$

$$= K\int_0^v \int_0^1 x^{\frac{1}{2}(n-1)-1}(1-x)^{\frac{1}{2}-1} y^{\frac{1}{2}(n-2)-1}\, dy\, dx$$

$$+ K\int_v^1 \int_0^{v/x} x^{\frac{1}{2}(n-1)-1}(1-x)^{\frac{1}{2}-1} y^{\frac{1}{2}(n-2)-1}\, dy\, dx$$

$$= K\frac{2}{n-2}\left\{\int_0^v x^{\frac{1}{2}(n-1)-1}(1-x)^{\frac{1}{2}-1}\, dx\right.$$

$$\left. + v^{\frac{1}{2}(n-2)}\int_v^1 x^{\frac{1}{2}-1}(1-x)^{\frac{1}{2}-1}\, dx\right\}$$

$$= K\frac{2}{n-2}\{B[\tfrac{1}{2}(n-1); \tfrac{1}{2}]I_v[\tfrac{1}{2}(n-1), \tfrac{1}{2}]$$

$$+ 2v^{\frac{1}{2}(n-2)} \arcsin\sqrt{1-v}\}$$

$$= I_v[\tfrac{1}{2}(n-1), \tfrac{1}{2}] + 2B^{-1}[\tfrac{1}{2}(n-1), \tfrac{1}{2}]v^{\frac{1}{2}n-1} \arcsin\sqrt{1-v}.$$

CASE 3: $p_1 = \cdots = p_q = 2$. From Section 9.4.1 we see that in this case V is distributed as $\prod_{i=1}^{q-1} X_i^2$, where the X_i are independent and X_i has density $\beta(x;\ n-1-2i, 2i)$.

We consider now the case of $q = 3$. Let $K^{-1} = B(n-3, 2)B(n-5, 4)$. Then

$$
\begin{aligned}
(6)\quad \Pr\{V \le v\} &= \Pr\{X_1X_2 \le \sqrt{v}\} \\
&= K\int_0^{\sqrt{v}}\int_0^1 x^{n-6}(1-x)^3y^{n-4}(1-y)\,dy\,dx \\
&\quad + K\int_{\sqrt{v}}^1\int_0^{\sqrt{v}/x} x^{n-6}(1-x)^3y^{n-4}(1-y)\,dy\,dx \\
&= B^{-1}(n-5,4)\int_0^{\sqrt{v}} x^{n-6}(1-x)^3\,dx \\
&\quad + K\left\{\frac{v^{\frac{1}{2}(n-3)}}{n-3}\int_{\sqrt{v}}^1 x^{-3}(1-x)^3\,dx - \frac{v^{\frac{1}{2}(n-2)}}{n-2}\int_{\sqrt{v}}^1 x^{-4}(1-x)^3\,dx\right. \\
&= I_{\sqrt{v}}(n-5,4) + B^{-1}(n-5,4)v^{\frac{1}{2}(n-5)}\left\{\frac{n}{6} - \frac{3}{2}(n-1)\sqrt{v}\right. \\
&\quad - \frac{3}{2}(n-4)v + \left(\frac{17}{6}n - \frac{15}{2}\right)v^{\frac{3}{2}} - \frac{3}{2}(n-2)v\log v \\
&\quad \left. - \frac{1}{2}(n-3)v^{\frac{3}{2}}\log v\right\}.
\end{aligned}
$$

Wilks (1935a) has given this distribution. The case of $p_i = 1, i = 1, 2, 3,$ is a special case of Wilks' distribution* for $p_1 = 1, p_2 = 1$, and p_3. Wilks also gives the distributions for $p_1 = 1, p_2 = p_3 = 2$; for $p_1 = 1, p_2 = 2, p_3 = 3$; for $p_1 = 1, p_2 = 2, p_3 = 4$; and for $p_1 = p_2 = 2, p_3 = 3$.

Wald and Brookner (1941) have given a method for deriving the distribution if not more than one p_i is odd. It can be seen that the same result can be obtained by integration of products of beta functions after using the duplication formula to reduce the moments.

9.5. AN ASYMPTOTIC EXPANSION OF THE DISTRIBUTION OF THE LIKELIHOOD RATIO CRITERION

The hth moment of $\lambda = V^{\frac{1}{2}N}$ is

$$
(1)\qquad \mathscr{E}\lambda^h = K\frac{\prod_{i=1}^{p}\Gamma\{\frac{1}{2}[N(1+h)-i]\}}{\prod_{i=1}^{q}\left\{\prod_{j=1}^{p_i}\Gamma\{\frac{1}{2}[N(1+h)-j]\}\right\}},
$$

* In Wilks' formula $\Gamma[\frac{1}{2}(N-2-i)]$ should be $\Gamma[\frac{1}{2}(n-2-i)]$.

where K is chosen so that $\mathscr{E}\lambda^0 = 1$. This is of the form of (1) of Section 8.6 with

$$a = p, \qquad b = p, \qquad x_k = \frac{N}{2}, \qquad \xi_k = \frac{-k}{2},$$

$$(2) \qquad y_j = \frac{N}{2}, \qquad \eta_j = \frac{-j + p_1 \cdots + p_{i-1}}{2},$$

$$j = p_1 = \cdots + p_{i-1} + 1, \cdots, p_1 + \cdots + p_i; \qquad i = 1, \cdots, q.$$

Then $f = \frac{1}{2}[p(p+1) - \sum p_i(p_i + 1)] = \frac{1}{2}(p^2 - \sum p_i^2)$, $\beta_k = \varepsilon_j = \frac{1}{2}(1 - \rho)N$.

In order to make the second term in the expansion vanish we take ρ as

$$(3) \qquad \rho = 1 - \frac{2(p^3 - \sum p_i^3) + 9(p^2 - \sum p_i^2)}{6N(p^2 - \sum p_i^2)}.$$

Let

$$(4) \qquad m = \rho N = N - \frac{3}{2} - \frac{p^3 - \sum p_i^3}{3(p^2 - \sum p_i^2)}.$$

Then $\omega_2 = \gamma_2/m^2$, where [as shown by Box (1949)]

$$(5) \qquad \gamma_2 = \frac{p^4 - \sum p_i^4}{48} - \frac{5(p^2 - \sum p_i^2)}{96} - \frac{(p^3 - \sum p_i^3)^2}{72(p^2 - \sum p_i^2)}.$$

We obtain from Section 8.6 the following expansion:

$$(6) \qquad \Pr\{-m \log V \leq v\} = \Pr\{\chi_f^2 \leq v\} + \frac{\gamma_2}{m^2}[\Pr\{\chi_{f+4}^2 \leq v\} - \Pr\{\chi_f^2 \leq v\}] + O(m^{-3}).$$

If $q = 2$, we obtain further terms in the expansion by using the results of Section 8.6.

If $p_i = 1$, we have

$$f = \tfrac{1}{2}p(p-1),$$

$$m = N - \frac{2p + 11}{6},$$

$$(7) \qquad \gamma_2 = \frac{p(p-1)}{288}(2p^2 - 2p - 13),$$

$$\gamma_3 = \frac{p(p-1)}{3240}(p-2)(2p-1)(p+1);$$

other terms are given by Box (1949). If $p_i = 2$ $(p = 2q)$

$$f = 2q(q-1),$$

$$m = N - \frac{4q+13}{6}, \tag{8}$$

$$\gamma_2 = \frac{q(q-1)}{72}(8q^2 - 8q - 7).$$

Table 1 gives an indication of the order of approximation of (6) for $p_i = 1$. In each case v is chosen so that the first term is 0.95.

TABLE 1

p	f	v	γ_2	N	m	γ_2/m^2	Second Term
4	6	12.592	$\frac{11}{24}$	15	$\frac{71}{6}$	0.0033	−0.0007
5	10	18.307	$\frac{15}{8}$	15	$\frac{69}{6}$	0.0142	−0.0021
6	15	24.996	$\frac{235}{48}$	15	$\frac{67}{6}$	0.0393	−0.0043
				16	$\frac{73}{6}$	0.0331	−0.0036

9.6. AN EXAMPLE

We take the following example from an industrial time study [Abruzzi (1950)]. The purpose of the study was to investigate the length of time taken by various operators in a garment factory to do several elements of a pressing operation. The entire pressing operation was divided into the following six elements:

1. Pick up and position garment.
2. Press and repress short dart.
3. Reposition garment on ironing board.
4. Press three quarters of length of long dart.
5. Press balance of long dart.
6. Hang garment on rack.

In this case $\mathbf{x}_\alpha$ is the vector of measurements on individual α. The component $x_{i\alpha}$ is the time taken to do the ith element of the operation. N is 76. The data (in seconds) are summarized in the sample mean vector and covariance matrix:

$$\bar{\mathbf{x}} = \begin{pmatrix} 9.47 \\ 25.56 \\ 13.25 \\ 31.44 \\ 27.29 \\ 8.80 \end{pmatrix}, \tag{1}$$

$$S = \begin{pmatrix} 2.57 & 0.85 & 1.56 & 1.79 & 1.33 & 0.42 \\ 0.85 & 37.00 & 3.34 & 13.47 & 7.59 & 0.52 \\ 1.56 & 3.34 & 8.44 & 5.77 & 2.00 & 0.50 \\ 1.79 & 13.47 & 5.77 & 34.01 & 10.50 & 1.77 \\ 1.33 & 7.59 & 2.00 & 10.50 & 23.01 & 3.43 \\ 0.42 & 0.52 & 0.50 & 1.77 & 3.43 & 4.59 \end{pmatrix}. \tag{2}$$

The sample standard deviations are (1.604, 6.041, 2.903, 5.832, 4.798, 2.141). The sample correlation matrix is

$$R = \begin{pmatrix} 1.000 & 0.088 & 0.334 & 0.191 & 0.173 & 0.123 \\ 0.088 & 1.000 & 0.186 & 0.384 & 0.262 & 0.040 \\ 0.334 & 0.186 & 1.000 & 0.343 & 0.144 & 0.080 \\ 0.191 & 0.384 & 0.343 & 1.000 & 0.375 & 0.142 \\ 0.173 & 0.262 & 0.144 & 0.375 & 1.000 & 0.334 \\ 0.123 & 0.040 & 0.080 & 0.142 & 0.334 & 1.000 \end{pmatrix}. \tag{3}$$

The investigators are interested in testing the hypothesis that the six variates are mutually independent. It often happens in time studies that a new operation is proposed in which the elements are combined in a different way; the new operation may use some of the elements several times and some elements may be omitted. If the times for the different elements in the operation for which data are available are independent, it may reasonably be assumed that they will be independent in a new operation. Then the distribution of time for the new operation can be estimated by using the means and variances of the individual items.

In this problem the criterion V is $V = |\boldsymbol{R}| = 0.472$. Since the sample size is large we can use asymptotic theory: $m = 433/6$, $f = 15$, and $-m \log V = 54.1$. Since the significance point for the χ^2-distribution with 15 degrees of freedom is 30.6 at the 0.01 significance level, we find the result significant. We reject the hypothesis of independence; we cannot consider the times of the elements independent.

9.7. THE CASE OF TWO SETS OF VARIATES

In the case of two sets of variates ($q = 2$), the random vector $\boldsymbol{X}$, the observation vector $\boldsymbol{x}_\alpha$, the mean vector $\boldsymbol{\mu}$, and the covariance matrix $\boldsymbol{\Sigma}$ are partitioned as follows:

$$\begin{aligned} \boldsymbol{X} &= \begin{pmatrix} \boldsymbol{X}^{(1)} \\ \boldsymbol{X}^{(2)} \end{pmatrix}, & \boldsymbol{x}_\alpha &= \begin{pmatrix} \boldsymbol{x}_\alpha^{(1)} \\ \boldsymbol{x}_\alpha^{(2)} \end{pmatrix}, \\ \boldsymbol{\mu} &= \begin{pmatrix} \boldsymbol{\mu}^{(1)} \\ \boldsymbol{\mu}^{(2)} \end{pmatrix}, & \boldsymbol{\Sigma} &= \begin{pmatrix} \boldsymbol{\Sigma}_{11} & \boldsymbol{\Sigma}_{12} \\ \boldsymbol{\Sigma}_{21} & \boldsymbol{\Sigma}_{22} \end{pmatrix}. \end{aligned} \tag{1}$$

The null hypothesis of independence specifies that $\mathbf{\Sigma}_{12} = \mathbf{0}$; that is, that $\mathbf{\Sigma}$ is of the form

$$\mathbf{\Sigma}_0 = \begin{pmatrix} \mathbf{\Sigma}_{11} & \mathbf{0} \\ \mathbf{0} & \mathbf{\Sigma}_{22} \end{pmatrix}. \tag{2}$$

The test criterion is

$$V = \frac{|A|}{|A_{11}| \cdot |A_{22}|}. \tag{3}$$

It was shown in Section 9.4.2 that when the null hypothesis is true, this criterion is distributed as $U_{p_1, p_2, N-1-p_2}$, the criterion for testing a hypothesis about regression coefficients (Chapter 8). We now wish to study further the relationship between testing the hypothesis of independence of two sets and testing the hypothesis that regression of one set on the other is zero.

The conditional distribution of $X_\alpha^{(1)}$ given $X_\alpha^{(2)} = x_\alpha^{(2)}$ is $N[\boldsymbol{\mu}^{(1)} + \boldsymbol{\beta}(x_\alpha^{(2)} - \boldsymbol{\mu}^{(2)}), \mathbf{\Sigma}_{11\cdot 2}] = N[\boldsymbol{\beta}(x_\alpha^{(2)} - \bar{x}^{(2)}) + \boldsymbol{\nu}, \mathbf{\Sigma}_{11\cdot 2}]$, where $\boldsymbol{\beta} = \mathbf{\Sigma}_{12}\mathbf{\Sigma}_{22}^{-1}$, $\mathbf{\Sigma}_{11\cdot 2} = \mathbf{\Sigma}_{11} - \mathbf{\Sigma}_{12}\mathbf{\Sigma}_{22}^{-1}\mathbf{\Sigma}_{21}$, and $\boldsymbol{\nu} = \boldsymbol{\mu}^{(1)} + \boldsymbol{\beta}(\bar{x}^{(2)} - \boldsymbol{\mu}^{(2)})$. Let $X_\alpha^* = X_\alpha^{(1)}$, $z_\alpha^{*\prime} = [(x_\alpha^{(2)} - \bar{x}^{(2)})' \; 1]$, $\boldsymbol{\beta}^* = (\boldsymbol{\beta} \; \boldsymbol{\nu})$, and $\mathbf{\Sigma}^* = \mathbf{\Sigma}_{11\cdot 2}$. Then the conditional distribution of X_α^* is $N(\boldsymbol{\beta}^* z_\alpha^*, \mathbf{\Sigma}^*)$. This is exactly the kind of distribution studied in Chapter 8.

The null hypothesis that $\mathbf{\Sigma}_{12} = \mathbf{0}$ is equivalent to the null hypothesis $\boldsymbol{\beta} = \mathbf{0}$. Considering $x_\alpha^{(2)}$ fixed, we know from Chapter 8 that the criterion (based on the likelihood ratio criterion) for testing this hypothesis is

$$U = \frac{|\sum (x_\alpha^* - \hat{\boldsymbol{\beta}}_\Omega^* z_\alpha^*)(x_\alpha^* - \hat{\boldsymbol{\beta}}_\Omega^* z_\alpha^*)'|}{|\sum (x_\alpha^* - \hat{\boldsymbol{\beta}}_{2\omega}^* z_\alpha^{*(2)})(x_\alpha^* - \hat{\boldsymbol{\beta}}_{2\omega}^* z_\alpha^{*(2)})'|}, \tag{4}$$

where

$$z_\alpha^{*(2)} = 1,$$

$$\hat{\boldsymbol{\beta}}_{2\omega}^* = \hat{\boldsymbol{\nu}} = \bar{x}^* = \bar{x}^{(1)},$$

$$\begin{aligned} \hat{\boldsymbol{\beta}}_\Omega^* &= (\hat{\boldsymbol{\beta}}_{1\Omega}^* \; \hat{\boldsymbol{\beta}}_{2\Omega}^*) \\ &= \left(\sum x_\alpha^* z_\alpha^{*(1)\prime} \; \sum x_\alpha^* z_\alpha^{*(2)\prime}\right) \begin{pmatrix} \sum z_\alpha^{*(1)} z_\alpha^{*(1)\prime} & \sum z_\alpha^{*(1)} z_\alpha^{*(2)\prime} \\ \sum z_\alpha^{*(2)} z_\alpha^{*(1)\prime} & \sum z_\alpha^{*(2)} z_\alpha^{*(2)\prime} \end{pmatrix}^{-1} \\ &= (A_{12} \; N\bar{x}^{(1)}) \begin{pmatrix} A_{22} & \mathbf{0} \\ \mathbf{0} & N \end{pmatrix}^{-1} \\ &= (A_{12}A_{22}^{-1} \; \bar{x}^{(1)}). \end{aligned} \tag{5}$$

The matrix in the denominator of U is

$$\sum (x_\alpha^{(1)} - \bar{x}^{(1)})(x_\alpha^{(1)} - \bar{x}^{(1)})' = A_{11}. \tag{6}$$

The matrix in the numerator is

$$(7)\quad \sum[\boldsymbol{x}_\alpha^{(1)} - \bar{\boldsymbol{x}}^{(1)} - A_{12}A_{22}^{-1}(\boldsymbol{x}_\alpha^{(2)} - \bar{\boldsymbol{x}}^{(2)})][\boldsymbol{x}_\alpha^{(1)} - \bar{\boldsymbol{x}}^{(1)} - A_{12}A_{22}^{-1}(\boldsymbol{x}_\alpha^{(2)} - \bar{\boldsymbol{x}}^{(2)})]' = A_{11} - A_{12}A_{22}^{-1}A_{21}.$$

Therefore

$$(8)\qquad U = \frac{|A_{11} - A_{12}A_{22}^{-1}A_{21}|}{|A_{11}|} = \frac{|A|}{|A_{11}|\cdot|A_{22}|}$$

which is exactly V.

Now let us see why it is that when the null hypothesis is true the distribution of $U = V$ does not depend on whether the $X_\alpha^{(2)}$ are held fixed. It was shown in Chapter 8 that when the null hypothesis is true the distribution of U depends only on p, q_1 and $N - q_2$, but not on $\boldsymbol{z}_\alpha$. Thus the conditional distribution of V given $X_\alpha^{(2)} = \boldsymbol{x}_\alpha^{(2)}$ does not depend on $\boldsymbol{x}_\alpha^{(2)}$; the joint distribution of V and $X_\alpha^{(2)}$ is the product of the distribution of V and the distribution of $X_\alpha^{(2)}$, and the marginal distribution of V is this conditional distribution. This shows that the distribution of V (under the null hypothesis) does not depend on whether the $X_\alpha^{(2)}$ are fixed or have any distribution (normal or not).

Let us extend this result to show that if $q > 2$, the distribution of V under the null hypothesis of independence does not depend on the distribution of one set of variates, say $X_\alpha^{(q)}$. We have

$$(9)\quad V = \frac{|A|}{|A_{11}|\begin{vmatrix} A_{22} & \cdots & A_{2q} \\ \vdots & & \vdots \\ A_{q2} & \cdots & A_{qq}\end{vmatrix}} \cdot \frac{\begin{vmatrix} A_{22} & \cdots & A_{2q} \\ \vdots & & \vdots \\ A_{q2} & \cdots & A_{qq}\end{vmatrix}}{|A_{22}|\begin{vmatrix} A_{33} & \cdots & A_{3q} \\ \vdots & & \vdots \\ A_{q3} & \cdots & A_{qq}\end{vmatrix}} \cdots \frac{\begin{vmatrix} A_{q-1,q-1} & A_{q-1,q} \\ A_{q,q-1} & A_{qq}\end{vmatrix}}{|A_{q-1,q-1}|\cdot|A_{qq}|}$$

$$= V_1V_2\cdots V_{q-1}.$$

When the null hypothesis is true, V_1 is distributed independently of $X_\alpha^{(2)}, \cdots, X_\alpha^{(q)}$ by the previous result. In turn we argue that V_j is distributed independently of $X_\alpha^{(j+1)}, \cdots, X_\alpha^{(q)}$. Thus $V_1 \cdots V_{q-1}$ is distributed independently of $X_\alpha^{(q)}$.

THEOREM 9.7.1. *Under the null hypothesis of independence, the distribution of V is that given earlier in this chapter if $q - 1$ sets are jointly normally distributed, even though one set is not normally distributed.*

Now let us return briefly to the case of $q = 2$. It is of some interest to generalize the correlation coefficient relating two variables to a "vector correlation coefficient" relating two sets of variates. One such generalization is based on determinants (in a way analogous to generalizing variance to generalized variance).

In the case of two scalar variables X_1 and X_2 the "coefficient of alienation" is

$$\frac{\sigma_{1\cdot 2}^2}{\sigma_1^2}, \tag{10}$$

where

$$\sigma_{1\cdot 2}^2 = \mathscr{E}(X_1 - \beta X_2)^2 \tag{11}$$

is the variance of X_1 about its regression on X_2 when

$$\mathscr{E}X_1 = \mathscr{E}X_2 = 0 \tag{12}$$

and

$$\mathscr{E}(X_1|X_2) = \beta X_2. \tag{13}$$

In the case of two vectors $\boldsymbol{X}^{(1)}$ and $\boldsymbol{X}^{(2)}$, the regression matrix is

$$\boldsymbol{\beta} = \boldsymbol{\Sigma}_{12}\boldsymbol{\Sigma}_{22}^{-1} \tag{14}$$

and the generalized variance of $\boldsymbol{X}^{(1)}$ about its regression on $\boldsymbol{X}^{(2)}$ is

$$|\mathscr{E}\{(\boldsymbol{X}^{(1)} - \boldsymbol{\beta}\boldsymbol{X}^{(2)})(\boldsymbol{X}^{(1)} - \boldsymbol{\beta}\boldsymbol{X}^{(2)})'\}| = |\boldsymbol{\Sigma}_{11} - \boldsymbol{\Sigma}_{12}\boldsymbol{\Sigma}_{22}^{-1}\boldsymbol{\Sigma}_{21}| = \frac{|\boldsymbol{\Sigma}|}{|\boldsymbol{\Sigma}_{22}|}. \tag{15}$$

Since the generalized variance of $\boldsymbol{X}^{(1)}$ is

$$|\mathscr{E}\boldsymbol{X}^{(1)}\boldsymbol{X}^{(1)\prime}| = |\boldsymbol{\Sigma}_{11}| \tag{16}$$

the vector coefficient of alienation is

$$\frac{|\boldsymbol{\Sigma}_{11} - \boldsymbol{\Sigma}_{12}\boldsymbol{\Sigma}_{22}^{-1}\boldsymbol{\Sigma}_{21}|}{|\boldsymbol{\Sigma}_{11}|} = \frac{|\boldsymbol{\Sigma}|}{|\boldsymbol{\Sigma}_{11}| \cdot |\boldsymbol{\Sigma}_{22}|}. \tag{17}$$

The sample equivalent of (17) is simply V.

The square of the correlation between two scalars X_1 and X_2 can be written as

$$\frac{\sigma^2(b)}{\sigma_1^2} \tag{18}$$

where $\sigma^2(b)$ is the variance of the regression function. The generalized variance of $\boldsymbol{\beta}\boldsymbol{X}^{(2)}$ is

$$|\mathscr{E}\boldsymbol{\beta}\boldsymbol{X}^{(2)}(\boldsymbol{\beta}\boldsymbol{X}^{(2)})'| = |\boldsymbol{\Sigma}_{12}\boldsymbol{\Sigma}_{22}^{-1}\boldsymbol{\Sigma}_{21}|. \tag{19}$$

Thus the square of the vector correlation coefficient is

$$\frac{|\mathbf{\Sigma}_{12}\mathbf{\Sigma}_{22}^{-1}\mathbf{\Sigma}_{21}|}{|\mathbf{\Sigma}_{11}|}(-1)^{p_1} = \frac{\begin{vmatrix} \mathbf{0} & \mathbf{\Sigma}_{12} \\ \mathbf{\Sigma}_{21} & \mathbf{\Sigma}_{22} \end{vmatrix}}{|\mathbf{\Sigma}_{11}| \cdot |\mathbf{\Sigma}_{22}|}. \tag{20}$$

If $p_1 = p_2$, (20) is

$$(-1)^{p_1}\frac{|\mathbf{\Sigma}_{12}|^2}{|\mathbf{\Sigma}_{11}| \cdot |\mathbf{\Sigma}_{22}|}. \tag{21}$$

Of course, measures such as these are not completely satisfactory as generalizations because they omit information relating the two sets; that is, the relationship between $\boldsymbol{X}^{(1)}$ and $\boldsymbol{X}^{(2)}$ cannot be expressed in a single number. In Chapter 11 we will investigate this subject much more deeply. We will find that the "canonical correlations" express the dependence between $\boldsymbol{X}^{(1)}$ and $\boldsymbol{X}^{(2)}$ independent of the coordinate systems of $\boldsymbol{X}^{(1)}$ and $\boldsymbol{X}^{(2)}$.

Other tests of independence of sets can be based on procedures considered in Chapter 8. The hypothesis that the subvectors $\boldsymbol{X}^{(1)}, \cdots, \boldsymbol{X}^{(q)}$ are mutually independent can be considered as the hypothesis that $\boldsymbol{X}^{(1)}$ is independent of $\boldsymbol{X}^{(2)}, \cdots, \boldsymbol{X}^{(q)}$, that $\boldsymbol{X}^{(2)}$ is independent of $\boldsymbol{X}^{(3)}, \cdots, \boldsymbol{X}^{(q)}$, and so forth, and this can be rephrased as the hypothesis that regression of $\boldsymbol{X}^{(1)}$ on $\boldsymbol{X}^{(2)}, \cdots, \boldsymbol{X}^{(q)}$ is zero, and so on.

REFERENCES

Abruzzi (1950); Box (1949); Daly (1940); Girshick (1941); Gumbel and Littauer (1952); Kelley (1928); Narain (1950); Wald and Brookner (1941); Wilks (1935a), (1943), 242–245.

PROBLEMS

1. Let x_1 = arithmetic speed, x_2 = arithmetic power, x_3 = intellectual interest, x_4 = social interest, x_5 = activity interest. T. L. Kelley [(1928), p. 114] observed the following correlations between batteries of tests identified as above, based on 109 pupils:

$$\begin{pmatrix} 1.0000 & 0.4249 & -0.0552 & -0.0031 & 0.1927 \\ 0.4249 & 1.0000 & -0.0416 & 0.0495 & 0.0687 \\ -0.0552 & -0.0416 & 1.0000 & 0.7474 & 0.1691 \\ -0.0031 & 0.0495 & 0.7474 & 1.0000 & 0.2653 \\ 0.1927 & 0.0687 & 0.1691 & 0.2653 & 1.0000 \end{pmatrix}.$$

Let $\boldsymbol{x}^{(1)\prime} = (x_1, x_2)$ and $\boldsymbol{x}^{(2)\prime} = (x_3, x_4, x_5)$. Test the hypothesis that $\boldsymbol{x}^{(1)}$ is independent of $\boldsymbol{x}^{(2)}$ at the 1% significance level.

2. Another set of time-study data [Abruzzi (1950)] is summarized by the correlation matrix based on 188 observations:

$$\begin{pmatrix} 1.00 & -0.27 & 0.06 & 0.07 & 0.02 \\ -0.27 & 1.00 & -0.01 & -0.02 & -0.02 \\ 0.06 & -0.01 & 1.00 & -0.07 & -0.04 \\ 0.07 & -0.02 & -0.07 & 1.00 & -0.10 \\ 0.02 & -0.02 & -0.04 & -0.10 & 1.00 \end{pmatrix}.$$

Test the hypothesis that $\sigma_{ij} = 0$ $(i \neq j)$ at the 5% significance level.

3. (Sec. 9.4) Derive some of the distributions obtained by Wilks (1935a) and referred to at the end of Section 9.4. [*Hint:* Use the results of Section 9.4.1.]

4. (Sec. 9.7) Prove that the criterion V of this chapter can be expressed as the product of U's given in Chapter 8.

5. (Sec. 9.7) Give the sample vector coefficient of alienation and the vector correlation coefficient.

6. (Sec. 9.7) If y is the sample vector coefficient of alienation and z the square of the vector correlation coefficient, find $\mathscr{E} y^g z^h$ when $\mathbf{\Sigma}_{12} = \mathbf{0}$.

7. (Sec. 9.5) For the case $p_i = 2$, express m and γ_2. Compute the second term of (6) when v is chosen so that the first term is 0.95 for $p = 4$ and 6 and $N = 15$.

CHAPTER 10

Testing Hypotheses of Equality of Covariance Matrices and Equality of Mean Vectors and Covariance Matrices

10.1. INTRODUCTION

In this chapter we study problems of testing hypotheses of equality of covariance matrices and equality of both covariance matrices and mean vectors. The tests discussed are likelihood ratio tests or adaptations of likelihood ratio tests. In each case the problem and test considered here are multivariate generalizations of a univariate problem and test. First we consider equality of covariance matrices and equality of covariance matrices and mean vectors of several populations without specifying the common covariance matrix or the common covariance matrix and mean vector. Later we consider the equality of a covariance matrix to a given matrix and also simultaneous equality of a covariance matrix to a given matrix and equality of a mean vector to a given vector. One other hypothesis considered, the equality of a covariance matrix to a given matrix except for a proportionality factor, has only a trivial corresponding univariate hypothesis. In each case the class of tests for a class of hypotheses leads to a confidence region.

10.2. CRITERIA FOR TESTING EQUALITY OF SEVERAL COVARIANCE MATRICES

In this section we study several normal distributions and consider using a set of samples, one from each population, to test the hypothesis that the covariance matrices of these populations are equal. Let $x_\alpha^{(g)}$ ($\alpha = 1, \cdots, N_g$; $g = 1, \cdots, q$) be an observation from the gth population $N(\boldsymbol{\mu}^{(g)}, \boldsymbol{\Sigma}_g)$. We wish to test the hypothesis

$$H_1:\ \boldsymbol{\Sigma}_1 = \cdots = \boldsymbol{\Sigma}_q. \tag{1}$$

Let

$$\sum_{g=1}^{q} N_g = N, \tag{2}$$

$$A_g = \sum_{\alpha=1}^{N_g} (\boldsymbol{x}_\alpha^{(g)} - \bar{\boldsymbol{x}}^{(g)})(\boldsymbol{x}_\alpha^{(g)} - \bar{\boldsymbol{x}}^{(g)})', \qquad g = 1, \cdots, q,$$

$$A = \sum_{g=1}^{q} A_g.$$

First we shall obtain the likelihood ratio criterion. The likelihood function is

$$L = \prod_{g=1}^{q} \frac{1}{(2\pi)^{\frac{1}{2}pN_g} |\boldsymbol{\Sigma}_g|^{\frac{1}{2}N_g}} \exp\left[-\tfrac{1}{2} \sum_{\alpha=1}^{N_g} (\boldsymbol{x}_\alpha^{(g)} - \boldsymbol{\mu}^{(g)})' \boldsymbol{\Sigma}_g^{-1} (\boldsymbol{x}_\alpha^{(g)} - \boldsymbol{\mu}^{(g)})\right]. \tag{3}$$

The space Ω is the parameter space in which each $\boldsymbol{\Sigma}_g$ is positive definite and $\boldsymbol{\mu}^{(g)}$ any vector. The space ω is the parameter space in which $\boldsymbol{\Sigma}_1 = \boldsymbol{\Sigma}_2 = \cdots = \boldsymbol{\Sigma}_q$ and $\boldsymbol{\mu}^{(g)}$ any vector. The maximum likelihood estimates of $\boldsymbol{\mu}^{(g)}$ and $\boldsymbol{\Sigma}_g$ in Ω are given by

$$\hat{\boldsymbol{\mu}}_\Omega^{(g)} = \bar{\boldsymbol{x}}^{(g)}, \qquad \hat{\boldsymbol{\Sigma}}_{g\Omega} = \frac{1}{N_g} A_g. \tag{4}$$

The maximum likelihood estimates of $\boldsymbol{\mu}^{(g)}$ in ω are given by (4), $\hat{\boldsymbol{\mu}}_\omega^{(g)} = \bar{\boldsymbol{x}}^{(g)}$, since the maximizing values of $\boldsymbol{\mu}^{(g)}$ are the same regardless of $\boldsymbol{\Sigma}_g$. The function to be maximized with respect to $\boldsymbol{\Sigma}_1 = \cdots = \boldsymbol{\Sigma}_q = \boldsymbol{\Sigma}$, say, is

$$\frac{1}{(2\pi)^{\frac{1}{2}pN} |\boldsymbol{\Sigma}|^{\frac{1}{2}N}} \exp\left[-\tfrac{1}{2} \sum_g \sum_\alpha (\boldsymbol{x}_\alpha^{(g)} - \bar{\boldsymbol{x}}^{(g)})' \boldsymbol{\Sigma}^{-1} (\boldsymbol{x}_\alpha^{(g)} - \bar{\boldsymbol{x}}^{(g)})\right]. \tag{5}$$

By Lemma 3.2.2 the maximizing value of $\boldsymbol{\Sigma}$ is

$$\hat{\boldsymbol{\Sigma}}_\omega = \frac{1}{N} A, \tag{6}$$

and the maximum of the likelihood function is

$$\frac{1}{(2\pi)^{\frac{1}{2}pN} |\hat{\boldsymbol{\Sigma}}_\omega|^{\frac{1}{2}N}} e^{-\frac{1}{2}pN}. \tag{7}$$

The likelihood ratio criterion for testing (1) is

$$\lambda_1 = \frac{\prod_{g=1}^{q} |\hat{\boldsymbol{\Sigma}}_{g\Omega}|^{\frac{1}{2}N_g}}{|\hat{\boldsymbol{\Sigma}}_\omega|^{\frac{1}{2}N}} \tag{8}$$

$$= \frac{\prod_{g=1}^{q} |A_g|^{\frac{1}{2}N_g}}{|A|^{\frac{1}{2}N}} \cdot \frac{N^{\frac{1}{2}pN}}{\prod_{g=1}^{q} N_g^{\frac{1}{2}pN_g}}.$$

The critical region is

$$\lambda_1 \le \lambda_1(\alpha), \tag{9}$$

where $\lambda_1(\alpha)$ is defined so that (9) holds with probability α when (1) is true.

Bartlett (1937a) has suggested modifying λ_1 in the univariate case by replacing sample numbers by the numbers of degrees of freedom of the A_g. Except for a numerical constant, the statistic he proposes is

$$V_1 = \frac{\prod_{g=1}^{q} |A_g|^{\frac{1}{2}n_g}}{|A|^{\frac{1}{2}n}}, \tag{10}$$

where $n_g = N_g - 1$ and $n = \sum n_g = N - q$. The numerator is proportional to a power of a weighted geometric mean of the sample generalized variances, and the denominator is proportional to a power of the determinant of a weighted arithmetic mean of the sample covariance matrices.

In the scalar case ($p = 1$) of two samples the criterion (10) is

$$\frac{(n_1)^{\frac{1}{2}n_1}(n_2)^{\frac{1}{2}n_2}(s_1^2)^{\frac{1}{2}n_1}(s_2^2)^{\frac{1}{2}n_2}}{(n_1 s_1^2 + n_2 s_2^2)^{\frac{1}{2}(n_1+n_2)}} = \frac{(n_1)^{\frac{1}{2}n_1}(n_2)^{\frac{1}{2}n_2}F^{\frac{1}{2}n_1}}{(n_1 F + n_2)^{\frac{1}{2}(n_1+n_2)}}, \tag{11}$$

where s_1^2 and s_2^2 are the usual unbiased estimates of σ_1^2 and σ_2^2 (the two population variances) and

$$F = \frac{s_1^2}{s_2^2}. \tag{12}$$

Thus the critical region

$$V_1 \le V_1(\alpha) \tag{13}$$

is based on the F-statistic with n_1 and n_2 degrees of freedom, and the inequality (13) implies a particular method of choosing $F_1(\alpha)$ and $F_2(\alpha)$ for the critical region

$$\begin{aligned} F &\le F_1(\alpha), \\ F &\ge F_2(\alpha). \end{aligned} \tag{14}$$

Brown (1939) and Scheffé (1942) have shown that (14) yields an unbiased test.

Bartlett gave a more intuitive argument for the use of V_1 in place of λ_1. He argues that if N_1, say, is small, A_1 is given too much weight in λ_1, and other effects may be missed.

It is clear that if one assumes

$$\mathscr{E} X_\alpha^{(g)} = \mathbf{B}_g z_\alpha^{(g)}, \tag{15}$$

where $\mathbf{z}_\alpha^{(g)}$ consists of k_g components, and if one estimates the matrix $\boldsymbol{\beta}_g$, defining

$$A_g = \sum_{\alpha=1}^{N_g} (\mathbf{x}_\alpha^{(g)} - \hat{\boldsymbol{\beta}}_g \mathbf{z}_\alpha^{(g)})(\mathbf{x}_\alpha - \hat{\boldsymbol{\beta}}_g \mathbf{z}_\alpha^{(g)})' \tag{16}$$

one uses (10) with $n_g = N_g - k_g$.

10.3. CRITERIA FOR TESTING THAT SEVERAL NORMAL DISTRIBUTIONS ARE IDENTICAL

In Section 8.8 we considered testing the equality of mean vectors when we assumed the covariance matrices were the same; that is, we tested

$$H_2: \quad \boldsymbol{\mu}^{(1)} = \boldsymbol{\mu}^{(2)} = \cdots = \boldsymbol{\mu}^{(q)} \qquad \text{given } \boldsymbol{\Sigma}_1 = \boldsymbol{\Sigma}_2 \cdots = \boldsymbol{\Sigma}_q. \tag{1}$$

The test of the assumption in H_2 was considered in Section 10.2. Now let us consider the hypothesis that both means and covariances are the same; this is a combination of H_1 and H_2. We test

$$H: \quad \boldsymbol{\mu}^{(1)} = \boldsymbol{\mu}^{(2)} = \cdots = \boldsymbol{\mu}^{(q)}, \qquad \boldsymbol{\Sigma}_1 = \boldsymbol{\Sigma}_2 = \cdots = \boldsymbol{\Sigma}_q. \tag{2}$$

As in Section 10.2, let $\mathbf{x}_\alpha^{(g)}$ $(\alpha = 1, \cdots, N_g)$ be an observation from $N(\boldsymbol{\mu}^{(g)}, \boldsymbol{\Sigma}_g)$ $(g = 1, \cdots, q)$. Then Ω is the unrestricted parameter space of $\{\boldsymbol{\mu}^{(g)}, \boldsymbol{\Sigma}_g\}$ $(g = 1, \cdots, q)$, where $\boldsymbol{\Sigma}_g$ is positive definite, and ω^* consists of the space restricted by (2).

The likelihood function is given by (3) of Section 10.2. The hypothesis H_1 of Section 10.2 is that the parameter point falls in ω; the hypothesis H_2 of Section 8.8 is that the parameter point falls in ω^* given it falls in $\omega \supset \omega^*$; and the hypothesis H here is that the parameter point falls in ω^* given that it is in Ω.

We use the following lemma:

LEMMA 10.3.1. *Let $\mathbf{y}$ be an observation vector on a random vector with density $f(\mathbf{z}, \boldsymbol{\theta})$, where $\boldsymbol{\theta}$ is a parameter vector in a space Ω. Let H_a be the hypothesis $\boldsymbol{\theta} \in \Omega_a \subset \Omega$, let H_b be the hypothesis $\boldsymbol{\theta} \in \Omega_b \subset \Omega_a$, given $\boldsymbol{\theta} \in \Omega_a$, and let H_{ab} be the hypothesis $\boldsymbol{\theta} \in \Omega_b$, given $\boldsymbol{\theta} \in \Omega$. If λ_a, the likelihood ratio criterion for testing H_a, λ_b for H_b, and λ_{ab} for H_{ab} are uniquely defined for the observation vector $\mathbf{y}$,*

$$\lambda_{ab} = \lambda_a \lambda_b. \tag{3}$$

Proof. Since

$$\lambda_a = \frac{\max_{\boldsymbol{\theta} \in \Omega_a} f(\mathbf{y}, \boldsymbol{\theta})}{\max_{\boldsymbol{\theta} \in \Omega} f(\mathbf{y}, \boldsymbol{\theta})}, \tag{4}$$

$$\lambda_b = \frac{\max_{\theta \in \Omega_b} f(y, \theta)}{\max_{\theta \in \Omega_a} f(y, \theta)}, \tag{5}$$

$$\lambda_{ab} = \frac{\max_{\theta \in \Omega_b} f(y, \theta)}{\max_{\theta \in \Omega} f(y, \theta)}, \tag{6}$$

the equality (3) is obvious.

Thus the likelihood ratio criterion for the hypothesis H is the product of the likelihood ratio criteria for H_1 and H_2,

$$\lambda = \lambda_1 \lambda_2 = \left(\prod_{g=1}^{q} \frac{|A_g|^{\frac{1}{2}N_g}}{N_g^{\frac{1}{2}pN_g}} \right) \frac{N^{\frac{1}{2}pN}}{|B|^{\frac{1}{2}N}}, \tag{7}$$

where

$$\begin{aligned} B &= \sum_{g=1}^{q} \sum_{\alpha=1}^{N_g} (x_\alpha^{(g)} - \bar{x})(x_\alpha^{(g)} - \bar{x})' \\ &= A + \sum_{g=1}^{q} N_g(\bar{x}^{(g)} - \bar{x})(\bar{x}^{(g)} - \bar{x})'. \end{aligned} \tag{8}$$

The critical region is defined by

$$\lambda \leq \lambda(\alpha), \tag{9}$$

where $\lambda(\alpha)$ is chosen so that the probability of (9) under H is α.

Let

$$V_2 = \frac{|A|^{\frac{1}{2}n}}{|B|^{\frac{1}{2}n}} = \lambda_2^{n/N}; \tag{10}$$

this is clearly equivalent to λ_2 for testing H_2. It seems reasonable that

$$V = V_1 V_2 = \frac{\prod_{g=1}^{q} |A_g|^{\frac{1}{2}n_g}}{|B|^{\frac{1}{2}n}} \tag{11}$$

would be preferable to λ for testing H.

10.4. THE MOMENTS OF THE CRITERIA

Here we shall find the joint moments of V_1 and V_2 when H is true. From these we find the moments of V_1 alone when H_1 is true and the moments of V when H is true. It follows from Section 8.8 that when $\mu_g = \mu$, the matrix $\sum_g N_g(\bar{X}^{(g)} - \bar{X})(\bar{X}^{(g)} - \bar{X})'$ is distributed as $\sum_{f=1}^{q-1} Y_f Y_f'$, where the Y_f are independently distributed, each according to $N(0, \Sigma)$ and independently of A_g.

We can express the joint moment of V_1 and V_2 as

$$(1)\quad \mathscr{E}V_1^hV_2^k = \mathscr{E}\left(\frac{\prod_g |A_g|^{\frac{1}{2}hn_g}}{|\sum_g A_g|^{\frac{1}{2}hn}} \cdot \frac{|\sum_g A_g|^{\frac{1}{2}kn}}{|\sum_g A_g + \sum_f Y_f Y_f'|^{\frac{1}{2}kn}}\right)$$

$$= \mathscr{E}\left(\prod_g |A_g|^{\frac{1}{2}hn_g} |\sum_g A_g|^{\frac{1}{2}(k-h)n} |\sum_g A_g + \sum_f Y_f Y_f'|^{-\frac{1}{2}kn}\right)$$

$$= \int \cdots \int \prod_g |A_g|^{\frac{1}{2}hn_g} |\sum_g A_g|^{\frac{1}{2}(k-h)n} |\sum_g A_g + \sum_f y_f y_f'|^{-\frac{1}{2}kn}$$

$$\cdot \prod_g w(A_g|\mathbf{\Sigma}, n_g) \prod_f n(y_f|\mathbf{0}, \mathbf{\Sigma}) \prod_g dA_g \prod_f dy_f$$

$$= \prod_{g=1}^q \frac{K(\mathbf{\Sigma}, n_g)}{K(\mathbf{\Sigma}, n_g + hn_g)} \int \cdots \int |\sum_g A_g|^{\frac{1}{2}(k-h)n} |\sum_g A_g + \sum_f y_f y_f'|^{-\frac{1}{2}kn}$$

$$\cdot \prod_g w(A_g|\mathbf{\Sigma}, n_g + hn_g) \prod_f n(y_f|\mathbf{0}, \mathbf{\Sigma}) \prod_g dA_g \prod_f dy_f,$$

by the argument used in Sections 8.4 and 9.3. The integration over A_g positive definite ($g = 1, \cdots, q$) can be considered (with a suitable change of variables) as an integration for $\sum A_g = A$ and an integration over A positive definite. The integration over $\sum A_g = A$ of $\prod_g w(A_g|\mathbf{\Sigma}, n_g + hn_g)$ gives $w(A|\mathbf{\Sigma}, n + hn)$ by Theorem 7.3.2. Thus

$$(2)\quad \mathscr{E}V_1^hV_2^k = \prod_g \frac{K(\mathbf{\Sigma}, n_g)}{K(\mathbf{\Sigma}, n_g + hn_g)} \int \cdots \int |A|^{\frac{1}{2}(k-h)n} |A + \sum y_f y_f'|^{-\frac{1}{2}kn}$$

$$\cdot w(A|\mathbf{\Sigma}, n + hn) \prod_f n(y_f|\mathbf{0}, \mathbf{\Sigma})\, dA \prod_f dy_f$$

$$= \prod_{g=1}^q \frac{K(\mathbf{\Sigma}, n_g)}{K(\mathbf{\Sigma}, n_g + hn_g)} \frac{K(\mathbf{\Sigma}, n + hn)}{K(\mathbf{\Sigma}, n)} \int \cdots \int |A|^{\frac{1}{2}kn}$$

$$|A + \sum y_f y_f'|^{-\frac{1}{2}kn} w(A|\mathbf{\Sigma}, n) \prod_f n(y_f|\mathbf{0}, \mathbf{\Sigma})\, dA \prod_f dy_f$$

$$= \prod_{g=1}^q \frac{K(\mathbf{\Sigma}, n_g)}{K(\mathbf{\Sigma}, n_g + hn_g)} \frac{K(\mathbf{\Sigma}, n + hn)}{K(\mathbf{\Sigma}, n)} \mathscr{E}V_2^k,$$

where $\mathscr{E}V_2^k$ is given by (9) of Section 8.4 with $q_1 = q - 1$. Substituting $K^{-1}(\mathbf{\Sigma}, m) = 2^{\frac{1}{2}pm}\pi^{p(p-1)/4}|\mathbf{\Sigma}|^{\frac{1}{2}m}\prod_{i=1}^p \Gamma[\frac{1}{2}(m + 1 - i)]$ into (2) we obtain

$$(3)\quad \mathscr{E}V_1^hV_2^k = \prod_{i=1}^p \left\{\prod_{g=1}^q \frac{\Gamma[\frac{1}{2}(n_g + hn_g + 1 - i)]}{\Gamma[\frac{1}{2}(n_g + 1 - i)]}\right\}$$

$$\cdot \frac{\Gamma[\frac{1}{2}(n + kn + 1 - i)]}{\Gamma[\frac{1}{2}(n + hn + 1 - i)]} \frac{\Gamma[\frac{1}{2}(n + q - i)]}{\Gamma[\frac{1}{2}(n + kn + q - i)]}.$$

The hth moment of V_1 when H is true is obtained from (3) by setting $k = 0$; this is also the hth moment of V_1 when H_1 is true because the distribution of A_g does not depend on $\boldsymbol{\mu}_g$. Thus

$$(4)\quad \mathscr{E}V_1^h = \prod_{i=1}^{p}\left\{\prod_{g=1}^{q}\frac{\Gamma[\frac{1}{2}(n_g + hn_g + 1 - i)]}{\Gamma[\frac{1}{2}(n_g + 1 - i)]}\right\}\frac{\Gamma[\frac{1}{2}(n + 1 - i)]}{\Gamma[\frac{1}{2}(n + hn + 1 - i)]}.$$

We find the hth moment of V by setting $k = h$,

$$(5)\quad \mathscr{E}V^h = \prod_{i=1}^{p}\left\{\prod_{g=1}^{q}\frac{\Gamma[\frac{1}{2}(n_g + hn_g + 1 - i)]}{\Gamma[\frac{1}{2}(n_g + 1 - i)]}\right\}\frac{\Gamma[\frac{1}{2}(n + q - i)]}{\Gamma[\frac{1}{2}(n + hn + q - i)]}.$$

Since $0 \leq V_1 \leq 1$, $0 \leq V_2 \leq 1$, $0 \leq V \leq 1$, the moments determine the distributions uniquely. Thus in principle at least, $V_1(\alpha)$, $V_2(\alpha)$, and $V(\alpha)$ can be determined. It should be noticed that these distributions involve no nuisance parameters.

We see from (2) that

$$(6)\quad \mathscr{E}V_1^hV_2^k = \mathscr{E}V_1^h\mathscr{E}V_2^k.$$

Thus V_1 and V_2 are independently distributed. It is possible, therefore, to use V_1 to test H_1 followed by V_2 to test H_2. Suppose we want a test of H at level α. We can use V_1 and V_2 by choosing $V_1(\beta)$ and $V_2(\gamma)$ so that

$$(7)\quad (1 - \beta)(1 - \gamma) = 1 - \alpha,$$

that is,

$$(8)\quad \beta + \gamma - \beta\gamma = \alpha.$$

It seems reasonable to guess that by choosing β and γ satisfying (8), one can emphasize the mean or covariance part of H. However, it is difficult to put this precisely because we know so little about the power of the tests.

We summarize in the following theorem:

THEOREM 10.4.1. *Let V_1 be the criterion defined by* (10) *of Section* 10.2 *for testing the hypothesis that H_1: $\boldsymbol{\Sigma}_1 = \cdots = \boldsymbol{\Sigma}_q$, where A_g is n_g times the sample covariance matrix and $n_g + 1$ is the size of the sample from the* gth *population; let V_2 be the criterion defined by* (10) *of Section* 10.3 *for testing the hypothesis H_2: $\boldsymbol{\mu}_1 = \cdots = \boldsymbol{\mu}_q$, given H_1 is true, where $B = A + \sum_g N_g(\bar{x}^{(g)} - \bar{x})(\bar{x}^{(g)} - \bar{x})'$. When H_1 and H_2 are true, V_1 and V_2 are independently distributed. The* hth *moment of V_1 when H_1 is true is given by* (4). *The* kth *moment of $V = V_1V_2$, the criterion for testing H_1 and H_2 simultaneously, is given by* (5).

This theorem was first proved by Wilks (1932a).

If p is even, say $p = 2r$, we can use the duplication formula for the gamma-function $[\Gamma(\alpha + \frac{1}{2})\Gamma(\alpha + 1) = \sqrt{\pi}\Gamma(2\alpha + 1)2^{-2\alpha}]$. Then

$$(9)\quad \mathscr{E}V_1^h = \prod_{j=1}^{r}\left\{\left[\prod_{g=1}^{q}\frac{\Gamma(n_g + hn_g + 1 - 2j)}{\Gamma(n_g + 1 - 2j)}\right]\frac{\Gamma(n + 1 - 2j)}{\Gamma(n + hn + 1 - 2j)}\right\}$$

and

$$(10)\quad \mathscr{E}V^h = \prod_{j=1}^{r}\left\{\left[\prod_{g=1}^{q}\frac{\Gamma(n_g + hn_g + 1 - 2j)}{\Gamma(n_g + 1 - 2j)}\right]\frac{\Gamma(n + q - 2j)}{\Gamma(n + q + hn - 2j)}\right\}.$$

On the basis of these moments we can express V_1 and V_2 as products of variables $X^a(1 - X)^b$ where the X's are independently distributed with β-densities.* In principle these can be integrated to obtain the distributions of V_1 and V. In Section 10.6 we consider V_1 when $p = 2, q = 2$ (the case of $p = 1$, $q = 2$ being a function of an F-statistic). In other cases the integrals become unmanageable. To find probabilities we use the asymptotic expansion given in the next section. Box (1949) has given some other approximate distributions.

10.5. ASYMPTOTIC EXPANSIONS OF THE DISTRIBUTIONS OF THE CRITERIA

Again we make use of Theorem 8.6.1 to obtain asymptotic expansions of the distributions of V_1 and of V. We assume that $n_g = k_g n$, where $\sum k_g = 1$. The asymptotic expansion is in terms of n increasing with $k_1, \cdots, k_g$ fixed. (We could assume only $\lim n_g/n = k_g > 0$.)

The hth moment of

$$(1)\quad W_1 = V_1 \cdot \frac{n^{\frac{1}{2}pn}}{\prod_{g=1}^{q} n_g^{\frac{1}{2}pn_g}} = V_1 \cdot \prod_{g=1}^{q}\left(\frac{n}{n_g}\right)^{\frac{1}{2}pn_g} = \left[\prod_{g=1}^{q}\left(\frac{1}{k_g}\right)^{k_g}\right]^{\frac{1}{2}pn} V_1$$

is

$$(2)\quad \mathscr{E}W_1^h = K\left(\frac{\prod_{j=1}^{p}(\frac{1}{2}n)^{\frac{1}{2}n}}{\prod_{g=1}^{q}\prod_{i=1}^{p}(\frac{1}{2}n_g)^{\frac{1}{2}n_g}}\right)^h \frac{\prod_{g=1}^{q}\prod_{i=1}^{p}\Gamma[\frac{1}{2}n_g(1 + h) + \frac{1}{2}(1 - i)]}{\prod_{j=1}^{p}\Gamma[\frac{1}{2}n(1 + h) + \frac{1}{2}(1 - j)]}.$$

This is of the form of (1) of Section 8.6 with

$$b = p, \qquad y_j = \tfrac{1}{2}n, \qquad \eta_j = \tfrac{1}{2}(1 - j), \qquad j = 1, \cdots, p,$$

$$(3)\quad a = pq, \qquad x_k = \tfrac{1}{2}n_g, \qquad k = (g - 1)p + 1, \cdots, gp, \; g = 1, \cdots, q,$$

$$\xi_k = \tfrac{1}{2}(1 - i), \qquad k = i, p + i, \cdots, (q - 1)p + i, \; i = 1, \cdots, p.$$

* Wilks (1932a) has given some other integral representations.

Then

$$(4)\qquad \begin{aligned} f &= -2[\sum \xi_k - \sum \eta_j - \tfrac{1}{2}(a-b)] \\ &= -[q\sum_{i=1}^{p}(1-i) - \sum_{j=1}^{p}(1-j) - (qp-p)] \\ &= -[-q\tfrac{1}{2}p(p-1) + \tfrac{1}{2}p(p-1) - (q-1)p] \\ &= \tfrac{1}{2}(q-1)p(p+1), \end{aligned}$$

and $\varepsilon_j = \frac{1}{2}(1-\rho)n$ and $\beta_k = \frac{1}{2}(1-\rho)n_g = \frac{1}{2}(1-\rho)k_g n$ $(k = (g-1)p + 1, \cdots, gp)$.

In order to make the second term in the expansion vanish, we take ρ as

$$(5)\qquad \rho = 1 - \left(\sum \frac{1}{n_g} - \frac{1}{n}\right)\frac{2p^2+3p-1}{6(p+1)(q-1)}.$$

Then

$$(6)\qquad \omega_2 = \frac{p(p+1)\left[(p-1)(p+2)\left(\sum \frac{1}{n_g^2} - \frac{1}{n^2}\right) - 6(q-1)(1-\rho)^2\right]}{48\rho^2}.$$

Thus

$$(7)\qquad \Pr\{-2\rho \log W_1 \le z\} = \Pr\{\chi_f^2 \le z\} + \omega_2[\Pr\{\chi_{f+4}^2 \le z\} - \Pr\{\chi_f^2 \le z\}] + O(n^{-3}).$$

Let $W = Vn^{\frac{1}{2}pn}\prod_{g=1}^{q} n_g^{-\frac{1}{2}pn_g}$. The hth moment is

$$(8)\qquad \mathscr{E}W^h = K\left(\frac{\prod_{j=1}^{p}(\frac{1}{2}n)^{\frac{1}{2}n}}{\prod_{g=1}^{q}\prod_{i=1}^{p}(\frac{1}{2}n_g)^{\frac{1}{2}n_g}}\right)^h \frac{\prod_{g=1}^{q}\prod_{i=1}^{p}\Gamma[\frac{1}{2}n_g(1+h) + \frac{1}{2}(1-i)]}{\prod_{j=1}^{p}\Gamma[\frac{1}{2}n(1+h) + \frac{1}{2}(q-j)]}.$$

Here a, b, x_k, y_j, and ξ_k are the same as before, but $\eta_j = \frac{1}{2}(q-j)$. We find $f = \frac{1}{2}(q-1)p(p+3)$. For the second term in the expansion to vanish we choose ρ so that

$$(9)\qquad 1-\rho = \left(\sum \frac{1}{n_g} - \frac{1}{n}\right)\frac{2p^2+3p-1}{6(q-1)(p+3)} + \frac{1}{n}\cdot\frac{p-q+2}{p+3}.$$

Then

$$(10)\qquad \begin{aligned} \omega_2 = \frac{p}{288\rho^2}\Bigg[&6\left(\sum \frac{1}{n_g^2} - \frac{1}{n^2}\right)(p+1)(p-1)(p+2) \\ &- \left(\sum \frac{1}{n_g} - \frac{1}{n}\right)^2 \frac{(2p^2+3p-1)^2}{(q-1)(p+3)} - 12\left(\sum \frac{1}{n_g} - \frac{1}{n}\right)\frac{(2p^2+3p-1)(p-q+2)}{n(p+3)} \\ &- 36\frac{(q-1)(p-q+2)^2}{n^2(p+3)} - \frac{12(q-1)}{n^2}(-2q^2+7q+3pq-2p^2-6p-4)\Bigg]. \end{aligned}$$

The asymptotic expansion of the distribution of $-2\rho \log W$ is

$$(11)\quad \Pr\{-2\rho \log W \le z\} = \Pr\{\chi_f^2 \le z\} + \omega_2[\Pr\{\chi_{f+4}^2 \le z\} - \Pr\{\chi_f^2 \le z\}] + O(n^{-3}).$$

Box (1949) has considered the case of W_1 in considerable detail. In addition to this expansion he has considered the use of (13) of Section 8.6. He also has given an F-approximation.

As an example, we use one given by E. S. Pearson and Wilks (1933). The measurements are made on tensile strength (X_1) and hardness (X_2) of aluminum die-castings. There are 12 observations in each of five samples. The observed sums of squares and cross-products in the five samples are

$$(12)\quad \begin{aligned} A_1 &= \begin{pmatrix} 78.948 & 214.18 \\ 214.18 & 1247.18 \end{pmatrix}, \\ A_2 &= \begin{pmatrix} 223.695 & 657.62 \\ 657.62 & 2519.31 \end{pmatrix}, \\ A_3 &= \begin{pmatrix} 57.448 & 190.63 \\ 190.63 & 1241.78 \end{pmatrix}, \\ A_4 &= \begin{pmatrix} 187.618 & 375.91 \\ 375.91 & 1473.44 \end{pmatrix}, \\ A_5 &= \begin{pmatrix} 88.456 & 259.18 \\ 259.18 & 1171.73 \end{pmatrix}, \end{aligned}$$

and the sum of these is

$$(13)\quad \sum A_i = \begin{pmatrix} 636.165 & 1697.52 \\ 1697.52 & 7653.44 \end{pmatrix}.$$

Then $-\log W_1$ is 5.399. To use the asymptotic expansion we find $\rho = 152/165 = 0.9212$ and $\omega_2 = 0.0022$. Since ω_2 is small, we can consider $-2\rho \log W_1$ as χ^2 with 12 degrees of freedom. Our observed criterion is, therefore, clearly not significant.

10.6. THE CASE OF TWO POPULATIONS

In the case of two populations we can obtain some results which cannot be obtained more generally. If $p = 1$, the criteria are .

$$(1)\quad \begin{aligned} V_1 &= \frac{A_1^{\frac{1}{2}n_1} A_2^{\frac{1}{2}n_2}}{(A_1 + A_2)^{\frac{1}{2}(n_1+n_2)}} \\ &= \left(\frac{A_1}{A_2}\right)^{\frac{1}{2}n_1} \left(1 + \frac{A_1}{A_2}\right)^{-\frac{1}{2}(n_1+n_2)}, \end{aligned}$$

$$(2) \qquad V_2 = \left(\frac{A}{B}\right)^{\frac{1}{2}n}$$

$$= \left(1 + \frac{B - A}{A}\right)^{-\frac{1}{2}n}.$$

The ratios $(n_2/n_1)A_1/A_2$ and $(n_1 + n_2)(B - A)/A$ are distributed independently as F_{n_1, n_2} and F_{1, n_1+n_2} respectively.

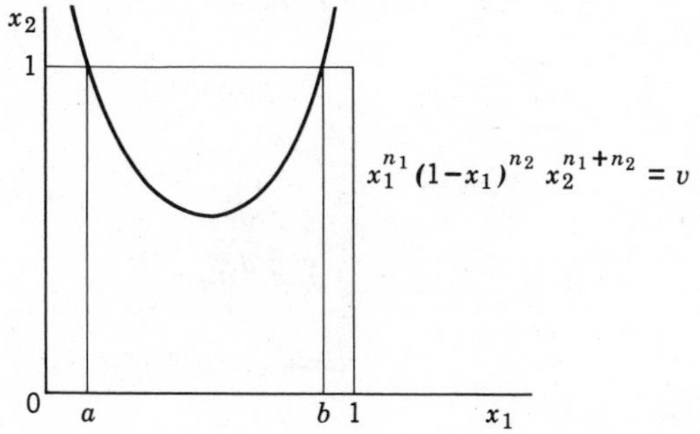

FIGURE 1

If $p = 2$, we can obtain a closed expression for the distribution of V_1. We have

$$(3) \quad \mathscr{E} V_1^h = \frac{\Gamma(n_1 + hn_1 - 1)\Gamma(n_2 + hn_2 - 1)\Gamma(n_1 + n_2 - 1)}{\Gamma(n_1 - 1)\Gamma(n_2 - 1)\Gamma[n_1 + n_2 + h(n_1 + n_2) - 1]}$$

$$= \left\{\frac{\Gamma(n_1 + hn_1 - 1)\Gamma(n_2 + hn_2 - 1)\Gamma(n_1 + n_2 - 2)}{\Gamma(n_1 - 1)\Gamma(n_2 - 1)\Gamma[n_1 + n_2 - 2 + h(n_1 + n_2)]}\right\}$$

$$\cdot \left\{\frac{\Gamma[n_1 + n_2 - 2 + h(n_1 + n_2)]\Gamma(n_1 + n_2 - 1)\Gamma(1)}{\Gamma(n_1 + n_2 - 2)\Gamma[n_1 + n_2 + h(n_1 + n_2) - 1]\Gamma(1)}\right\}$$

$$= \int_0^1 \int_0^1 [x_1^{n_1}(1 - x_1)^{n_2} x_2^{n_1+n_2}]^h \, x_1^{n_1-2}(1 - x_1)^{n_2-2} x_2^{n_1+n_2-3} \, dx_1 \, dx_2$$

$$\cdot B^{-1}(n_1 - 1, n_2 - 1) B^{-1}(n_1 + n_2 - 2, 1).$$

Thus V_1 is distributed as $X_1^{n_1}(1 - X_1)^{n_2} X_2^{n_1+n_2}$, where X_1 and X_2 are independently distributed accordingly to $\beta(x;\ n_1 - 1, n_2 - 1)$ and $\beta(x;\ n_1 + n_2 - 2, 1)$ respectively. Thus

$$(4) \qquad \Pr\{V_1 \le v\} = \Pr\{X_1^{n_1}(1 - X_1)^{n_2} X_2^{n_1+n_2} \le v\}$$

where

$$0 \le v \le \left(\frac{n_1}{n_1 + n_2}\right)^{n_1} \left(\frac{n_2}{n_1 + n_2}\right)^{n_2}.$$

Let $a \le b$ be the two roots of $x_1^{n_1}(1 - x_1)^{n_2} = v$; see Figure 1.

Then (4) is

$$
\begin{aligned}
(5)\quad \Pr\{V_1 \le v\} &= \Pr\{X_1 \le a\} + \Pr\{X_1 \ge b\} \\
&\quad + \Pr\{0 \le X_2^{n_1+n_2} \le v/[X_1^{n_1}(1-X_1)^{n_2}],\ a \le X_1 \le b\} \\
&= B^{-1}(n_1-1, n_2-1)\int_0^a x_1^{n_1-2}(1-x_1)^{n_2-2}\,dx_1 \\
&\quad + B^{-1}(n_1-1, n_2-1)(n_1+n_2-2)\int_a^b \int_0^{\{v/[x_1^{n_1}(1-x_1)^{n_2}]\}^{1/(n_1+n_2)}} \\
&\qquad x_1^{n_1-2}(1-x_1)^{n_2-2} x_2^{n_1+n_2-3}\,dx_2\,dx_1 \\
&\quad + B^{-1}(n_1-1, n_2-1)\int_b^1 x_1^{n_1-2}(1-x_1)^{n_2-2}\,dx_1 \\
&= I_a(n_1-1, n_2-1) + B^{-1}(n_1-1, n_2-1)v^{(n_1+n_2-2)/(n_1+n_2)} \\
&\qquad \cdot \int_a^b x_1^{2n_1/(n_1+n_2)-2}(1-x_1)^{2n_2/(n_1+n_2)-2}\,dx_1 \\
&\quad + 1 - I_b(n_1-1, n_2-1).
\end{aligned}
$$

In general, the integral above is not easy to evaluate.

If $n_1 = n_2 = m$, say, then $a = \frac{1}{2}(1 - \sqrt{1 - 4v^{1/m}})$, $b = \frac{1}{2}(1 + \sqrt{1 - 4v^{1/m}}) = 1 - a$. Thus $I_a(m-1, m-1) = 1 - I_b(m-1, m-1)$ and

$$
\begin{aligned}
(6)\quad \Pr\{V_1 \le v\} &= 2I_a(m-1, m-1) \\
&\quad + B^{-1}(m-1, m-1)v^{1-(1/m)}\int_a^b x_1^{-1}(1-x_1)^{-1}\,dx_1 \\
&= 2I_{\frac{1}{2}[1-\sqrt{1-4v^{(1/m)}}]}(m-1, m-1) \\
&\quad + 2B^{-1}(m-1, m-1)v^{1-(1/m)}\log\left(\frac{1+\sqrt{1-4v^{(1/m)}}}{1-\sqrt{1-4v^{(1/m)}}}\right).
\end{aligned}
$$

E. S. Pearson and Wilks (1933) have given this in another form.

For $p > 2$, the matter of integration is much more complicated.

We can learn something about the criterion by putting it in terms of roots of determinantal equations. We have

$$
\begin{aligned}
(7)\qquad V_1 &= \frac{|A_1|^{\frac{1}{2}n_1}|A_2|^{\frac{1}{2}n_2}}{|A_1 + A_2|^{\frac{1}{2}(n_1+n_2)}} \\
&= \prod_{i=1}^{p} \theta_i^{\frac{1}{2}n_1} \prod_{i=1}^{p}(1+\theta_i)^{-\frac{1}{2}(n_1+n_2)}
\end{aligned}
$$

where $\theta_1 \geq \theta_2 \geq \cdots \geq \theta_p$ are the roots of

$$|A_1 - \theta A_2| = 0. \tag{8}$$

This follows because (8) can be written as $|A_1 A_2^{-1} - \theta I| = 0$ or $|A_1 A_2^{-1} + I - (1 + \theta)I| = 0$ and therefore $\prod \theta_i = |A_1 A_2^{-1}|$ and $\prod(1 + \theta_i) = |A_1 A_2^{-1} + I|$.

It is clear that V_1 and ω of H_1 are invariant with regard to nonsingular transformations

$$Y = CX + \boldsymbol{\nu}. \tag{9}$$

We can choose C so that

$$\begin{aligned} \mathscr{C}(Y^{(2)}, Y^{(2)\prime}) &= C\boldsymbol{\Sigma}_2 C' = I, \\ \mathscr{C}(Y^{(1)}, Y^{(1)\prime}) &= C\boldsymbol{\Sigma}_1 C' = \boldsymbol{\Lambda}, \end{aligned} \tag{10}$$

where $\boldsymbol{\Lambda}$ is a diagonal matrix, the diagonal elements $\lambda_1 \geq \lambda_2 \geq \cdots \geq \lambda_p$ being the roots of

$$|\boldsymbol{\Sigma}_1 - \lambda \boldsymbol{\Sigma}_2| = 0. \tag{11}$$

Thus, the distribution of V_1 when H_1 is not true depends only on the p parameters $\lambda_1, \cdots, \lambda_p$. (Note that the means do not affect A_1 and A_2.)

The hypothesis H_1 is that $\lambda_1 = \cdots = \lambda_p = 1$. The criterion V_1 is in a way a measure of how close the θ_i are to n_1/n_2. Any other measure of how close the θ_i are to n_1/n_2 also yields a test of H_1. Another test, for example [suggested by Roy (1953)], is $\theta_1 \geq \theta_1^0$, $\theta_p \leq \theta_p^0$. Any test based on the θ's is invariant under transformations leaving the problem invariant.

For $q = 2$

$$V = \frac{|A_1|^{\frac{1}{2}n_1} |A_2|^{\frac{1}{2}n_2}}{\left| A_1 + A_2 + \dfrac{N_1 N_2}{N_1 + N_2} (\bar{\boldsymbol{x}}^{(1)} - \bar{\boldsymbol{x}}^{(2)})(\bar{\boldsymbol{x}}^{(1)} - \bar{\boldsymbol{x}}^{(2)})' \right|^{\frac{1}{2}(n_1+n_2)}}. \tag{12}$$

If we make the transformation (9), we find that when H is not true, the distribution of V depends on $\lambda_1, \cdots, \lambda_p$ and $C(\boldsymbol{\mu}^{(1)} - \boldsymbol{\mu}^{(2)})$. If some of the roots λ_i are equal, we can reduce the number of nonzero elements in $C(\boldsymbol{\mu}^{(1)} - \boldsymbol{\mu}^{(2)})$ to less than p.

10.7. TESTING THE HYPOTHESIS THAT A COVARIANCE MATRIX IS PROPORTIONAL TO A GIVEN MATRIX; THE SPHERICITY TEST

10.7.1. The Hypothesis

In many statistical analyses that are considered univariate, the assumption is made that a set of random variables are independent and have a common variance. In this section we consider a test of these assumptions based on repeated sets of observations.

More precisely, we use a sample of p-component vectors $x_1, \cdots, x_N$ from $N(\mu, \Sigma)$ to test the hypothesis H: $\Sigma = \sigma^2 I$, where σ^2 is not specified. The hypothesis can be given an algebraic interpretation in terms of the characteristic roots of Σ, that is, the roots of

$$|\Sigma - \phi I| = 0. \tag{1}$$

The hypothesis is true, if and only if, all the roots of (1) are equal.* Another way of putting it is that the arithmetic mean of roots $\phi_1, \cdots, \phi_p$ is equal to the geometric mean, that is,

$$\frac{\prod \phi_i^{1/p}}{\frac{\sum \phi_i}{p}} = \frac{|\Sigma|^{1/p}}{\frac{\operatorname{tr} \Sigma}{p}} = 1. \tag{2}$$

The lengths squared of the principal axes of the ellipsoids of constant density are proportional to the roots ϕ_i (see Chapter 11); the hypothesis specifies that these are equal, that is, that the ellipsoids are spheres.

The hypothesis H is equivalent to the more general form, $\Psi = \sigma^2 \Psi_0$, with Ψ_0 specified, having observation vectors $y_1, \cdots, y_N$ from $N(\nu, \Psi)$. Let C be a matrix such that

$$C\Psi_0 C' = I, \tag{3}$$

and let $\mu^* = C\nu$, $\Sigma^* = C\Psi C'$, $x_\alpha^* = Cy_\alpha$. Then $x_1^*, \cdots, x_N^*$ are observations from $N(\mu^*, \Sigma^*)$ and the hypothesis is transformed into H: $\Sigma^* = \sigma^2 I$.

10.7.2. The Criterion

In the canonical form the hypothesis H is a combination of the hypothesis H_1: Σ is diagonal or the components of X are independent and H_2: the diagonal elements of Σ are equal given Σ is diagonal or the variances of the components of X are equal given the components are independent. Thus by Lemma 10.3.1 the likelihood ratio criterion λ for H is the product of the criterion λ_1 for H_1 and λ_2 for H_2. From Section 9.2 we see that the criterion for H_1 is

$$\lambda_1 = \frac{|A|^{\frac{1}{2}N}}{\prod a_{ii}^{\frac{1}{2}N}} = |r_{ij}|^{\frac{1}{2}N}, \tag{4}$$

* This follows from the fact that $\Sigma = O'\Phi O$, where Φ is a diagonal matrix with roots as diagonal elements and O is an orthogonal matrix.

where

(5) $$A = \sum_{\alpha=1}^{N} (\boldsymbol{x}_\alpha - \bar{\boldsymbol{x}})(\boldsymbol{x}_\alpha - \bar{\boldsymbol{x}})' = (a_{ij})$$

and $r_{ij} = a_{ij}/\sqrt{a_{ii}a_{jj}}$. We use the results of Section 10.2 to obtain λ_2 by considering the ith component of $\boldsymbol{x}_\alpha$ as the αth observation from the ith population (p here is q in Section 10.2, N here is N_g there, pN here is N there). Thus

(6) $$\lambda_2 = \frac{\prod_i [\sum_\alpha (x_{i\alpha} - \bar{x}_i)^2]^{\frac{1}{2}N}}{[\sum_{i,\alpha} (x_{i\alpha} - \bar{x}_i)^2/p]^{\frac{1}{2}pN}}$$

$$= \frac{\prod a_{ii}^{\frac{1}{2}N}}{(\operatorname{tr} A/p)^{\frac{1}{2}pN}}.$$

Thus the criterion for H is

(7) $$\lambda = \lambda_1 \lambda_2 = \frac{|A|^{\frac{1}{2}N}}{(\operatorname{tr} A/p)^{\frac{1}{2}pN}}.$$

It will be observed that λ resembles (2). If $\theta_1, \cdots, \theta_p$ are the roots of

(8) $$|A - \theta I| = 0,$$

the criterion is a power of the ratio of the geometric mean and the arithmetic mean

(9) $$\lambda = \left(\frac{\prod \theta_i^{1/p}}{\sum \theta_i/p}\right)^{\frac{1}{2}pN}.$$

Now let us go back to the hypothesis $\boldsymbol{\Psi} = \sigma^2 \boldsymbol{\Psi}_0$, given observation vectors $\boldsymbol{y}_1, \cdots, \boldsymbol{y}_N$ from $N(\boldsymbol{\nu}, \boldsymbol{\Psi})$. In the transformed variables $\{\boldsymbol{x}_\alpha^*\}$ the criterion is $|A^*|^{\frac{1}{2}N}(\operatorname{tr} A^*/p)^{-\frac{1}{2}pN}$, where

(10) $$A^* = \sum_{\alpha=1}^{N} (\boldsymbol{x}_\alpha^* - \bar{\boldsymbol{x}}^*)(\boldsymbol{x}_\alpha^* - \bar{\boldsymbol{x}}^*)'$$

$$= C \sum_{\alpha=1}^{N} (\boldsymbol{y}_\alpha - \bar{\boldsymbol{y}})(\boldsymbol{y}_\alpha - \bar{\boldsymbol{y}})' C'$$

$$= CBC',$$

where

(11) $$B = \sum_{\alpha=1}^{N} (\boldsymbol{y}_\alpha - \bar{\boldsymbol{y}})(\boldsymbol{y}_\alpha - \bar{\boldsymbol{y}})'.$$

From (3) we have $\boldsymbol{\Psi}_0 = C^{-1}(C')^{-1} = (C'C)^{-1}$. Thus

$$|A^*| = \frac{|B|}{|\boldsymbol{\Psi}_0|} = |B\boldsymbol{\Psi}_0^{-1}|, \tag{12}$$

$$\begin{aligned} \operatorname{tr} A^* &= \operatorname{tr} CBC' = \operatorname{tr} BC'C \\ &= \operatorname{tr} B'\boldsymbol{\Psi}_0^{-1}. \end{aligned}$$

The results can be summarized.

THEOREM 10.7.1. *Given a set of p component observation vectors $y_1, \cdots, y_N$ from $N(\boldsymbol{\nu}, \boldsymbol{\Psi})$, the likelihood ratio criterion for testing the hypothesis H: $\boldsymbol{\Psi} = \sigma^2\boldsymbol{\Psi}_0$,* where *$\boldsymbol{\Psi}_0$ is specified and σ^2 is not specified, is*

$$\frac{|B\boldsymbol{\Psi}_0^{-1}|^{\frac{1}{2}N}}{(\operatorname{tr} B\boldsymbol{\Psi}_0^{-1}/p)^{\frac{1}{2}pN}}. \tag{13}$$

Mauchly (1940b) gives this criterion and its moments under the null hypothesis.

One can consider $\operatorname{tr} B\boldsymbol{\Psi}_0^{-1}/(pN)$ or $\operatorname{tr} B\boldsymbol{\Psi}_0^{-1}/[p(N-1)]$ as an estimate of σ^2 [Hotelling (1951)]. It follows from the preceding discussion that $\operatorname{tr} B\boldsymbol{\Psi}_0^{-1}$ has the χ^2-distribution with $p(N-1)$ degrees of freedom.

10.7.3. The Moments of the Criterion

As usual, we identify the distribution of λ by finding its moments. As was observed in Section 7.6, when $\boldsymbol{\Sigma}$ is diagonal the correlation coefficients $\{r_{ij}\}$ are distributed independently of the variances $\{a_{ii}/(N-1)\}$. Since λ_1 depends only on $\{r_{ij}\}$ and λ_2 depends only on $\{a_{ii}\}$, they are independently distributed when the null hypothesis is true, and therefore $\mathscr{E}\lambda^g = \mathscr{E}\lambda_1^g \mathscr{E}\lambda_2^g$. Let $W = \lambda^{2/N}$, $W_1 = \lambda_1^{2/N}$, $W_2 = \lambda_2^{2/N}$. From Section 9.3 we see that

$$\mathscr{E}W_1^h = \frac{\Gamma^p(\frac{1}{2}n)}{\Gamma^p(\frac{1}{2}n+h)} \prod_{i=1}^{p} \frac{\Gamma[\frac{1}{2}(n+1-i)+h]}{\Gamma[\frac{1}{2}(n+1-i)]}, \tag{14}$$

where* $n = N-1$. The hth moment of W_2 is obtained from Section 10.4 by letting p of that section be 1, n_g of that section be n of this, and q of that section be p of this. Then $W_2 = p^p V_1^{2/n}$. Thus

$$\mathscr{E}W_2^h = p^{hp} \frac{\Gamma^p(\frac{1}{2}n+h)\Gamma(\frac{1}{2}pn)}{\Gamma^p(\frac{1}{2}n)\Gamma(\frac{1}{2}pn+ph)}. \tag{15}$$

It follows that

$$\mathscr{E}W^h = p^{hp} \frac{\Gamma(\frac{1}{2}pn)}{\Gamma(\frac{1}{2}pn+ph)} \prod_{i=1}^{p} \frac{\Gamma[\frac{1}{2}(n+1-i)+h]}{\Gamma[\frac{1}{2}(n+1-i)]}. \tag{16}$$

* If the mean of $\boldsymbol{X}_\alpha$ is a regression on $\boldsymbol{z}_\alpha$, n is the number of degrees of freedom of $\boldsymbol{A}$.

It can be seen from the moments that W can be expressed as p^p times a polynomial in independent variables with β-distributions since W_1 and W_2 can be represented in such a fashion. For $p = 2$, we have a particularly simple case, for then

$$\mathscr{E}W^h = 4^h \frac{\Gamma(n)}{\Gamma(n+2h)} \prod_{i=1}^{2} \frac{\Gamma[\frac{1}{2}(n+1-i)+h]}{\Gamma[\frac{1}{2}(n+1-i)]} \tag{17}$$

$$= \frac{\Gamma(n)\Gamma(n-1+2h)}{\Gamma(n+2h)\Gamma(n-1)} = \frac{n-1}{n-1+2h}$$

$$= (n-1)\int_0^1 z^{n-2+2h}\,dz,$$

by use of the duplication formula for the gamma function. Thus W is distributed as Z^2 where Z has the density $(n-1)z^{n-2}$ and W has the density $\frac{1}{2}(n-1)w^{\frac{1}{2}(n-3)}$. The cdf is

$$\Pr\{W \le w\} = F(w) = w^{\frac{1}{2}(n-1)}. \tag{18}$$

This result can also be found from the joint distribution of θ_1, θ_2, the roots of (8).

10.7.4. Asymptotic Expansion of the Distribution

From (16) we see that the rth moment of $W^{\frac{1}{2}n} = Z$, say, is

$$\mathscr{E}Z^r = Kp^{\frac{1}{2}npr} \frac{\prod_{i=1}^{p} \Gamma[\frac{1}{2}n(1+r)+\frac{1}{2}(1-i)]}{\Gamma[\frac{1}{2}pn(1+r)]}. \tag{19}$$

This is of the form of (1), Section 8.6, with

$$a = p, \quad x_k = \tfrac{1}{2}n, \quad \xi_k = \tfrac{1}{2}(1-k), \quad k = 1, \cdots, p,$$
$$b = 1, \quad y_1 = \tfrac{1}{2}np, \quad \eta_1 = 0.$$

Thus the expansion of Section 8.6 is valid with $f = \frac{1}{2}p(p+1) - 1$. To make the second term in the expansion zero we take ρ so

$$1 - \rho = \frac{2p^2 + p + 2}{6pn}. \tag{20}$$

Then

$$\omega_2 = \frac{(p+2)(p-1)(p-2)(2p^3+6p^2+3p+2)}{288p^2n^2\rho^2}. \tag{21}$$

Thus the cdf of W is found from

$$\Pr\{-2\rho \log Z \le z\} = \Pr\{-n\rho \log W \le z\} \tag{22}$$
$$= \Pr\{\chi_f^2 \le z\} + \omega_2(\Pr\{\chi_{f+4}^2 \le z\} - \Pr\{\chi_f^2 \le z\}) + O(n^{-3}).$$

10.7.5. Confidence Regions

Given observations $\mathbf{y}_1, \cdots, \mathbf{y}_N$ from $N(\boldsymbol{\nu}, \boldsymbol{\Psi})$, we can test $\boldsymbol{\Psi} = \sigma^2\boldsymbol{\Psi}_0$ for any specified $\boldsymbol{\Psi}_0$. From this family of tests we can set up a confidence region for $\boldsymbol{\Psi}$. If any matrix is in the confidence region, all multiples of it are. This kind of confidence region is of interest if all components of $\mathbf{y}_\alpha$ are measured in the same unit, but the investigator wants a region independent of this common unit. The confidence region of confidence $1 - \varepsilon$ consists of all matrices $\boldsymbol{\Psi}^*$ satisfying

$$\frac{|B\boldsymbol{\Psi}^{*-1}|}{[(\operatorname{tr} B\boldsymbol{\Psi}^{*-1})/p]^p} \geq \lambda^{2/N}(\varepsilon), \tag{23}$$

where $\lambda(\varepsilon)$ is the ε significance level for the criterion.

Consider the case of $p = 2$. If the common unit of measurement is irrelevant, the investigator is interested in $\tau = \psi_{11}/\psi_{22}$ and $\rho = \psi_{12}/\sqrt{\psi_{11}\psi_{22}}$. In this case

$$\boldsymbol{\Psi}^{-1} = \frac{1}{\psi_{11}\psi_{22}(1-\rho^2)}\begin{pmatrix} \psi_{22} & -\rho\sqrt{\psi_{11}\psi_{22}} \\ -\rho\sqrt{\psi_{11}\psi_{22}} & \psi_{11} \end{pmatrix} \tag{24}$$

$$= \frac{1}{\psi_{11}(1-\rho^2)}\begin{pmatrix} 1 & -\rho\sqrt{\tau} \\ -\rho\sqrt{\tau} & \tau \end{pmatrix}.$$

The region in terms of τ and ρ is

$$4\frac{(b_{11}b_{22} - b_{12}^2)(1-\rho^2)\tau}{(b_{11} + \tau b_{22} - 2\rho\sqrt{\tau}b_{12})^2} \geq \lambda^{2/N}(\varepsilon) \tag{25}$$

Hickman (1953) has given an example of such a confidence region.

10.8. TESTING THE HYPOTHESIS THAT A COVARIANCE MATRIX IS EQUAL TO A GIVEN MATRIX

If Y is distributed according to $N(\boldsymbol{\nu}, \boldsymbol{\Psi})$, we wish to test H_1 that $\boldsymbol{\Psi} = \boldsymbol{\Psi}_0$, where $\boldsymbol{\Psi}_0$ is a given positive definite matrix. By the argument of the preceding section we see that this is equivalent to testing the hypothesis H_1: $\boldsymbol{\Sigma} = I$, where $\boldsymbol{\Sigma}$ is the covariance matrix of a vector X distributed according to $N(\boldsymbol{\mu}, \boldsymbol{\Sigma})$. Given a sample $\mathbf{x}_1, \cdots, \mathbf{x}_N$, the likelihood ratio criterion is

$$\lambda_1 = \frac{\max_{\boldsymbol{\mu}} L(\boldsymbol{\mu}, I)}{\max_{\boldsymbol{\mu}, \boldsymbol{\Sigma}} L(\boldsymbol{\mu}, \boldsymbol{\Sigma})}, \tag{1}$$

where the likelihood function is

$$L(\boldsymbol{\mu}, \boldsymbol{\Sigma}) = (2\pi)^{-\frac{1}{2}pN}|\boldsymbol{\Sigma}|^{-\frac{1}{2}N}e^{-\frac{1}{2}\sum_{\alpha}(\boldsymbol{x}_\alpha-\boldsymbol{\mu})'\boldsymbol{\Sigma}^{-1}(\boldsymbol{x}_\alpha-\boldsymbol{\mu})}. \tag{2}$$

Results in Chapter 3 show that

$$\lambda_1 = \frac{(2\pi)^{-\frac{1}{2}pN}e^{-\frac{1}{2}\sum_{\alpha}(\boldsymbol{x}_\alpha-\bar{\boldsymbol{x}})'(\boldsymbol{x}_\alpha-\bar{\boldsymbol{x}})}}{(2\pi)^{-\frac{1}{2}pN}\left|\frac{1}{N}A\right|^{-\frac{1}{2}N}e^{-\frac{1}{2}pN}} = \left(\frac{e}{N}\right)^{\frac{1}{2}pN}|A|^{\frac{1}{2}N}e^{-\frac{1}{2}\operatorname{tr}A}, \tag{3}$$

where

$$A = \sum_{\alpha}(\boldsymbol{x}_\alpha - \bar{\boldsymbol{x}})(\boldsymbol{x}_\alpha - \bar{\boldsymbol{x}})'. \tag{4}$$

In terms of the roots of

$$|A - \theta I| = 0, \tag{5}$$

the criterion is

$$\lambda_1 = \left(\frac{e}{N}\right)^{\frac{1}{2}pN}\prod_{i=1}^{p}\theta_i^{\frac{1}{2}N}e^{-\frac{1}{2}\Sigma\theta_i}. \tag{6}$$

Using the algebra of the previous section, we prove:

THEOREM 10.8.1. *Given $\boldsymbol{y}_1, \cdots, \boldsymbol{y}_N$ as observation vectors of p components from $N(\boldsymbol{\nu}, \boldsymbol{\Psi})$, the likelihood ratio criterion for testing the hypothesis H_1: $\boldsymbol{\Psi} = \boldsymbol{\Psi}_0$, where $\boldsymbol{\Psi}_0$ is specified, is*

$$\lambda_1 = \left(\frac{e}{N}\right)^{\frac{1}{2}pN}|B\boldsymbol{\Psi}_0^{-1}|^{\frac{1}{2}N}e^{-\frac{1}{2}\operatorname{tr}(B\boldsymbol{\Psi}_0^{-1})}, \tag{7}$$

where

$$B = \sum(\boldsymbol{y}_\alpha - \bar{\boldsymbol{y}})(\boldsymbol{y}_\alpha - \bar{\boldsymbol{y}})'. \tag{8}$$

The criterion is a function of the sum and the product of the roots. This is intuitively reasonable because the hypothesis can be stated as the sum of the roots of $\boldsymbol{\Sigma}$ being p and the product of the roots being 1.

Now let us find the hth moment of the criterion, under the null and the alternative hypotheses. Let the distribution of A be $W(\boldsymbol{\Sigma}, n)$ where $n = N - 1$. We wish to find

$$\begin{aligned} \mathscr{E}\lambda_1^h &= \int\cdots\int\left(\frac{e^{\frac{1}{2}pN}}{N^{\frac{1}{2}pN}}|A|^{\frac{1}{2}N}e^{-\frac{1}{2}\operatorname{tr}A}\right)^h w(A|\boldsymbol{\Sigma}, n)\,dA \\ &= \frac{e^{\frac{1}{2}phN}}{N^{\frac{1}{2}phN}}\int\cdots\int|A|^{\frac{1}{2}Nh}e^{-\frac{1}{2}h\operatorname{tr}A}w(A|\boldsymbol{\Sigma}, n)\,dA. \end{aligned} \tag{9}$$

Since

$$(10)\quad |A|^{\frac{1}{2}Nh}e^{-\frac{1}{2}h\operatorname{tr}A}w(A|\Sigma, n)$$

$$= \frac{|A|^{\frac{1}{2}(n+Nh-p-1)}e^{-\frac{1}{2}(\operatorname{tr}\Sigma^{-1}A+\operatorname{tr}hA)}}{2^{\frac{1}{2}pn}\pi^{\frac{1}{4}p(p-1)}|\Sigma|^{\frac{1}{2}n}\prod_{i=1}^{p}\Gamma[\frac{1}{2}(n+1-i)]}$$

$$= \frac{2^{\frac{1}{2}pNh}\prod\Gamma[\frac{1}{2}(n+Nh+1-i)]}{|\Sigma^{-1}+hI|^{\frac{1}{2}(n+Nh)}|\Sigma|^{\frac{1}{2}n}\prod\Gamma[\frac{1}{2}(n+1-i)]}$$

$$\cdot\frac{|\Sigma^{-1}+hI|^{\frac{1}{2}(n+Nh)}|A|^{\frac{1}{2}(n+Nh-p-1)}e^{-\frac{1}{2}\operatorname{tr}(\Sigma^{-1}+hI)A}}{2^{\frac{1}{2}p(n+Nh)}\pi^{\frac{1}{4}p(p-1)}\prod_{i=1}^{p}\Gamma[\frac{1}{2}(n+Nh+1-i)]}$$

$$= \frac{2^{\frac{1}{2}pNh}|\Sigma|^{\frac{1}{2}Nh}\prod\Gamma[\frac{1}{2}(n+Nh+1-i)]}{|I+h\Sigma|^{\frac{1}{2}(n+Nh)}\prod\Gamma[\frac{1}{2}(n+1-i)]}\, w(A|(\Sigma^{-1}+hI)^{-1}, n+Nh),$$

the hth moment of λ_1 is

$$(11)\qquad \mathscr{E}\lambda_1^h = \left(\frac{2e}{N}\right)^{\frac{1}{2}phN}\frac{|\Sigma|^{\frac{1}{2}Nh}\prod\Gamma[\frac{1}{2}(n+Nh+1-i)]}{|I+h\Sigma|^{\frac{1}{2}(n+Nh)}\prod\Gamma[\frac{1}{2}(n+1-i)]}.$$

It can be shown that the characteristic function of $-2\log\lambda_1$ is

$$(12)\quad \mathscr{E}e^{-2it\log\lambda_1} = \mathscr{E}\lambda_1^{-2it}$$

$$= \left(\frac{2e}{N}\right)^{-ipNt}\frac{|\Sigma|^{-iNt}}{|I-2it\Sigma|^{\frac{1}{2}n-iNt}}\cdot\prod_{j=1}^{p}\frac{\Gamma[\frac{1}{2}(N-j)-iNt]}{\Gamma[\frac{1}{2}(N-j)]}.$$

When the null hypothesis is true, $\Sigma = I$, and

$$(13)\quad \mathscr{E}e^{-2it\log\lambda_1} = \left(\frac{2e}{N}\right)^{-ipNt}(1-2it)^{-\frac{1}{2}p(n-2iNt)}\prod_j\frac{\Gamma[\frac{1}{2}(N-j)-iNt]}{\Gamma[\frac{1}{2}(N-j)]}.$$

This characteristic function is the product of p terms such as

$$(14)\qquad \phi_j(t) = \left(\frac{2e}{N}\right)^{-iNt}(1-2it)^{-\frac{1}{2}(n-2iNt)}\frac{\Gamma[\frac{1}{2}(N-j)-iNt]}{\Gamma[\frac{1}{2}(N-j)]}.$$

Thus $-2\log\lambda_1$ is distributed as the sum of p independent variates, the characteristic function of the jth being (14). Using Stirling's approximation for the gamma function and letting $n = N - 1$, we have

$$(15)\quad \phi_j(t) \sim 2^{-iNt}e^{-iNt}N^{iNt}(1-2it)^{\frac{1}{2}(2iNt-N+1)} \cdot \frac{e^{-[\frac{1}{2}(N-j)-iNt]}[\frac{1}{2}(N-j-2)-iNt]^{\frac{1}{2}(N-j-1)-iNt}}{e^{-[\frac{1}{2}(N-j)]}[\frac{1}{2}(N-j-2)]^{\frac{1}{2}(N-j-1)}}$$

$$= (1-2it)^{-\frac{1}{2}j}\left(1 - \frac{it(j+2)}{\frac{1}{2}(N-j-2)(1-2it)}\right)^{\frac{1}{2}(N-j)-\frac{1}{2}} \cdot \left(1 - \frac{it(j+2)}{itN(1-2it)}\right)^{-iNt}.$$

As $N \to \infty$, $\phi_j(t) \to (1-2it)^{-\frac{1}{2}j}$, the characteristic function of χ_j^2, χ^2 with j degrees of freedom. Thus $-2\log\lambda_1$ is asymptotically distributed as $\sum_{j=1}^{p}\chi_j^2$, which is χ^2 with $\sum j = \frac{1}{2}p(p+1)$ degrees of freedom.

10.9. TESTING THE HYPOTHESIS THAT A MEAN VECTOR AND A COVARIANCE MATRIX ARE EQUAL TO A GIVEN VECTOR AND MATRIX

In Chapter 3 we pointed out that if $\boldsymbol{\Psi}$ is known, $(\bar{\boldsymbol{y}} - \boldsymbol{\nu}_0)'\boldsymbol{\Psi}_0^{-1}(\bar{\boldsymbol{y}} - \boldsymbol{\nu}_0)$ is suitable for testing

$$(1)\qquad H_2:\ \boldsymbol{\nu} = \boldsymbol{\nu}_0, \qquad \text{given } \boldsymbol{\Psi} = \boldsymbol{\Psi}_0.$$

Now let us combine H_1 of Section 10.8 and H_2 and test

$$(2)\qquad H:\ \boldsymbol{\nu} = \boldsymbol{\nu}_0, \qquad \boldsymbol{\Psi} = \boldsymbol{\Psi}_0,$$

on the basis of a sample $\boldsymbol{y}_1, \cdots, \boldsymbol{y}_N$ from $N(\boldsymbol{\nu}, \boldsymbol{\Psi})$.

Let

$$(3)\qquad \boldsymbol{X} = \boldsymbol{C}(\boldsymbol{Y} - \boldsymbol{\nu}_0),$$

where

$$(4)\qquad \boldsymbol{C}\boldsymbol{\Psi}_0\boldsymbol{C}' = \boldsymbol{I}.$$

Then $\boldsymbol{x}_1, \cdots, \boldsymbol{x}_N$ constitutes a sample from $N(\boldsymbol{\mu}, \boldsymbol{\Sigma})$ and the hypothesis is

$$(5)\qquad H:\ \boldsymbol{\mu} = \boldsymbol{0}, \qquad \boldsymbol{\Sigma} = \boldsymbol{I}.$$

The likelihood ratio criterion for H_2: $\boldsymbol{\mu} = \boldsymbol{0}$, given $\boldsymbol{\Sigma} = \boldsymbol{I}$, is

$$(6)\qquad \lambda_2 = e^{-\frac{1}{2}N\bar{\boldsymbol{x}}'\bar{\boldsymbol{x}}}.$$

The likelihood ratio criterion for H is (by Lemma 10.3.1)

$$\lambda = \lambda_1 \lambda_2 = \left(\frac{e}{N}\right)^{\frac{1}{2}pN} |A|^{\frac{1}{2}N} e^{-\frac{1}{2}\operatorname{tr} A} e^{-\frac{1}{2}N\bar{x}'\bar{x}} \tag{7}$$

$$= \left(\frac{e}{N}\right)^{\frac{1}{2}pN} |A|^{\frac{1}{2}N} e^{-\frac{1}{2}\operatorname{tr}(A+N\bar{x}\bar{x}')}$$

$$= \left(\frac{e}{N}\right)^{\frac{1}{2}pN} |A|^{\frac{1}{2}N} e^{-\frac{1}{2}\Sigma x_\alpha' x_\alpha}.$$

The two factors λ_1 and λ_2 are independent because λ_1 is a function of A and λ_2 is a function of $\bar{x}$, and A and $\bar{x}$ are independent. Since

$$\mathscr{E}\lambda_2^h = \mathscr{E}e^{-\frac{1}{2}hN\Sigma\bar{x}_i^2} = \mathscr{E}e^{-\frac{1}{2}h\chi_p^2} \tag{8}$$

$$= (1+h)^{-\frac{1}{2}p},$$

the hth moment of λ is

$$\mathscr{E}\lambda^h = \mathscr{E}\lambda_1^h \mathscr{E}\lambda_2^h = \left(\frac{2e}{N}\right)^{\frac{1}{2}pNh} \frac{1}{(1+h)^{\frac{1}{2}pN(1+h)}} \prod_{i=1}^{p} \frac{\Gamma[\frac{1}{2}(N-i+Nh)]}{\Gamma[\frac{1}{2}(N-i)]} \tag{9}$$

under the null hypothesis. Clearly

$$-2\log\lambda = -2\log\lambda_1 - 2\log\lambda_2 \tag{10}$$

has asymptotically the χ^2-distribution with $p(p+1)/2 + p$ degrees of freedom.

Now let us return to the observations $y_1, \cdots, y_N$. Then

$$\sum_\alpha x_\alpha' x_\alpha = \sum_\alpha (y_\alpha - \nu_0)' C' C (y_\alpha - \nu_0) \tag{11}$$

$$= \sum_\alpha (y_\alpha - \nu_0)' \Psi_0^{-1} (y_\alpha - \nu_0)$$

$$= \operatorname{tr} A + N\bar{x}'\bar{x}$$

$$= \operatorname{tr}(B\Psi_0^{-1}) + N(\bar{y} - \nu_0)' \Psi_0^{-1} (\bar{y} - \nu_0)$$

and

$$|A| = |B\Psi_0^{-1}|. \tag{12}$$

THEOREM 10.9.1. *Given the p-component observation vectors* $y_1, \cdots, y_N$ *from* $N(\nu, \Psi)$, *the likelihood ratio criterion for testing the hypothesis* H: $\nu = \nu_0$, $\Psi = \Psi_0$, *is*

$$\lambda = \left(\frac{e}{N}\right)^{\frac{1}{2}pN} |B\Psi_0^{-1}|^{\frac{1}{2}N} e^{-\frac{1}{2}[\operatorname{tr} B\Psi_0 + N(\bar{y}-\nu_0)'\Psi_0^{-1}(\bar{y}-\nu_0)]}. \tag{13}$$

When the null hypothesis is true, $-2\log\lambda$ *is asymptotically distributed as* χ^2 *with* $\frac{1}{2}p(p+1) + p$ *degrees of freedom.*

REFERENCES

Sections 10.2–10.6. Bartlett (1937a); Bishop (1939); Box (1949); Brown (1939); Nair (1939); E. S. Pearson and Wilks (1933); Pillai (1955); Plackett (1947); S. N. Roy (1952a), (1953); Scheffé (1942); Wilks (1932a).

Section 10.7. Girshick (1941); Hickman (1953); Hotelling (1951); Ihm (1955); Mauchly (1940a), (1940b).

PROBLEMS

1. (Sec. 10.2) Sums of squares and cross-products of deviations from the means of four measurements are given below (see Sec. 5.3). The populations are Iris versicolor (1), Iris setosa (2), and Iris virginica (3); each sample consists of 50 observations.

$$A_1 = \begin{pmatrix} 13.0552 & 4.1740 & 8.9620 & 2.7332 \\ 4.1740 & 4.8250 & 4.0500 & 2.0190 \\ 8.9620 & 4.0500 & 10.8200 & 3.5820 \\ 2.7332 & 2.0190 & 3.5820 & 1.9162 \end{pmatrix},$$

$$A_2 = \begin{pmatrix} 6.0882 & 4.8616 & 0.8014 & 0.5062 \\ 4.8616 & 7.0408 & 0.5732 & 0.4556 \\ 0.8014 & 0.5732 & 1.4778 & 0.2974 \\ 0.5062 & 0.4556 & 0.2974 & 0.5442 \end{pmatrix},$$

$$A_3 = \begin{pmatrix} 19.8128 & 4.5944 & 14.8612 & 2.4056 \\ 4.5944 & 5.0962 & 3.4976 & 2.3338 \\ 14.8612 & 3.4976 & 14.9248 & 2.3924 \\ 2.4056 & 2.3338 & 2.3924 & 3.6962 \end{pmatrix}.$$

(*a*) Test the hypothesis $\boldsymbol{\Sigma}_1 = \boldsymbol{\Sigma}_2$ at the 5% significance level.

(*b*) Test the hypothesis $\boldsymbol{\Sigma}_1 = \boldsymbol{\Sigma}_2 = \boldsymbol{\Sigma}_3$ at the 5% significance level.

2. (Sec. 10.2) Let $x_\alpha^{(g)}$, $\alpha = 1, \cdots, N$, be observations from $N(\boldsymbol{\mu}^{(g)}, \boldsymbol{\Sigma}_g)$, $g = 1, 2, 3$ ($\boldsymbol{\Sigma}_g$ of order p). Give a criterion for testing the hypothesis $\boldsymbol{\Sigma}_1 = \boldsymbol{\Sigma}_2$; give a criterion for testing the hypothesis that $\boldsymbol{\Sigma}_3$ is the same as $\boldsymbol{\Sigma}_1$ and $\boldsymbol{\Sigma}_2$, when $\boldsymbol{\Sigma}_1 = \boldsymbol{\Sigma}_2$ is assumed; and verify that these two criteria are statistically independent when $\boldsymbol{\Sigma}_1 = \boldsymbol{\Sigma}_2 = \boldsymbol{\Sigma}_3$.

3. (Sec. 10.2) (*a*) Let $\boldsymbol{Y}^{(g)}$ $(g = 1, \cdots, q)$ be a set of random vectors each with p components. Suppose

$$\mathscr{E}\boldsymbol{Y}^{(g)} = \boldsymbol{0}, \qquad \mathscr{E}\boldsymbol{Y}^{(g)}\boldsymbol{Y}^{(h)\prime} = \delta_{gh}\boldsymbol{\Sigma}_g.$$

Let $\boldsymbol{C}$ be an orthogonal matrix of order q such that each element of the last row is

$$c_{qh} = 1/\sqrt{q}.$$

Define

$$\boldsymbol{Z}^{(g)} = \sum_{h=1}^{q} c_{gh}\boldsymbol{Y}^{(h)}, \qquad g = 1, \cdots, q.$$

Show that

$$\mathscr{E}\boldsymbol{Z}^{(q)}\boldsymbol{Z}^{(g)\prime} = \boldsymbol{0}, \qquad g = 1, \cdots, q-1,$$

if and only if

$$\boldsymbol{\Sigma}_1 = \boldsymbol{\Sigma}_2 = \cdots = \boldsymbol{\Sigma}_q.$$

(*b*) Let $\boldsymbol{X}_\alpha^{(g)}$ $(\alpha = 1, \cdots, N)$ be a random sample from $N(\boldsymbol{\mu}^{(g)}, \boldsymbol{\Sigma}_g)$ $(g = 1, \cdots, q)$. Use the result from (*a*) to construct a test of the hypothesis

$$H\colon \boldsymbol{\Sigma}_1 = \cdots = \boldsymbol{\Sigma}_q,$$

based on a test of independence of $\boldsymbol{Z}^{(q)}$ and the set $\boldsymbol{Z}^{(1)}, \cdots, \boldsymbol{Z}^{(q-1)}$. Find the exact distribution of the criterion for the case $p = 2$.

4. (Sec. 10.2) Show that H_1 and H_2 are invariant under transformations $\boldsymbol{X}^{*(g)} = \boldsymbol{C}\boldsymbol{X}^{(g)} + \boldsymbol{c}$. Verify that V_1 and V_2 are invariant under transformations $\boldsymbol{x}_\alpha^{*(g)} = \boldsymbol{C}\boldsymbol{x}_\alpha^{(g)} + \boldsymbol{c}$.

5. (Sec. 10.4) Express V_1 and V_2 as products of variables $X^a(1 - X)^b$, where the X's are independently distributed with β-densities.

6. (Sec. 10.6) Let $\boldsymbol{x}_1^{(\nu)}, \cdots, \boldsymbol{x}_N^{(\nu)}$ be observations from $N(\boldsymbol{\mu}^{(\nu)}, \boldsymbol{\Sigma}_\nu)$, $\nu = 1, 2$, and let $\boldsymbol{A}_\nu = \sum(\boldsymbol{x}_\alpha^{(\nu)} - \bar{\boldsymbol{x}}^{(\nu)})(\boldsymbol{x}_\alpha^{(\nu)} - \bar{\boldsymbol{x}}^{(\nu)})'$.

(*a*) Prove that the likelihood ratio test for H: $\boldsymbol{\Sigma}_1 = \boldsymbol{\Sigma}_2$ is equivalent to rejecting H if

$$T = \frac{|\boldsymbol{A}_1| \cdot |\boldsymbol{A}_2|}{|\boldsymbol{A}_1 + \boldsymbol{A}_2|^2} \leq C.$$

(*b*) Let $d_1^2, d_2^2, \cdots, d_p^2$ be the roots of $|\boldsymbol{\Sigma}_1 - \lambda\boldsymbol{\Sigma}_2| = 0$, and let

$$\boldsymbol{D} = \begin{pmatrix} d_1 & 0 \cdots & 0 \\ 0 & d_2 \cdots & 0 \\ \cdot & \cdot & \cdot \\ \cdot & \cdot & \cdot \\ \cdot & \cdot & \cdot \\ 0 & 0 \cdots & d_p \end{pmatrix}.$$

Show that T is distributed as $|\boldsymbol{B}_1| \cdot |\boldsymbol{B}_2| / |\boldsymbol{B}_1 + \boldsymbol{B}_2|^2$, where $\boldsymbol{B}_1$ is distributed according to $W(\boldsymbol{D}^2, N - 1)$ and $\boldsymbol{B}_2$ is distributed according to $W(\boldsymbol{I}, N - 1)$. Show that T is distributed as $|\boldsymbol{D}\boldsymbol{C}_1\boldsymbol{D}| \cdot |\boldsymbol{C}_2| / |\boldsymbol{D}\boldsymbol{C}_1\boldsymbol{D} + \boldsymbol{C}_2|^2$, where $\boldsymbol{C}_i$ is distributed according to $W(\boldsymbol{I}, N - 1)$.

7. (Sec. 10.7) Consider the test that $\boldsymbol{\Sigma} = \sigma^2\boldsymbol{I}$, where σ^2 is unspecified.

(*a*) What group of transformations of $\boldsymbol{x}$ leaves the null hypothesis invariant and leaves the set of alternative hypotheses invariant? Give the largest group.

(*b*) The power function of the test depends on functions of $\boldsymbol{\mu}$ and $\boldsymbol{\Sigma}$. Give the minimum number of such functions.

8. (Sec. 10.7) Find the distribution of W for $p = 2$ under the null hypothesis (*a*) directly from the distribution of $\boldsymbol{A}$ and (*b*) from the distribution of the characteristic roots (Chapter 13).

9. (Sec. 10.7) Let $\boldsymbol{x}_1, \cdots, \boldsymbol{x}_N$ be a sample from $N(\boldsymbol{\mu}, \boldsymbol{\Sigma})$. What is the likelihood ratio criterion for testing the hypothesis $\boldsymbol{\mu} = k\boldsymbol{\mu}_0$, $\boldsymbol{\Sigma} = k^2\boldsymbol{\Sigma}_0$, where $\boldsymbol{\mu}_0$ and $\boldsymbol{\Sigma}_0$ are specified and k is unspecified?

10. (Sec. 10.7) Let $\boldsymbol{x}_1^{(1)}, \cdots, \boldsymbol{x}_{N_1}^{(1)}$ be a sample from $N(\boldsymbol{\mu}^{(1)}, \boldsymbol{\Sigma}_1)$ and $\boldsymbol{x}_1^{(2)}, \cdots, \boldsymbol{x}_{N_2}^{(2)}$ be a sample from $N(\boldsymbol{\mu}^{(2)}, \boldsymbol{\Sigma}_2)$. What is the likelihood ratio criterion for testing the hypothesis that $\boldsymbol{\Sigma}_1 = k^2\boldsymbol{\Sigma}_2$, where k is unspecified? What is the likelihood ratio criterion for testing the hypothesis that $\boldsymbol{\mu}^{(1)} = k\boldsymbol{\mu}^{(2)}$ and $\boldsymbol{\Sigma}_1 = k^2\boldsymbol{\Sigma}_2$, where k is unspecified?

11. (Sec. 10.7) Let $\boldsymbol{x}_\alpha$ (of p components, $\alpha = 1, \cdots, N$) be observations from $N(\boldsymbol{\mu}, \boldsymbol{\Sigma})$. We define the following hypotheses

$$H\colon\ \boldsymbol{\mu} = \boldsymbol{0}, \qquad \boldsymbol{\Sigma} = k^2\boldsymbol{\Sigma}_0,$$
$$H_1\colon\ \boldsymbol{\Sigma} = k^2\boldsymbol{\Sigma}_0,$$
$$H_2\colon\ \boldsymbol{\mu} = \boldsymbol{0}, \qquad \text{given that } \boldsymbol{\Sigma} = k^2\boldsymbol{\Sigma}_0.$$

In each case k^2 is unspecified, but $\mathbf{\Sigma}_0$ is specified. Find the likelihood ratio criterion λ_2 for testing H_2. Give the asymptotic distribution of $-2 \log \lambda_2$ under H_2. Obtain the exact distribution of a suitable monotonic function of λ_2 under H_2.

12. (Sec. 10.7) Find the likelihood ratio criterion λ for testing H of Problem 11 (given $\boldsymbol{x}_1, \cdots, \boldsymbol{x}_N$). What is the asymptotic distribution of $-2 \log \lambda$ under H?

13. (Sec. 10.7) Show that $\lambda = \lambda_1 \lambda_2$, where λ is defined in Problem 12, λ_2 is defined in Problem 11, and λ_1 is the likelihood ratio criterion for H_1 in Problem 11. Are λ_1 and λ_2 independently distributed under H? Prove your answer.

14. (Sec. 10.7) Verify that $\operatorname{tr} \boldsymbol{B} \boldsymbol{\Psi}_0^{-1}$ has the χ^2-distribution with $p(N-1)$ degrees of freedom.

CHAPTER 11

Principal Components

11.1. INTRODUCTION

Principal components are linear combinations of random or statistical variables which have special properties in terms of variances. For example, the first principal component is the normalized linear combination (that is, the sum of squares of the coefficients being one) with maximum variance. In effect, transforming the original vector variable to the vector of principal components amounts to a rotation of coordinate axes to a new coordinate system that has inherent statistical properties. This choosing of a coordinate system is to be contrasted with the many problems treated previously where the coordinate system is irrelevant.

The principal components turn out to be the characteristic vectors of the covariance matrix. Thus the study of principal components can be considered as putting into statistical terms the usual developments of characteristic roots and vectors (for positive semidefinite matrices).

From the point of view of statistical theory, the set of principal components yields a convenient set of coordinates, and the accompanying variances of the components characterize their statistical properties. In statistical practice, the method of principal components is used to find the linear combinations with large variance. In many exploratory studies the number of variables under consideration is too large to handle. Since it is the deviations in these studies which are of interest, a way of reducing the number of variables to be treated is to discard the linear combinations which have small variances and study only those with large variances. For example, a physical anthropologist may make dozens of measurements of lengths and breadths of each of a number of individuals, such measurements as ear length, ear breadth, facial length, facial breadth, and so forth. He may be interested in describing and analyzing how individuals differ in these kinds of physiological characteristics. Eventually he will want to "explain" these differences, but first he wants to know what measurements or combinations of measurements show considerable variation; that is, which should need further study. The principal components give a new set of linearly combined measurements. It may be that most of the variation from individual to individual resides in three linear combinations; then the anthropologist can direct his study to these three quantities; the other linear combinations vary so little from one person to the next that study of them will tell little of individual variation.

Hotelling, who developed many of these ideas, has given a rather thorough discussion (1933).

11.2. DEFINITION OF PRINCIPAL COMPONENTS IN THE POPULATION

Suppose the random vector X of p components has the covariance matrix $\mathbf{\Sigma}$. Since we shall be interested only in variances and covariances in this chapter, we shall assume that the mean vector is $\mathbf{0}$. Moreover, in developing the ideas and algebra here, the actual distribution of X is irrelevant except for the covariance matrix; however, if X is normally distributed, more meaning can be given the principal components.

In the following treatment we shall not use the usual theory of characteristic roots and vectors; as a matter of fact, that theory will be implicitly derived. The treatment will include the cases where $\mathbf{\Sigma}$ is singular (that is, positive semidefinite) and where $\mathbf{\Sigma}$ has multiple roots.

Let $\boldsymbol{\beta}$ be a p-component column vector such that $\boldsymbol{\beta}'\boldsymbol{\beta} = 1$. The variance of $\boldsymbol{\beta}'X$ is

$$\mathscr{E}(\boldsymbol{\beta}'X)^2 = \mathscr{E}\boldsymbol{\beta}'XX'\boldsymbol{\beta} = \boldsymbol{\beta}'\mathbf{\Sigma}\boldsymbol{\beta}. \tag{1}$$

To determine the normalized linear combination $\boldsymbol{\beta}'X$ with maximum variance, we must find a vector $\boldsymbol{\beta}$ satisfying $\boldsymbol{\beta}'\boldsymbol{\beta} = 1$ which maximizes (1). Let

$$\phi = \boldsymbol{\beta}'\mathbf{\Sigma}\boldsymbol{\beta} - \lambda(\boldsymbol{\beta}'\boldsymbol{\beta} - 1) = \sum_{i,j} \beta_i\sigma_{ij}\beta_j - \lambda(\sum_i \beta_i^2 - 1), \tag{2}$$

where λ is a Lagrange multiplier. The vector of partial derivatives $(\partial\phi/\partial\beta_i)$ is

$$\frac{\partial\phi}{\partial\boldsymbol{\beta}} = 2\mathbf{\Sigma}\boldsymbol{\beta} - 2\lambda\boldsymbol{\beta} \tag{3}$$

(by Theorem 8 of Appendix 1). Since $\boldsymbol{\beta}'\mathbf{\Sigma}\boldsymbol{\beta}$ and $\boldsymbol{\beta}'\boldsymbol{\beta}$ have derivatives everywhere in a region containing $\boldsymbol{\beta}'\boldsymbol{\beta} = 1$, a vector $\boldsymbol{\beta}$ maximizing $\boldsymbol{\beta}'\mathbf{\Sigma}\boldsymbol{\beta}$ must satisfy the expression (3) set equal to $\mathbf{0}$; that is

$$(\mathbf{\Sigma} - \lambda I)\boldsymbol{\beta} = \mathbf{0}. \tag{4}$$

In order to get a solution of (4) with $\boldsymbol{\beta}'\boldsymbol{\beta} = 1$ we must have $\mathbf{\Sigma} - \lambda I$ singular; in other words, λ must satisfy

$$|\mathbf{\Sigma} - \lambda I| = 0. \tag{5}$$

The function $|\boldsymbol{\Sigma} - \lambda \boldsymbol{I}|$ is a polynomial in λ of degree p. Therefore (5) has p roots; let these be $\lambda_1 \geq \lambda_2 \geq \cdots \geq \lambda_p$. ($\boldsymbol{\beta}'$ complex conjugate in (6) proves λ real.) If we multiply (4) on the left by $\boldsymbol{\beta}'$ we obtain

$$\boldsymbol{\beta}'\boldsymbol{\Sigma}\boldsymbol{\beta} = \lambda\boldsymbol{\beta}'\boldsymbol{\beta} = \lambda. \tag{6}$$

This shows that if $\boldsymbol{\beta}$ satisfies (4) (and $\boldsymbol{\beta}'\boldsymbol{\beta} = 1$), then the variance of $\boldsymbol{\beta}'X$ [given by (1)] is λ. Thus for the maximum variance we should use in (4) the largest root λ_1. Let $\boldsymbol{\beta}^{(1)}$ be a normalized solution of $(\boldsymbol{\Sigma} - \lambda_1 \boldsymbol{I})\boldsymbol{\beta} = \mathbf{0}$. Then $U_1 = \boldsymbol{\beta}^{(1)\prime}X$ is a normalized linear combination with maximum variance. [If $\boldsymbol{\Sigma} - \lambda_1 \boldsymbol{I}$ is of rank $p - 1$, then there is only one solution to $(\boldsymbol{\Sigma} - \lambda_1 \boldsymbol{I})\boldsymbol{\beta} = \mathbf{0}$ and $\boldsymbol{\beta}'\boldsymbol{\beta} = 1$.]

Now let us find a normalized combination $\boldsymbol{\beta}'X$ that has maximum variance of all linear combinations uncorrelated with U_1. Lack of correlation is

$$\begin{aligned} 0 &= \mathscr{E}\boldsymbol{\beta}'XU_1 = \mathscr{E}\boldsymbol{\beta}'XX'\boldsymbol{\beta}^{(1)} \\ &= \boldsymbol{\beta}'\boldsymbol{\Sigma}\boldsymbol{\beta}^{(1)} = \lambda_1\boldsymbol{\beta}'\boldsymbol{\beta}^{(1)}, \end{aligned} \tag{7}$$

since $\boldsymbol{\Sigma}\boldsymbol{\beta}^{(1)} = \lambda_1\boldsymbol{\beta}^{(1)}$. Thus $\boldsymbol{\beta}'X$ is orthogonal to U in both the statistical sense (of lack of correlation) and in the geometric sense (of the inner product of the vectors $\boldsymbol{\beta}$ and $\boldsymbol{\beta}^{(1)}$ being zero). (That is, $\lambda_1\boldsymbol{\beta}'\boldsymbol{\beta}^{(1)} = 0$ only if $\boldsymbol{\beta}'\boldsymbol{\beta}^{(1)} = 0$ when $\lambda_1 \neq 0$, and $\lambda_1 \neq 0$ if $\boldsymbol{\Sigma} \neq \mathbf{0}$; the case of $\boldsymbol{\Sigma} = \mathbf{0}$ is obviously trivial and is not treated.) We now want to maximize

$$\phi_2 = \boldsymbol{\beta}'\boldsymbol{\Sigma}\boldsymbol{\beta} - \lambda(\boldsymbol{\beta}'\boldsymbol{\beta} - 1) - 2\nu_1\boldsymbol{\beta}'\boldsymbol{\Sigma}\boldsymbol{\beta}^{(1)}, \tag{8}$$

where λ and ν_1 are Lagrange multipliers. The vector of partial derivatives is

$$\frac{\partial\phi_2}{\partial\boldsymbol{\beta}} = 2\boldsymbol{\Sigma}\boldsymbol{\beta} - 2\lambda\boldsymbol{\beta} - 2\nu_1\boldsymbol{\Sigma}\boldsymbol{\beta}^{(1)}, \tag{9}$$

and we set this equal to $\mathbf{0}$. From (9) we obtain by multiplying on the left by $\boldsymbol{\beta}^{(1)\prime}$

$$\begin{aligned} 0 &= 2\boldsymbol{\beta}^{(1)\prime}\boldsymbol{\Sigma}\boldsymbol{\beta} - 2\lambda\boldsymbol{\beta}^{(1)\prime}\boldsymbol{\beta} - 2\nu_1\boldsymbol{\beta}^{(1)\prime}\boldsymbol{\Sigma}\boldsymbol{\beta}^{(1)} \\ &= -2\nu_1\lambda_1, \end{aligned} \tag{10}$$

by (7). Therefore, $\nu_1 = 0$ and $\boldsymbol{\beta}$ must satisfy (4), and therefore λ must satisfy (5). Let $\lambda_{(2)}$ be the maximum of $\lambda_1, \cdots, \lambda_p$ such that there is a vector $\boldsymbol{\beta}$ satisfying $(\boldsymbol{\Sigma} - \lambda_{(2)}\boldsymbol{I})\boldsymbol{\beta} = 0$, $\boldsymbol{\beta}'\boldsymbol{\beta} = 1$, and (7); call this vector $\boldsymbol{\beta}^{(2)}$ and the corresponding linear combination $U_2 = \boldsymbol{\beta}^{(2)\prime}X$. (It will be shown eventually that $\lambda_{(2)} = \lambda_2$. We define $\lambda_{(1)} = \lambda_1$.)

This procedure is continued; at the $(r + 1)$st step we want to find a vector $\boldsymbol{\beta}$ such that $\boldsymbol{\beta}'X$ has maximum variance of all normalized linear combinations which are uncorrelated with $U_1, \cdots, U_r$, that is, such that

$$(11) \qquad \begin{aligned} 0 &= \mathscr{E}\boldsymbol{\beta}'XU_i = \mathscr{E}\boldsymbol{\beta}'XX'\boldsymbol{\beta}^{(i)} \\ &= \boldsymbol{\beta}'\boldsymbol{\Sigma}\boldsymbol{\beta}^{(i)} = \lambda_{(i)}\boldsymbol{\beta}'\boldsymbol{\beta}^{(i)}, \qquad i = 1, \cdots, r. \end{aligned}$$

We want to maximize

$$(12) \qquad \phi_{r+1} = \boldsymbol{\beta}'\boldsymbol{\Sigma}\boldsymbol{\beta} - \lambda(\boldsymbol{\beta}'\boldsymbol{\beta} - 1) - 2\sum_{i=1}^{r} \nu_i\boldsymbol{\beta}'\boldsymbol{\Sigma}\boldsymbol{\beta}^{(i)},$$

where λ and $\nu_1, \cdots, \nu_r$ are Lagrange multipliers. The vector of partial derivatives is

$$(13) \qquad \frac{\partial\phi_{r+1}}{\partial\boldsymbol{\beta}} = 2\boldsymbol{\Sigma}\boldsymbol{\beta} - 2\lambda\boldsymbol{\beta} - 2\sum_{i=1}^{r} \nu_i\boldsymbol{\Sigma}\boldsymbol{\beta}^{(i)},$$

and we set this equal to $\mathbf{0}$. Multiplying (13) on the left by $\boldsymbol{\beta}^{(j)\prime}$, we obtain

$$(14) \qquad 0 = 2\boldsymbol{\beta}^{(j)\prime}\boldsymbol{\Sigma}\boldsymbol{\beta} - 2\lambda\boldsymbol{\beta}^{(j)\prime}\boldsymbol{\beta} - 2\nu_j\boldsymbol{\beta}^{(j)\prime}\boldsymbol{\Sigma}\boldsymbol{\beta}^{(j)}.$$

If $\lambda_{(j)} \neq 0$, this gives $-2\nu_j\lambda_{(j)} = 0$ and $\nu_j = 0$. If $\lambda_{(j)} = 0$, $\boldsymbol{\Sigma}\boldsymbol{\beta}^{(j)} = \lambda_{(j)}\boldsymbol{\beta}^{(j)} = \mathbf{0}$ and the jth term in the sum in (13) vanishes. Thus $\boldsymbol{\beta}$ must satisfy (4) and therefore λ must satisfy (5).

Let $\lambda_{(r+1)}$ be the maximum of $\lambda_1, \cdots, \lambda_p$ such that there is a vector $\boldsymbol{\beta}$ satisfying $(\boldsymbol{\Sigma} - \lambda_{(r+1)}I)\boldsymbol{\beta} = \mathbf{0}$, $\boldsymbol{\beta}'\boldsymbol{\beta} = 1$ and (11); call this vector $\boldsymbol{\beta}^{(r+1)}$, and the corresponding linear combination $U_{r+1} = \boldsymbol{\beta}^{(r+1)\prime}X$. If $\lambda_{(r+1)} = 0$ and $\lambda_{(j)} = 0\, (j \neq r+1)$, then $\boldsymbol{\beta}^{(j)\prime}\boldsymbol{\Sigma}\boldsymbol{\beta}^{(r+1)} = 0$ does not imply $\boldsymbol{\beta}^{(j)\prime}\boldsymbol{\beta}^{(r+1)} = 0$. However, $\boldsymbol{\beta}^{(r+1)}$ can be replaced by a linear combination of $\boldsymbol{\beta}^{(r+1)}$ and the $\boldsymbol{\beta}^{(j)}$'s with $\lambda_{(j)}$'s being 0 so that the new $\boldsymbol{\beta}^{(r+1)}$ is orthogonal to all $\boldsymbol{\beta}^{(j)} (j = 1, \cdots, r)$. This procedure is carried on until at the $(m+1)$st stage one cannot find a vector $\boldsymbol{\beta}$ satisfying $\boldsymbol{\beta}'\boldsymbol{\beta} = 1$, (4), and (11). Either $m = p$ or $m < p$ since $\boldsymbol{\beta}^{(1)}, \cdots, \boldsymbol{\beta}^{(m)}$ must be linearly independent.

We shall now show that the inequality $m < p$ leads to a contradiction. If $m < p$ there exist $p - m$ vectors, say $e_{m+1}, \cdots, e_p$, such that $\boldsymbol{\beta}^{(i)\prime}e_j = 0$, $e_i'e_j = \delta_{ij}$. (This follows from Lemma 2 in Appendix 1.) Let $(e_{m+1}, \cdots, e_p) = E$. Now we shall show that there exists a $(p - m)$-component vector c and a number θ such that $Ec = \sum c_i e_i$ is a solution to (4) with $\lambda = \theta$. Consider a root of $|E'\boldsymbol{\Sigma}E - \theta I| = 0$ and a corresponding vector c satisfying $E'\boldsymbol{\Sigma}Ec = \theta c$. The vector $\boldsymbol{\Sigma}Ec$ is orthogonal to $\boldsymbol{\beta}^{(1)}, \cdots, \boldsymbol{\beta}^{(m)}$ (since $\boldsymbol{\beta}^{(i)\prime}\boldsymbol{\Sigma}Ec = \lambda_{(i)}\boldsymbol{\beta}^{(i)\prime}\sum c_j e_j = \lambda_{(i)}\sum c_j\boldsymbol{\beta}^{(i)\prime}e_j = 0$) and therefore is a vector in the space spanned by $e_{m+1}, \cdots, e_p$ and can be written as Eg (where g is a $(p - m)$-component vector). Multiplying $\boldsymbol{\Sigma}Ec = Eg$ on the left by E', we obtain $E'\boldsymbol{\Sigma}Ec = E'Eg = g$. Thus $g = \theta c$ and we have $\boldsymbol{\Sigma}(Ec) = \theta(Ec)$. Then $(Ec)'X$ is uncorrelated with $\boldsymbol{\beta}^{(j)\prime}X (j = 1, \cdots, m)$ and thus leads to a new $\boldsymbol{\beta}^{(m+1)}$. Since this contradicts the assumption that $m < p$, we must have $m = p$.

Let $\boldsymbol{\beta} = (\boldsymbol{\beta}^{(1)} \cdots \boldsymbol{\beta}^{(p)})$ and

$$\boldsymbol{\Lambda} = \begin{pmatrix} \lambda_{(1)} & 0 & \cdots & 0 \\ 0 & \lambda_{(2)} & \cdots & 0 \\ \cdot & \cdot & & \cdot \\ \cdot & \cdot & & \cdot \\ \cdot & \cdot & & \cdot \\ 0 & 0 & \cdots & \lambda_{(p)} \end{pmatrix}. \tag{15}$$

The equations $\boldsymbol{\Sigma\beta}^{(r)} = \lambda_{(r)}\boldsymbol{\beta}^{(r)}$ can be written in matrix form as

$$\boldsymbol{\Sigma\beta} = \boldsymbol{\beta\Lambda}, \tag{16}$$

and the equations $\boldsymbol{\beta}^{(r)\prime}\boldsymbol{\beta}^{(r)} = 1$ and $\boldsymbol{\beta}^{(r)\prime}\boldsymbol{\beta}^{(s)} = 0$, $r \neq s$, can be written as

$$\boldsymbol{\beta}'\boldsymbol{\beta} = I. \tag{17}$$

From (16) and (17) we obtain

$$\boldsymbol{\beta}'\boldsymbol{\Sigma\beta} = \boldsymbol{\Lambda}. \tag{18}$$

From the fact that

$$\begin{aligned} |\boldsymbol{\Sigma} - \lambda I| &= |\boldsymbol{\beta}'| \cdot |\boldsymbol{\Sigma} - \lambda I| \cdot |\boldsymbol{\beta}| \\ &= |\boldsymbol{\beta}'\boldsymbol{\Sigma\beta} - \lambda\boldsymbol{\beta}'\boldsymbol{\beta}| = |\boldsymbol{\Lambda} - \lambda I| \\ &= \prod(\lambda_{(i)} - \lambda) \end{aligned} \tag{19}$$

we see that the roots of (19) are the diagonal elements of $\boldsymbol{\Lambda}$; that is, $\lambda_{(1)} = \lambda_1, \lambda_{(2)} = \lambda_2, \cdots, \lambda_{(p)} = \lambda_p$.

We have proved the following theorem:

THEOREM 11.2.1. *Let the p-component random vector X have $\mathscr{E}X = \mathbf{0}$ and $\mathscr{E}XX' = \boldsymbol{\Sigma}$. Then there exists an orthogonal linear transformation*

$$U = \boldsymbol{\beta}'X \tag{20}$$

such that the covariance matrix of U is $\mathscr{E}UU' = \boldsymbol{\Lambda}$ and

$$\boldsymbol{\Lambda} = \begin{pmatrix} \lambda_1 & 0 & \cdots & 0 \\ 0 & \lambda_2 & \cdots & 0 \\ \cdot & \cdot & & \cdot \\ \cdot & \cdot & & \cdot \\ \cdot & \cdot & & \cdot \\ 0 & 0 & \cdots & \lambda_p \end{pmatrix}, \tag{21}$$

where $\lambda_1 \geq \lambda_2 \geq \cdots \geq \lambda_p \geq 0$ are the roots of (5). The rth column of $\boldsymbol{\beta}$, $\boldsymbol{\beta}^{(r)}$, satisfies $(\boldsymbol{\Sigma} - \lambda_r I)\boldsymbol{\beta} = \mathbf{0}$. The rth component of U, $U_r = \boldsymbol{\beta}^{(r)\prime}X$, has maximum variance of all normalized linear combinations uncorrelated with $U_1, \cdots, U_{r-1}$.

The vector $\boldsymbol{U}$ is defined as the vector of principal components of $\boldsymbol{X}$. It will be observed that we have proved Theorem 2 of Appendix 1 for $\boldsymbol{B}$ positive semidefinite, and indeed, the proof holds for any symmetric $\boldsymbol{B}$. It might be noted that once the transformation to $U_1, \cdots, U_p$ has been made, it is obvious that U_1 is the normalized linear combination with maximum variance, for if $U^* = \sum c_i U_i$, where $\sum c_i^2 = 1$ (U^* also being a normalized linear combination of the X's), then $\operatorname{Var}(U^*) = \sum c_i^2 \lambda_i = \lambda_1 + \sum_{i=2}^p c_i^2(\lambda_i - \lambda_1)$ (since $c_1^2 = 1 - \sum_2^p c_i^2$), which is clearly maximum for $c_i^2 = 0$, $i = 2, \cdots, p$. Similarly, U_2 is the normalized linear combination uncorrelated with U_1 which has maximum variance ($U^* = \sum c_i U_i$ being uncorrelated with U_1 implying $c_1 = 0$); in turn the maximal properties of $U_3, \cdots, U_p$ are verified.

Some other consequences can be derived.

COROLLARY 11.2.1. *Suppose* $\lambda_{r+1} = \cdots = \lambda_{r+m} = \nu$ (*that is,* ν *is a root of multiplicity* m); *then* $(\boldsymbol{\Sigma} - \nu \boldsymbol{I})$ *is of rank* $p - m$. *Furthermore* $\boldsymbol{\beta}^* = (\boldsymbol{\beta}^{(r+1)} \cdots \boldsymbol{\beta}^{(r+m)})$ *is uniquely determined except for multiplication on the right by an orthogonal matrix.*

Proof. From the derivation of the theorem we have $(\boldsymbol{\Sigma} - \nu\boldsymbol{I})\boldsymbol{\beta}^{(i)} = \boldsymbol{0}$, $i = r+1, \cdots, r+m$; that is, $\boldsymbol{\beta}^{(r+1)}, \cdots, \boldsymbol{\beta}^{(r+m)}$ are m linearly independent solutions of $(\boldsymbol{\Sigma} - \nu\boldsymbol{I})\boldsymbol{\beta} = \boldsymbol{0}$. To show that there cannot be another linearly independent solution, take $\sum_{i=1}^p x_i \boldsymbol{\beta}^{(i)}$, where the x_i are scalars. If it is a solution we have $\nu \sum x_i \boldsymbol{\beta}^{(i)} = \boldsymbol{\Sigma}(\sum x_i \boldsymbol{\beta}^{(i)}) = \sum x_i \boldsymbol{\Sigma}\boldsymbol{\beta}^{(i)} = \sum x_i \lambda_i \boldsymbol{\beta}^{(i)}$. Since $\nu x_i = \lambda_i x_i$, we must have $x_i = 0$ unless $i = r+1, \cdots, r+m$. Thus the rank is $p - m$.

If $\boldsymbol{\beta}^*$ is one set of solutions to $(\boldsymbol{\Sigma} - \nu\boldsymbol{I})\boldsymbol{\beta} = \boldsymbol{0}$, then any other set of solutions are linear combinations of the others, that is, are $\boldsymbol{\beta}^*\boldsymbol{A}$ for $\boldsymbol{A}$ nonsingular. However, the orthogonality conditions $\boldsymbol{\beta}^{*\prime}\boldsymbol{\beta}^* = \boldsymbol{I}$ applied to the linear combinations give $\boldsymbol{I} = (\boldsymbol{\beta}^*\boldsymbol{A})'(\boldsymbol{\beta}^*\boldsymbol{A}) = \boldsymbol{A}'\boldsymbol{\beta}^{*\prime}\boldsymbol{\beta}^*\boldsymbol{A} = \boldsymbol{A}'\boldsymbol{A}$, and thus $\boldsymbol{A}$ must be orthogonal. Q.E.D.

THEOREM 11.2.2. *An orthogonal transformation* $\boldsymbol{V} = \boldsymbol{C}\boldsymbol{X}$ *of a random vector* $\boldsymbol{X}$ *leaves invariant the generalized variance and the sum of the variances of the components.*

Proof. Let $\mathscr{E}\boldsymbol{X} = \boldsymbol{0}$ and $\mathscr{E}\boldsymbol{X}\boldsymbol{X}' = \boldsymbol{\Sigma}$. Then $\mathscr{E}\boldsymbol{V} = \boldsymbol{0}$ and $\mathscr{E}\boldsymbol{V}\boldsymbol{V}' = \boldsymbol{C}\boldsymbol{\Sigma}\boldsymbol{C}'$. The generalized variance of $\boldsymbol{V}$ is

$$|\boldsymbol{C}\boldsymbol{\Sigma}\boldsymbol{C}'| = |\boldsymbol{C}| \cdot |\boldsymbol{\Sigma}| \cdot |\boldsymbol{C}'| = |\boldsymbol{\Sigma}| \cdot |\boldsymbol{C}\boldsymbol{C}'| = |\boldsymbol{\Sigma}|, \tag{22}$$

which is the generalized variance of $\boldsymbol{X}$. The sum of the variances of the components of $\boldsymbol{V}$ is

$$\sum \mathscr{E} V_i^2 = \operatorname{tr}(\boldsymbol{C}\boldsymbol{\Sigma}\boldsymbol{C}') = \operatorname{tr}(\boldsymbol{\Sigma}\boldsymbol{C}'\boldsymbol{C}) = \operatorname{tr}(\boldsymbol{\Sigma}\boldsymbol{I}) = \operatorname{tr}\boldsymbol{\Sigma} = \sum \mathscr{E} X_i^2. \tag{23}$$

COROLLARY 11.2.2. *The generalized variance of the vector of principal components is the generalized variance of the original vector, and the sum of the variances of the principal components is the sum of the variances of the original variates.*

Another approach to the above theory can be based on the surfaces of constant density of the normal distribution with mean vector $\mathbf{0}$ and covariance $\mathbf{\Sigma}$ (nonsingular). The density is

$$\frac{1}{(2\pi)^{\frac{1}{2}p}|\mathbf{\Sigma}|^{\frac{1}{2}}} e^{-\frac{1}{2}\mathbf{x}'\mathbf{\Sigma}^{-1}\mathbf{x}}, \tag{24}$$

and surfaces of constant density are ellipsoids

$$\mathbf{x}'\mathbf{\Sigma}^{-1}\mathbf{x} = C. \tag{25}$$

A principal axis of this ellipsoid is defined as the line from $-\mathbf{y}$ to $\mathbf{y}$, where $\mathbf{y}$ is a point on the ellipsoid where the squared distance $\mathbf{x}'\mathbf{x}$ has a stationary point. Using the method of Lagrange multipliers, we determine the stationary points by considering

$$\psi = \mathbf{x}'\mathbf{x} - \lambda\mathbf{x}'\mathbf{\Sigma}^{-1}\mathbf{x}, \tag{26}$$

where λ is a Lagrange multiplier. We differentiate ψ with respect to the components of $\mathbf{x}$, and the derivatives set equal to 0 are

$$\frac{\partial\psi}{\partial\mathbf{x}} = 2\mathbf{x} - 2\lambda\mathbf{\Sigma}^{-1}\mathbf{x} = \mathbf{0}, \tag{27}$$

or

$$\mathbf{x} = \lambda\mathbf{\Sigma}^{-1}\mathbf{x}. \tag{28}$$

Multiplication by $\mathbf{\Sigma}$ gives

$$\mathbf{\Sigma}\mathbf{x} = \lambda\mathbf{x}. \tag{29}$$

This equation is the same as (4) and the same algebra can be developed. Thus the vectors $\boldsymbol{\beta}^{(1)}, \cdots, \boldsymbol{\beta}^{(p)}$ give the principal axes of the ellipsoid. The transformation $\mathbf{u} = \boldsymbol{\beta}\mathbf{x}$ is a rotation of the coordinates axes so that the new axes are in the direction of the principal axes of the ellipsoid. In the new coordinates the ellipsoid is

$$\mathbf{u}'\mathbf{\Lambda}^{-1}\mathbf{u} = \sum \frac{u_i^2}{\lambda_i} = C. \tag{30}$$

Thus the length of the ith principal axis is $2\sqrt{\lambda_i C}$.

A third approach to the same results is in terms of "planes of closest fit" [Pearson (1901)]. Consider a plane through the origin $\boldsymbol{\alpha}'\mathbf{x} = 0$, where $\boldsymbol{\alpha}'\boldsymbol{\alpha} = 1$. The distance of a point $\mathbf{x}$ from this plane is $\boldsymbol{\alpha}'\mathbf{x}$. Let us find the coefficients of a plane such that the expected distance squared of a random point $\mathbf{X}$ from the plane is a minimum, where $\mathscr{E}\mathbf{X} = \mathbf{0}$ and $\mathscr{E}\mathbf{X}\mathbf{X}'$

$= \mathbf{\Sigma}$. Thus we wish to minimize $\mathscr{E}(\boldsymbol{\alpha}'X)^2 = \mathscr{E}\boldsymbol{\alpha}'XX'\boldsymbol{\alpha} = \boldsymbol{\alpha}'\mathbf{\Sigma}\boldsymbol{\alpha}$, subject to the restriction $\boldsymbol{\alpha}'\boldsymbol{\alpha} = 1$. Comparison with the first approach immediately shows that the solution is $\boldsymbol{\alpha} = \boldsymbol{\beta}^{(p)}$.

Analysis into principal components is most suitable where all the components of X are measured in the same units. If they are not measured in the same units, the rationale of maximizing $\boldsymbol{\beta}'\mathbf{\Sigma}\boldsymbol{\beta}$ relative to $\boldsymbol{\beta}'\boldsymbol{\beta}$ is questionable; in fact, the analysis will depend on the various units of measurement. Suppose $\mathbf{\Delta}$ is a diagonal matrix and let $Y = \mathbf{\Delta}X$. For example, one component of X may be measured in inches and the corresponding component of Y may be measured in feet; another component of X may be in pounds and the corresponding one of Y in ounces. The covariance matrix of Y is $\mathscr{E}YY' = \mathscr{E}\mathbf{\Delta}XX'\mathbf{\Delta} = \mathbf{\Delta}\mathbf{\Sigma}\mathbf{\Delta} = \mathbf{\Psi}$, say. Then analysis of Y into principal components involves maximizing $\mathscr{E}(\boldsymbol{\gamma}'Y)^2 = \boldsymbol{\gamma}'\mathbf{\Psi}\boldsymbol{\gamma}$ relative to $\boldsymbol{\gamma}'\boldsymbol{\gamma}$ and leads to the equation $\mathbf{0} = (\mathbf{\Psi} - \nu I)\boldsymbol{\gamma} = (\mathbf{\Delta}\mathbf{\Sigma}\mathbf{\Delta} - \nu I)\boldsymbol{\gamma}$, where ν must satisfy $|\mathbf{\Psi} - \nu I| = 0$. Multiplication on the left by $\mathbf{\Delta}^{-1}$ gives

$$\mathbf{0} = (\mathbf{\Sigma} - \nu\mathbf{\Delta}^{-2})(\mathbf{\Delta}\boldsymbol{\gamma}). \tag{31}$$

Let $\mathbf{\Delta}\boldsymbol{\gamma} = \boldsymbol{\alpha}$; that is, $\boldsymbol{\gamma}'Y = \boldsymbol{\gamma}'\mathbf{\Delta}X = \boldsymbol{\alpha}'X$. Then (31) results from maximizing $\mathscr{E}(\boldsymbol{\alpha}'X)^2 = \boldsymbol{\alpha}'\mathbf{\Sigma}\boldsymbol{\alpha}$ relative to $\boldsymbol{\alpha}'\mathbf{\Delta}^{-2}\boldsymbol{\alpha}$. This last quadratic form is a weighted sum of squares, the weights being the diagonal elements of $\mathbf{\Delta}^{-2}$. It might be noted that if $\mathbf{\Delta}^{-2}$ is taken to be the matrix

$$\mathbf{\Delta}^{-2} = \begin{pmatrix} \sigma_{11} & 0 & \cdots & 0 \\ 0 & \sigma_{22} & \cdots & 0 \\ \cdot & \cdot & & \cdot \\ \cdot & \cdot & & \cdot \\ \cdot & \cdot & & \cdot \\ 0 & 0 & \cdots & \sigma_{pp} \end{pmatrix}, \tag{32}$$

then $\mathbf{\Psi}$ is the matrix of correlations.

11.3. MAXIMUM LIKELIHOOD ESTIMATES OF THE PRINCIPAL COMPONENTS AND THEIR VARIANCES

The primary problem of statistical inference in principal component analysis is to estimate the vectors $\boldsymbol{\beta}^{(1)}, \cdots, \boldsymbol{\beta}^{(p)}$ and the scalars $\lambda_1, \cdots, \lambda_p$. We apply the algebra of the preceding section to an estimate of the covariance matrix.

THEOREM 11.3.1. *Let $x_1, \cdots, x_N$ be N ($> p$) observations from $N(\boldsymbol{\mu}, \mathbf{\Sigma})$, where $\mathbf{\Sigma}$ is a matrix with p different characteristic roots. Then a set of maximum likelihood estimates of $\lambda_1, \cdots, \lambda_p$ and $\boldsymbol{\beta}^{(1)}, \cdots, \boldsymbol{\beta}^{(p)}$ defined in Theorem* 11.2.1 *are the roots $k_1 > \cdots > k_p$ of*

$$|\hat{\mathbf{\Sigma}} - kI| = 0, \tag{1}$$

and a set of corresponding vectors $\boldsymbol{b}^{(1)}, \cdots, \boldsymbol{b}^{(p)}$ *satisfying*

$$(\hat{\boldsymbol{\Sigma}} - k_i \boldsymbol{I})\boldsymbol{b}^{(i)} = \boldsymbol{0}, \tag{2}$$

$$\boldsymbol{b}^{(i)\prime}\boldsymbol{b}^{(i)} = 1, \tag{3}$$

where $\hat{\boldsymbol{\Sigma}}$ *is the maximum likelihood estimate of* $\boldsymbol{\Sigma}$.

Proof. Since the roots of $|\boldsymbol{\Sigma} - \lambda \boldsymbol{I}| = 0$ are different, each vector $\boldsymbol{\beta}^{(i)}$ is uniquely defined except that $\boldsymbol{\beta}^{(i)}$ can be replaced by $-\boldsymbol{\beta}^{(i)}$. If we require that the first nonzero component of $\boldsymbol{\beta}^{(i)}$ be positive, then $\boldsymbol{\beta}^{(i)}$ is uniquely defined, and $\boldsymbol{\mu}, \boldsymbol{\Lambda}, \boldsymbol{\beta}$ is a single-valued function of $\boldsymbol{\mu}, \boldsymbol{\Sigma}$. By Corollary 3.2.1, the set of maximum likelihood estimates of $\boldsymbol{\mu}, \boldsymbol{\Lambda}, \boldsymbol{\beta}$ is the same function of $\hat{\boldsymbol{\mu}}, \hat{\boldsymbol{\Sigma}}$. This function is defined by equations (1), (2), and (3) with the corresponding restriction that the first nonzero component of $\boldsymbol{b}^{(i)}$ be positive. [It can be shown that if $|\boldsymbol{\Sigma}| \neq 0$, the probability is 1 that the roots of (1) are different because the conditions on $\hat{\boldsymbol{\Sigma}}$ for the roots to have multiplicities higher than 1 determine a region in the space of $\hat{\boldsymbol{\Sigma}}$ of dimensionality less than $\frac{1}{2}p(p+1)$]. From (18) of Section 11.2 we see that

$$\boldsymbol{\Sigma} = \boldsymbol{\beta}\boldsymbol{\Lambda}\boldsymbol{\beta}' = \sum \lambda_i \boldsymbol{\beta}^{(i)}\boldsymbol{\beta}^{(i)\prime} \tag{4}$$

and by the same algebra

$$\hat{\boldsymbol{\Sigma}} = \sum k_i \boldsymbol{b}^{(i)}\boldsymbol{b}^{(i)\prime}. \tag{5}$$

Replacing $\boldsymbol{b}^{(i)}$ by $-\boldsymbol{b}^{(i)}$ clearly does not change $\sum k_i \boldsymbol{b}^{(i)}\boldsymbol{b}^{(i)\prime}$. Since the likelihood function depends only on $\hat{\boldsymbol{\Sigma}}$ (see Section 3.2), the maximum of the likelihood function is attained by taking any set of solutions to (2) and (3). Q.E.D.

It is possible to assume explicitly arbitrary multiplicities of roots of $\boldsymbol{\Sigma}$. If these multiplicities are not all unity, the maximum likelihood estimates are not defined as in Theorem 11.3.1. We shall not go into this matter, but shall consider only an extreme example. Suppose that we assume that the equation $|\boldsymbol{\Sigma} - \lambda \boldsymbol{I}| = 0$ has one root of multiplicity p. Let this root be λ_1. Then by Corollary 11.2.1, $\boldsymbol{\Sigma} - \lambda_1 \boldsymbol{I}$ is of rank 0; that is, $\boldsymbol{\Sigma} - \lambda_1 \boldsymbol{I} = \boldsymbol{0}$ or $\boldsymbol{\Sigma} = \lambda_1 \boldsymbol{I}$. If $\boldsymbol{X}$ is distributed according to $N(\boldsymbol{\mu}, \boldsymbol{\Sigma}) = N(\boldsymbol{\mu}, \lambda_1 \boldsymbol{I})$, the components of $\boldsymbol{X}$ are independently distributed with variance λ_1. Thus the maximum likelihood estimate of λ_1 is

$$\hat{\lambda}_1 = \frac{1}{pN} \sum_{i=1}^{p} \sum_{\alpha=1}^{N} (x_{i\alpha} - \bar{x}_i)^2, \tag{6}$$

and $\hat{\boldsymbol{\Sigma}} = \hat{\lambda}_1 \boldsymbol{I}$, and $\hat{\boldsymbol{\beta}}$ can be any orthogonal matrix. It might be pointed out that in Section 10.7 we considered a test of the hypothesis that $\boldsymbol{\Sigma} = \lambda_1 \boldsymbol{I}$ (with λ_1 unspecified); that is, the hypothesis is that $\boldsymbol{\Sigma}$ has one characteristic root of multiplicity p.

In most applications of principal component analysis it can be assumed that the roots of $\boldsymbol{\Sigma}$ are different. It might also be pointed out that in some uses of this method the algebra is applied to the matrix of correlation coefficients rather than to the covariance matrix. In general this leads to different roots and vectors.

11.4. COMPUTATION OF THE MAXIMUM LIKELIHOOD ESTIMATES OF THE PRINCIPAL COMPONENTS

There are several ways of computing the characteristic roots and characteristic vectors (principal components) of a matrix $\hat{\boldsymbol{\Sigma}}$. We shall indicate two of these methods.

One method involves expanding the determinantal equation

$$0 = |\hat{\boldsymbol{\Sigma}} - k\boldsymbol{I}| \tag{1}$$

and solving the resulting pth degree equation in k (for example, by Horner's method) for the roots $k_1 > k_2 > \cdots > k_p$. Then $\hat{\boldsymbol{\Sigma}} - k_i\boldsymbol{I}$ is of rank $p - 1$, and a solution of $(\hat{\boldsymbol{\Sigma}} - k_i\boldsymbol{I})\boldsymbol{b}^{(i)} = \boldsymbol{0}$ can be obtained by taking $b_j^{(i)}$ as the cofactor of the element in the first (or any other fixed) column and jth row of $\hat{\boldsymbol{\Sigma}} - k_i\boldsymbol{I}$.

The second method is iterative. The equation for a characteristic root and the corresponding characteristic vector can be written

$$\boldsymbol{\Sigma}\boldsymbol{x} = \lambda\boldsymbol{x}, \tag{2}$$

where we have written the equation for the population. Let $\boldsymbol{x}_{(0)}$ be any vector not orthogonal to the first characteristic vector, and define

$$\boldsymbol{x}_{(i)} = \boldsymbol{\Sigma}\boldsymbol{y}_{(i-1)}, \qquad i = 1, 2, \cdots, \tag{3}$$

$$\boldsymbol{y}_{(i)} = \frac{1}{\sqrt{\boldsymbol{x}_{(i)}'\boldsymbol{x}_{(i)}}}\boldsymbol{x}_{(i)}, \qquad i = 0, 1, 2, \cdots. \tag{4}$$

It will be shown that

$$\lim_{i \to \infty} \boldsymbol{y}_{(i)} = \pm\boldsymbol{\beta}^{(1)}, \tag{5}$$

$$\lim_{i \to \infty} \boldsymbol{x}_{(i)}'\boldsymbol{x}_{(i)} = \lambda_1^2. \tag{6}$$

We have $\boldsymbol{\Sigma} = \boldsymbol{\beta}\boldsymbol{\Lambda}\boldsymbol{\beta}'$, and thus, by induction

$$\boldsymbol{\Sigma}^i = (\boldsymbol{\beta}\boldsymbol{\Lambda}\boldsymbol{\beta}')\boldsymbol{\Sigma}^{i-1} = (\boldsymbol{\beta}\boldsymbol{\Lambda}\boldsymbol{\beta}')\boldsymbol{\beta}\boldsymbol{\Lambda}^{i-1\prime}\boldsymbol{\beta} = \boldsymbol{\beta}\boldsymbol{\Lambda}^i\boldsymbol{\beta}'. \tag{7}$$

Let $s_i = 1/\sqrt{\boldsymbol{x}_{(i)}'\boldsymbol{x}_{(i)}}$. From (3) and (4) we have

$$\boldsymbol{y}_{(i)} = s_i\boldsymbol{\Sigma}\boldsymbol{y}_{(i-1)}. \tag{8}$$

By repeated use of (8) we have

$$y_{(i)} = \left(\prod_{j=0}^{i} s_j\right) \Sigma^i x_{(0)} \tag{9}$$
$$= t_i \beta \Lambda^i \beta' x_{(0)},$$

where $t_i = \prod_{j=0}^{i} s_j$. From (4) we have

$$1 = y'_{(i)} y_{(i)} = t_i^2 x'_{(0)} \beta \Lambda^i \beta' \beta \Lambda^i \beta' x_{(0)}. \tag{10}$$

We can write

$$y_{(i)} = t_i \lambda_1^i \beta \left(\frac{1}{\lambda_1} \Lambda\right)^i \beta' x_{(0)}. \tag{11}$$

The limit of $\left(\frac{1}{\lambda_1} \Lambda\right)^i$ is

$$\lim_{i\to\infty} \left(\frac{1}{\lambda_1} \Lambda\right)^i = \lim_{i\to\infty} \begin{pmatrix} 1 & 0 & \cdots & 0 \\ 0 & \left(\frac{\lambda_2}{\lambda_1}\right)^i & \cdots & 0 \\ \cdot & \cdot & & \cdot \\ \cdot & \cdot & & \cdot \\ \cdot & \cdot & & \cdot \\ 0 & 0 & \cdots & \left(\frac{\lambda_p}{\lambda_1}\right)^i \end{pmatrix} \tag{12}$$

$$= \begin{pmatrix} 1 & 0 \cdots 0 \\ 0 & 0 \cdots 0 \\ \cdot & \cdot \quad\ \cdot \\ \cdot & \cdot \quad\ \cdot \\ \cdot & \cdot \quad\ \cdot \\ 0 & 0 \cdots 0 \end{pmatrix}$$

since $\lambda_j/\lambda_1 < 1$ for $j > 1$. Thus

$$\lim_{i\to\infty} \beta \left(\frac{1}{\lambda_1} \Lambda\right)^i \beta' x_0 = \beta \begin{pmatrix} 1 & 0 \cdots 0 \\ 0 & 0 \cdots 0 \\ \cdot & \cdot \quad\ \cdot \\ \cdot & \cdot \quad\ \cdot \\ \cdot & \cdot \quad\ \cdot \\ 0 & 0 \cdots 0 \end{pmatrix} \beta' x_0 \tag{13}$$
$$= (\beta^{(1)}\ \mathbf{0} \cdots \mathbf{0}) \beta' x_0$$
$$= \beta^{(1)} \beta^{(1)\prime} x_0$$
$$= (\beta^{(1)\prime} x_0) \beta^{(1)}.$$

From (10) and (13) we have

$$\lim_{i \to \infty} (t_i \lambda_1^i)^2 = \frac{1}{(\boldsymbol{\beta}^{(1)\prime} \boldsymbol{x}_0)^2}. \tag{14}$$

Thus (5) and (6) follow.

To find the second root and vector we define

$$\boldsymbol{\Sigma}_2 = \boldsymbol{\Sigma} - \lambda_1 \boldsymbol{\beta}^{(1)} \boldsymbol{\beta}^{(1)\prime}. \tag{15}$$

Then

$$\begin{aligned} \boldsymbol{\Sigma}_2 \boldsymbol{\beta}^{(i)} &= \boldsymbol{\Sigma} \boldsymbol{\beta}^{(i)} - \lambda_1 \boldsymbol{\beta}^{(1)} \boldsymbol{\beta}^{(1)\prime} \boldsymbol{\beta}^{(i)} \\ &= \boldsymbol{\Sigma} \boldsymbol{\beta}^{(i)} = \lambda_i \boldsymbol{\beta}^{(i)} \end{aligned} \tag{16}$$

if $i \neq 1$ and

$$\boldsymbol{\Sigma}_2 \boldsymbol{\beta}^{(1)} = \mathbf{0}. \tag{17}$$

Thus λ_2 is the largest root of $\boldsymbol{\Sigma}_2$ and $\boldsymbol{\beta}^{(2)}$ is the corresponding vector. The iteration process is now applied to $\boldsymbol{\Sigma}_2$ to find λ_2 and $\boldsymbol{\beta}^{(2)}$. Defining $\boldsymbol{\Sigma}_3 = \boldsymbol{\Sigma}_2 - \lambda_2 \boldsymbol{\beta}^{(2)} \boldsymbol{\beta}^{(2)\prime}$, we can find λ_3 and $\boldsymbol{\beta}^{(3)}$, and so forth.

There are several ways in which the labor of the iteration procedure may be reduced. One is to raise $\boldsymbol{\Sigma}$ to a power before proceeding with the iteration. Thus one can use $\boldsymbol{\Sigma}^2$ defining

$$\boldsymbol{x}_{(i)} = \boldsymbol{\Sigma}^2 \boldsymbol{y}_{(i-1)}, \tag{18}$$

$$\boldsymbol{y}_{(i)} = \frac{\boldsymbol{x}_{(i)}}{\sqrt{\boldsymbol{x}_{(i)}' \boldsymbol{x}_{(i)}}}. \tag{19}$$

This procedure will give twice as rapid convergence as the use of (3) and (4). Using $\boldsymbol{\Sigma}^4 = \boldsymbol{\Sigma}^2 \boldsymbol{\Sigma}^2$ will lead to convergence four times as rapid, and so on. It should be noted that since $\boldsymbol{\Sigma}^2$ is symmetric, there are only $p(p + 1)/2$ elements to be found.

Another acceleration procedure has been suggested by Aitken (1937). Suppose one requires accuracy to q significant figures. If the components of $\boldsymbol{y}_{(i)}$ agree with those of $\boldsymbol{y}_{(i-1)}$ to q significant figures, define $\boldsymbol{z}^{(0)} = \boldsymbol{y}_{(i)}$, $\boldsymbol{z}^{(1)} = \boldsymbol{\Sigma} \boldsymbol{z}^{(0)}$, $\boldsymbol{z}^{(2)} = \boldsymbol{\Sigma} \boldsymbol{z}^{(1)}$, $\boldsymbol{z}^{(3)} = \boldsymbol{\Sigma} \boldsymbol{z}^{(2)}$,

$$r_i^{(k)} = \frac{z_i^{(k)}}{z_i^{(k-1)}} \qquad (i = 1, \cdots, p;\ k = 1, 2, 3). \tag{20}$$

Then

$$R_i = \frac{r_i^{(1)} r_i^{(3)} - [r_i^{(2)}]^2}{r_i^{(1)} + r_i^{(3)} - 2 r_i^{(2)}} \qquad (i = 1, \cdots, p) \tag{21}$$

should agree to the desired number of significant figures. Take the common value of the R_i as the estimate of $\lambda_{(1)}$. The components of the estimate of $\boldsymbol{\beta}^{(1)}$ are proportional to

$$(22) \qquad Z_i = \frac{z_i^{(1)} z_i^{(3)} - [z_i^{(2)}]^2}{R_i z_i^{(1)} + z_i^{(3)}/R_i - 2z_i^{(2)}} \qquad (i = 1, \cdots, p).$$

11.5. AN EXAMPLE

In Section 5.3 we considered two samples of observations on varieties of iris [Fisher (1936b)]; as an example of principal component analysis we use one of those samples, namely Iris versicolor. There are 50 observations ($N = 50$, $n = N - 1 = 49$). Each observation consists of four measurements on a plant; x_1 is sepal length, x_2 is sepal width, x_3 is petal length, and x_4 is petal width. The observed sums of squares and cross products of deviations from means is

$$(1) \quad \boldsymbol{A} = \sum_{\alpha=1}^{50} (\boldsymbol{x}_\alpha - \bar{\boldsymbol{x}})(\boldsymbol{x}_\alpha - \bar{\boldsymbol{x}})' = \begin{pmatrix} 13.0552 & 4.1740 & 8.9620 & 2.7332 \\ 4.1740 & 4.8250 & 4.0500 & 2.0190 \\ 8.9620 & 4.0500 & 10.8200 & 3.5820 \\ 2.7332 & 2.0190 & 3.5820 & 1.9162 \end{pmatrix},$$

and the estimate of $\boldsymbol{\Sigma}$ is

$$(2) \quad \boldsymbol{S} = \tfrac{1}{49} \boldsymbol{A} = \begin{pmatrix} 0.266,433 & 0.085,184 & 0.182,899 & 0.055,780 \\ 0.085,184 & 0.098,469 & 0.082,653 & 0.041,204 \\ 0.182,899 & 0.082,653 & 0.220,816 & 0.073,102 \\ 0.055,780 & 0.041,204 & 0.073,102 & 0.039,106 \end{pmatrix}.$$

We use the iterative procedure to find the first principal component, by computing in turn $\boldsymbol{z}^{(j)} = \boldsymbol{S}\boldsymbol{z}^{(j-1)}$. As an initial approximation, we use $\boldsymbol{z}^{(0)\prime} = (1, 0, 1, 0)$. It is not necessary to normalize the vector at each iteration, but to compare successive vectors, we compute $z_i^{(j)}/z_i^{(j-1)} = r_i^{(j)}$, which is an approximation to l_1, the largest root of $\boldsymbol{S}$. After seven iterations, $r_i^{(7)}$ agree to within two units in the fifth decimal place (fifth significant figure). This vector is normalized, and $\boldsymbol{S}$ is applied to the normalized vector. The ratios, $r_i^{(8)}$, agree to within two units in the sixth place; the value of l_1 is (nearly accurate to the sixth place) $l_1 = 0.487,875$. The normalized eighth iterated vector is our estimate of $\boldsymbol{\beta}^{(1)}$, namely,

$$(3) \qquad \boldsymbol{b}^{(1)} = \begin{pmatrix} 0.686,724,4 \\ 0.305,346,3 \\ 0.623,662,8 \\ 0.214,983,7 \end{pmatrix}.$$

This vector agrees with the normalized seventh iterate to about one unit in the sixth place. It should be pointed out that l_1 and $\boldsymbol{b}^{(1)}$ have to be calculated more accurately than l_2 and $\boldsymbol{b}^{(2)}$, and so forth. The trace of $\boldsymbol{S}$ is

0.624,824, which is the sum of the roots. Thus l_1 is more than three times the sum of the other roots.

We next compute

$$(4)\quad S_2 = S - l_1 b^{(1)} b^{(1)\prime}$$

$$= \begin{pmatrix} 0.036{,}355{,}9 & -0.017{,}117{,}9 & -0.026{,}050{,}2 & -0.016{,}247{,}2 \\ -0.017{,}117{,}9 & 0.052{,}981{,}3 & -0.010{,}254{,}6 & 0.009{,}177{,}7 \\ -0.026{,}050{,}2 & -0.010{,}254{,}6 & 0.031{,}054{,}4 & 0.007{,}689{,}0 \\ -0.016{,}247{,}2 & 0.009{,}177{,}7 & 0.007{,}689{,}0 & 0.016{,}557{,}4 \end{pmatrix},$$

and iterate $z^{(j)} = S_2 z^{(j-1)}$, using $z^{(0)\prime} = (0, 1, 0, 0)$. (In the actual computation S_2 was multiplied by 10 and the first row and column were multiplied by -1). In this case the iteration does not proceed as rapidly; as will be seen, the ratio of l_2 to l_3 is approximately 1.32. After 15 iterations, $r_i^{(15)}$ agree to three decimal places or two significant figures. On iterations 15, 16, and 17, the $r_i^{(j)}$ are computed to ten significant figures. The R_i for the acceleration can then be computed to about six figures, and the accelerated vector to about three places. (In this case the number of valid significant figures in the accelerated vector is only one-third the number in the $r_i^{(j)}$.) The accelerated vector is iterated once, normalized, and iterated again. On this last iteration, the ratios agree to within four units in the fifth significant figure. We obtain $l_2 = 0.072{,}382{,}8$ and

$$(5)\qquad b^{(2)} = \begin{pmatrix} -0.669{,}033 \\ 0.567{,}484 \\ 0.343{,}309 \\ 0.335{,}307 \end{pmatrix}.$$

We find

$$(6)\quad S_3 = S_2 - l_2 b^{(2)} b^{(2)\prime}$$

$$= \begin{pmatrix} 0.003{,}955 & 0.010{,}363 & -0.009{,}425 & -0.000{,}010 \\ 0.010{,}363 & 0.029{,}671 & -0.024{,}356 & -0.004{,}595 \\ -0.009{,}425 & -0.024{,}356 & 0.022{,}523 & -0.000{,}643 \\ -0.000{,}010 & -0.004{,}595 & -0.000{,}643 & 0.008{,}419 \end{pmatrix}.$$

Using an initial approximation of $z^{(0)\prime} = (0, 1, 1, 0)$, we iterate six times to obtain $l_3 = 0.054{,}775$ and

$$(7)\qquad b^{(3)} = \begin{pmatrix} -0.265{,}105 \\ -0.729{,}589 \\ 0.627{,}178 \\ 0.063{,}676 \end{pmatrix}.$$

Next we compute

(8) $S_4 = S_3 - l_3 b^{(3)} b^{(3)\prime}$

$$= \begin{pmatrix} 0.000,105 & -0.000,232 & -0.000,318 & 0.000,915 \\ -0.000,232 & 0.000,514 & 0.000,708 & -0.002,050 \\ -0.000,318 & 0.000,708 & 0.000,977 & -0.002,831 \\ 0.000,915 & -0.002,050 & -0.002,831 & 0.008,197 \end{pmatrix}.$$

This matrix is approximately of rank one; the characteristic vector is proportional to any column. We find $l_4 = 0.009,793$ and

$$(9) \qquad b^{(4)} = \begin{pmatrix} 0.102,31 \\ -0.228,90 \\ -0.316,02 \\ 0.915,02 \end{pmatrix}.$$

We compute $S_4 - l_4 b^{(4)} b^{(4)\prime}$ as a check, and we find that the entries of this matrix are within 0.000,003 of being zero. Because the entries in S_4 are so small, the elements of $b^{(4)}$ can be correct to only four places and probably are correct to only three places. Since the other vectors are probably accurate to four or five places, it would likely be more accurate to compute $b^{(4)}$ to be orthogonal to $b^{(1)}$, $b^{(2)}$, and $b^{(3)}$.

The results may be summarized as follows:

$$(10) \qquad (l_1, l_2, l_3, l_4) = (0.4879 \quad 0.0724 \quad 0.0548 \quad 0.0098),$$

$$(11) \qquad B = \begin{pmatrix} 0.6867 & -0.6690 & -0.2651 & 0.1023 \\ 0.3053 & 0.5675 & -0.7296 & -0.2289 \\ 0.6237 & 0.3433 & 0.6272 & -0.3160 \\ 0.2150 & 0.3353 & 0.0637 & 0.9150 \end{pmatrix}.$$

It will be observed that the first component accounts for 78% of the total variance in the four measurements; the last component accounts for a little more than 1% of the total variance. In fact, the variance of $0.7x_1 + 0.3x_2 + 0.6x_3 + 0.2x_4$ (an approximation to the first principal component) is 0.478, which is almost 77% of the total variance. If one is interested in studying the variations in conditions that lead to variations of (x_1, x_2, x_3, x_4), one can look for variations in conditions that lead to variations of $0.7x_1 + 0.3x_2 + 0.6x_3 + 0.2x_4$. It is not very important if the other variations in (x_1, x_2, x_3, x_4) are neglected in exploratory investigations.

REFERENCES

Aitken (1937); Fisher (1936); Girshick (1936); Hotelling (1933), (1936a); Lorge and Morrison (1938); K. Pearson (1901); Stone (1947); Wilks (1943), 252–257.

PROBLEMS

1. In the example, Section 9.6, consider the three pressing operations (x_2, x_4, x_5). Find the first principal component of this estimated covariance matrix. [*Hint:* Start with the vector (1, 1, 1) and iterate.]

2. Prove that the characteristic vectors of $\begin{pmatrix} 1 & \rho \\ \rho & 1 \end{pmatrix}$ are $\begin{pmatrix} 1/\sqrt{2} \\ 1/\sqrt{2} \end{pmatrix}$ and $\begin{pmatrix} 1/\sqrt{2} \\ -1/\sqrt{2} \end{pmatrix}$ corresponding to roots $1 + \rho$ and $1 - \rho$.

3. Verify that the proof of Theorem 11.2.1 yields a proof of Theorem 2 of Appendix 1 for any real symmetric matrix.

4. Let $\boldsymbol{z} = \boldsymbol{y} + \boldsymbol{x}$, where $\mathscr{E}\boldsymbol{y} = \mathscr{E}\boldsymbol{x} = \boldsymbol{0}$, $\mathscr{E}\boldsymbol{y}\boldsymbol{y}' = \boldsymbol{\Phi}$, $\mathscr{E}\boldsymbol{x}\boldsymbol{x}' = \sigma^2\boldsymbol{I}$, $\mathscr{E}\boldsymbol{y}\boldsymbol{x}' = \boldsymbol{0}$. The p components of $\boldsymbol{y}$ can be called systematic parts and the components of $\boldsymbol{x}$ errors.

(*a*) Find the linear combination $\boldsymbol{\gamma}'\boldsymbol{z}$ of unit variance that has minimum error variance (that is, $\boldsymbol{\gamma}'\boldsymbol{x}$ has minimum variance).

(*b*) Under the assumptions $\mathscr{E}z_i^2 = 1$, find the linear function $\boldsymbol{\gamma}'\boldsymbol{z}$ of unit variance that maximizes the sum of squares of the correlations between z_i and $\boldsymbol{\gamma}'\boldsymbol{z}$.

(*c*) Relate these results to principal components.

5. Prove directly the sample analogue of Theorem 11.2.1, where $\sum \boldsymbol{x}_\alpha = \boldsymbol{0}$, $\sum \boldsymbol{x}_\alpha \boldsymbol{x}_\alpha' = \boldsymbol{A}$.

6. Let $\sigma_{ii} = 1$, $\sigma_{ij} = \rho$, $i \neq j$. Prove that one characteristic root of $\boldsymbol{\Sigma}$ is $1 + (p - 1)\rho$ with a vector proportional to $(1, 1, \cdots, 1)$, and that the other roots are all equal to $1 - \rho$.

7. Let $\boldsymbol{\Sigma} = \boldsymbol{\Phi} + \sigma^2\boldsymbol{I}$, where $\boldsymbol{\Phi}$ is positive semidefinite of rank r. Prove that each characteristic vector of $\boldsymbol{\Phi}$ is a vector of $\boldsymbol{\Sigma}$ and each root of $\boldsymbol{\Sigma}$ is a root of $\boldsymbol{\Phi}$ plus σ^2.

CHAPTER 12

Canonical Correlations and Canonical Variables

12.1. INTRODUCTION

In this section we consider two sets of variates with a joint distribution, and we analyze the correlations between the variables of one set and those of the other set. We find a new coordinate system in the space of each set of variates in such a way that the new coordinates display unambiguously the system of correlation. More precisely, we find the linear combinations of variables in each set that have maximum correlation; these linear combinations are the first coordinates in the new systems. Then a second linear combination in each set is sought such that the correlation between these is the maximum of correlations between such linear combinations as are uncorrelated with the first linear combinations. The procedure is continued until the two new coordinate systems are completely specified.

The statistical method outlined is of particular usefulness in exploratory studies. The investigator may have two large sets of variates and may wish to study the interrelations. If the two sets are very large he may wish to consider only a few linear combinations of each set. Then he will want to study those linear combinations most highly correlated. For example, one set of variables may be measurements of physical characteristics, such as various lengths and breadths of skulls; the other variables may be measurements of mental characteristics, such as scores on intelligence tests. If the investigator is interested in relating these, he may find that the interrelation is almost completely described by the correlation between the first few canonical variates.

This theory was developed by Hotelling (1936b).

12.2. CANONICAL CORRELATIONS AND VARIATES IN THE POPULATION

Suppose the random vector X of p components has the covariance matrix $\mathbf{\Sigma}$ (which is assumed to be positive definite). Since we are only interested in variances and covariances in this chapter, we shall assume

$\mathscr{E}\boldsymbol{X} = \mathbf{0}$. In developing the concepts and algebra we do not need to assume that $\boldsymbol{X}$ is normally distributed, though this latter assumption will be made to develop sampling theory.

We partition $\boldsymbol{X}$ into two subvectors of p_1 and p_2 components respectively,

$$\boldsymbol{X} = \begin{pmatrix} \boldsymbol{X}^{(1)} \\ \boldsymbol{X}^{(2)} \end{pmatrix}. \tag{1}$$

For convenience we shall assume $p_1 \leq p_2$. The covariance matrix is partitioned similarly into p_1 and p_2 rows and columns,

$$\boldsymbol{\Sigma} = \begin{pmatrix} \boldsymbol{\Sigma}_{11} & \boldsymbol{\Sigma}_{12} \\ \boldsymbol{\Sigma}_{21} & \boldsymbol{\Sigma}_{22} \end{pmatrix}. \tag{2}$$

In the previous chapter we developed a rotation of coordinate axes to a new system in which the variance properties were clearly exhibited. Here we shall develop a transformation of the first p_1 coordinate axes and a transformation of the last p_2 coordinate axes to a new $(p_1 + p_2)$-system that will exhibit clearly the intercorrelations between $\boldsymbol{X}^{(1)}$ and $\boldsymbol{X}^{(2)}$.

Consider an arbitrary linear combination, $U = \boldsymbol{\alpha}'\boldsymbol{X}^{(1)}$, of the components of $\boldsymbol{X}^{(1)}$ and an arbitrary linear function, $V = \boldsymbol{\gamma}'\boldsymbol{X}^{(2)}$, of the components of $\boldsymbol{X}^{(2)}$. We first ask for the linear functions that have maximum correlation. Since the correlation of a multiple of U and a multiple of V is the same as the correlation of U and V, we can make an arbitrary normalization of $\boldsymbol{\alpha}$ and $\boldsymbol{\gamma}$. We therefore require $\boldsymbol{\alpha}$ and $\boldsymbol{\gamma}$ to be such that U and V have unit variance, that is,

$$1 = \mathscr{E}U^2 = \mathscr{E}\boldsymbol{\alpha}'\boldsymbol{X}^{(1)}\boldsymbol{X}^{(1)\prime}\boldsymbol{\alpha} = \boldsymbol{\alpha}'\boldsymbol{\Sigma}_{11}\boldsymbol{\alpha}, \tag{3}$$

$$1 = \mathscr{E}V^2 = \mathscr{E}\boldsymbol{\gamma}'\boldsymbol{X}^{(2)}\boldsymbol{X}^{(2)\prime}\boldsymbol{\gamma} = \boldsymbol{\gamma}'\boldsymbol{\Sigma}_{22}\boldsymbol{\gamma}. \tag{4}$$

We note that $\mathscr{E}U = \mathscr{E}\boldsymbol{\alpha}'\boldsymbol{X}^{(1)} = \boldsymbol{\alpha}'\mathscr{E}\boldsymbol{X}^{(1)} = 0$ and similarly $\mathscr{E}V = 0$. Then the correlation between U and V is

$$\mathscr{E}UV = \mathscr{E}\boldsymbol{\alpha}'\boldsymbol{X}^{(1)}\boldsymbol{X}^{(2)\prime}\boldsymbol{\gamma} = \boldsymbol{\alpha}'\boldsymbol{\Sigma}_{12}\boldsymbol{\gamma}. \tag{5}$$

Thus the algebraic problem is to find $\boldsymbol{\alpha}$ and $\boldsymbol{\gamma}$ to maximize (5) subject to (3) and (4).

Let

$$\psi = \boldsymbol{\alpha}'\boldsymbol{\Sigma}_{12}\boldsymbol{\gamma} - \tfrac{1}{2}\lambda(\boldsymbol{\alpha}'\boldsymbol{\Sigma}_{11}\boldsymbol{\alpha} - 1) - \tfrac{1}{2}\mu(\boldsymbol{\gamma}'\boldsymbol{\Sigma}_{22}\boldsymbol{\gamma} - 1), \tag{6}$$

where λ and μ are Lagrange multipliers. We differentiate ψ with respect to the elements of $\boldsymbol{\alpha}$ and $\boldsymbol{\gamma}$. The vectors of derivatives set equal to zero are

$$\frac{\partial \psi}{\partial \boldsymbol{\alpha}} = \boldsymbol{\Sigma}_{12}\boldsymbol{\gamma} - \lambda\boldsymbol{\Sigma}_{11}\boldsymbol{\alpha} = \mathbf{0}, \tag{7}$$

$$\frac{\partial \psi}{\partial \boldsymbol{\gamma}} = \boldsymbol{\Sigma}_{12}'\boldsymbol{\alpha} - \mu\boldsymbol{\Sigma}_{22}\boldsymbol{\gamma} = \mathbf{0}. \tag{8}$$

Multiplication of (7) on the left by $\boldsymbol{\alpha}'$ and (8) on the left by $\boldsymbol{\gamma}'$ gives

$$\boldsymbol{\alpha}'\boldsymbol{\Sigma}_{12}\boldsymbol{\gamma} - \lambda\boldsymbol{\alpha}'\boldsymbol{\Sigma}_{11}\boldsymbol{\alpha} = 0, \tag{9}$$

$$\boldsymbol{\gamma}'\boldsymbol{\Sigma}_{12}'\boldsymbol{\alpha} - \mu\boldsymbol{\gamma}'\boldsymbol{\Sigma}_{22}\boldsymbol{\gamma} = 0. \tag{10}$$

Since $\boldsymbol{\alpha}'\boldsymbol{\Sigma}_{11}\boldsymbol{\alpha} = 1$ and $\boldsymbol{\gamma}'\boldsymbol{\Sigma}_{22}\boldsymbol{\gamma} = 1$, this shows that $\lambda = \mu = \boldsymbol{\alpha}'\boldsymbol{\Sigma}_{12}\boldsymbol{\gamma}$. Thus (7) and (8) can be written as

$$-\lambda\boldsymbol{\Sigma}_{11}\boldsymbol{\alpha} + \boldsymbol{\Sigma}_{12}\boldsymbol{\gamma} = \mathbf{0}, \tag{11}$$

$$\boldsymbol{\Sigma}_{21}\boldsymbol{\alpha} - \lambda\boldsymbol{\Sigma}_{22}\boldsymbol{\gamma} = \mathbf{0}, \tag{12}$$

since $\boldsymbol{\Sigma}_{12}' = \boldsymbol{\Sigma}_{21}$. In one matrix equation this is

$$\begin{pmatrix} -\lambda\boldsymbol{\Sigma}_{11} & \boldsymbol{\Sigma}_{12} \\ \boldsymbol{\Sigma}_{21} & -\lambda\boldsymbol{\Sigma}_{22} \end{pmatrix} \begin{pmatrix} \boldsymbol{\alpha} \\ \boldsymbol{\gamma} \end{pmatrix} = \mathbf{0}. \tag{13}$$

In order that there be a nontrivial solution [which is necessary for a solution satisfying (3) and (4)], the matrix on the left must be singular; that is,

$$\begin{vmatrix} -\lambda\boldsymbol{\Sigma}_{11} & \boldsymbol{\Sigma}_{12} \\ \boldsymbol{\Sigma}_{21} & -\lambda\boldsymbol{\Sigma}_{22} \end{vmatrix} = 0. \tag{14}$$

The determinant on the left is a polynomial of degree p. To demonstrate this, consider a Laplace expansion by minors of the first p_1 columns. One term is $|-\lambda\boldsymbol{\Sigma}_{11}| \cdot |-\lambda\boldsymbol{\Sigma}_{22}| = (-\lambda)^{p_1+p_2}|\boldsymbol{\Sigma}_{11}| \cdot |\boldsymbol{\Sigma}_{22}|$. The other terms in the expansion are of lower degree in λ because one or more rows of each minor in the first p_1 columns does not contain λ. Since $\boldsymbol{\Sigma}$ is positive definite, $|\boldsymbol{\Sigma}_{11}| \cdot |\boldsymbol{\Sigma}_{22}| \neq 0$ (Corollary 3 of Appendix 1). This shows that (14) is a polynomial equation of degree p and has p roots, say $\lambda_1 \geq \lambda_2 \geq \cdots \geq \lambda_p$. ($\boldsymbol{\alpha}'$ and $\boldsymbol{\gamma}'$ complex conjugate in (9) and (10) prove λ real.)

From (9) we see that $\lambda = \boldsymbol{\alpha}'\boldsymbol{\Sigma}_{12}\boldsymbol{\gamma}$ is the correlation between $U = \boldsymbol{\alpha}'\boldsymbol{X}^{(1)}$ and $V = \boldsymbol{\gamma}'\boldsymbol{X}^{(2)}$ when $\boldsymbol{\alpha}$ and $\boldsymbol{\gamma}$ satisfy (13) for some value of λ. Since we want the maximum correlation, we take $\lambda = \lambda_1$. Let a solution to (13) for $\lambda = \lambda_1$ be $\boldsymbol{\alpha}^{(1)}$, $\boldsymbol{\gamma}^{(1)}$ and let $U_1 = \boldsymbol{\alpha}^{(1)\prime}\boldsymbol{X}^{(1)}$ and $V_1 = \boldsymbol{\gamma}^{(1)\prime}\boldsymbol{X}^{(2)}$. Then U_1 and V_1 are normalized linear combinations of $\boldsymbol{X}^{(1)}$ and $\boldsymbol{X}^{(2)}$ respectively, with maximum correlation.

We now consider finding a second linear combination of $\boldsymbol{X}^{(1)}$, say $U = \boldsymbol{\alpha}'\boldsymbol{X}^{(1)}$, and a second linear combination of $\boldsymbol{X}^{(2)}$, say $V = \boldsymbol{\gamma}'\boldsymbol{X}^{(2)}$, such that of all linear combinations uncorrelated with U_1 and V_1 these have maximum correlation. This procedure is continued. At the rth step we have obtained linear combinations $U_1 = \boldsymbol{\alpha}^{(1)\prime}\boldsymbol{X}^{(1)}$, $V_1 = \boldsymbol{\gamma}^{(1)\prime}\boldsymbol{X}^{(2)}, \cdots, U_r = \boldsymbol{\alpha}^{(r)\prime}\boldsymbol{X}^{(1)}$, $V_r = \boldsymbol{\gamma}^{(r)\prime}\boldsymbol{X}^{(2)}$ with corresponding correlations [roots of (14)]

$\lambda^{(1)} = \lambda_1, \lambda^{(2)}, \cdots, \lambda^{(r)}$. We ask for a linear combination of $\boldsymbol{X}^{(1)}$, $U = \boldsymbol{\alpha}'\boldsymbol{X}^{(1)}$, and a linear combination $\boldsymbol{\gamma}'\boldsymbol{X}^{(2)}$ which of all linear combinations uncorrelated with $U_1, V_1, \cdots, U_r, V_r$ have maximum correlation. The condition that U be uncorrelated with U_i is

$$0 = \mathscr{E}UU_i = \mathscr{E}\boldsymbol{\alpha}'\boldsymbol{X}^{(1)}\boldsymbol{X}^{(1)\prime}\boldsymbol{\alpha}^{(i)} = \boldsymbol{\alpha}'\boldsymbol{\Sigma}_{11}\boldsymbol{\alpha}^{(i)}. \tag{15}$$

If $\lambda^{(i)} \neq 0$, $\boldsymbol{\Sigma}_{11}\boldsymbol{\alpha}^{(i)} = (1/\lambda^{(i)})\boldsymbol{\Sigma}_{12}\boldsymbol{\gamma}^{(i)}$, and, therefore,

$$0 = \boldsymbol{\alpha}'\boldsymbol{\Sigma}_{12}\boldsymbol{\gamma}^{(i)} = \mathscr{E}UV_i. \tag{16}$$

If $\lambda^{(i)} = 0$, $\boldsymbol{\Sigma}_{12}\boldsymbol{\gamma}^{(i)} = 0$ and (16) holds.

The condition that V be uncorrelated with V_i is

$$0 = \mathscr{E}VV_i = \boldsymbol{\gamma}'\boldsymbol{\Sigma}_{22}\boldsymbol{\gamma}^{(i)}. \tag{17}$$

By the same argument we have

$$0 = \boldsymbol{\gamma}'\boldsymbol{\Sigma}_{21}\boldsymbol{\alpha}^{(i)} = \mathscr{E}VU_i. \tag{18}$$

We now maximize $\mathscr{E}U_{r+1}V_{r+1}$, choosing $\boldsymbol{\alpha}$ and $\boldsymbol{\gamma}$ to satisfy (3), (4), (15), and (17) for $i = 1, 2, \cdots, r$. Consider

$$\psi_{r+1} = \boldsymbol{\alpha}'\boldsymbol{\Sigma}_{12}\boldsymbol{\gamma} - \tfrac{1}{2}\lambda(\boldsymbol{\alpha}'\boldsymbol{\Sigma}_{11}\boldsymbol{\alpha} - 1) - \tfrac{1}{2}\mu(\boldsymbol{\gamma}'\boldsymbol{\Sigma}_{22}\boldsymbol{\gamma} - 1) + \sum_{i=1}^{r} \nu_i\boldsymbol{\alpha}'\boldsymbol{\Sigma}_{11}\boldsymbol{\alpha}^{(i)} + \sum_{i=1}^{r} \theta_i\boldsymbol{\gamma}'\boldsymbol{\Sigma}_{22}\boldsymbol{\gamma}^{(i)}, \tag{19}$$

where $\lambda, \mu, \nu_1, \cdots, \nu_r, \theta_1, \cdots, \theta_r$ are Lagrange multipliers. The vectors of partial derivatives of ψ_{r+1} with respect to the elements of $\boldsymbol{\alpha}$ and $\boldsymbol{\gamma}$ are set equal to zero, giving

$$\frac{\partial\psi_{r+1}}{\partial\boldsymbol{\alpha}} = \boldsymbol{\Sigma}_{12}\boldsymbol{\gamma} - \lambda\boldsymbol{\Sigma}_{11}\boldsymbol{\alpha} + \sum\nu_i\boldsymbol{\Sigma}_{11}\boldsymbol{\alpha}^{(i)} = \boldsymbol{0}, \tag{20}$$

$$\frac{\partial\psi_{r+1}}{\partial\boldsymbol{\gamma}} = \boldsymbol{\Sigma}_{21}\boldsymbol{\alpha} - \mu\boldsymbol{\Sigma}_{22}\boldsymbol{\gamma} + \sum\theta_i\boldsymbol{\Sigma}_{22}\boldsymbol{\gamma}^{(i)} = \boldsymbol{0}. \tag{21}$$

Multiplication of (20) on the left by $\boldsymbol{\alpha}^{(j)\prime}$ and (21) on the left by $\boldsymbol{\gamma}^{(j)\prime}$ gives

$$0 = \nu_j\boldsymbol{\alpha}^{(j)\prime}\boldsymbol{\Sigma}_{11}\boldsymbol{\alpha}^{(j)} = \nu_j, \tag{22}$$

$$0 = \theta_j\boldsymbol{\gamma}^{(j)\prime}\boldsymbol{\Sigma}_{22}\boldsymbol{\gamma}^{(j)} = \theta_j. \tag{23}$$

Thus equations (20) and (21) are simply (11) and (12) or alternatively (13). We therefore take the largest λ_i, say, $\lambda^{(r+1)}$, such that there is a solution to (13) satisfying (3), (4), (15), and (17) for $i = 1, \cdots, r$. Let this solution be $\boldsymbol{\alpha}^{(r+1)}, \boldsymbol{\gamma}^{(r+1)}$, and let $U_{r+1} = \boldsymbol{\alpha}^{(r+1)\prime}\boldsymbol{X}^{(1)}$ and $V_{r+1} = \boldsymbol{\gamma}^{(r+1)\prime}\boldsymbol{X}^{(2)}$.

This procedure is continued step by step as long as successive solutions can be found which satisfy the conditions, namely, (13) for some λ_i, (3), (4), (15), and (17). Let m be the number of steps for which this can be done. Now we shall show that $m = p_1$ $(\leq p_2)$. Let $\mathbf{A} = (\boldsymbol{\alpha}^{(1)} \cdots \boldsymbol{\alpha}^{(m)})$, $\boldsymbol{\Gamma}_1 = (\boldsymbol{\gamma}^{(1)} \cdots \boldsymbol{\gamma}^{(m)})$ and

$$\boldsymbol{\Lambda} = \begin{pmatrix} \lambda^{(1)} & 0 & \cdots & 0 \\ 0 & \lambda^{(2)} & \cdots & 0 \\ \cdot & \cdot & & \cdot \\ \cdot & \cdot & & \cdot \\ \cdot & \cdot & & \cdot \\ 0 & 0 & \cdots & \lambda^{(m)} \end{pmatrix}. \tag{24}$$

Conditions (3) and (15) can be summarized as

$$\mathbf{A}'\boldsymbol{\Sigma}_{11}\mathbf{A} = I. \tag{25}$$

Since $\boldsymbol{\Sigma}_{11}$ is of rank p_1 and I is of rank m, $m \leq p_1$. Now let us show that $m < p_1$ leads to a contradiction by showing that in this case there is another vector satisfying the conditions. Since $\mathbf{A}'\boldsymbol{\Sigma}_{11}$ is $m \times p_1$, there exists a $p_1 \times (p_1 - m)$ matrix E (of rank $p_1 - m$) such that $\mathbf{A}'\boldsymbol{\Sigma}_{11}E = \mathbf{0}$. Similarly there is a $p_2 \times (p_2 - m)$ matrix F (of rank $p_2 - m$) such that $\boldsymbol{\Gamma}_1'\boldsymbol{\Sigma}_{22}F = \mathbf{0}$. We also have $\boldsymbol{\Gamma}_1'\boldsymbol{\Sigma}_{21}E = \boldsymbol{\Lambda}\mathbf{A}'\boldsymbol{\Sigma}_{11}E = \mathbf{0}$ and $\mathbf{A}'\boldsymbol{\Sigma}_{12}F = \boldsymbol{\Lambda}\boldsymbol{\Gamma}_1'\boldsymbol{\Sigma}_{22}F = \mathbf{0}$. Since E is of rank $p_1 - m$, $E'\boldsymbol{\Sigma}_{11}E$ is nonsingular, and similarly $F'\boldsymbol{\Sigma}_{22}F$ is nonsingular. Thus there is at least one root of

$$\begin{vmatrix} -\nu E'\boldsymbol{\Sigma}_{11}E & E'\boldsymbol{\Sigma}_{12}F \\ F'\boldsymbol{\Sigma}_{21}E & -\nu F'\boldsymbol{\Sigma}_{22}F \end{vmatrix} = 0, \tag{26}$$

because $|E'\boldsymbol{\Sigma}_{11}E| \cdot |F'\boldsymbol{\Sigma}_{22}F| \neq 0$. From the preceding algebra we see that there exist vectors $\boldsymbol{a}$ and $\boldsymbol{b}$ so that

$$E'\boldsymbol{\Sigma}_{12}F\boldsymbol{b} = \nu E'\boldsymbol{\Sigma}_{11}E\boldsymbol{a}, \tag{27}$$

$$F'\boldsymbol{\Sigma}_{21}E\boldsymbol{a} = \nu F'\boldsymbol{\Sigma}_{22}F\boldsymbol{b}. \tag{28}$$

Let $E\boldsymbol{a} = \boldsymbol{g}$ and $F\boldsymbol{b} = \boldsymbol{h}$. We now want to show that ν, $\boldsymbol{g}$, and $\boldsymbol{h}$ form a new solution $\lambda^{(m+1)}$, $\boldsymbol{\alpha}^{(m+1)}$, $\boldsymbol{\gamma}^{(m+1)}$. Let $\boldsymbol{\Sigma}_{11}^{-1}\boldsymbol{\Sigma}_{12}\boldsymbol{h} = \boldsymbol{k}$. Since $\mathbf{A}'\boldsymbol{\Sigma}_{11}\boldsymbol{k} = \mathbf{A}'\boldsymbol{\Sigma}_{12}F\boldsymbol{b} = \mathbf{0}$, $\boldsymbol{k}$ is orthogonal to the rows of $\mathbf{A}'\boldsymbol{\Sigma}_{11}$ and therefore is a linear combination of the columns of E, say $E\boldsymbol{c}$. Thus the equation $\boldsymbol{\Sigma}_{12}\boldsymbol{h} = \boldsymbol{\Sigma}_{11}\boldsymbol{k}$ can be written

$$\boldsymbol{\Sigma}_{12}F\boldsymbol{b} = \boldsymbol{\Sigma}_{11}E\boldsymbol{c}. \tag{29}$$

Multiplication by E' on the left gives

$$E'\boldsymbol{\Sigma}_{12}F\boldsymbol{b} = E'\boldsymbol{\Sigma}_{11}E\boldsymbol{c}. \tag{30}$$

Since $E'\boldsymbol{\Sigma}_{11}E$ is nonsingular, comparison of (27) and (30) shows that $\boldsymbol{c} = \nu\boldsymbol{a}$, and therefore $\boldsymbol{k} = \nu\boldsymbol{g}$. Thus

$$\boldsymbol{\Sigma}_{12}\boldsymbol{h} = \nu\boldsymbol{\Sigma}_{11}\boldsymbol{g}. \tag{31}$$

In a similar fashion we show that

$$\boldsymbol{\Sigma}_{21}\boldsymbol{g} = \nu\boldsymbol{\Sigma}_{22}\boldsymbol{h}. \tag{32}$$

Therefore $\nu = \lambda^{(m+1)}$, $\boldsymbol{g} = \boldsymbol{\alpha}^{(m+1)}$, $\boldsymbol{h} = \boldsymbol{\gamma}^{(m+1)}$ is another solution. But this is contrary to the assumption that $\lambda^{(m)}$, $\boldsymbol{\alpha}^{(m)}$, $\boldsymbol{\gamma}^{(m)}$ was the last possible solution. Thus $m = p_1$.

The conditions on the λ's, $\boldsymbol{\alpha}$'s and $\boldsymbol{\gamma}$'s can be summarized as

$$\mathbf{A}'\boldsymbol{\Sigma}_{11}\mathbf{A} = I, \tag{33}$$

$$\mathbf{A}'\boldsymbol{\Sigma}_{12}\boldsymbol{\Gamma}_1 = \boldsymbol{\Lambda}. \tag{34}$$

$$\boldsymbol{\Gamma}_1'\boldsymbol{\Sigma}_{22}\boldsymbol{\Gamma}_1 = I. \tag{35}$$

Let $\boldsymbol{\Gamma}_2 = (\boldsymbol{\gamma}^{(p_1+1)} \cdots \boldsymbol{\gamma}^{(p_2)})$ be a $p_2 \times (p_2 - p_1)$ matrix satisfying

$$\boldsymbol{\Gamma}_2'\boldsymbol{\Sigma}_{22}\boldsymbol{\Gamma}_1 = \mathbf{0}, \tag{36}$$

$$\boldsymbol{\Gamma}_2'\boldsymbol{\Sigma}_{22}\boldsymbol{\Gamma}_2 = I. \tag{37}$$

This matrix can be formed one column at a time; $\boldsymbol{\gamma}^{(p_1+1)}$ is a vector orthogonal to $\boldsymbol{\Sigma}_{22}\boldsymbol{\Gamma}_1$ and normalized so $\boldsymbol{\gamma}^{(p_1+1)'}\boldsymbol{\Sigma}_{22}\boldsymbol{\gamma}^{(p_1+1)} = 1$; $\boldsymbol{\gamma}^{(p_1+2)}$ is a vector orthogonal to $\boldsymbol{\Sigma}_{22}(\boldsymbol{\Gamma}_1\ \boldsymbol{\gamma}^{(p_1+1)})$ and normalized so $\boldsymbol{\gamma}^{(p_1+2)'}\boldsymbol{\Sigma}_{22}\boldsymbol{\gamma}^{(p_1+2)} = 1$, and so forth. Let $\boldsymbol{\Gamma} = (\boldsymbol{\Gamma}_1\ \boldsymbol{\Gamma}_2)$; this square matrix is nonsingular since $\boldsymbol{\Gamma}'\boldsymbol{\Sigma}_{22}\boldsymbol{\Gamma} = I$. Consider the determinant

$$\begin{aligned}
&\begin{vmatrix} \mathbf{A}' & \mathbf{0} \\ \mathbf{0} & \boldsymbol{\Gamma}_1' \\ \mathbf{0} & \boldsymbol{\Gamma}_2' \end{vmatrix} \begin{vmatrix} -\lambda\boldsymbol{\Sigma}_{11} & \boldsymbol{\Sigma}_{12} \\ \boldsymbol{\Sigma}_{21} & -\lambda\boldsymbol{\Sigma}_{22} \end{vmatrix} \begin{vmatrix} \mathbf{A} & \mathbf{0} & \mathbf{0} \\ \mathbf{0} & \boldsymbol{\Gamma}_1 & \boldsymbol{\Gamma}_2 \end{vmatrix} \\
&= \begin{vmatrix} -\lambda I & \boldsymbol{\Lambda} & \mathbf{0} \\ \boldsymbol{\Lambda} & -\lambda I & \mathbf{0} \\ \mathbf{0} & \mathbf{0} & -\lambda I \end{vmatrix} \\
&= (-\lambda)^{p_2-p_1} \begin{vmatrix} -\lambda I & \boldsymbol{\Lambda} \\ \boldsymbol{\Lambda} & -\lambda I \end{vmatrix} \\
&= (-\lambda)^{p_2-p_1}|-\lambda I| \cdot |-\lambda I - \boldsymbol{\Lambda}(-\lambda I)^{-1}\boldsymbol{\Lambda}| \\
&= (-\lambda)^{p_2-p_1}|\lambda^2 I - \boldsymbol{\Lambda}^2| \\
&= (-\lambda)^{p_2-p_1}\prod(\lambda^2 - \lambda^{(i)2}).
\end{aligned} \tag{38}$$

Except for a constant factor the above polynomial is

$$\begin{vmatrix} -\lambda\boldsymbol{\Sigma}_{11} & \boldsymbol{\Sigma}_{12} \\ \boldsymbol{\Sigma}_{21} & -\lambda\boldsymbol{\Sigma}_{22} \end{vmatrix}. \tag{39}$$

Thus the roots of (14) are the roots of (38) set equal to zero, namely, $\lambda = \pm\lambda^{(i)}$, $i = 1, \cdots, p_1$, and $\lambda = 0$ (of multiplicity $p_2 - p_1$). Thus $(\lambda_1, \cdots, \lambda_p) = (\lambda_1, \cdots, \lambda_{p_1}, 0, \cdots, 0, -\lambda_{p_1}, \cdots, -\lambda_1)$. The set $\{\lambda^{(i)2}\}$ $(i = 1, \cdots, p_1)$ is the set $\{\lambda_i^2\}$ $(i = 1, \cdots, p_1)$. To show that the set $\{\lambda^{(i)}\}$ $(i = 1, \cdots, p_1)$ is the set $\{\lambda_i\}$ $(i = 1, \cdots, p_1)$ we only need to show that $\lambda^{(i)}$ is nonnegative [and therefore is one of the λ_i $(i = 1, \cdots, p_1)$]. We observe that

$$\boldsymbol{\Sigma}_{12}\boldsymbol{\gamma}^{(r)} = -\lambda^{(r)}\boldsymbol{\Sigma}_{11}(-\boldsymbol{\alpha}^{(r)}), \tag{40}$$

$$\boldsymbol{\Sigma}_{21}(-\boldsymbol{\alpha}^{(r)}) = -\lambda^{(r)}\boldsymbol{\Sigma}_{22}\boldsymbol{\gamma}^{(r)}; \tag{41}$$

thus, if $\lambda^{(r)}$, $\boldsymbol{\alpha}^{(r)}$, $\boldsymbol{\gamma}^{(r)}$ is a solution, so is $-\lambda^{(r)}$, $-\boldsymbol{\alpha}^{(r)}$, $\boldsymbol{\gamma}^{(r)}$. If $\lambda^{(r)}$ were negative, then $-\lambda^{(r)}$ would be nonnegative and $-\lambda^{(r)} \geq \lambda^{(r)}$. But since $\lambda^{(r)}$ was to be maximum, we must have $\lambda^{(r)} \geq -\lambda^{(r)}$ and therefore $\lambda^{(r)} \geq 0$. Since the set $\{\lambda^{(i)}\}$ is the same as $\{\lambda_i\}$ $(i = 1, \cdots, p_1)$, we must have $\lambda^{(i)} = \lambda_i$.

Let

$$\boldsymbol{U} = \begin{pmatrix} U_1 \\ \cdot \\ \cdot \\ \cdot \\ U_{p_1} \end{pmatrix} = \mathbf{A}'\boldsymbol{X}^{(1)}, \tag{42}$$

$$\boldsymbol{V}^{(1)} = \begin{pmatrix} V_1 \\ \cdot \\ \cdot \\ \cdot \\ V_{p_1} \end{pmatrix} = \boldsymbol{\Gamma}_1'\boldsymbol{X}^{(2)}, \tag{43}$$

$$\boldsymbol{V}^{(2)} = \begin{pmatrix} V_{p_1+1} \\ \cdot \\ \cdot \\ \cdot \\ V_{p_2} \end{pmatrix} = \boldsymbol{\Gamma}_2'\boldsymbol{X}^{(2)}. \tag{44}$$

The components of $\boldsymbol{U}$ are one set of canonical variates and the components of $\boldsymbol{V} = \begin{pmatrix} \boldsymbol{V}^{(1)} \\ \boldsymbol{V}^{(2)} \end{pmatrix}$ are the other set. We have

$$\mathscr{E}\begin{pmatrix} \boldsymbol{U} \\ \boldsymbol{V}^{(1)} \\ \boldsymbol{V}^{(2)} \end{pmatrix}(\boldsymbol{U}' \ \boldsymbol{V}^{(1)\prime} \ \boldsymbol{V}^{(2)\prime}) = \begin{pmatrix} \mathbf{A}' & \mathbf{0} \\ \mathbf{0} & \boldsymbol{\Gamma}_1' \\ \mathbf{0} & \boldsymbol{\Gamma}_2' \end{pmatrix}\begin{pmatrix} \boldsymbol{\Sigma}_{11} & \boldsymbol{\Sigma}_{12} \\ \boldsymbol{\Sigma}_{21} & \boldsymbol{\Sigma}_{22} \end{pmatrix}\begin{pmatrix} \mathbf{A} & \mathbf{0} & \mathbf{0} \\ \mathbf{0} & \boldsymbol{\Gamma}_1 & \boldsymbol{\Gamma}_2 \end{pmatrix} = \begin{pmatrix} \boldsymbol{I} & \boldsymbol{\Lambda} & \mathbf{0} \\ \boldsymbol{\Lambda} & \boldsymbol{I} & \mathbf{0} \\ \mathbf{0} & \mathbf{0} & \boldsymbol{I} \end{pmatrix}, \tag{45}$$

where

$$\mathbf{\Lambda} = \begin{pmatrix} \lambda_1 & 0 & \cdots & 0 \\ 0 & \lambda_2 & \cdots & 0 \\ \cdot & \cdot & & \cdot \\ \cdot & \cdot & & \cdot \\ \cdot & \cdot & & \cdot \\ 0 & 0 & \cdots & \lambda_{p_1} \end{pmatrix}. \tag{46}$$

DEFINITION 12.2.1. *Let* $\boldsymbol{X} = \begin{pmatrix} \boldsymbol{X}^{(1)} \\ \boldsymbol{X}^{(2)} \end{pmatrix}$, *where* $\boldsymbol{X}^{(1)}$ *has* p_1 *components and* $\boldsymbol{X}^{(2)}$ *has* p_2 $(= p - p_1 \geq p_1)$ *components. The* rth *pair of canonical variates are the pair of linear combinations,* $U_r = \boldsymbol{\alpha}^{(r)\prime}\boldsymbol{X}^{(1)}$ *and* $V_r = \boldsymbol{\gamma}^{(r)\prime}\boldsymbol{X}^{(2)}$, *each of unit variance and uncorrelated with the first* $r - 1$ *pairs of canonical variates and having maximum correlation. The correlation is the* rth *canonical correlation.*

THEOREM 12.2.1. *Let* $\boldsymbol{X} = \begin{pmatrix} \boldsymbol{X}^{(1)} \\ \boldsymbol{X}^{(2)} \end{pmatrix}$ *be a random vector with covariance matrix* $\mathbf{\Sigma}$. *The* rth *canonical correlation between* $\boldsymbol{X}^{(1)}$ *and* $\boldsymbol{X}^{(2)}$ *is the* rth *largest root of* (14). *The coefficients of* $\boldsymbol{\alpha}^{(r)\prime}\boldsymbol{X}^{(1)}$ *and* $\boldsymbol{\gamma}^{(r)\prime}\boldsymbol{X}^{(2)}$ *defining the* rth *pair of canonical variates satisfy* (13) *for* $\lambda = \lambda_r$ *and* (3) *and* (4).

We can now verify (without differentiation) that U_1, V_1 have maximum correlation. The linear combinations $\boldsymbol{a}'\boldsymbol{U} = (\boldsymbol{a}'\mathbf{A}')\boldsymbol{X}^{(1)}$ and $\boldsymbol{b}'\boldsymbol{V} = (\boldsymbol{b}'\mathbf{\Gamma}')\boldsymbol{X}^{(2)}$ are normalized by $\boldsymbol{a}'\boldsymbol{a} = 1$ and $\boldsymbol{b}'\boldsymbol{b} = 1$. Since $\mathbf{A}$ and $\mathbf{\Gamma}$ are nonsingular any vector $\boldsymbol{\alpha}$ can be written as $\mathbf{A}\boldsymbol{a}$ and any vector $\boldsymbol{\gamma}$ can be written as $\mathbf{\Gamma}\boldsymbol{b}$ and hence any linear combinations $\boldsymbol{\alpha}'\boldsymbol{X}^{(1)}$ and $\boldsymbol{\gamma}'\boldsymbol{X}^{(2)}$ can be written as $\boldsymbol{a}'\boldsymbol{U}$ and $\boldsymbol{b}'\boldsymbol{V}$. The correlation between them is

$$\boldsymbol{a}'(\mathbf{\Lambda}\ \mathbf{0})\boldsymbol{b} = \sum_{i=1}^{p_1} \lambda_i a_i b_i. \tag{47}$$

Let $\lambda_i a_i / \sqrt{\sum(\lambda_i a_i)^2} = c_i$. Then the maximum of $\boldsymbol{a}'(\mathbf{\Lambda}\ \mathbf{0})\boldsymbol{b} = \sqrt{\sum(\lambda_i a_i)^2}\ \sum c_i b_i$ with respect to $\boldsymbol{b}$ is for $b_i = c_i$, for $\sum c_i b_i$ is the cosine of the angle between the vector $\boldsymbol{b}$ and $(c_1, \cdots, c_{p_1}, 0, \cdots, 0)$. Then (47) is

$$\sqrt{\sum \lambda_i^2 a_i^2} = \sqrt{\sum_2^{p_1}(\lambda_i^2 - \lambda_1^2)a_i^2 + \lambda_1^2}$$

and this is maximized by taking $a_i = 0$, $i = 2, \cdots, p_1$. Thus the maximized linear combinations are U_1 and V_1. In verifying that U_2 and V_2 form the second pair of canonical variates we note that lack of correlation between U_1 and a linear combination $\boldsymbol{a}'\boldsymbol{U}$ is $0 = \mathscr{E}U_1\boldsymbol{a}'\boldsymbol{U} = \mathscr{E}U_1\sum a_i U_i = a_1$ and lack of correlation between V_1 and $\boldsymbol{b}'\boldsymbol{V}$ is $0 = b_1$. The algebra used above gives the desired result with sums starting with $i = 2$.

We can derive a single matrix equation for $\boldsymbol{\alpha}$ or $\boldsymbol{\gamma}$. If we multiply (11) by λ and (12) by $\boldsymbol{\Sigma}_{22}^{-1}$, we have

$$\lambda\boldsymbol{\Sigma}_{12}\boldsymbol{\gamma} = \lambda^2\boldsymbol{\Sigma}_{11}\boldsymbol{\alpha}, \tag{48}$$

$$\boldsymbol{\Sigma}_{22}^{-1}\boldsymbol{\Sigma}_{21}\boldsymbol{\alpha} = \lambda\boldsymbol{\gamma}. \tag{49}$$

Substitution from (49) into (48) gives

$$\boldsymbol{\Sigma}_{12}\boldsymbol{\Sigma}_{22}^{-1}\boldsymbol{\Sigma}_{21}\boldsymbol{\alpha} = \lambda^2\boldsymbol{\Sigma}_{11}\boldsymbol{\alpha} \tag{50}$$

or

$$(\boldsymbol{\Sigma}_{12}\boldsymbol{\Sigma}_{22}^{-1}\boldsymbol{\Sigma}_{21} - \lambda^2\boldsymbol{\Sigma}_{11})\boldsymbol{\alpha} = 0. \tag{51}$$

The quantities $\lambda_1^2, \cdots, \lambda_{p_1}^2$ satisfy

$$|\boldsymbol{\Sigma}_{12}\boldsymbol{\Sigma}_{22}^{-1}\boldsymbol{\Sigma}_{21} - \nu\boldsymbol{\Sigma}_{11}| = 0 \tag{52}$$

and $\boldsymbol{\alpha}^{(1)}, \cdots, \boldsymbol{\alpha}^{(p_1)}$ satisfy (50) for $\lambda^2 = \lambda_1^2, \cdots, \lambda_{p_1}^2$ respectively. The similar equations for $\boldsymbol{\gamma}^{(1)}, \cdots, \boldsymbol{\gamma}^{(p_2)}$ occur when $\lambda^2 = \lambda_1^2, \cdots, \lambda_{p_2}^2$ are substituted with

$$(\boldsymbol{\Sigma}_{21}\boldsymbol{\Sigma}_{11}^{-1}\boldsymbol{\Sigma}_{12} - \lambda^2\boldsymbol{\Sigma}_{22})\boldsymbol{\gamma} = \mathbf{0}. \tag{53}$$

Another approach to canonical variates is used if the two sets of variates are not random. Suppose we have a set of variates $\boldsymbol{X}_\phi^{(1)}$ with expected values $\boldsymbol{\beta}\boldsymbol{x}_\phi^{(2)}$ for $\phi = 1, \cdots, n$, and let

$$\mathscr{E}(\boldsymbol{X}_\phi^{(1)} - \boldsymbol{\beta}\boldsymbol{x}_\phi^{(2)})(\boldsymbol{X}_\phi^{(1)} - \boldsymbol{\beta}\boldsymbol{x}_\phi^{(2)})' = \boldsymbol{\Psi}. \tag{54}$$

This distribution arises as the conditional distribution of $\boldsymbol{X}^{(1)}$ given $\boldsymbol{X}^{(2)} = \boldsymbol{x}_\phi^{(2)}$ if the joint distribution of $\boldsymbol{X}$ is $N(\mathbf{0}, \boldsymbol{\Sigma})$; then $\boldsymbol{\beta} = \boldsymbol{\Sigma}_{12}\boldsymbol{\Sigma}_{22}^{-1}$ and $\boldsymbol{\Psi} = \boldsymbol{\Sigma}_{11} - \boldsymbol{\Sigma}_{12}\boldsymbol{\Sigma}_{22}^{-1}\boldsymbol{\Sigma}_{21}$. Now consider a linear combination $U_\phi = \boldsymbol{\alpha}'\boldsymbol{X}_\phi^{(1)}$. This has expected value $\mathscr{E}U_\phi = \boldsymbol{\alpha}'\boldsymbol{\beta}\boldsymbol{x}_\phi^{(2)}$ and variance $\operatorname{Var}(U_\phi) = \boldsymbol{\alpha}'\boldsymbol{\Psi}\boldsymbol{\alpha}$; the mean sum of squares of the expected values is

$$\begin{aligned}\frac{1}{n}\sum_{\phi=1}^{n}(\mathscr{E}U_\phi)^2 &= \frac{1}{n}\sum\boldsymbol{\alpha}'\boldsymbol{\beta}\boldsymbol{x}_\phi^{(2)}\boldsymbol{x}_\phi^{(2)\prime}\boldsymbol{\beta}'\boldsymbol{\alpha} \\ &= \boldsymbol{\alpha}'\boldsymbol{\beta}S_{22}\boldsymbol{\beta}'\boldsymbol{\alpha},\end{aligned} \tag{55}$$

where

$$S_{22} = \frac{1}{n}\sum\boldsymbol{x}_\phi^{(2)}\boldsymbol{x}_\phi^{(2)\prime}. \tag{56}$$

To maximize the mean sum of squares relative to the variance, we maximize (55) for $\boldsymbol{\alpha}'\boldsymbol{\Psi}\boldsymbol{\alpha} = 1$. This leads to the vector equation

$$(\boldsymbol{\beta}S_{22}\boldsymbol{\beta}' - \nu\boldsymbol{\Psi})\boldsymbol{\alpha} = \mathbf{0} \tag{57}$$

for ν satisfying

$$|\boldsymbol{\beta}S_{22}\boldsymbol{\beta}' - \nu\boldsymbol{\Psi}| = 0. \tag{58}$$

Multiplication of (57) on the left by $\boldsymbol{\alpha}'$ shows that for $\boldsymbol{\alpha}$ and ν satisfying $\boldsymbol{\alpha}'\boldsymbol{\Psi}\boldsymbol{\alpha} = 1$ and (57), $\boldsymbol{\alpha}'\boldsymbol{\beta}S_{22}\boldsymbol{\beta}'\boldsymbol{\alpha} = \nu$; to obtain the maximum we take the largest root of (58). The linear combination of $\boldsymbol{X}^{(1)}$ that has a maximum mean sum of squares of all linear combinations uncorrelated with the first corresponds to the solution of (57) for the second largest root of (58).

The entire relationship between this second approach and the first will require consideration of estimation. We can hint at it by observing that if the second model is derived from the first by making $\boldsymbol{X}^{(2)} = \boldsymbol{x}_\phi^{(2)}$, then $\boldsymbol{\Psi} = \boldsymbol{\Sigma}_{11} - \boldsymbol{\Sigma}_{12}\boldsymbol{\Sigma}_{22}^{-1}\boldsymbol{\Sigma}_{21}$, $\boldsymbol{\beta} = \boldsymbol{\Sigma}_{12}\boldsymbol{\Sigma}_{22}^{-1}$; S_{22} corresponds to $\boldsymbol{\Sigma}_{22}$ so $\boldsymbol{\beta}S_{22}\boldsymbol{\beta}'$ corresponds to $\boldsymbol{\Sigma}_{12}\boldsymbol{\Sigma}_{22}^{-1}\boldsymbol{\Sigma}_{21}$ and $\boldsymbol{\beta}S_{22}\boldsymbol{\beta}' - \nu\boldsymbol{\Psi}$ corresponds to

$$\boldsymbol{\Sigma}_{12}\boldsymbol{\Sigma}_{22}^{-1}\boldsymbol{\Sigma}_{21} - \nu(\boldsymbol{\Sigma}_{11} - \boldsymbol{\Sigma}_{12}\boldsymbol{\Sigma}_{22}^{-1}\boldsymbol{\Sigma}_{21}) = (1+\nu)\left(\boldsymbol{\Sigma}_{12}\boldsymbol{\Sigma}_{22}^{-1}\boldsymbol{\Sigma}_{21} - \frac{\nu}{1+\nu}\boldsymbol{\Sigma}_{11}\right).$$

Let $U_\phi^{(1)} = \boldsymbol{\alpha}^{(1)\prime}\boldsymbol{X}_\phi^{(1)}$, where $\boldsymbol{\alpha}^{(1)}$ is a solution of (57) for $\nu = \nu_1$, the largest root of (58). Then $\mathscr{E}U_\phi^{(1)} = \boldsymbol{\alpha}^{(1)\prime}\boldsymbol{\beta}\boldsymbol{x}_\phi^{(2)}$. Let $\boldsymbol{\alpha}^{(1)\prime}\boldsymbol{\beta} = k\boldsymbol{\gamma}^{(1)\prime}$, where k is determined so

$$1 = \frac{1}{n}\sum_\phi \boldsymbol{\gamma}^{(1)\prime}\boldsymbol{x}_\phi^{(2)}\boldsymbol{x}_\phi^{(2)\prime}\boldsymbol{\gamma}^{(1)} \tag{59}$$
$$= \boldsymbol{\gamma}^{(1)\prime}S_{22}\boldsymbol{\gamma}^{(1)}.$$

Then $k = \sqrt{\nu_1}$. Let $v_\phi^{(1)} = \boldsymbol{\gamma}^{(1)\prime}\boldsymbol{x}_\phi^{(2)}$. Thus $\mathscr{E}U_\phi^{(1)} = \sqrt{\nu_1}v_\phi^{(1)}$.

In a similar fashion we define $U_\phi^{(2)}, \cdots, U_\phi^{(p_1)}$ and $v_\phi^{(2)}, \cdots, v_\phi^{(p_1)}$. Then

$$\mathscr{E}U_\phi^{(i)} = \sqrt{\nu_i}v_\phi^{(i)}, \tag{60}$$

and $U_\phi^{(i)}$ is uncorrelated with $U_\phi^{(j)}$, $i \neq j$. The variance of $U_\phi^{(i)}$ is 1. Similarly,

$$\frac{1}{n}\sum v_\phi^{(i)}v_\phi^{(j)} = \delta_{ij}. \tag{61}$$

Thus we get a canonical resolution of the regression structure.

We can make another interpretation of these developments in terms of prediction. Consider two random variables U and V with means 0 and variances σ_u^2 and σ_v^2 and correlation ρ. Consider approximating U by a multiple of V, say bV; then the mean square error of approximation is

$$\mathscr{E}(U - bV)^2 = \sigma_u^2 - 2b\sigma_u\sigma_v\rho + b^2\sigma_v^2 \tag{62}$$
$$= \sigma_u^2(1-\rho^2) + (b\sigma_v - \rho\sigma_u)^2.$$

This is minimized by taking $b = \sigma_u\rho/\sigma_v$. We can consider bV as a linear prediction of U from V; then $\sigma_u^2(1-\rho^2)$ is the mean square error of prediction. The ratio of the mean square error of prediction to the variance of U is $\sigma_u^2(1-\rho^2)/\sigma_u^2 = 1 - \rho^2$; this is a measure of the relative effect of V on U or the relative effectiveness of V in predicting U. Thus the greater ρ^2 or $|\rho|$ is, the more effective is V in predicting U.

Now consider the random vector $\boldsymbol{X}$ partitioned according to (1), and consider using a linear combination $V = \boldsymbol{\gamma}'\boldsymbol{X}^{(2)}$ to predict a linear combination $U = \boldsymbol{\alpha}'\boldsymbol{X}^{(1)}$. Then V predicts U best if the correlation between U and V is a maximum. Thus we can say that $\boldsymbol{\alpha}^{(1)'}\boldsymbol{X}^{(1)}$ is the linear combination of $\boldsymbol{X}^{(1)}$ that can be predicted best and $\boldsymbol{\gamma}^{(1)'}\boldsymbol{X}^{(2)}$ is the best predictor.

The mean square effect of V on U can be measured as

$$\mathscr{E}(bV)^2 = \rho^2 \frac{\sigma_u^2}{\sigma_v^2} \mathscr{E}V^2 = \rho^2\sigma_u^2, \tag{63}$$

and the relative mean square effect can be measured by the ratio $\mathscr{E}(bV)^2/\mathscr{E}U^2 = \rho^2$. Thus maximum effect of a linear combination of $\boldsymbol{X}^{(2)}$ on a linear combination of $\boldsymbol{X}^{(1)}$ is made by $\boldsymbol{\gamma}^{(1)'}\boldsymbol{X}^{(2)}$ on $\boldsymbol{\alpha}^{(1)'}\boldsymbol{X}^{(1)}$.

A similar interpretation can be made in the case that $\boldsymbol{X}^{(2)}$ is not random. Here we take expectation with respect to $\boldsymbol{X}_\phi^{(1)}$ and average over ϕ.

It should be pointed out that in the special case of $p_1 = 1$, the one canonical correlation is the multiple correlation between $\boldsymbol{X}^{(1)} = X_1$ and $\boldsymbol{X}^{(2)}$.

The definition of canonical variates and correlations was made in terms of the covariance matrix $\boldsymbol{\Sigma} = \mathscr{E}(\boldsymbol{X} - \mathscr{E}\boldsymbol{X})(\boldsymbol{X} - \mathscr{E}\boldsymbol{X})'$. We could extend this treatment by starting with a normally distributed vector $\boldsymbol{Y}$ with $p + p_3$ components and define $\boldsymbol{X}$ as the vector having the conditional distribution of the first p components of $\boldsymbol{Y}$ given the value of the last p_3 components. This would mean treating $\boldsymbol{X}_\phi$ with mean $\mathscr{E}\boldsymbol{X}_\phi = \boldsymbol{\Theta}\boldsymbol{y}_\phi^{(3)}$; the elements of the covariance matrix would be the partial covariances of the first p elements of $\boldsymbol{Y}$.

12.3. ESTIMATION OF CANONICAL CORRELATIONS AND VARIATES

Let $\boldsymbol{x}_1, \cdots, \boldsymbol{x}_N$ be N observations from $N(\boldsymbol{\mu}, \boldsymbol{\Sigma})$. Let $\boldsymbol{x}_\alpha$ be partitioned into two subvectors of p_1 and p_2 components respectively,

$$\boldsymbol{x}_\alpha = \begin{pmatrix} \boldsymbol{x}_\alpha^{(1)} \\ \boldsymbol{x}_\alpha^{(2)} \end{pmatrix}. \tag{1}$$

The maximum likelihood estimate of $\boldsymbol{\Sigma}$ [partitioned as in (2) of Section 12.2] is

$$\begin{aligned} \hat{\boldsymbol{\Sigma}} &= \begin{pmatrix} \hat{\boldsymbol{\Sigma}}_{11} & \hat{\boldsymbol{\Sigma}}_{12} \\ \hat{\boldsymbol{\Sigma}}_{21} & \hat{\boldsymbol{\Sigma}}_{22} \end{pmatrix} = \frac{1}{N}\sum(\boldsymbol{x}_\alpha - \bar{\boldsymbol{x}})(\boldsymbol{x}_\alpha - \bar{\boldsymbol{x}})' \\ &= \frac{1}{N}\begin{pmatrix} \sum(\boldsymbol{x}_\alpha^{(1)} - \bar{\boldsymbol{x}}^{(1)})(\boldsymbol{x}_\alpha^{(1)} - \bar{\boldsymbol{x}}^{(1)})' & \sum(\boldsymbol{x}_\alpha^{(1)} - \bar{\boldsymbol{x}}^{(1)})(\boldsymbol{x}_\alpha^{(2)} - \bar{\boldsymbol{x}}^{(2)})' \\ \sum(\boldsymbol{x}_\alpha^{(2)} - \bar{\boldsymbol{x}}^{(2)})(\boldsymbol{x}_\alpha^{(1)} - \bar{\boldsymbol{x}}^{(1)})' & \sum(\boldsymbol{x}_\alpha^{(2)} - \bar{\boldsymbol{x}}^{(2)})(\boldsymbol{x}_\alpha^{(2)} - \bar{\boldsymbol{x}}^{(2)})' \end{pmatrix}. \end{aligned} \tag{2}$$

The maximum likelihood estimates of the canonical correlations $\mathbf{\Lambda}$ and the canonical variates defined by $\mathbf{A}$ and $\mathbf{\Gamma}$ involve applying the algebra of the previous section to $\hat{\mathbf{\Sigma}}$. The matrices $\mathbf{\Lambda}$, $\mathbf{A}$, and $\mathbf{\Gamma}_1$ are uniquely defined if we assume the canonical correlations different and that the first nonzero element of each column of $\mathbf{A}$ is positive. The indeterminacy in $\mathbf{\Gamma}_2$ allows multiplication on the right by a $(p_2 - p_1) \times (p_2 - p_1)$ orthogonal matrix; this indeterminacy can be removed by various types of requirements, for example, that the submatrix formed by the lower $p_2 - p_1$ rows be triangular with positive diagonal elements. Application of Corollary 3.2.1 then shows that the maximum likelihood estimates of $\lambda_1, \cdots, \lambda_p$ are the roots of

$$\begin{vmatrix} -\lambda\hat{\mathbf{\Sigma}}_{11} & \hat{\mathbf{\Sigma}}_{12} \\ \hat{\mathbf{\Sigma}}_{21} & -\lambda\hat{\mathbf{\Sigma}}_{22} \end{vmatrix} = 0, \tag{3}$$

and the jth columns of $\hat{\mathbf{A}}$ and $\hat{\mathbf{\Gamma}}_1$ satisfy

$$\begin{pmatrix} -\hat{\lambda}_j\hat{\mathbf{\Sigma}}_{11} & \hat{\mathbf{\Sigma}}_{12} \\ \hat{\mathbf{\Sigma}}_{21} & -\hat{\lambda}_j\hat{\mathbf{\Sigma}}_{22} \end{pmatrix}\begin{pmatrix} \hat{\boldsymbol{\alpha}}^{(j)} \\ \hat{\boldsymbol{\gamma}}^{(j)} \end{pmatrix} = 0, \tag{4}$$

$$\hat{\boldsymbol{\alpha}}^{(j)\prime}\hat{\mathbf{\Sigma}}_{11}\hat{\boldsymbol{\alpha}}^{(j)} = 1, \tag{5}$$

$$\hat{\boldsymbol{\gamma}}^{(j)\prime}\hat{\mathbf{\Sigma}}_{22}\hat{\boldsymbol{\gamma}}^{(j)} = 1. \tag{6}$$

$\hat{\mathbf{\Gamma}}_2$ satisfies

$$\hat{\mathbf{\Gamma}}_2'\hat{\mathbf{\Sigma}}_{22}\hat{\mathbf{\Gamma}}_1 = \mathbf{0}, \tag{7}$$

$$\hat{\mathbf{\Gamma}}_2'\hat{\mathbf{\Sigma}}_{22}\hat{\mathbf{\Gamma}}_2 = I. \tag{8}$$

When the other restrictions on $\mathbf{A}$ and $\mathbf{\Gamma}_2$ are made, $\hat{\mathbf{A}}$, $\hat{\mathbf{\Gamma}}$, and $\hat{\mathbf{\Lambda}}$ are uniquely defined.

THEOREM 12.3.1. *Let $\boldsymbol{x}_1, \cdots, \boldsymbol{x}_N$ be N observations from $N(\boldsymbol{\mu}, \mathbf{\Sigma})$. Let $\mathbf{\Sigma}$ be partitioned into p_1 and p_2 $(p_1 \leq p_2)$ rows and columns as in* (2) *in Section* 12.2 *and let $\boldsymbol{x}_\alpha$ be similarly partitioned as in* (1). *The maximum likelihood estimates of the canonical correlations are the roots of* (3) *where $\hat{\mathbf{\Sigma}}_{ij}$ are defined by* (2). *The maximum likelihood estimates of the coefficients of the* jth *canonical components satisfy* (4), (5), *and* (6), $j = 1, \cdots, p_1$; *the remaining components satisfy* (7) *and* (8).

In the population the canonical correlations and canonical variates were found in terms of maximizing correlations of linear combinations of two sets of variates. The entire argument can be carried out in terms of the sample. Thus $\hat{\boldsymbol{\alpha}}^{(1)\prime}\boldsymbol{x}_\alpha^{(1)}$ and $\hat{\boldsymbol{\gamma}}^{(1)\prime}\boldsymbol{x}_\alpha^{(2)}$ have maximum sample correlation between any linear combinations of $\boldsymbol{x}_\alpha^{(1)}$ and $\boldsymbol{x}_\alpha^{(2)}$, and this correlation is $\hat{\lambda}_1$. Similarly, $\hat{\boldsymbol{\alpha}}^{(2)\prime}\boldsymbol{x}_\alpha^{(1)}$ and $\hat{\boldsymbol{\gamma}}^{(2)\prime}\boldsymbol{x}_\alpha^{(2)}$ have the second maximum sample correlation, and so forth.

It may also be observed that we could define the sample canonical variates and correlations in terms of S, the unbiased estimate of $\mathbf{\Sigma}$. Then $a^{(j)} = \sqrt{(N-1)/N}\,\hat{\alpha}^{(j)}$, $c^{(j)} = \sqrt{(N-1)/N}\,\hat{\gamma}^{(j)}$, and $l_j = \hat{\lambda}_j$ satisfy

$$S_{12}c^{(j)} = l_j S_{11} a^{(j)}, \tag{9}$$

$$S_{21}a^{(j)} = l_j S_{22} c^{(j)}, \tag{10}$$

$$a^{(j)\prime} S_{11} a^{(j)} = 1, \tag{11}$$

$$c^{(j)\prime} S_{22} c^{(j)} = 1. \tag{12}$$

We shall call the linear combinations $a^{(j)\prime}x_\alpha^{(1)}$ and $c^{(j)\prime}x_\alpha^{(2)}$ the sample canonical variates.

We can also derive the sample canonical variates from the sample correlation matrix,

$$R = \left(\frac{\hat{\sigma}_{ij}}{\sqrt{\hat{\sigma}_{ii}}\sqrt{\hat{\sigma}_{jj}}}\right) = \left(\frac{s_{ij}}{\sqrt{s_{ii}s_{jj}}}\right). \tag{13}$$

Let

$$S_1 = \begin{pmatrix} \sqrt{s_{11}} & 0 & \cdots & 0 \\ 0 & \sqrt{s_{22}} & \cdots & 0 \\ \cdot & \cdot & & \cdot \\ \cdot & \cdot & & \cdot \\ \cdot & \cdot & & \cdot \\ 0 & 0 & \cdots & \sqrt{s_{p_1p_1}} \end{pmatrix}, \tag{14}$$

$$S_2 = \begin{pmatrix} \sqrt{s_{p_1+1,\,p_1+1}} & 0 & \cdots & 0 \\ 0 & \sqrt{s_{p_1+2,\,p_1+2}} & \cdots & 0 \\ \cdot & \cdot & & \cdot \\ \cdot & \cdot & & \cdot \\ \cdot & \cdot & & \cdot \\ 0 & 0 & \cdots & \sqrt{s_{pp}} \end{pmatrix}. \tag{15}$$

Then we can write (9) through (12) as

$$R_{12}(S_2 c^{(j)}) = l_j R_{11}(S_1 a^{(j)}), \tag{16}$$

$$R_{21}(S_1 a^{(j)}) = l_j R_{22}(S_2 c^{(j)}), \tag{17}$$

$$(S_1 a^{(j)})' R_{11} (S_1 a^{(j)}) = 1, \tag{18}$$

$$(S_2 c^{(j)})' R_{22} (S_2 c^{(j)}) = 1. \tag{19}$$

Now let us consider the model where $\mathscr{E}X_\phi^{(1)} = \mu^{(1)} + \mathbf{\beta}(x_\phi^{(2)} - \bar{x}^{(2)})$ and $\mathscr{E}(X_\phi^{(1)} - \mathscr{E}X_\phi^{(1)})(X_\phi^{(1)} - \mathscr{E}X_\phi^{(1)})' = \mathbf{\Psi}$. Then $\mathbf{\beta}$ and $\mathbf{\Psi}$ are estimated by

$$B = S_{12}S_{22}^{-1}, \tag{20}$$

$$S_{11} - BS_{22}B' = S_{11} - S_{12}S_{22}^{-1}S_{21}. \tag{21}$$

The roots of

$$|S_{12}S_{22}^{-1}S_{21} - \nu(S_{11} - S_{12}S_{22}^{-1}S_{21})| = 0, \tag{22}$$

say, $\bar{\nu}_1, \cdots, \bar{\nu}_{p_1}$, are the estimates of $\nu_1, \cdots, \nu_{p_1}$. Let $\bar{\boldsymbol{a}}^{(i)}$ be the solution of

$$[S_{12}S_{22}^{-1}S_{21} - \bar{\nu}_i(S_{11} - S_{12}S_{22}^{-1}S_{21})]\bar{\boldsymbol{a}} = \boldsymbol{0}, \tag{23}$$

$$\bar{\boldsymbol{a}}'(S_{11} - S_{12}S_{22}^{-1}S_{21})\bar{\boldsymbol{a}} = 1. \tag{24}$$

Then $\bar{\nu}_i = l_i^2/(1 - l_i^2)$ and $\bar{\boldsymbol{a}}^{(i)} = [1/(1 - l_i^2)]^{\frac{1}{2}}\boldsymbol{a}^{(i)}$. Then $(1/\sqrt{\bar{\nu}_i})B'\bar{\boldsymbol{a}}_i = \bar{\boldsymbol{c}}_i$ is the same as $\boldsymbol{c}_i$.

We can give these developments a geometric interpretation. The rows of the matrix $(\boldsymbol{x}_1, \cdots, \boldsymbol{x}_N)$ can be interpreted as p vectors in an N-dimensional space and the rows of $(\boldsymbol{x}_1 - \bar{\boldsymbol{x}}, \cdots, \boldsymbol{x}_N - \bar{\boldsymbol{x}})$ are the p vectors projected on the $(N - 1)$-dimensional subspace orthogonal to the equiangular line. Denote these as $\boldsymbol{x}_1^*, \cdots, \boldsymbol{x}_p^*$. Any vector $\boldsymbol{u}^*$ with components $\boldsymbol{\alpha}'(\boldsymbol{x}_1^{(1)} - \bar{\boldsymbol{x}}^{(1)}, \cdots, \boldsymbol{x}_N^{(1)} - \bar{\boldsymbol{x}}^{(1)}) = \alpha_1\boldsymbol{x}_1^* + \cdots + \alpha_{p_1}\boldsymbol{x}_{p_1}^*$ is in the p_1-space spanned by $\boldsymbol{x}_1^*, \cdots, \boldsymbol{x}_{p_1}^*$, and a vector $\boldsymbol{v}^*$ with components $\boldsymbol{\gamma}'(\boldsymbol{x}_1^{(2)} - \bar{\boldsymbol{x}}^{(2)}, \cdots, \boldsymbol{x}_N^{(2)} - \bar{\boldsymbol{x}}^{(2)}) = \gamma_1\boldsymbol{x}_{p_1+1}^* + \cdots + \gamma_{p_2}\boldsymbol{x}_p^*$ is in the p_2-space spanned by $\boldsymbol{x}_{p_1+1}^*, \cdots, \boldsymbol{x}_p^*$. The cosine of the angle between these two vectors is the correlation between $u_\alpha = \boldsymbol{\alpha}'\boldsymbol{x}_\alpha^{(1)}$ and $v_\alpha = \boldsymbol{\gamma}'\boldsymbol{x}_\alpha^{(2)}$ $(\alpha = 1, \cdots, N)$. Finding $\boldsymbol{\alpha}$ and $\boldsymbol{\gamma}$ to maximize the correlation is equivalent to finding the vectors in the p_1-space and the p_2-space such that the angle between them is least (that is, has the greatest cosine). This gives the first canonical variates and the first canonical correlation is the cosine of the angle. Similarly, the second canonical variates correspond to vectors orthogonal to the first canonical variates and with the angle minimized.

12.4. COMPUTATION

We will discuss briefly computation in terms of the population quantities. Usually (50), (51), or (52) of Section 12.2 will be used. The computation of $\boldsymbol{\Sigma}_{12}\boldsymbol{\Sigma}_{22}^{-1}\boldsymbol{\Sigma}_{21}$ is described in Section 8.2.3. It can also be accomplished by solving $\boldsymbol{\Sigma}_{21} = \boldsymbol{\Sigma}_{22}F$ for $\boldsymbol{\Sigma}_{22}^{-1}\boldsymbol{\Sigma}_{21}$ and then multiplying by $\boldsymbol{\Sigma}_{12}$. If p_1 is sufficiently small, the determinant $|\boldsymbol{\Sigma}_{12}\boldsymbol{\Sigma}_{22}^{-1}\boldsymbol{\Sigma}_{21} - \nu\boldsymbol{\Sigma}_{11}|$ can be expanded into a polynomial in ν, and the polynomial equation may be solved for ν. The solutions are then inserted into (51) to arrive at the vectors $\boldsymbol{\alpha}$.

In many cases p_1 is too large for this procedure to be efficient. Then one uses an iterative procedure

$$\boldsymbol{\Sigma}_{12}\boldsymbol{\Sigma}_{22}^{-1}\boldsymbol{\Sigma}_{21}\boldsymbol{\alpha}(i) = \lambda^2(i + 1)\boldsymbol{\Sigma}_{11}\boldsymbol{\alpha}(i + 1), \tag{1}$$

starting with an initial approximation $\boldsymbol{\alpha}(0)$; the vector $\boldsymbol{\alpha}(i+1)$ may be normalized by

$$\boldsymbol{\alpha}(i+1)'\boldsymbol{\Sigma}_{11}\boldsymbol{\alpha}(i+1) = 1. \tag{2}$$

Usually (1) is replaced by

$$\boldsymbol{\Sigma}_{11}^{-1}\boldsymbol{\Sigma}_{12}\boldsymbol{\Sigma}_{22}^{-1}\boldsymbol{\Sigma}_{21}\boldsymbol{\alpha}(i) = \lambda^2(i+1)\boldsymbol{\alpha}(i+1) \tag{3}$$

by solving $\boldsymbol{\Sigma}_{12}\boldsymbol{\Sigma}_{22}^{-1}\boldsymbol{\Sigma}_{21} = \boldsymbol{\Sigma}_{11}E$. Then $\lambda^2(i+1)$ converges to λ_1^2 and $\boldsymbol{\alpha}(i+1)$ converges to $\boldsymbol{\alpha}^{(1)}$ (if $\lambda_1 > \lambda_2$). This can be demonstrated in a fashion similar to that used for principal components. From (45) of Section 12.2 we deduce that

$$\boldsymbol{\Sigma}_{11}^{-1}\boldsymbol{\Sigma}_{12}\boldsymbol{\Sigma}_{22}^{-1}\boldsymbol{\Sigma}_{21} = \mathbf{A}\boldsymbol{\Lambda}^2\mathbf{A}^{-1}. \tag{4}$$

The result of $i+1$ iterations is

$$\begin{aligned} \boldsymbol{\alpha}(i+1) &= t_{i+1}(\mathbf{A}\boldsymbol{\Lambda}^2\mathbf{A}^{-1})^{i+1}\boldsymbol{\alpha}(0) \\ &= t_{i+1}\mathbf{A}\boldsymbol{\Lambda}^{2(i+1)}\mathbf{A}^{-1}\boldsymbol{\alpha}(0), \end{aligned} \tag{5}$$

where t_{i+1} accomplishes the normalization. Then $t_{i+1}\boldsymbol{\Lambda}^{2(i+1)}$ converges to the matrix with 1 in the upper left-hand corner and 0's elsewhere. It follows that $\boldsymbol{\alpha}(i+1)$ converges to a multiple of the first column of $\mathbf{A}$ which is $\boldsymbol{\alpha}^{(1)}$.

Equation (4) is of the form

$$\sum_{i=1}^{p_1} \boldsymbol{\alpha}^{(i)}\lambda_i^2\tilde{\boldsymbol{\alpha}}^{(i)\prime}, \tag{6}$$

where $\tilde{\boldsymbol{\alpha}}^{(i)\prime}$ is the ith row of $\mathbf{A}^{-1}$. From the fact that $\mathbf{A}'\boldsymbol{\Sigma}_{11}\mathbf{A} = I$, we find that $\mathbf{A}'\boldsymbol{\Sigma}_{11} = \mathbf{A}^{-1}$ and thus

$$\boldsymbol{\alpha}^{(i)\prime}\boldsymbol{\Sigma}_{11} = \tilde{\boldsymbol{\alpha}}^{(i)\prime}. \tag{7}$$

Now

$$\begin{aligned} \boldsymbol{\Sigma}_{11}^{-1}\boldsymbol{\Sigma}_{12}\boldsymbol{\Sigma}_{22}^{-1}\boldsymbol{\Sigma}_{21} - \lambda_1^2\boldsymbol{\alpha}^{(1)}\tilde{\boldsymbol{\alpha}}^{(1)\prime} &= \sum_{i=2}^{p_1} \boldsymbol{\alpha}^{(i)}\lambda_i^2\tilde{\boldsymbol{\alpha}}^{(i)\prime} \\ &= \mathbf{A}\begin{pmatrix} 0 & 0 & \cdots & 0 \\ 0 & \lambda_2^2 & \cdots & 0 \\ \cdot & \cdot & & \cdot \\ \cdot & \cdot & & \cdot \\ \cdot & \cdot & & \cdot \\ 0 & 0 & \cdots & \lambda_{p_1}^2 \end{pmatrix}\mathbf{A}^{-1}. \end{aligned} \tag{8}$$

The maximum characteristic root of this matrix is λ_2^2. If we now use this matrix for iteration, we will obtain λ_2^2 and $\boldsymbol{\alpha}^{(2)}$. The procedure is continued to find as many λ_i^2 and $\boldsymbol{\alpha}^{(i)}$ as desired.

Given λ_i and $\boldsymbol{\alpha}^{(i)}$, we find $\boldsymbol{\gamma}^{(i)}$ from $\boldsymbol{\Sigma}_{21}\boldsymbol{\alpha}^{(i)} = \lambda_i\boldsymbol{\Sigma}_{22}\boldsymbol{\gamma}^{(i)}$ or $(1/\lambda_i)\boldsymbol{\Sigma}_{22}^{-1}\boldsymbol{\Sigma}_{21}\boldsymbol{\alpha}^{(i)} = \boldsymbol{\gamma}^{(i)}$. A check on the computations is provided by comparing $\boldsymbol{\Sigma}_{12}\boldsymbol{\gamma}^{(i)}$ and $\lambda_i\boldsymbol{\Sigma}_{11}\boldsymbol{\alpha}^{(i)}$.

For the sample we perform these calculations with $\hat{\boldsymbol{\Sigma}}_{ij}$ or $\boldsymbol{S}_{ij}$ substituted for $\boldsymbol{\Sigma}_{ij}$. It is often convenient to use $\boldsymbol{R}_{ij}$ in the computation (because $-1 < r_{ij} < 1$) to obtain $\boldsymbol{S}_1\boldsymbol{a}^{(j)}$ and $\boldsymbol{S}_2\boldsymbol{c}^{(j)}$; from these $\boldsymbol{a}^{(j)}$ and $\boldsymbol{c}^{(j)}$ are easy to compute.

12.5. AN EXAMPLE

In this section we consider a simple illustrative example. C. R. Rao [(1952), p. 245] gives some measurements on the first and second adult sons in a sample of 25 families. (These have been used in Problem 1 of Chapter 3 and Problem 17 of Chapter 4.) Let $x_{1\alpha}$ be the head length of the first son in the αth family, $x_{2\alpha}$ be the head breadth of the first son, $x_{3\alpha}$ be the head length of the second son, and $x_{4\alpha}$ be the head breadth of the second son. We shall investigate the relations between the measurements for the first son and for the second. Thus $\boldsymbol{x}_\alpha^{(1)\prime} = (x_{1\alpha}, x_{2\alpha})$ and $\boldsymbol{x}_\alpha^{(2)\prime} = (x_{3\alpha}, x_{4\alpha})$. The data can be summarized as*

$$\bar{\boldsymbol{x}}' = (185.72, 151.12, 183.84, 149.24),$$

$$S = \frac{1}{24}A = \begin{pmatrix} 95.2933 & 52.8683 & 69.6617 & 46.1117 \\ 52.8683 & 54.3600 & 51.3117 & 35.0533 \\ 69.6617 & 51.3117 & 100.8067 & 56.5400 \\ 46.1117 & 35.0533 & 56.5400 & 45.0233 \end{pmatrix}. \tag{1}$$

The matrix of correlations is

$$R = \left(\begin{array}{cc|cc} 1.0000 & 0.7346 & 0.7108 & 0.7040 \\ 0.7346 & 1.0000 & 0.6932 & 0.7086 \\ \hline 0.7108 & 0.6932 & 1.0000 & 0.8392 \\ 0.7040 & 0.7086 & 0.8392 & 1.0000 \end{array}\right) \tag{2}$$

$$= \begin{pmatrix} \boldsymbol{R}_{11} & \boldsymbol{R}_{12} \\ \boldsymbol{R}_{21} & \boldsymbol{R}_{22} \end{pmatrix}.$$

All of the correlations are about 0.7 except for the correlation between the two measurements on second sons. In particular, $\boldsymbol{R}_{12}$ is nearly of rank one, and hence the second canonical correlation will be near zero.

We compute

$$\boldsymbol{R}_{22}^{-1}\boldsymbol{R}_{21} = \begin{pmatrix} 0.405{,}769 & 0.333{,}205 \\ 0.363{,}480 & 0.428{,}976 \end{pmatrix}, \tag{3}$$

* Rao's computations are in error; his last "difference" is incorrect.

$$(4) \qquad R_{12}R_{22}^{-1}R_{21} = \begin{pmatrix} 0.544{,}311 & 0.538{,}841 \\ 0.538{,}841 & 0.534{,}950 \end{pmatrix}.$$

The determinantal equation is

$$(5) \qquad 0 = \begin{vmatrix} 0.544{,}311 - \quad \nu & 0.538{,}841 - 0.7346\nu \\ 0.538{,}841 - 0.7346\nu & 0.534{,}950 - \quad \nu \end{vmatrix}$$

$$= 0.460{,}363\nu^2 - 0.287{,}596\nu + 0.000{,}830.$$

The roots are 0.621,816 and 0.002,900; thus $l_1 = 0.788{,}553$ and $l_2 = 0.053{,}852$. Corresponding to these roots are the vectors

$$(6) \qquad S_1 a^{(1)} = \begin{pmatrix} 0.552{,}166 \\ 0.521{,}548 \end{pmatrix}, \qquad S_1 a^{(2)} = \begin{pmatrix} 1.366{,}501 \\ -1.378{,}467 \end{pmatrix},$$

where S_1 is the diagonal matrix with diagonal elements $\sqrt{s_{11}} = 9.7618$, $\sqrt{s_{22}} = 7.3729$. We apply $(1/l_i)R_{22}^{-1}R_{21}$ to $S_1 a^{(i)}$ to obtain

$$(7) \qquad S_2 c^{(1)} = \begin{pmatrix} 0.504{,}511 \\ 0.538{,}242 \end{pmatrix}, \qquad S_2 c^{(2)} = \begin{pmatrix} 1.767{,}281 \\ -1.757{,}288 \end{pmatrix},$$

where S_2 is the diagonal matrix with diagonal elements $\sqrt{s_{33}} = 10.0402$ and $\sqrt{s_{44}} = 6.7099$. We check these computations by calculating

$$(8) \qquad \frac{1}{l_1} R_{11}^{-1}R_{12}(S_2 c^{(1)}) = \begin{pmatrix} 0.552{,}157 \\ 0.521{,}560 \end{pmatrix}, \qquad \frac{1}{l_2} R_{11}^{-1}R_{12}(S_2 c^{(2)}) = \begin{pmatrix} 1.365{,}151 \\ -1.376{,}741 \end{pmatrix}.$$

The first vector in (8) corresponds closely to the first vector in (6); in fact, it is a slight improvement for the computation is equivalent to an iteration on $S_1 a^{(1)}$. The second vector in (8) does not correspond as closely to the second vector in (6). One reason is that l_2 is correct to only four or five significant figures (as is $\nu_2 = l_2^2$) and thus the components of $S_2 c^{(2)}$ can be correct to only as many significant figures; secondly, the fact that $S_2 c^{(2)}$

corresponds to the smaller root means that the iteration decreases the accuracy instead of increasing it. Our final results are

$$\begin{array}{lrcc} & & (1) & (2) \\ & l_i = & 0.789, & 0.054, \\ (9) & a^{(i)} = & \begin{pmatrix} 0.0566 \\ 0.0707 \end{pmatrix}, & \begin{pmatrix} 0.1400 \\ -0.1870 \end{pmatrix}, \\ & c^{(i)} = & \begin{pmatrix} 0.0502 \\ 0.0802 \end{pmatrix}, & \begin{pmatrix} 0.1760 \\ -0.2619 \end{pmatrix}. \end{array}$$

The larger of the two canonical correlations, 0.789, is higher than any of the individual correlations of a variable of the first set with a variable of the other. The second canonical correlation is very near zero. This means that to study the relation between two head dimensions of first sons and second sons we can confine our attention to the first canonical variates; the second canonical variates are correlated only slightly. The first canonical variate in each set is approximately proportional to the sum of the two measurements divided by their respective standard deviations; the second canonical variate in each set is approximately proportional to the difference of the two standardized measurements.

REFERENCES

Bartlett (1939a); Finney (1956); Hotelling (1935), (1936b); Kendall (1946), 348–354; Krull (1951); Quenouille (1950); C. R. Rao (1952); S. N. Roy (1947); Steel (1951), (1955); Vinograde (1950); Waugh (1942); Wilks (1943), 257–260.

PROBLEMS

1. Find the canonical correlations and canonical variates between the first two variables and the last three in Problem 18 of Chapter 4.

2. Prove directly the sample analogue of Theorem 12.2.1.

3. Give explicitly a normalization factor t_{i+1} in (5) of Section 12.4, and prove that $t_{i+1}\mathbf{\Lambda}^{2(i+1)}$ converges to a matrix with 1 in the upper left-hand corner and 0's elsewhere.

4. Let $z_\alpha = z_{1\alpha} = 1$, $\alpha = 1, \cdots, n$, and $\boldsymbol{\beta} = \boldsymbol{\beta}$. Verify that $\boldsymbol{\alpha}^{(1)} = \mathbf{\Sigma}^{-1}\boldsymbol{\beta}$. Relate this result to the discriminant function (Chapter 6).

5. (*a*) Let $\boldsymbol{X}' = (\boldsymbol{X}^{(1)\prime}\ \boldsymbol{X}^{(2)\prime})$, $\mathscr{E}\boldsymbol{X} = \mathbf{0}$,

$$\mathscr{E}\boldsymbol{X}\boldsymbol{X}' = \begin{pmatrix} \mathbf{\Sigma}_{11} & \mathbf{\Sigma}_{12} \\ \mathbf{\Sigma}_{21} & \mathbf{\Sigma}_{22} \end{pmatrix}.$$

$U = \boldsymbol{\alpha}'\boldsymbol{X}^{(1)}$, $V = \boldsymbol{\gamma}'\boldsymbol{X}^{(2)}$, $\mathscr{E}U^2 = 1 = \mathscr{E}V^2$, where $\boldsymbol{\alpha}$ and $\boldsymbol{\gamma}$ are vectors. Show that choosing $\boldsymbol{\alpha}$ and $\boldsymbol{\gamma}$ to maximize $\mathscr{E}UV$ is equivalent to choosing $\boldsymbol{\alpha}$ and $\boldsymbol{\gamma}$ to minimize the generalized variance of $(U\ V)$.

(*b*) Let $\boldsymbol{X}' = (\boldsymbol{X}^{(1)\prime}\ \boldsymbol{X}^{(2)\prime}\ \boldsymbol{X}^{(3)\prime})$, $\mathscr{E}\boldsymbol{X} = \boldsymbol{0}$,

$$\mathscr{E}\boldsymbol{X}\boldsymbol{X}' = \boldsymbol{\Sigma} = \begin{pmatrix} \boldsymbol{\Sigma}_{11} & \boldsymbol{\Sigma}_{12} & \boldsymbol{\Sigma}_{13} \\ \boldsymbol{\Sigma}_{21} & \boldsymbol{\Sigma}_{22} & \boldsymbol{\Sigma}_{23} \\ \boldsymbol{\Sigma}_{31} & \boldsymbol{\Sigma}_{32} & \boldsymbol{\Sigma}_{33} \end{pmatrix},$$

$U = \boldsymbol{\alpha}'\boldsymbol{X}^{(1)}$, $V = \boldsymbol{\gamma}'\boldsymbol{X}^{(2)}$, $W = \boldsymbol{\beta}'\boldsymbol{X}^{(3)}$, $\mathscr{E}U^2 = \mathscr{E}V^2 = \mathscr{E}W^2 = 1$. Consider finding $\boldsymbol{\alpha}, \boldsymbol{\gamma}, \boldsymbol{\beta}$ to minimize the generalized variance of (U, V, W). Show that this minimum is invariant with respect to transformations $\boldsymbol{X}^{*(i)} = \boldsymbol{A}_i\boldsymbol{X}^{(i)}, |\boldsymbol{A}_i| \neq 0$.

(*c*) By using such transformations, transform $\boldsymbol{\Sigma}$ into the simplest possible form.

(*d*) In the case of $\boldsymbol{X}^{(i)}$ consisting of two components, reduce the problem (of minimizing the generalized variance) to its simplest form.

(*e*) In this case give the derivative equations.

(*f*) Show that the minimum generalized variance is 1 if and only if $\boldsymbol{\Sigma}_{12} = \boldsymbol{0}$, $\boldsymbol{\Sigma}_{13} = \boldsymbol{0}$, $\boldsymbol{\Sigma}_{23} = \boldsymbol{0}$. (*Note:* This extension of the notion of canonical variates does not yield to a "nice" explicit treatment.)

CHAPTER 13

The Distribution of Certain Characteristic Roots and Vectors that Do Not Depend on Parameters

13.1. INTRODUCTION

In this chapter we find the distribution of the sample principal component vectors and their sample variances when all population variances are 1 (Section 13.3). We also find the distribution of the sample canonical correlations and one set of canonical vectors when the two sets of original variates are independent. This second distribution will be shown to be equivalent to the distribution of roots and vectors obtained in the next section. The distribution of the roots is particularly of interest because many invariant tests are functions of these roots. For example, invariant tests of the general linear hypothesis (Section 8.10) depend on the sample only through the roots of the determinantal equation

$$|(\hat{\boldsymbol{\beta}}_{1\Omega} - \boldsymbol{\beta}_1^*)A_{11\cdot 2}(\hat{\boldsymbol{\beta}}_{1\Omega} - \boldsymbol{\beta}_1^*)' - \lambda N\hat{\boldsymbol{\Sigma}}_\Omega| = 0. \tag{1}$$

If the hypothesis is true, the roots have the distribution given in Theorem 13.2.2 or 13.2.3. Thus the significance level of any invariant test of the general linear hypothesis can be obtained from the distribution derived in the next section. If the test criterion is one of the ordered roots (for example, the largest root), then the desired distribution is a marginal distribution of the joint distribution of roots.

The other distributions of roots given in this chapter are also useful for other invariant tests. These same distributions arise as limiting distributions when one treats more complicated problems (such as the distributions of roots when the general linear hypothesis is not true).

13.2. THE CASE OF TWO WISHART MATRICES

13.2.1. The Transformation

Let us consider A^* and B^* $(p \times p)$ distributed independently according to $W(\boldsymbol{\Sigma}, m)$ and $W(\boldsymbol{\Sigma}, n)$ respectively $(m, n \geq p)$. We shall call the roots of

$$|A^* - \lambda B^*| = 0 \tag{1}$$

the *characteristic roots of* A^* *in the metric of* B^* and the vectors satisfying

$$(A^* - \lambda B^*)x^* = 0 \tag{2}$$

the *characteristic vectors of* A^* *in the metric of* B^*. In this section we shall consider the distribution of these roots and vectors. Later it will be shown that the squares of canonical correlation coefficients have this distribution if the population canonical correlations are all zero.

First we shall transform A^* and B^* so that the distributions do not involve an arbitrary matrix Σ. Let C be a matrix such that

$$C\Sigma C' = I. \tag{3}$$

Let

$$\begin{aligned} A &= CA^*C', \\ B &= CB^*C'. \end{aligned} \tag{4}$$

Then A and B are independently distributed according to $W(I, m)$ and $W(I, n)$ respectively (Section 7.3.3). Since

$$\begin{aligned} |A - \lambda B| &= |CA^*C' - \lambda CB^*C'| \\ &= |C(A^* - \lambda B^*)C'| = |C| \cdot |A^* - \lambda B^*| \cdot |C'|, \end{aligned}$$

the roots of (1) are the roots of

$$|A - \lambda B| = 0. \tag{5}$$

The corresponding vectors satisfying

$$(A - \lambda B)x = 0 \tag{6}$$

satisfy

$$\begin{aligned} 0 &= C^{-1}(A - \lambda B)x \\ &= C^{-1}(CA^*C' - \lambda CB^*C')x \\ &= (A^* - \lambda B^*)C'x. \end{aligned} \tag{7}$$

Thus the vectors x^* are the vectors $C'x$.

It will be convenient to consider the roots of

$$|A - f(A + B)| = 0 \tag{8}$$

and the vectors Y satisfying

$$[A - f(A + B)]y = 0. \tag{9}$$

The latter equation can be written

$$
\begin{aligned}
\mathbf{0} &= (A - fA - fB)y \\
&= [(1-f)A - fB]y.
\end{aligned}
\tag{10}
$$

Since the probability that $f = 1$ is 0 (that is, that $|-B| = 0$), the above equation is

$$\left(A - \frac{f}{1-f}B\right) y = \mathbf{0}. \tag{11}$$

Thus the roots of (5) are related to the roots of (8) by $\lambda = f/(1-f)$ or $f = \lambda/(1+\lambda)$ and the vectors satisfying (6) are equal (or proportional) to those satisfying (9).

We now consider finding the distribution of the roots and vectors satisfying (8) and (9). Let the roots be ordered $f_1 > f_2 > \cdots > f_p > 0$ (since the probability of two roots being equal is 0). Let

$$F = \begin{pmatrix} f_1 & 0 \cdots & 0 \\ 0 & f_2 \cdots & 0 \\ \cdot & \cdot & \cdot \\ \cdot & \cdot & \cdot \\ \cdot & \cdot & \cdot \\ 0 & 0 \cdots & f_p \end{pmatrix}. \tag{12}$$

Suppose the corresponding vector solutions of (9) normalized by

$$y'(A + B)y = 1 \tag{13}$$

are $y_1, \cdots, y_p$. These vectors must satisfy

$$y_i'(A + B)y_j = 0, \qquad i \neq j, \tag{14}$$

because $y_i'Ay_j = f_jy_i'(A + B)y_j$ and $y_i'Ay_j = f_iy_i'(A + B)y_j$ and this can be only if (14) holds ($f_i \neq f_j$).

Let the $p \times p$ matrix Y be

$$Y = (y_1 \cdots y_p). \tag{15}$$

Equation (9) can be summarized as

$$AY = (A + B)YF, \tag{16}$$

and equations (13) and (14) give

$$Y'(A + B)Y = I. \tag{17}$$

From (16) we have

$$Y'AY = Y'(A + B)YF = F. \tag{18}$$

Multiplication of (17) and (18) on the left by $(Y')^{-1}$ and on the right by Y^{-1} gives

$$\begin{aligned} A + B &= (Y')^{-1}Y^{-1}, \\ A &= (Y')^{-1}FY^{-1}. \end{aligned} \tag{19}$$

Now let $Y^{-1} = E$. Then

$$\begin{aligned} A + B &= E'E, \\ A &= E'FE, \\ B &= E'(I - F)E. \end{aligned} \tag{20}$$

We now consider the joint distribution of E and F. From (20) we see that E and F define A and B uniquely. From (8) and (9) and the ordering $f_1 > \cdots > f_p$ we see that A and B define F uniquely. Equations (8) for $f = f_i$ and (9) define y_i uniquely except for multiplication by -1 (that is, replacing y_i by $-y_i$). Since $YE = I$, this means that E is defined uniquely except that rows of E can be multiplied by -1. To remove this indeterminacy we require that $e_{i1} \geq 0$ (the probability that $e_{i1} = 0$ is 0). Thus E and F are uniquely defined in terms of A and B.

13.2.2. The Jacobian

To find the density of E and F we substitute in the density of A and B according to (20) and multiply by the Jacobian of the transformation. We devote this subsection to finding the Jacobian

$$\left| \frac{\partial(A, B)}{\partial(E, F)} \right|. \tag{21}$$

Since the transformation from A and B to A and $G = A + B$ has the Jacobian unity, we shall find

$$\left| \frac{\partial(A, G)}{\partial(E, F)} \right| = \left| \frac{\partial(A, B)}{\partial(E, F)} \right|. \tag{22}$$

First we notice that if $x_\alpha = f_\alpha(y_1, \cdots, y_n)$, $\alpha = 1, \cdots, n$, is a one-to-one transformation, the Jacobian is the determinant of the linear transformation

$$dx_\alpha = \sum_\beta \frac{\partial f_\alpha}{\partial y_\beta} dy_\beta, \tag{23}$$

where dx_α and dy_β are only formally differentials (that is, we write these as a mnemonic device). If $f_\alpha(y_1, \cdots, y_n)$ is a polynomial, then $\partial f_\alpha / \partial y_\beta$ is the coefficient of y_β^* in the expansion of $f_\alpha(y_1 + y_1^*, \cdots, y_n + y_n^*)$ [in fact the coefficient in the expansion of $f_\alpha(y_1, \cdots, y_{\beta-1}, y_\beta + y_\beta^*, y_{\beta+1}, \cdots, y_n)$]. The elements of A and G are polynomials in E and F. Thus the derivative of an element of A is the coefficient of an element of E^* or F^* in the expansion of $(E + E^*)'(F + F^*)(E + E^*)$ and the derivative of an element of G is the coefficient of an element of E^* or F^* in the expansion of $(E + E^*)'(E + E^*)$. Thus the Jacobian of the transformation from A, G to E, F is the determinant of the linear transformation

$$dA = (dE)'FE + E'(dF)E + E'F(dE), \tag{24}$$

$$dG = (dE)'E + E'(dE). \tag{25}$$

Since A and G (dA and dG) are symmetric, only the functionally independent component equations above are used.

Multiply (24) and (25) on the left by E'^{-1} and on the right by E^{-1} to obtain

$$E'^{-1}(dA)E^{-1} = E'^{-1}(dE)'F + dF + F(dE)E^{-1}, \tag{26}$$

$$E'^{-1}(dG)E^{-1} = E'^{-1}(dE)' + (dE)E^{-1}. \tag{27}$$

It should be kept in mind that (24) and (25) are now considered as a linear transformation without regard to how the equations were obtained.

Let

$$E'^{-1}(dA)E^{-1} = d\bar{A}, \tag{28}$$

$$E'^{-1}(dG)E^{-1} = d\bar{G}, \tag{29}$$

$$(dE)E^{-1} = dW. \tag{30}$$

Then

$$d\bar{A} = (dW)'F + dF + F(dW), \tag{31}$$

$$d\bar{G} = dW' + dW. \tag{32}$$

The linear transformation from dE, dF to dA, dG is considered as the linear transformation from dE, dF to dW, dF with determinant $|E^{-1}|^p = |E|^{-p}$ (because each row of dE is transformed by E^{-1}), the linear transformation from dW, dF to $d\bar{A}$, $d\bar{G}$, and the linear transformation from $d\bar{A}$, $d\bar{G}$ to $dA = E'(d\bar{A})E$, $dG = E'd\bar{G}E$ with determinant $|E|^{p+1} \cdot |E|^{p+1}$ (from Section 7.3.3); and the determinant of the linear transformation from dE, dF to dA, dG is the product of the determinants of the three component transformations. The transformation (31), (32) is written in components as

$$
\begin{aligned}
d\bar{a}_{ii} &= df_{ii} + 2f_i dw_{ii}, \\
d\bar{a}_{ij} &= f_j dw_{ji} + f_i dw_{ij}, \qquad i<j, \\
d\bar{g}_{ii} &= 2dw_{ii}, \\
d\bar{g}_{ij} &= dw_{ji} + dw_{ij}, \qquad i<j.
\end{aligned} \tag{33}
$$

The determinant is

$$
\begin{array}{r|cccc|}
 & df_{ii} & dw_{ii} & dw_{ij}\,(i<j) & dw_{ij}\,(i>j) \\
d\bar{a}_{ii} & \mathbf{I} & \mathbf{2F} & \mathbf{0} & \mathbf{0} \\
d\bar{g}_{ii} & \mathbf{0} & \mathbf{2I} & \mathbf{0} & \mathbf{0} \\
d\bar{a}_{ij}\,(i<j) & \mathbf{0} & \mathbf{0} & \mathbf{M} & \mathbf{N} \\
d\bar{g}_{ij}\,(i<j) & \mathbf{0} & \mathbf{0} & \mathbf{I} & \mathbf{I}
\end{array} \tag{34}
$$

$$
= \begin{vmatrix} \mathbf{I} & \mathbf{2F} \\ \mathbf{0} & \mathbf{2I} \end{vmatrix} \cdot \begin{vmatrix} \mathbf{M} & \mathbf{N} \\ \mathbf{I} & \mathbf{I} \end{vmatrix} = 2^p\,|\mathbf{M} - \mathbf{N}|,
$$

where

$$
\mathbf{M} = \begin{array}{c}
 \\ d\bar{a}_{12} \\ \vdots \\ d\bar{a}_{1p} \\ d\bar{a}_{23} \\ \vdots \\ d\bar{a}_{2p} \\ \vdots \\ d\bar{a}_{p-1,\,p}
\end{array}
\begin{array}{c}
\begin{array}{cccccccc} dw_{12} & \cdots & dw_{1p} & dw_{23} & \cdots & dw_{2p} & \cdots & dw_{p-1,\,p} \end{array} \\
\left[\begin{array}{ccc|ccc|c|c}
f_1 & \cdots & 0 & 0 & \cdots & 0 & & 0 \\
\vdots & & \vdots & \vdots & & \vdots & & \vdots \\
0 & \cdots & f_1 & 0 & \cdots & 0 & & 0 \\
\hline
0 & \cdots & 0 & f_2 & \cdots & 0 & & 0 \\
\vdots & & \vdots & \vdots & & \vdots & & \vdots \\
0 & \cdots & 0 & 0 & \cdots & f_2 & & 0 \\
\hline
 & & & & & & & \\
\hline
0 & \cdots & 0 & 0 & \cdots & 0 & & f_{p-1}
\end{array}\right]
\end{array} \tag{35}
$$

and

$$
(36)\qquad
N = \begin{array}{c} \\ d\bar{a}_{12} \\ \cdot \\ \cdot \\ \cdot \\ d\bar{a}_{1p} \\ d\bar{a}_{23} \\ \cdot \\ \cdot \\ \cdot \\ d\bar{a}_{2p} \\ \cdot \\ \cdot \\ \cdot \\ d\bar{a}_{p-1,\,p} \end{array}
\begin{array}{c}
\begin{array}{ccc|ccc|c|c} dw_{21} & \cdots & dw_{p1} & dw_{32} & \cdots & dw_{p2} & \cdots & dw_{p,\,p-1} \end{array} \\
\left[\begin{array}{ccc|ccc|c|c}
f_2 & \cdots & 0 & 0 & \cdots & 0 & & 0 \\
\cdot & & \cdot & \cdot & & \cdot & & \cdot \\
\cdot & & \cdot & \cdot & & \cdot & & \cdot \\
\cdot & & \cdot & \cdot & & \cdot & & \cdot \\
0 & \cdots & f_p & 0 & \cdots & 0 & & 0 \\
\hline
0 & \cdots & 0 & f_3 & \cdots & 0 & & 0 \\
\cdot & & \cdot & \cdot & & \cdot & & \cdot \\
\cdot & & \cdot & \cdot & & \cdot & & \cdot \\
\cdot & & \cdot & \cdot & & \cdot & & \cdot \\
0 & \cdots & 0 & 0 & \cdots & f_p & & 0 \\
\hline
 & & & & & & & \\
 & & & & & & & \\
 & & & & & & & \\
\hline
0 & \cdots & 0 & 0 & \cdots & 0 & & f_p
\end{array}\right]
\end{array}
$$

Then

$$(37)\qquad |N - M| = \prod_{i<j} (f_i - f_j).$$

The determinant of the linear transformation (24), (25) is

$$(38)\qquad |E|^{-p}|E|^{p+1}|E|^{p+1}2^p \prod_{i<j} (f_i - f_j) = 2^p|E|^{p+2} \prod_{i<j} (f_i - f_j).$$

THEOREM 13.2.1. *The Jacobian of the transformation* (20) *is the absolute value of* (38).

13.2.3. The Joint Distribution of the Matrix E and the Roots

The joint density of A and B is

$$(39)\qquad w(A|I, m)w(B|I, n) = C_1|A|^{\frac{1}{2}(m-p-1)}|B|^{\frac{1}{2}(n-p-1)}e^{-\frac{1}{2}\operatorname{tr}(A+B)},$$

where

$$
\begin{aligned}
(40)\qquad C_1 &= K(I, m)K(I, n) \\
&= [2^{\frac{1}{2}p(n+m)}\pi^{\frac{1}{4}p(p-1)}]^{-1} \prod_{i=1}^{p} \{\Gamma[\tfrac{1}{2}(n+1-i)]\Gamma[\tfrac{1}{2}(m+1-i)]\}^{-1}.
\end{aligned}
$$

Therefore the joint density of E and F is

$$
\begin{aligned}
(41)\qquad & C_1|E'FE|^{\frac{1}{2}(m-p-1)}|E'(I-F)E|^{\frac{1}{2}(n-p-1)} \\
&\qquad\qquad e^{-\frac{1}{2}\operatorname{tr}E'E}2^p|E'E|^{\frac{1}{2}(p+2)} \prod_{i<j} (f_i - f_j).
\end{aligned}
$$

Since $|E'FE| = |E'|\cdot|F|\cdot|E| = |F|\cdot|E'E| = \prod f_i|E'E|$ and $|E'(I-F)E| = |I-F|\cdot|E'E| = \prod(1-f_i)|E'E|$, the density of E and F is

$$(42)\quad 2^pC_1|E'E|^{\frac{1}{2}(m+n-p)}e^{-\frac{1}{2}\operatorname{tr}E'E}\prod f_i^{\frac{1}{2}(m-p-1)}\prod(1-f_i)^{\frac{1}{2}(n-p-1)}\prod_{i<j}(f_i-f_j).$$

Clearly, E and F are statistically independent because the density factors into a function of E and a function of F. To determine the marginal densities we have only to find the two normalizing constants (the product of which is 2^pC_1).

Let us evaluate

$$(43)\qquad 2^p\int|E'E|^{\frac{1}{2}(m+n-p)}e^{-\frac{1}{2}\operatorname{tr}E'E}\,dE,$$

where the integration is $0 < e_{i1} < \infty$, $-\infty < e_{ij} < \infty, j \neq 1$. The value of (43) is unchanged if we let $-\infty < e_{i1} < \infty$ and multiply by 2^{-p}. Thus (43) is

$$(44)\quad (2\pi)^{\frac{1}{2}p^2}\int_{-\infty}^{\infty}\cdots\int_{-\infty}^{\infty}|E'E|^{\frac{1}{2}(m+n-p)}\left[\frac{1}{(2\pi)^{\frac{1}{2}p^2}}\exp\left(-\tfrac{1}{2}\sum_{i,j}e_{ij}^2\right)\right]\prod de_{ij}.$$

Except for the constant $(2\pi)^{\frac{1}{2}p^2}$, (44) is a definition of the expectation of the $\frac{1}{2}(m+n-p)$ power of $|E'E|$ when the e_{ij} have as density the function within brackets. This expected value is the $\frac{1}{2}(m+n-p)$-th moment of the generalized variance $|E'E|$ when $E'E$ has the distribution $W(I, p)$. (See Section 7.5.) Thus (44) is

$$(45)\qquad (2\pi)^{\frac{1}{2}p^2}\prod_{i=1}^{p}\frac{\Gamma[\frac{1}{2}(m+n+1-i)]}{\Gamma[\frac{1}{2}(p+1-i)]}2^{\frac{1}{2}p(n+m-p)}.$$

Thus the density of E is

$$(46)\qquad \frac{\prod_{i=1}^{p}\Gamma[\frac{1}{2}(p+1-i)]}{2^{\frac{1}{2}p(m+n-2)}\pi^{\frac{1}{2}p^2}\prod_{i=1}^{p}\Gamma[\frac{1}{2}(m+n+1-i)]}|E'E|^{\frac{1}{2}(m+n-p)}e^{-\frac{1}{2}\operatorname{tr}E'E}.$$

The density of f_i is (42) divided by (46); that is, the density of f_i is

$$(47)\qquad C_2\prod_{i=1}^{p}f_i^{\frac{1}{2}(m-p-1)}\prod_{i=1}^{p}(1-f_i)^{\frac{1}{2}(n-p-1)}\prod_{i<j}(f_i-f_j),$$

for $0 \leq f_p \leq \cdots \leq f_1 \leq 1$, where

$$(48)\quad C_2 = \frac{\pi^{\frac{1}{2}p}\prod_{i=1}^{p}\Gamma[\frac{1}{2}(m+n+1-i)]}{\prod_{i=1}^{p}\{\Gamma[\frac{1}{2}(n+1-i)]\Gamma[\frac{1}{2}(m+1-i)]\Gamma[\frac{1}{2}(p+1-i)]\}}.$$

The density of λ_i is obtained from (47) by letting

$$f_i = \frac{\lambda_i}{\lambda_i + 1};$$

we have

$$\frac{df_i}{d\lambda_i} = \frac{1}{(\lambda_i + 1)^2},$$

$$f_i - f_j = \frac{\lambda_i - \lambda_j}{(\lambda_i + 1)(\lambda_j + 1)}, \tag{49}$$

$$1 - f_i = \frac{1}{\lambda_i + 1}.$$

Thus the density of λ_i is

$$C_2 \prod_{i=1}^{p} \lambda_i^{\frac{1}{2}(m-p-1)} \prod_{i=1}^{p} (\lambda_i + 1)^{-\frac{1}{2}(m+n)} \prod_{i<j} (\lambda_i - \lambda_j) \tag{50}$$

for $0 \leq \lambda_p \leq \cdots \leq \lambda_1$.

THEOREM 13.2.2. *If A and B are distributed independently according to $W(\Sigma, m)$ and $W(\Sigma, n)$ respectively ($m \geq p, n \geq p$), the joint density of the roots of $|A - \lambda B| = 0$ is (50) where C_2 is defined by (48).*

The joint density of Y can be found from (46) and the fact that the Jacobian is $|Y|^{-2p}$ (see Theorem 11 of Appendix 1).

13.2.4. The Distribution for A Singular

The matrix A above can be represented as

$$A = \sum_{\alpha=1}^{m} Y_\alpha Y_\alpha', \tag{51}$$

where the Y_α are independent, each with the distribution $N(\mathbf{0}, I)$. Now let us treat the case of $m < p$. We shall find the distribution of the nonzero roots of

$$|A - f(A + B)| = 0 \tag{52}$$

for $m < p$.

To derive this, let $A + B = G$. The joint density function of B and Y_α is

$$\frac{|B|^{\frac{1}{2}(n-p-1)} e^{-\frac{1}{2}\operatorname{tr} B}}{2^{\frac{1}{2}np} \pi^{p(p-1)/4} \prod_{i=1}^{p} \Gamma[\frac{1}{2}(n + 1 - i)]} \cdot \frac{e^{-\frac{1}{2}\operatorname{tr} A}}{(2\pi)^{\frac{1}{2}mp}}. \tag{53}$$

The Jacobian of the transformation from the $y_{i\alpha}$ and B to the $y_{i\alpha}$ and $G = A + B$ is clearly 1, so that the joint density function of the $y_{i\alpha}$ and G is

$$(54)\qquad \frac{|G - A|^{\frac{1}{2}(n-p-1)} e^{-\frac{1}{2}\operatorname{tr} G}}{2^{\frac{1}{2}p(m+n)} \pi^{\frac{1}{2}mp + p(p-1)/4} \prod_{i=1}^{p} \Gamma[\frac{1}{2}(n+1-i)]}.$$

Let $G = CC'$, where C is a square matrix chosen in some unique way. This is merely a definition of the matrix C which will be used for transforming the $y_{i\alpha}$; that is, we are not replacing G by its expression in terms of C. Let

$$(55)\qquad Y = (y_{i\alpha}) = CU.$$

Here $U = (u_{i\alpha})$ has the same dimensions as Y $(i = 1, \cdots, p,\ \alpha = 1, \cdots, m)$. The Jacobian of the transformation relating the αth column of Y to the αth column of U is $|C|$, so that the Jacobian of the whole transformation is

$$(56)\qquad J = \operatorname{mod} |C|^m = |G|^{\frac{1}{2}m}.$$

Then

$$(57)\qquad A = YY' = CUU'C',$$

$$(58)\qquad \begin{aligned} |G - A| &= |CC' - CUU'C'| \\ &= |C(I - UU')C'| = |C|^2 \cdot |I - UU'| \\ &= |G| \cdot |I - UU'|. \end{aligned}$$

Thus, by (54), the joint density function of G and U is

$$(59)\qquad \frac{|G|^{\frac{1}{2}(n+m-p-1)} |I - UU'|^{\frac{1}{2}(n-p-1)} e^{-\frac{1}{2}\operatorname{tr} G}}{2^{\frac{1}{2}p(m+n)} \pi^{\frac{1}{2}mp + p(p-1)/4} \prod_{i=1}^{p} \Gamma[\frac{1}{2}(n+1-i)]}.$$

We integrate out G and make use of the fact that it is distributed according to $W(I, m + n)$, which can be derived from (59) and is also obvious when the definition of G as $A + B$ is recalled. The density function of the elements of U is found to be

$$(60)\qquad \frac{1}{\pi^{\frac{1}{2}mp}} \prod_{i=1}^{p} \left\{ \frac{\Gamma[\frac{1}{2}(m+n+1-i)]}{\Gamma[\frac{1}{2}(n+1-i)]} \right\} |I - UU'|^{\frac{1}{2}(n-p-1)}.$$

The f_i whose distribution we want to find are the nonzero roots of

$$(61)\qquad \begin{aligned} 0 = |A - f(A + B)| = |A - fG| &= |CUU'C' - fCC'| \\ &= |C| \cdot |UU' - fI| \cdot |C|. \end{aligned}$$

Since C is nonsingular, the f_i satisfy

$$|UU' - fI| = 0. \tag{62}$$

We shall show that the nonzero roots f_i of (62) are the same as the nonzero roots of

$$|U'U - fI| = 0. \tag{63}$$

For each root $f \neq 0$ of (62) there is a vector x satisfying

$$(UU' - fI)x = 0. \tag{64}$$

Multiplication by U' on the left gives

$$\begin{aligned} 0 &= U'(UU' - fI)x \\ &= (U'U - fI)U'x. \end{aligned} \tag{65}$$

Thus $U'x$ is a characteristic vector of $U'U$ and f is the corresponding characteristic root.

Let $f_1 > \cdots > f_m$ be the nonzero roots of (62). Then $|UU' - fI| = \prod_{i=1}^m (f_i - f)(-f)^{p-m}$ and $|U'U - fI| = \prod_{i=1}^m (f_i - f)$. Setting $f = 1$ shows

$$|I - U'U| = |I - UU'|. \tag{66}$$

Putting (66) in (60), the problem becomes to find the distribution of the m nonzero roots $1 > f_1 > \cdots > f_m > 0$ of (63), where U is distributed with density function

$$K|I - U'U|^{\frac{1}{2}(n-p-1)}. \tag{67}$$

Applying the argument leading to (59) and (62) to the case of the number of dimensions of the vector being not greater than the number of vectors in A, we can show that (47) (with asterisks inserted) is the distribution of the roots of

$$|U_* U_*' - fI| = 0, \tag{68}$$

where U_* (with p^* rows and m^* columns, $m^* \geq p^*$) is distributed with density function

$$K|I - U_* U_*'|^{\frac{1}{2}(n^* - p^* - 1)}. \tag{69}$$

If now we put $m^* = p$, $p^* = m$, $n^* = n + m - p$, and $U_* = U'$, then (68) and (69) are the same as are (63) and (67) and thus $f_1, \cdots, f_m$ must be distributed according to (47), with p replaced by m, m by p, and n by $n + m - p$, that is,

$$\pi^{\frac{1}{2}m} \prod_{i=1}^{m} \left\{ \frac{\Gamma[\frac{1}{2}(m+n+1-i)]}{\Gamma[\frac{1}{2}(m+1-i)]\Gamma[\frac{1}{2}(m+n-p+1-i)]\Gamma[\frac{1}{2}(p+1-i)]} \right\} \cdot \prod_{i=1}^{m} \{ f_i^{\frac{1}{2}(p-m-1)} (1-f_i)^{\frac{1}{2}(n-p-1)} \} \prod_{i<j} (f_i - f_j). \tag{70}$$

THEOREM 13.2.3. *If A is distributed as $\sum_{\alpha=1}^{m} Y_\alpha Y'_\alpha$, where the Y_α are independent, each with the distribution $N(0, \Sigma)$, $m \le p$, and B is independently distributed according to $W(\Sigma, n)$, $n \ge p$, then the density of the nonzero roots of $|A - f(A + B)| = 0$ is given by* (70).

It is of interest to note that these distributions of roots were found independently and at about the same time by Fisher (1939), Girshick (1939a), P. L. Hsu (1939a), Mood (1951), and Roy (1939). The development of the Jacobian in Section 13.2.2 is due mainly to Hsu [as reported by Deemer and Olkin (1951)].

13.3. THE CASE OF ONE NONSINGULAR WISHART MATRIX

In this section we shall find the distribution of the roots of

$$|A - \lambda I| = 0, \tag{1}$$

where the matrix A has the distribution $W(I, n)$. It will be observed that the variances of the principal components of a sample of $n + 1$ from $N(\mu, I)$ are $1/n$ times the roots of (1). We shall find the following theorem useful:

THEOREM 13.3.1. *If the symmetric matrix B has a density of the form $g(\lambda_1, \cdots, \lambda_p)$, where $\lambda_1 > \cdots > \lambda_p$ are the characteristic roots of B, then the joint distribution of the roots is*

$$\frac{\pi^{p(p+1)/4} g(\lambda_1, \cdots, \lambda_p) \prod_{i<j} (\lambda_i - \lambda_j)}{\prod_{i=1}^{p} \Gamma[\frac{1}{2}(p - i + 1)]}. \tag{2}$$

Proof. From Theorem 2 of Appendix 1 we know that there exists an orthogonal matrix C such that

$$B = C' \Lambda C, \tag{3}$$

where

$$\Lambda = \begin{pmatrix} \lambda_1 & 0 \cdots 0 \\ 0 & \lambda_2 \cdots 0 \\ \cdot & \cdot \quad\quad \cdot \\ \cdot & \cdot \quad\quad \cdot \\ \cdot & \cdot \quad\quad \cdot \\ 0 & 0 \cdots \lambda_p \end{pmatrix}. \tag{4}$$

If the λ's are numbered in descending order of magnitude and if $c_{i1} \geq 0$, then (with probability 1) the transformation from $\boldsymbol{B}$ to $\boldsymbol{\Lambda}$ and $\boldsymbol{C}$ is unique. Let the matrix $\boldsymbol{C}$ be given the coordinates $C_1, \cdots, C_{p(p-1)/2}$, and let the Jacobian of the transformation be $f(\boldsymbol{\Lambda}, \boldsymbol{C})$. Then the joint density of $\boldsymbol{\Lambda}$ and $\boldsymbol{C}$ is $g(\lambda_1, \cdots, \lambda_p)f(\boldsymbol{\Lambda}, \boldsymbol{C})$. To prove the theorem we must show that

$$(5)\qquad \int \cdots \int f(\boldsymbol{\Lambda}, \boldsymbol{C})\, dc_1 \cdots dc_{p(p-1)/2} = \frac{\pi^{p(p+1)/4} \prod_{i<j} (\lambda_i - \lambda_j)}{\prod \Gamma[\frac{1}{2}(p+1-i)]}.$$

We show this by taking a special case where $\boldsymbol{B} = \boldsymbol{UU'}$ and $\boldsymbol{U}\,(p \times m, m \geq p)$ has the density

$$(6)\qquad \pi^{-\frac{1}{2}mp} \prod_{i=1}^{p} \frac{\Gamma[\frac{1}{2}(m+n+1-i)]}{\Gamma[\frac{1}{2}(n+1-i)]} |\boldsymbol{I} - \boldsymbol{UU'}|^{\frac{1}{2}(n-p-1)}.$$

Then by Lemma 13.3.1, which will be stated below, $\boldsymbol{B}$ has the density

$$(7)\qquad \begin{aligned} &\pi^{-\frac{1}{4}p(p-1)} \prod_{i=1}^{p} \left\{\frac{\Gamma[\frac{1}{2}(m+n+1-i)]}{\Gamma[\frac{1}{2}(m+1-i)]\Gamma[\frac{1}{2}(n+1-i)]}\right\} \\ &\qquad \cdot |\boldsymbol{I} - \boldsymbol{B}|^{\frac{1}{2}(n-p-1)} |\boldsymbol{B}|^{\frac{1}{2}(m-p-1)} \\ &= \pi^{-\frac{1}{4}p(p-1)} \prod_{i=1}^{p} \left\{\frac{\Gamma[\frac{1}{2}(m+n+1-i)]}{\Gamma[\frac{1}{2}(m+1-i)]\Gamma[\frac{1}{2}(n+1-i)]}\right\} \\ &\qquad \prod (1-\lambda_i)^{\frac{1}{2}(n-p-1)} \prod \lambda_i^{\frac{1}{2}(m-p-1)} \\ &= g^*(\lambda_1, \cdots, \lambda_p). \end{aligned}$$

The joint density of $\boldsymbol{\Lambda}$ and $\boldsymbol{C}$ is $f(\boldsymbol{\Lambda}, \boldsymbol{C})g^*(\lambda_1, \cdots, \lambda_p)$. In the preceding section we proved that the marginal distribution of $\boldsymbol{\Lambda}$ is (50). Thus

$$(8)\qquad \begin{aligned} &\int \cdots \int g^*(\lambda_1, \cdots, \lambda_p) f(\boldsymbol{\Lambda}, \boldsymbol{C})\, d\boldsymbol{C} \\ &= g^*(\lambda_1, \cdots, \lambda_p) \int \cdots \int f(\boldsymbol{\Lambda}, \boldsymbol{C})\, d\boldsymbol{C} \\ &= \frac{\pi^{\frac{1}{4}p(p+1)} \prod (\lambda_i - \lambda_j)}{\prod_{i=1}^{p} \Gamma[\frac{1}{2}(p+1-i)]} g^*(\lambda_1, \cdots, \lambda_p). \end{aligned}$$

This proves (5) and hence the theorem.

The statement above (7) is based on the following lemma:

LEMMA 13.3.1. *If the density of* $\boldsymbol{Y}(p \times m)$ *is* $f(\boldsymbol{YY'})$, *then the density of* $\boldsymbol{B} = \boldsymbol{YY'}$ *is*

$$(9)\qquad \frac{|\boldsymbol{B}|^{\frac{1}{2}(m-p-1)} f(\boldsymbol{B}) \pi^{\frac{1}{2}p[m-\frac{1}{2}(p-1)]}}{\prod_{i=1}^{p} \Gamma[\frac{1}{2}(m+1-i)]}.$$

The proof of this, like that of Theorem 13.3.1, depends on exhibiting a special case; let $f(YY') = (2\pi)^{-\frac{1}{2}pm}e^{-\frac{1}{2}\operatorname{tr} YY'}$, then (9) is $w(B|I, m)$.

Now let us find the density of the roots of (1). The density of A is

$$(10)\qquad \frac{|A|^{\frac{1}{2}(n-p-1)}e^{-\frac{1}{2}\operatorname{tr} A}}{2^{\frac{1}{2}pn}\pi^{\frac{1}{4}p(p-1)}\prod_{i=1}^{p}\Gamma[\frac{1}{2}(n+1-i)]} = \frac{\prod_{i=1}^{p}\lambda_i^{\frac{1}{2}(n-p-1)}\exp(-\frac{1}{2}\sum_{i=1}^{p}\lambda_i)}{2^{\frac{1}{2}pn}\pi^{\frac{1}{4}p(p-1)}\prod_{i=1}^{p}\Gamma[\frac{1}{2}(n+1-i)]}.$$

Thus by the theorem we obtain as the distribution of the roots of A

$$(11)\qquad \frac{\pi^{\frac{1}{2}p}\prod_{i=1}^{p}\lambda_i^{\frac{1}{2}(n-p-1)}\exp\left(-\frac{1}{2}\sum_{i=1}^{p}\lambda_i\right)\prod_{i<j}(\lambda_i-\lambda_j)}{2^{\frac{1}{2}pn}\prod_{i=1}^{p}\{\Gamma[\frac{1}{2}(n+1-i)]\Gamma[\frac{1}{2}(p+1-i)]\}}.$$

THEOREM 13.3.2. *If $A(p \times p)$ has the distribution $W(I, n)$, then the characteristic roots ($\lambda_1 \geq \lambda_2 \geq \cdots \geq \lambda_p \geq 0$) have the density* (11) *over the range where the density is not* 0.

COROLLARY 13.3.1. *Let $v_1 \geq \cdots \geq v_p$ be the sample variances of the sample principal components of a sample of size $N = n + 1$ from $N(\mu, \sigma^2 I)$. Then $(n/\sigma^2)\, v_i$ are distributed with density* (11).

The characteristic vectors of A are uniquely defined (except for multiplication by -1) with probability 1 by

$$(12)\qquad \begin{aligned}(A - \lambda I)y &= 0,\\ y'y &= 1,\end{aligned}$$

since the roots are different with probability 1. Let the vectors with $y_{1j} \geq 0$ be

$$(13)\qquad Y = (y_1, \cdots, y_p).$$

Then

$$(14)\qquad AY = Y\Lambda.$$

From Section 11.2 we know that

$$(15)\qquad Y'Y = I.$$

Multiplication of (14) on the right by $Y^{-1} = Y'$ gives

$$(16)\qquad A = Y\Lambda Y'.$$

Thus $Y' = C$, defined above.

Now let us consider the joint distribution of $\boldsymbol{\Lambda}$ and $\boldsymbol{C}$. The matrix $\boldsymbol{A}$ has the distribution of

(17) $$\boldsymbol{A} = \sum_{\alpha=1}^{n} \boldsymbol{X}_\alpha \boldsymbol{X}_\alpha',$$

where the $\boldsymbol{X}_\alpha$ are independently distributed, each according to $N(\mathbf{0}, \boldsymbol{I})$. Let

(18) $$\boldsymbol{X}_\alpha^* = \boldsymbol{Q}\boldsymbol{X}_\alpha,$$

where $\boldsymbol{Q}$ is any orthogonal matrix. Then the $\boldsymbol{X}_\alpha^*$ are independently distributed according to $N(\mathbf{0}, \boldsymbol{I})$ and

(19) $$\boldsymbol{A}^* = \sum_{\alpha=1}^{n} \boldsymbol{X}_\alpha^* \boldsymbol{X}_\alpha^{*\prime} = \boldsymbol{Q}\boldsymbol{A}\boldsymbol{Q}'$$

is distributed according to $W(\boldsymbol{I}, n)$. The roots of $\boldsymbol{A}^*$ are the roots of $\boldsymbol{A}$; thus

(20) $$\boldsymbol{A}^* = \boldsymbol{C}^{**\prime}\boldsymbol{\Lambda}\boldsymbol{C}^{**},$$

(21) $$\boldsymbol{C}^{**\prime}\boldsymbol{C}^{**} = \boldsymbol{I}$$

define $\boldsymbol{C}^{**}$ if we require $c_{i1}^{**} \geq 0$. Let

(22) $$\boldsymbol{C}^* = \boldsymbol{C}\boldsymbol{Q}'.$$

Let

(23) $$J(\boldsymbol{C}^*) = \begin{pmatrix} \dfrac{c_{11}^*}{|c_{11}^*|} & 0 & \cdots & 0 \\ 0 & \dfrac{c_{21}^*}{|c_{21}^*|} & \cdots & 0 \\ \cdot & \cdot & & \cdot \\ \cdot & \cdot & & \cdot \\ \cdot & \cdot & & \cdot \\ 0 & 0 & \cdots & \dfrac{c_{p1}^*}{|c_{p1}^*|} \end{pmatrix},$$

with $c_{i1}^*/|c_{i1}^*| = 1$ if $c_{i1}^* = 0$. Thus $J(\boldsymbol{C}^*)$ is a diagonal matrix; the ith diagonal element is 1 if $c_{i1}^* \geq 0$ and is -1 if $c_{i1}^* < 0$. Thus

(24) $$\boldsymbol{C}^{**} = J(\boldsymbol{C}^*)\boldsymbol{C}^* = J(\boldsymbol{C}\boldsymbol{Q}')\boldsymbol{C}\boldsymbol{Q}'.$$

The distribution of $\boldsymbol{C}^{**}$ is the same as that of $\boldsymbol{C}$. We now shall show that this fact defines the distribution of $\boldsymbol{C}$.

DEFINITION 13.3.1. *If the random orthogonal matrix* $\mathbf{E}$ *of order* p *has a distribution such that* $\mathbf{E}\boldsymbol{Q}'$ *has the same distribution for every orthogonal* $\boldsymbol{Q}$, *the distribution of* $\mathbf{E}$ *is said to have the "Haar invariant" distribution (or normalized measure).*

The definition is possible because it has been proved that there is only one distribution with the required invariance property [Halmos (1950)]. It has also been shown that this distribution is the only one invariant under multiplication on the left by an orthogonal matrix (that is, the distribution of $\boldsymbol{QE}$ is the same as that of $\boldsymbol{E}$). From this it follows that the probability is $1/2^p$ that $\boldsymbol{E}$ be such that $e_{i1} \geq 0$. This can be seen as follows. Let $\boldsymbol{J}_1, \cdots, \boldsymbol{J}_{2^p}$ be the 2^p diagonal matrices with elements $+1$ and -1. Since the distribution of $\boldsymbol{J}_i\boldsymbol{E}$ is the same as $\boldsymbol{E}$, the probability that $e_{i1} \geq 0$ is the same as the probability that the elements in the first column of $\boldsymbol{J}_i\boldsymbol{E}$ are nonnegative. These events for $i = 1, \cdots, 2^p$ are mutually exclusive and exhaustive (except for elements being 0 which have probability 0) and thus the probability of any one is $1/2^p$.

The conditional distribution of $\boldsymbol{E}$ given $e_{i1} \geq 0$ is 2^p times the Haar invariant distribution over this part of the space. We shall call it the conditional Haar invariant distribution.

LEMMA 13.3.2. *If the orthogonal matrix* $\boldsymbol{E}$ *has a distribution such that* $e_{i1} \geq 0$ *and if* $\boldsymbol{E}^{**} = \boldsymbol{J}(\boldsymbol{EQ}')\boldsymbol{EQ}'$ *has the same distribution for every orthogonal* $\boldsymbol{Q}$, *then* $\boldsymbol{E}$ *has the conditional Haar invariant distribution.*

Proof. Let the space V of orthogonal matrices be partitioned into the subspaces $V_1, \cdots, V_{2^p}$ so that $\boldsymbol{J}_iV_i = V_1$, say, where $\boldsymbol{J}_1 = \boldsymbol{I}$ and V_1 is the set for which $e_{i1} \geq 0$. Let μ_1 be the measure in V_1 defined by the distribution of $\boldsymbol{E}$ assumed in the lemma. The measure $\mu(W)$ of a (measurable) set W in V_i is defined as $(1/2^p)\mu_1(\boldsymbol{J}_iW)$. Now we want to show that μ is the Haar invariant measure. Let W be any (measurable) set in V_1. The lemma assumes that $2^p\mu(W) = \mu_1(W) = \Pr\{\boldsymbol{E} \varepsilon W\} = \Pr\{\boldsymbol{E}^{**} \varepsilon W\} = \sum\mu_1(\boldsymbol{J}_i([W\boldsymbol{Q}' \cap V_i]) = 2^p\mu(W\boldsymbol{Q}')$. If U is any (meausurable) set in V, then $U = \cup_{j=1}^{2^p} (U \cap V_j)$. Since $\mu(U \cap V_j) = (1/2^p)\mu_1[\boldsymbol{J}_j(U \cap V_j)]$, by the above this is $\mu[(U \cap V_j)\boldsymbol{Q}']$. Thus $\mu(U) = \mu(U\boldsymbol{Q}')$. Thus μ is invariant and μ_1 is the conditional invariant distribution.

From the lemma we see that the matrix $\boldsymbol{C}$ has the conditional Haar invariant distribution. Since the distribution of $\boldsymbol{C}$ conditional on $\boldsymbol{\Lambda}$ is the same, $\boldsymbol{C}$ and $\boldsymbol{\Lambda}$ are independent.

THEOREM 13.3.3. *If* $\boldsymbol{C} = \boldsymbol{Y}'$, *where* $\boldsymbol{Y} = (\boldsymbol{Y}_1, \cdots, \boldsymbol{Y}_p)$ *are the normalized characteristic vectors of* $\boldsymbol{A}$ *with* $y_{1i} \geq 0$ *and where* $\boldsymbol{A}$ *is distributed according to* $W(\boldsymbol{I}, n)$, *then* $\boldsymbol{C}$ *has the conditional Haar invariant distribution and* $\boldsymbol{C}$ *is distributed independently of the characteristic roots.*

From the preceding work we can generalize Theorem 13.3.1.

THEOREM 13.3.4. *If the symmetric matrix* $\boldsymbol{B}$ *has a density of the form* $g(\lambda_1, \cdots, \lambda_p)$, *where* $\lambda_1 > \cdots > \lambda_p$ *are the characteristic roots of* $\boldsymbol{B}$, *then the joint distribution of the roots is* (3) *and the matrix of normalized characteristic vectors* $\boldsymbol{Y}$ $(y_{1i} \geq 0)$ *is independently distributed according to the conditional Haar invariant distribution.*

Proof. The density of QBQ', where $QQ' = I$, is the same as that of B (for the roots are invariant) and therefore the distribution of $J(Y'Q')Y'Q'$ is the same as that of Y'. Then Theorem 13.3.4 follows from Lemma 13.3.2.

We shall give an application of this theorem to the case where $B = B'$ is normally distributed with the (functionally independent) components of B independent with means 0 and variances $\mathscr{E}b_{ii}^2 = 1$ and $\mathscr{E}b_{ij}^2 = \frac{1}{2}\ (i < j)$.

THEOREM 13.3.5. *Let $B = B'$ have the density*

$$\pi^{-p(p+1)/4} 2^{-\frac{1}{2}p} e^{-\frac{1}{2}\operatorname{tr} B^2}. \tag{25}$$

Then the characteristic roots $\lambda_1 > \cdots > \lambda_p$ of B have the density

$$2^{-\frac{1}{2}p}\{\prod \Gamma[\tfrac{1}{2}(p+1-i)]\}^{-1} e^{-\frac{1}{2}\Sigma\lambda_i^2} \prod_{i<j} (\lambda_i - \lambda_j) \tag{26}$$

and the matrix Y of the normalized characteristic vectors ($y_{1i} \geq 0$) is independently distributed according to the conditional Haar invariant distribution.

Proof. Since the characteristic roots of B^2 are $\lambda_1^2, \cdots, \lambda_p^2$ and $\operatorname{tr} B^2 = \Sigma\lambda_i^2$, the theorem follows directly.

13.4. CANONICAL CORRELATIONS

The sample canonical correlations were shown in Section 12.3 to be the square roots of the roots of

$$|A_{12}A_{22}^{-1}A_{21} - fA_{11}| = 0, \tag{1}$$

where

$$A_{ij} = \sum_{\alpha=1}^{N} (X_\alpha^{(i)} - \bar{X}^{(i)})(X_\alpha^{(j)} - \bar{X}^{(j)})', \tag{2}$$

and the distribution of

$$X = \begin{pmatrix} X^{(1)} \\ X^{(2)} \end{pmatrix} \tag{3}$$

is $N(\boldsymbol{\mu}, \boldsymbol{\Sigma})$, where

$$\boldsymbol{\Sigma} = \begin{pmatrix} \boldsymbol{\Sigma}_{11} & \boldsymbol{\Sigma}_{12} \\ \boldsymbol{\Sigma}_{21} & \boldsymbol{\Sigma}_{22} \end{pmatrix}. \tag{4}$$

From Section 3.3 we know that the distribution of A_{ij} is the same as that of

$$A_{ij} = \sum_{\alpha=1}^{n} Y_\alpha^{(i)} Y_\alpha^{(j)\prime}, \tag{5}$$

where $n = N - 1$ and

$$Y = \begin{pmatrix} Y^{(1)} \\ Y^{(2)} \end{pmatrix} \tag{6}$$

is distributed according to $N(\mathbf{0}, \boldsymbol{\Sigma})$. Let us assume that the dimensionality of $\boldsymbol{Y}^{(1)}$, say p_1, is not greater than the dimensionality of $\boldsymbol{Y}^{(2)}$, say p_2. Then there are p_1 nonzero roots of (1), say

$$f_1 > f_2 > \cdots > f_{p_1}. \tag{7}$$

Now we shall find the distribution of $\{f_i\}$ when

$$\boldsymbol{\Sigma}_{12} = \boldsymbol{0}. \tag{8}$$

For the moment assume $\{\boldsymbol{Y}_\alpha^{(2)}\}$ to be fixed. Then A_{22} is fixed, and

$$\boldsymbol{B} = \boldsymbol{A}_{12}\boldsymbol{A}_{22}^{-1} \tag{9}$$

is the matrix of regression coefficients of $\boldsymbol{Y}^{(1)}$ on $\boldsymbol{Y}^{(2)}$. From Section 4.4 we know that

$$\begin{aligned} \boldsymbol{A}_{11\cdot 2} = \sum_{\alpha=1}^{n} (\boldsymbol{Y}_\alpha^{(1)} - \boldsymbol{B}\boldsymbol{Y}_\alpha^{(2)})(\boldsymbol{Y}_\alpha^{(1)} - \boldsymbol{B}\boldsymbol{Y}_\alpha^{(2)})' &= \boldsymbol{A}_{11} - \boldsymbol{B}\boldsymbol{A}_{22}\boldsymbol{B}' \\ &= \boldsymbol{A}_{11} - \boldsymbol{A}_{12}\boldsymbol{A}_{22}^{-1}\boldsymbol{A}_{21} \end{aligned} \tag{10}$$

and

$$\boldsymbol{Q} = \boldsymbol{B}\boldsymbol{A}_{22}\boldsymbol{B}' = \boldsymbol{A}_{12}\boldsymbol{A}_{22}^{-1}\boldsymbol{A}_{21} \tag{11}$$

$(\boldsymbol{\beta} = \boldsymbol{0})$ are independently distributed according to $W(\boldsymbol{\Sigma}_{11}, n - p_2)$ and $W(\boldsymbol{\Sigma}_{11}, p_2)$ respectively. In terms of $\boldsymbol{Q}$ equation (1) defining f is

$$|\boldsymbol{Q} - f(\boldsymbol{A}_{11\cdot 2} + \boldsymbol{Q})| = 0. \tag{12}$$

The distribution of f_i $(i = 1, \cdots, p_1)$ is the distribution of the nonzero roots of (12), and this distribution is given by (see Section 13.2)

$$\pi^{\frac{1}{2}p_1} \prod_{i=1}^{p_1} \frac{\Gamma[\frac{1}{2}(N-i)]}{\Gamma[\frac{1}{2}(N - p_2 - i)]\Gamma[\frac{1}{2}(p_1 + 1 - i)]\Gamma[\frac{1}{2}(p_2 + 1 - i)]} \prod_{i=1}^{p_1} \{f_i^{\frac{1}{2}(p_2-p_1-1)}(1 - f_i)^{\frac{1}{2}(N-p_2-p_1-2)}\} \prod_{i<j} (f_i - f_j). \tag{13}$$

Since the conditional distribution (13) does not depend upon $\boldsymbol{Y}^{(2)}$, (13) is the unconditional distribution of the squares of the sample canonical correlation coefficients of the two sets $\boldsymbol{X}_\alpha^{(1)}$ and $\boldsymbol{X}_\alpha^{(2)}$ $(\alpha = 1, \cdots, N)$. The distribution (13) also holds when the $\boldsymbol{X}^{(2)}$ are actually fixed variate vectors or have any distribution, so long as $\boldsymbol{X}^{(1)}$ and $\boldsymbol{X}^{(2)}$ are independently distributed and $\boldsymbol{X}^{(1)}$ has a multivariate normal distribution.

In the special case when $p_1 = 1$, $p_2 = p - 1$, (13) reduces to

$$\frac{\Gamma[\frac{1}{2}(N-1)]}{\Gamma[\frac{1}{2}(N-p)]\Gamma[\frac{1}{2}(p-1)]} f^{\frac{1}{2}(p-3)}(1 - f)^{\frac{1}{2}(N-p-2)} \tag{14}$$

which is the distribution of the sample multipie correlation coefficient between $\boldsymbol{X}^{(1)}$ $(p_1 = 1)$ and $\boldsymbol{X}^{(2)}$ $(p_2 = p - 1)$.

REFERENCES

Deemer and Olkin (1951); Fisher (1939); Girshick (1939a); Goldstine and von Neumann (1951); Halmos (1950); Hsu (1939a); Kendall (1946), 354–358; Laha (1954); Marriott (1952); Mood (1951); Nanda (1948a), (1948b), (1950), (1951); Olkin (1952); Olkin and Roy (1954); S. N. Roy (1939b), (1942a), (1945a), (1952); Sastry (1948); Wilks (1943), 260–270.

PROBLEMS

1. (Sec. 13.2) Prove Theorem 13.2.1 for $p = 2$ by calculating the Jacobian directly.

2. (Sec. 13.2) Prove Theorem 13.3.2 for $p = 2$ directly by representing the orthogonal matrix $\boldsymbol{C}$ in terms of cosine and sine of an angle.

3. (Sec. 13.3) Give the Haar invariant distribution explicitly for the orthogonal matrix represented in terms of cosine and sine of an angle.

4. (Sec. 13.3) Let $\boldsymbol{A}$ and $\boldsymbol{B}$ be distributed according to $W(\boldsymbol{\Sigma}, m)$ and $W(\boldsymbol{\Sigma}, n)$ respectively. Let $\lambda_1 > \cdots > \lambda_p$ be the roots of $|\boldsymbol{A} - \lambda\boldsymbol{B}| = 0$ and $\phi_1 > \cdots > \phi_p$ be the roots of $|\boldsymbol{A} - \phi\boldsymbol{\Sigma}| = 0$. Find the distribution of the ϕ's from that of the λ's by letting $n \to \infty$.

5. (Sec. 13.2) Consider the distribution of the roots of $|\boldsymbol{A} - \lambda\boldsymbol{B}| = 0$ when $\boldsymbol{A}$ and $\boldsymbol{B}$ are of order two and are distributed according to $W(\boldsymbol{\Sigma}, m)$ and $W(\boldsymbol{\Sigma}, n)$ respectively. (*a*) Find the distribution of the largest root. (*b*) Find the distribution of the smallest root. (*c*) Find the distribution of the sum of the roots.

6. Let $\boldsymbol{A}$ be distributed according to $W(\boldsymbol{\Sigma}, n)$. In case of $p = 2$ find the distribution of the characteristic roots of $\boldsymbol{A}$. [*Hint:* Transform so that $\boldsymbol{\Sigma}$ goes into a diagonal matrix.]

7. From the result in Problem 6 find the distribution of the sphericity criterion (when the null hypothesis is not true).

8. (Sec. 13.2) Prove that the Jacobian $|\partial(\boldsymbol{G}, \boldsymbol{A})/\partial(\boldsymbol{E}, \boldsymbol{F})|$ is $\prod(f_i - f_j)$ times a function of $\boldsymbol{E}$ by showing that the Jacobian vanishes for $f_i = f_j$ and that its degree in f_i is the same as that of $\prod(f_i - f_j)$.

9. (Sec. 13.3) Prove Lemma 13.3.1 in as much detail as Theorem 13.3.1.

CHAPTER 14

A Review of Some Other Work in Multivariate Analysis

14.1. INTRODUCTION

In this chapter we shall take a brief look at some more advanced topics of multivariate analysis. The preceding thirteen chapters give a good deal of the fundamentals, but, of course, many important aspects are not covered. We shall here consider some of the more important and more recent developments in statistical inference to be applied to multivariate normal distributions. We shall not treat all of these, and we shall not treat some other problems which generally are considered to apply to other distributions. The selection of the topics in this chapter are, of course, influenced by the author's own interests.

14.2. TESTING HYPOTHESES OF RANK AND ESTIMATING LINEAR RESTRICTIONS ON REGRESSION COEFFICIENTS; CANONICAL CORRELATIONS AND CANONICAL VARIATES

In Chapter 8 we considered some aspects of the regression problem. Let $\boldsymbol{x}_\alpha$ $(\alpha = 1, \cdots, N)$ be an observation from $N(\boldsymbol{\beta} \boldsymbol{z}_\alpha, \boldsymbol{\Sigma})$. We considered testing the hypothesis $\boldsymbol{\beta}_1 = \boldsymbol{\beta}_1^*$, where $\boldsymbol{\beta} = (\boldsymbol{\beta}_1\ \boldsymbol{\beta}_2)$. For the sake of simplifying the discussion in this section we will assume that $\boldsymbol{\beta} = \boldsymbol{\beta}_1$ and the hypothesis is $\boldsymbol{\beta} = \boldsymbol{0}$. (The more general case can be reduced to this by proper transformations; see Chapter 8.) The null hypothesis specifies that the $\boldsymbol{z}_\alpha$ have no effect on the $\boldsymbol{x}_\alpha$. If the null hypothesis is not true, we can ask whether the $\boldsymbol{z}_\alpha$ affect the $\boldsymbol{x}_\alpha$ in a certain number of directions; that is, we consider the means of the N normal distributions as N points in a p-dimensional space, and we ask whether the dimensionality of the linear subspace is some number. Suppose the dimensionality is q^*; then we can choose a new coordinate system in the p-space of the $\boldsymbol{x}_\alpha$ so that the linear subspace is spanned by the first q^* axes. In the new system we can say that the independent variables affect q^* of the dependent variables. The new coordinate system shows clearly the influence of the independent variables on the dependent variables.

If $c'x_\alpha$ is a linear combination whose expected value is not affected by z_α, then $\mathscr{E}c'x_\alpha = c'\beta z_\alpha = 0$ implies $c'\beta = 0$. If there are q^* linearly independent linear combinations of x_α that are affected and $p - q^*$ that are not, then there are $p - q^*$ linearly independent c's such that $c'\beta = \mathbf{0}$. Thus the rank of β is q^*.

Now let us put this in terms of canonical correlations and variates according to the second model treated in Section 12.2. With β of rank q^*, there are q^* roots of (58) of Section 12.2, different from zero; that is, roots of $|\beta C\beta' - \nu\Sigma| = 0$, where $C = (1/N)\sum z_\alpha z_\alpha'$. Any vector satisfying (57), $(\beta C\beta' - \nu\Sigma)\alpha = \mathbf{0}$, for a root $\nu = 0$ satisfies $\beta C\beta'\alpha = \mathbf{0}$ and $(\beta'\alpha)'C(\beta'\alpha) = 0$, and therefore $\alpha'\beta = \mathbf{0}$.

In the sample we estimate β by $\hat{\beta}$ and Σ by $\hat{\Sigma} = (1/N)(\sum x_\alpha x_\alpha' - \hat{\beta}D\hat{\beta}')$, where $D = \sum z_\alpha z_\alpha'$. Let $\bar{\nu}_1 \geq \cdots \geq \bar{\nu}_p \geq 0$ be the roots of

$$|\hat{\beta}D\hat{\beta}' - \nu N\hat{\Sigma}| = 0 \tag{1}$$

and let $\hat{\alpha}^{(1)}, \cdots, \hat{\alpha}^{(p)}$ be vectors satisfying

$$(\hat{\beta}D\hat{\beta}' - \bar{\nu}_i N\hat{\Sigma})\hat{\alpha}^{(i)} = \mathbf{0}, \tag{2}$$

$$\hat{\alpha}^{(i)\prime}\hat{\Sigma}\hat{\alpha}^{(i)} = 1. \tag{3}$$

If the roots $\bar{\nu}_i$ are all different, (2) and (3) specify the $\hat{\alpha}^{(i)}$ uniquely and

$$\hat{\alpha}^{(i)\prime}\hat{\Sigma}\hat{\alpha}^{(j)} = 0, \qquad i \neq j. \tag{4}$$

If several of the $\bar{\nu}_i$ are zero, we can impose (4), but there is still some freedom left. It has been shown that $\bar{\nu}_i$, $\hat{\alpha}^{(i)}$ are maximum likelihood estimates of ν_i, $\alpha^{(i)}$. If the restriction that β is of rank q^* is imposed this is also true, but then $(\hat{\alpha}^{(q^*+1)}, \cdots, \hat{\alpha}^{(p)})$, associated with the smallest $\bar{\nu}_i$, can be multiplied by an arbitrary orthogonal matrix on the right [as can $(\alpha^{(q^*+1)}, \cdots, \alpha^{(p)})$]. In case of restrictions, the derivation is not straightforward; also the maximum likelihood estimates of β and Σ must take the restrictions into account.

It would seem intuitively reasonable to reject the null hypothesis that β is of rank q^* against the alternative that it is greater if the $p - q^*$ smallest roots are not sufficiently small. In fact, the likelihood ratio criterion is $\lambda = \prod_{i=q^*+1}^{p}(1 + \bar{\nu}_i)^{-\frac{1}{2}N}$. Other functions of these roots can also be used. If the null hypothesis is true, $-2 \log \lambda$ (and $N\sum_{q^*+1}^{p}\bar{\nu}_i$) has an asymptotic χ^2-distribution with $(p - q^*)(q - q^*)$ degrees of freedom.

In many situations one has a multiple-decision problem of deciding the rank of β. A possible procedure is to test that the rank is 0 against the alternative that it is 1, then 1 against 2, and so forth. The first step would involve deciding whether $\bar{\nu}_1$ is too big, and so on.

The distribution theory involved here is complicated. For fixed sample size, the distributions will be discussed in Section 14.4, and the asymptotic theory will be considered in Section 14.5. Tests about the $\boldsymbol{\alpha}^{(i)}$ corresponding to $\nu_i = 0$ can be based on exact theory.

These considerations also go under the name of the theory of discriminant functions (see Problem 4 of Chapter 12). Methods have been given for tests about the canonical variates corresponding to nonzero roots.

14.3. THE NONCENTRAL WISHART DISTRIBUTION

In Chapter 13 we deduced the distribution of the roots of

$$|\hat{\boldsymbol{\beta}} D \hat{\boldsymbol{\beta}}' - \nu N \hat{\boldsymbol{\Sigma}}| = 0 \tag{1}$$

when $\boldsymbol{\beta} = \mathbf{0}$; this was based on $\hat{\boldsymbol{\beta}} D \hat{\boldsymbol{\beta}}'$ and $N\hat{\boldsymbol{\Sigma}}$ having independent Wishart distributions. The distributions needed for the inference outlined in the preceding section are the distributions of the roots of (1) and of the corresponding vectors when $\boldsymbol{\beta} \neq \mathbf{0}$; in principle the methods of Chapter 13 can be used to obtain these distributions from the distributions of $\hat{\boldsymbol{\beta}} D \hat{\boldsymbol{\beta}}'$ and $N\hat{\boldsymbol{\Sigma}}$ when $\boldsymbol{\beta} \neq \mathbf{0}$; the distribution of $\hat{\boldsymbol{\beta}} D \hat{\boldsymbol{\beta}}'$ is the noncentral Wishart distribution.

Let $A = \sum_{\alpha=1}^{n} Y_\alpha Y_\alpha'$, where Y_α is an observation from $N(\boldsymbol{\mu}_\alpha, \boldsymbol{\Sigma})$, and let $\mathbf{T} = \sum_{\alpha=1}^{n} \boldsymbol{\mu}_\alpha \boldsymbol{\mu}_\alpha'$, and let t be the rank of $\mathbf{T}$. Then the density of A is $w(A|\boldsymbol{\Sigma}, n)$ times a function that depends on the roots of $|\mathbf{T} - \lambda\boldsymbol{\Sigma}| = 0$ and on the roots of*

$$|\mathbf{T} - \lambda \boldsymbol{\Sigma} A^{-1} \boldsymbol{\Sigma}| = 0. \tag{2}$$

If $t = 1$, this function involves a Bessel function; if $t = 2$ it involves an infinite series of Bessel functions; if $t = 3$ it can be expressed in terms of a triple infinite series. For higher values of t it is expressed as a multiple integral.

Unfortunately, the noncentral Wishart distribution is so complicated that its usefulness is limited.

14.4. DISTRIBUTIONS OF CERTAIN CHARACTERISTIC ROOTS AND VECTORS THAT DEPEND ON PARAMETERS

To compute the powers of various tests and to carry out the inference outlined in Section 14.2, it is desirable to know the distributions of the roots and vectors of certain matrices under various conditions. Some of these distributions have been obtained, but in general they are very complicated. The distribution of the roots $\bar{\nu}_1 \geq \cdots \geq \bar{\nu}_p$ of $|\hat{\boldsymbol{\beta}} D \hat{\boldsymbol{\beta}}' - \nu N \hat{\boldsymbol{\Sigma}}| = 0$ depends on the roots $\nu_1 \geq \cdots \geq \nu_p$ of $|\boldsymbol{\beta} C \boldsymbol{\beta}' - \nu \boldsymbol{\Sigma}| = 0$. A formal procedure has been given for obtaining the distribution as a product

* (4) in Anderson and Girshick (1944) and (8) in Anderson (1946) should be written as (2) here.

of the distribution for all ν's being zero and a complicated factor. This factor can be given explicitly in case the rank of $\boldsymbol{\beta}$ is one.*

Another case to be treated is where the roots are of $|S_{12}S_{22}^{-1}S_{21} - \nu S_{11}| = 0$, where S has a Wishart distribution and $\boldsymbol{\Sigma}_{12} \neq \mathbf{0}$.

Still another case is $|A - \nu B| = 0$ when A and B have Wishart matrices with different covariance matrices. The evaluation of the power function of an invariant test of $\boldsymbol{\Sigma}_1 = \boldsymbol{\Sigma}_2$ (Chapter 10) is in terms of such distributions.

14.5. THE ASYMPTOTIC DISTRIBUTION OF CERTAIN CHARACTERISTIC ROOTS AND VECTORS

Since the distributions mentioned in Section 14.4 are extremely complicated, there is much interest in the asymptotic distributions. Consider first the roots of $|\hat{\boldsymbol{\beta}}D\hat{\boldsymbol{\beta}}' - \nu N\hat{\boldsymbol{\Sigma}}| = 0$. Since $\text{plim}\,(1/N)\hat{\boldsymbol{\beta}}D\hat{\boldsymbol{\beta}}' = \boldsymbol{\beta}\lim\,(1/N)D\boldsymbol{\beta}'$ and $\text{plim}\,\hat{\boldsymbol{\Sigma}} = \boldsymbol{\Sigma}$, we have $\text{plim}\,(\bar{\nu}_i - \nu_i) = 0$ (if $\lim\,(1/N)D$ exists). If $\nu_1 > \cdots > \nu_p > 0$ [and $\lim\sqrt{N}(\nu_i - \nu_j) \neq 0$], that is, if all the population roots are different, then $\sqrt{N}(\bar{\nu}_i - \nu_i)$ and $\sqrt{N}(\hat{\alpha}^{(i)} - \alpha^{(i)})$ have a limiting joint normal distribution. The asymptotic distributions are much more complicated if some of the roots are equal [or if $\lim\sqrt{N}(\nu_i - \nu_j) = 0$], but a very important situation is where some of the roots are zero; in particular, where $\boldsymbol{\beta}$ is of rank less than p (but $q \geq p$). Suppose the first q^* population roots are distinct (and stay distinct as $N \to \infty$) and the last $p - q^*$ population roots are zero. Then $\sqrt{N}(\bar{\nu}_1 - \nu_1), \cdots, \sqrt{N}(\bar{\nu}_{q^*} - \nu_{q^*})$ are asymptotically independently normally distributed; the limiting distribution of $N\bar{\nu}_{q^*+1}, \cdots, N\bar{\nu}_p$ is the distribution of the characteristic roots of A $(p - q^*) \times (p - q^*)$ with the distribution $W(\boldsymbol{I}, q - q^*)$; that is, the density (11) of Section 13.3 with n replaced by $q - q^*$ and p replaced by $p - q^*$. It follows that $N\sum_{i=q^*+1}^{p}\bar{\nu}_i$ has an asymptotic χ^2-distribution with $(p - q^*)\cdot(q - q^*)$ degrees of freedom (as tr A), although the individual roots do not have asymptotic χ^2-distributions.

14.6. PRINCIPAL COMPONENTS

In Chapter 11 we defined the principal components as the characteristic vectors of a covariance matrix $\boldsymbol{\Sigma}$, and in Chapter 13 we derived the distribution of the characteristic roots and vectors of a sample covariance matrix S, when $\boldsymbol{\Sigma} = \boldsymbol{I}$. We are interested in some hypotheses about the roots and vectors of $\boldsymbol{\Sigma}$ when $\boldsymbol{\Sigma} \neq \boldsymbol{I}$. Suppose $\boldsymbol{X} = \boldsymbol{Y} + \boldsymbol{Z}$; for convenience of discussion suppose $\mathscr{E}\boldsymbol{Y} = \mathscr{E}\boldsymbol{Z} = \mathbf{0}$ and $\mathscr{E}\boldsymbol{Y}\boldsymbol{Z}' = \mathbf{0}$. Let $\mathscr{E}\boldsymbol{Y}\boldsymbol{Y}' = \boldsymbol{\Phi}$ and $\mathscr{E}\boldsymbol{Z}\boldsymbol{Z}' = \boldsymbol{\Psi}$; then $\boldsymbol{\Sigma} = \boldsymbol{\Phi} + \boldsymbol{\Psi}$. We consider Y to be the

* Roy (1942b) incorrectly claims to treat a more general case.

real effects and in a lower dimensional space, say, of q dimensions; then $\mathbf{\Phi}$ is of rank q. We consider $\boldsymbol{Z}$ to consist of errors of measurement; it may therefore be reasonable to let $\mathbf{\Psi} = \sigma^2 \boldsymbol{I}$. Then $\mathbf{\Sigma} = \mathbf{\Phi} + \sigma^2 \boldsymbol{I}$ has $p - q$ roots equal to σ^2. In this situation we may want to test whether q is a given value q^* against the alternative that it is greater. One would expect to accept this hypothesis if the $p - q^*$ smaller roots of $\boldsymbol{S}$ are approximately the same. The asymptotic theory has been worked out. Given $q = q^*$, one is interested in the first q^* vectors of $\mathbf{\Sigma}$, which characterize $\mathbf{\Phi}$; the asymptotic theory has been developed.

These considerations are closely related to factor analysis (see Section 14.7). Sometimes $\boldsymbol{S}$ is replaced by the matrix of correlations. In this case, however, even the asymptotic theory is complicated.

14.7. FACTOR ANALYSIS

Suppose $\boldsymbol{X} = \mathbf{\Lambda}\boldsymbol{f} + \boldsymbol{\mu} + \boldsymbol{U}$, where $\boldsymbol{f}$ is an m-component vector of (nonobservable) factor scores, $\boldsymbol{\mu}$ is a fixed vector of means, and $\boldsymbol{U}$ is a vector of (nonobservable) errors (or errors plus specific factors); the $p \times m$ matrix $\mathbf{\Lambda}$ consists of factor loadings ($m < p$). When $\boldsymbol{f}$ is random we assume $\mathscr{E}\boldsymbol{f} = \mathbf{0}$, $\mathscr{E}\boldsymbol{U} = \mathbf{0}$, $\mathscr{E}\boldsymbol{f}\boldsymbol{f}' = \boldsymbol{M}$, $\mathscr{E}\boldsymbol{U}\boldsymbol{U}' = \mathbf{\Psi}$, diagonal, and $\mathscr{E}\boldsymbol{f}\boldsymbol{U}' = \mathbf{0}$. Then $\mathscr{E}\boldsymbol{X} = \boldsymbol{\mu}$ and the covariance matrix of the observable $\boldsymbol{X}$ is

$$\mathscr{E}(\boldsymbol{X} - \boldsymbol{\mu})(\boldsymbol{X} - \boldsymbol{\mu})' = \mathbf{\Lambda}\boldsymbol{M}\mathbf{\Lambda}' + \mathbf{\Psi}. \tag{1}$$

There are problems about the model, such as what covariance matrices $\mathbf{\Sigma}$ can be represented by (1) for a given m and, if there is such a representation, what restrictions shall be put on $\mathbf{\Lambda}$ and $\boldsymbol{M}$ to make them unique. In the way of statistical inference, there is the problem of estimating $\mathbf{\Lambda}$, $\boldsymbol{M}$, and $\mathbf{\Psi}$ from a set of observations on $\boldsymbol{X}$; the centroid method is the simplest; the notion of principal components can be used; and there is a maximum likelihood solution. Another problem is to test whether m is a given number; more important is to decide what number m is. Associated with each observation $\boldsymbol{x}_\alpha$ there is an unobserved $\boldsymbol{f}_\alpha$; in some cases one wishes to estimate this $\boldsymbol{f}_\alpha$.

A rather complete exposition of these problems and solutions is given by Anderson and Rubin (1956). Kendall and Smith (1950) and Bartlett (1953) have also given expository papers. The reference list includes only some standard textbooks and a few papers on statistical inference.

The factor analysis model can be put another way. There are $p - r$ linearly independent vectors $\boldsymbol{\alpha}$ such that $\boldsymbol{\alpha}'\mathbf{\Lambda} = \mathbf{0}$; that is, $\mathbf{\Lambda}\boldsymbol{f} = \boldsymbol{Z}$ satisfies $p - r$ linear equations. The problem of estimating such $\boldsymbol{\alpha}$'s often goes under the name of estimating linear structural relations. In case $p - r$ is small some assumption of nonnormality must be made (or strong assumptions about the distribution of $\boldsymbol{U}$ are made).

14.8. STOCHASTIC EQUATIONS

Closely related to the subject of Section 14.2 is the study of stochastic equations. Suppose X_α has a multivariate normal distribution such that $\mathscr{E}\mathbf{A}X_\alpha = \mathbf{\Gamma} z_\alpha$ and the covariance matrix of $\mathbf{A}X_\alpha$ is $\mathbf{\Sigma}$, where $\mathbf{A}$ is $(p \times p)$ nonsingular. We can express this as

$$\text{(1)} \qquad \mathbf{A}X_\alpha - \mathbf{\Gamma} z_\alpha = U_\alpha,$$

where the (unobserved) U_α is distributed according to $N(\mathbf{0}, \mathbf{\Sigma})$. (1) is considered as a set of linear equations in the components of X_α, and may be solved as

$$\text{(2)} \qquad X_\alpha = \mathbf{A}^{-1}\mathbf{\Gamma} z_\alpha + \mathbf{A}^{-1}U_\alpha;$$

the equations are stochastic because the right-hand side of (1) is a random vector. This model written in the form of (2) is simply a regression model; that is, X_α is distributed according to $N(\boldsymbol{\beta} z_\alpha, \mathbf{\Sigma}^*)$, where $\boldsymbol{\beta} = \mathbf{A}^{-1}\mathbf{\Gamma}$ and $\mathbf{\Sigma}^* = \mathbf{A}^{-1}\mathbf{\Sigma}(\mathbf{A}^{-1})'$. The form of (1) may have some intrinsic interest; for example, if (1) represents a model for the formation of economic quantities X_α, each component equation may represent the behavior of a given set of individuals.

From the regression model $N(\boldsymbol{\beta} z_\alpha, \mathbf{\Sigma}^*)$ we go to (1) by the equations $\mathbf{A}\boldsymbol{\beta} = \mathbf{\Gamma}$ and $\mathbf{\Sigma} = \mathbf{A}\mathbf{\Sigma}^*\mathbf{A}'$. Now in order to determine $\mathbf{A}$, $\mathbf{\Gamma}$, and $\mathbf{\Sigma}$ uniquely (that is, to identify these parameters) we need some restrictions on them. These may be linear restrictions on $\mathbf{A}$ and $\mathbf{\Gamma}$ (for example, that some coefficients are zero) or restrictions on $\mathbf{\Sigma}$ (for instance, that $\mathbf{\Sigma}$ is diagonal). If the restrictions are that one row of $\mathbf{\Gamma}$ is $\mathbf{0}$, then we have exactly the situation of Section 14.2. Other kinds of restrictions lead to more complicated problems. Maximum likelihood estimates have been obtained, computing methods formulated, and asymptotic theory developed.

The list of references to this section is very incomplete; the reader is referred to the reference list in Hood and Koopmans (1953). A number of papers are included in that volume as well as the volume edited by Koopmans (1950).

14.9. TIME SERIES ANALYSIS

One time series model specifies a vector X_t as $z_t + U_t$, where z_t is the (nonrandom) vector of systematic parts and U_t is the (random) vector of errors. Various assumptions can be made about z_t, such as being trigonometric functions of time, or polynomials; the U_t are considered identically and independently distributed. This model is in many cases treated as in Chapter 8 and Section 14.2.

Another model is the stochastic difference equation. Let $\cdots, \boldsymbol{U}_{-1}, \boldsymbol{U}_0, \boldsymbol{U}_1, \cdots$ be a sequence of variables independently distributed according to $N(\mathbf{0}, \boldsymbol{\Sigma})$. Suppose $\cdots, \boldsymbol{Y}_{-1}, \boldsymbol{Y}_0, \boldsymbol{Y}_1, \cdots$ satisfy

$$\boldsymbol{Y}_t = \sum_{i=1}^{r} \boldsymbol{\Gamma}_i \boldsymbol{Y}_{t-i} + \boldsymbol{U}_t. \tag{1}$$

Then we say the sequence $\boldsymbol{Y}_t$ satisfies a stochastic difference equation. Variations on the model are to let the sequence of $\boldsymbol{U}$'s be $\boldsymbol{U}_1, \boldsymbol{U}_2, \cdots$ and define $\boldsymbol{Y}_0, \boldsymbol{Y}_{-1}, \cdots, \boldsymbol{Y}_{-(r-1)}$ as fixed. One can also insert in (1) the term $\boldsymbol{\Gamma z}_t$. One method of inference is to treat $\boldsymbol{\Gamma z}_t + \sum_i \boldsymbol{\Gamma}_i \boldsymbol{Y}_{t-i}$ as if it were $\boldsymbol{\beta z}_t$. The methods of regression analysis can be used, and the asymptotic theory is similar. The model can also be modified by making the left-hand side of (1) $\mathbf{A}\boldsymbol{Y}_t$.

A model in some sense more general than that above is to consider $\boldsymbol{X}_t$ as a stationary Gaussian (normal) stochastic process. Let $\mathscr{E}\boldsymbol{X}_t = \boldsymbol{\mu}$, $\mathscr{E}(\boldsymbol{X}_t - \boldsymbol{\mu})(\boldsymbol{X}_s - \boldsymbol{\mu})' = \boldsymbol{\Sigma}_{t-s}$, and suppose that any set $\boldsymbol{X}_{t_1}, \cdots, \boldsymbol{X}_{t_n}$ has a joint normal distribution. In the case that $\boldsymbol{X}_t$ is complex-valued,

$$\sigma_{ij}(t-s) = \int_{-\pi}^{\pi} e^{i(t-s)\lambda}\, dF_{ij}(\lambda);$$

the case of real values is a little more complicated [see Cramér (1940)]. In this model we want to estimate $\boldsymbol{\Sigma}_{t-s}$ and the set of spectral functions $F_{ij}(\lambda)$. Not a great deal has been done in this area.

References to the first model are given in Section 14.2; references to the second in 14.8.

REFERENCES

Section 14.2. T. W. Anderson (1951b); T. W. Anderson and Rubin (1949); Bartlett (1939a), (1947a), (1948), (1951); Cochran (1943); Fisher (1938), (1940); Geary (1948), (1949); P. L. Hsu (1941a), (1941b); C. R. Rao (1949a), (1949b), (1949c), (1952), 370–378; S. N. Roy (1939b); Tintner (1945), (1950b), (1952); Williams (1952), (1955).

Section 14.3. T. W. Anderson (1946); T. W. Anderson and Girshick (1944); Herz (1955); A. T. James (1955a), (1955b); Weibull (1953).

Section 14.4. T. W. Anderson (1945); Bartlett (1947b); Girshick (1941); S. N. Roy (1942a), (1942b), (1946a), (1946b), (1946c).

Section 14.5. T. W. Anderson (1948), (1951c); T. W. Anderson and Rubin (1950); Girshick (1939); Hotelling (1936); P. L. Hsu (1941a), (1941b), (1941c); Nanda (1948a), (1948b).

Section 14.6. T. W. Anderson (1956); Bartlett (1950), (1951b), (1951c), (1954); Girshick (1939a); Lawley (1953), (1956).

Section 14.7. T. W. Anderson and Rubin (1956); Bartlett (1937b), (1953); Holzinger and Harmon (1941); Kendall and Smith (1950); Lawley (1940), (1942), (1953); C. R. Rao (1955b); Rasch (1953); Thomson (1951); Thurstone (1947); Whittle (1952).

Section 14.8. See references for Section 14.2 and in Hood and Koopmans (1953); Koopmans (1950); Wold (1953).

Section 14.9. Cramér (1940). See also references for Sections 14.2 and 14.8.

APPENDIX 1

Matrix Theory

1. DEFINITION OF A MATRIX AND OPERATIONS ON MATRICES

In this appendix we summarize some of the well-known definitions and theorems of matrix algebra. A number of results which are not always contained in books on matrix algebra are proved here.

An $m \times n$ matrix A is a rectangular array of real numbers

$$A = \begin{pmatrix} a_{11} & a_{12} & \cdots & a_{1n} \\ a_{21} & a_{22} & \cdots & a_{2n} \\ \cdot & \cdot & & \cdot \\ \cdot & \cdot & & \cdot \\ \cdot & \cdot & & \cdot \\ a_{m1} & a_{m2} & \cdots & a_{mn} \end{pmatrix}, \tag{1}$$

which may be abbreviated (a_{ij}), $i = 1, 2, \cdots, m$; $j = 1, 2, \cdots, n$. Capital boldface letters will be used to denote matrices whose elements are the corresponding lower-case letters with appropriate subscripts. The sum of two matrices A and B of the same numbers of rows and columns, respectively, is defined by

$$A + B = (a_{ij}) + (b_{ij}) = (a_{ij} + b_{ij}). \tag{2}$$

The product of a matrix by a real number λ is defined by

$$\lambda A = A\lambda = (\lambda a_{ij}). \tag{3}$$

It can be verified that these operations have the algebraic properties

$$A + B = B + A, \tag{4}$$

$$(A + B) + C = A + (B + C), \tag{5}$$

$$A + (-1)A = (0), \tag{6}$$

$$(\lambda + \mu)A = \lambda A + \mu A, \tag{7}$$

$$\lambda(A + B) = \lambda A + \lambda B. \tag{8}$$

(9) $$\lambda(\mu A) = (\lambda\mu)A.$$

The matrix (0) with all elements 0 can be denoted as $\mathbf{0}$.

If A has the same number of columns as B has rows, that is, $A = (a_{ij})$, $i = 1, \cdots, l$, $j = 1, \cdots, m$, $B = (b_{jk})$, $j = 1, \cdots, m$, $k = 1, \cdots, n$, then A and B can be multiplied according to the rule

(10) $$AB = (a_{ij})(b_{jk}) = \left(\sum_{j=1}^{m} a_{ij}b_{jk}\right), \qquad i = 1, \cdots, l;\ k = 1, \cdots, n;$$

that is, AB is a matrix with l rows and n columns, the element in the ith row and kth column being $\sum_{j=1}^{m} a_{ij}b_{jk}$. The matrix product has the properties

(11) $$(AB)C = A(BC),$$

(12) $$A(B + C) = AB + AC,$$

(13) $$(A + B)C = AC + BC.$$

The relationships (11) through (13) hold provided one side is meaningful (that is, the numbers of rows and columns are such that the operations can be performed) since it follows then that the other side is also meaningful. Because of (11) we can write

(14) $$(AB)C = A(BC) = ABC.$$

The product BA may be meaningless even if AB is meaningful, and even when both are meaningful they are not necessarily equal.

The *transpose* of the $l \times m$ matrix $A = (a_{ij})$ is defined to be the $m \times l$ matrix A' which has in the jth row and ith column the element that A has in the ith row and jth column. The operation of transposition has the properties

(15) $$(A')' = A,$$

(16) $$(A + B)' = A' + B',$$

(17) $$(AB)' = B'A',$$

again with the restriction which is understood throughout these notes that at least one side is meaningful.

A vector x with m components can be treated as a matrix with m rows and one column. Therefore, the above operations hold for vectors.

We shall now be concerned with square matrices of the same size, which can be added and multiplied at will. The number of rows and columns will be taken to be p. A is called *symmetric* if $A = A'$. A particular matrix of considerable interest is the *identity* matrix

$$I = \begin{pmatrix} 1 & 0 & 0 \cdots 0 \\ 0 & 1 & 0 \cdots 0 \\ 0 & 0 & 1 \cdots 0 \\ \cdot & \cdot & \cdot \quad\; \cdot \\ \cdot & \cdot & \cdot \quad\; \cdot \\ \cdot & \cdot & \cdot \quad\; \cdot \\ 0 & 0 & 0 \cdots 1 \end{pmatrix} = (\delta_{ij}), \tag{18}$$

where δ_{ij}, the Kronecker delta, is defined by

$$\delta_{ij} = \begin{cases} 1, & \text{if } i = j, \\ 0, & \text{if } i \neq j. \end{cases} \tag{19}$$

The identity matrix satisfies

$$IA = AI = A. \tag{20}$$

Associated with any square matrix A is the determinant $|A|$, defined by

$$|A| = \sum (-1)^{f(j_1, \cdots, j_p)} \prod_{i=1}^{p} a_{ij_i}, \tag{21}$$

where the summation is taken over all permutations $(j_1, \cdots, j_p)$ of the set of integers $(1, \cdots, p)$, and $f(j_1, \cdots, j_p)$ is the number of transpositions required to change $(1, \cdots, p)$ into $(j_1, \cdots, j_p)$. A transposition consists of interchanging two numbers, and it can be shown that, although one can transform $(1, \cdots, p)$ into $(j_1, \cdots, j_p)$ by transpositions in many different ways, the number of transpositions required is always even or always odd so that $(-1)^{f(j_1, \cdots, j_p)}$ is consistently defined. It can be shown that

$$|AB| = |A| \cdot |B|. \tag{22}$$

Also

$$|A| = |A'|. \tag{23}$$

A submatrix of A is the rectangular array obtained from A by deleting rows and columns. A *minor* is the determinant of a square submatrix of A. The minor of an element a_{ij} is the determinant of the submatrix of A obtained by deleting the ith row and jth column. The cofactor of a_{ij}, say A_{ij}, is $(-1)^{i+j}$ times the minor of a_{ij}. It can be shown that

$$|A| = \sum_{j=1}^{p} a_{ij} A_{ij} = \sum_{j=1}^{p} a_{jk} A_{jk}. \tag{24}$$

If $|A| \neq 0$, there exists a unique matrix B such that $AB = I$. B is called the inverse of A and is denoted by A^{-1}. Let a^{hk} be the element of A^{-1} in the hth row and kth column. Then

$$a^{hk} = \frac{A_{kh}}{|A|}. \tag{25}$$

The operation of taking the inverse satisfies

$$(AC)^{-1} = C^{-1}A^{-1}, \tag{26}$$

since

$$(AC)(C^{-1}A^{-1}) = A(CC^{-1})A^{-1} = AIA^{-1} = AA^{-1} = I. \tag{27}$$

Also $I^{-1} = I$ and $A^{-1}A = I$. Furthermore, since the transposition of (27) gives $(A^{-1})'A' = I$ we have $(A^{-1})' = (A')^{-1}$.

A matrix whose determinant is not zero is called *nonsingular*. If $|A| \neq 0$, then the only solution to

$$A\boldsymbol{z} = \boldsymbol{0} \tag{28}$$

is the trivial one of $\boldsymbol{z} = \boldsymbol{0}$ [by multiplication of (28) on the left by A^{-1}]. If $|A| = 0$, there is at least one nontrivial solution (that is, $\boldsymbol{z} \neq \boldsymbol{0}$). Thus an equivalent definition of A being nonsingular is that (28) have only the trivial solution.

A set of vectors $\boldsymbol{z}_1, \cdots, \boldsymbol{z}_r$ is said to be *linearly independent* if there exists no set of scalars $c_1, \cdots, c_r$, not all zero, such that $\sum c_i \boldsymbol{z}_i = \boldsymbol{0}$. A $q \times p$ matrix D is said to be of *rank* r if the maximum number of linearly independent columns is r. Then every minor of order $r + 1$ must be zero (from the remarks in the preceding paragraph applied to the relevant $r + 1$ order square matrix) and at least one minor of order r must be nonzero. Conversely, if there is at least one minor of order r that is nonzero, there is at least one set of r columns (or rows) which is linearly independent. If all minors of order $r + 1$ are zero, there cannot be any set of $r + 1$ columns (or rows) which are linearly independent for such linear independence would imply a nonzero minor of order $r + 1$, but this contradicts the assumption. Thus rank r is equivalently defined by the maximum number of linearly independent rows, by the maximum number of linearly independent columns, or the maximum order of nonzero minors.

We now consider the quadratic form

$$\boldsymbol{x}'A\boldsymbol{x} = \sum_{i,j=1}^{p} a_{ij}x_i x_j, \tag{29}$$

where $\boldsymbol{x}' = (x_1, \cdots, x_p)$ and $A = (a_{ij})$. The matrix A is symmetric. This matrix A and the quadratic form are called *positive semidefinite* if

$\boldsymbol{x}'\boldsymbol{A}\boldsymbol{x} \geq 0$ for all $\boldsymbol{x}$. If $\boldsymbol{x}'\boldsymbol{A}\boldsymbol{x} > 0$ for all $\boldsymbol{x} \neq \boldsymbol{0}$, $\boldsymbol{A}$ and the quadratic form are called *positive definite*.

THEOREM 1. *If* $\boldsymbol{C}$ *with* p *rows and columns is positive definite, and if* $\boldsymbol{B}$ *with* p *rows and* q *columns,* $q \leq p$, *is of rank* q, *then* $\boldsymbol{B}'\boldsymbol{C}\boldsymbol{B}$ *is positive definite.*

Proof. Given a vector $\boldsymbol{y} \neq \boldsymbol{0}$, let $\boldsymbol{x} = \boldsymbol{B}\boldsymbol{y}$. Since $\boldsymbol{B}$ is of rank q, $\boldsymbol{B}\boldsymbol{y} = \boldsymbol{x} \neq \boldsymbol{0}$. Then

$$\begin{aligned} y'(B'CB)y &= (By)'C(By) \\ &= x'Cx > 0. \end{aligned} \tag{30}$$

The proof is completed by observing that $\boldsymbol{B}'\boldsymbol{C}\boldsymbol{B}$ is symmetric. As a converse, we observe that $\boldsymbol{B}'\boldsymbol{C}\boldsymbol{B}$ is positive definite only if $\boldsymbol{B}$ is of rank q, for otherwise there exists $\boldsymbol{y} \neq \boldsymbol{0}$ such that $\boldsymbol{B}\boldsymbol{y} = \boldsymbol{0}$.

COROLLARY 1. *If* $\boldsymbol{C}$ *is positive definite and* $\boldsymbol{B}$ *is nonsingular, then* $\boldsymbol{B}'\boldsymbol{C}\boldsymbol{B}$ *is positive definite.*

COROLLARY 2. *If* $\boldsymbol{C}$ *is positive definite, then* $\boldsymbol{C}^{-1}$ *is positive definite.*

Proof. $\boldsymbol{C}$ must be nonsingular for if $\boldsymbol{C}\boldsymbol{x} = \boldsymbol{0}$ for $\boldsymbol{x} \neq \boldsymbol{0}$, then $\boldsymbol{x}'\boldsymbol{C}\boldsymbol{x} = 0$ for this $\boldsymbol{x}$, but this is contrary to the assumption that $\boldsymbol{C}$ is positive definite. Let $\boldsymbol{B}$ in Theorem 1 be $\boldsymbol{C}^{-1}$. Then $\boldsymbol{B}'\boldsymbol{C}\boldsymbol{B} = (\boldsymbol{C}^{-1})'\boldsymbol{C}\boldsymbol{C}^{-1} = (\boldsymbol{C}^{-1})'$. Transposing $\boldsymbol{C}\boldsymbol{C}^{-1} = \boldsymbol{I}$, we have $(\boldsymbol{C}^{-1})'\boldsymbol{C}' = (\boldsymbol{C}^{-1})'\boldsymbol{C} = \boldsymbol{I}$. Thus $\boldsymbol{C}^{-1} = (\boldsymbol{C}^{-1})'$. Q.E.D.

COROLLARY 3. *Let* $\boldsymbol{D}$ *be a* $q \times q$ *matrix formed by deleting* $p - q$ *rows of a positive definite matrix* $\boldsymbol{C}$ *and the corresponding* $p - q$ *columns of* $\boldsymbol{C}$. *Then* $\boldsymbol{D}$ *is positive definite.*

Proof. This follows from Theorem 1 by forming $\boldsymbol{B}$ by taking the $p \times p$ identity matrix and deleting the columns corresponding to those deleted from $\boldsymbol{C}$.

The *trace* of a square matrix $\boldsymbol{A}$ is defined as $\operatorname{tr}\boldsymbol{A} = \sum_{i=1}^{p} a_{ii}$. The following properties are verified directly:

$$\operatorname{tr}(A + B) = \operatorname{tr}A + \operatorname{tr}B, \tag{31}$$

$$\operatorname{tr}AB = \operatorname{tr}BA. \tag{32}$$

A square matrix $\boldsymbol{A}$ is said to be diagonal if $a_{ij} = 0$, $i \neq j$. Then $|\boldsymbol{A}| = \prod_{i=1}^{p} a_{ii}$, for in (24) $|\boldsymbol{A}| = a_{11}A_{11}$, and in turn A_{11} is evaluated similarly.

A square matrix $\boldsymbol{A}$ is said to be triangular if $a_{ij} = 0$ for $i > j$ or alternatively for $i < j$. In either case the product of two such matrices is again triangular in the same way, for the i, jth term $(i > j)$ of $\boldsymbol{AB}$ is $\sum_{k=1}^{p} a_{ik}b_{kj} = 0$ if $a_{ik} = 0$ for $k < i$ and $b_{kj} = 0$ for $k > j$ with $j < i$.

2. CHARACTERISTIC ROOTS AND VECTORS

The *characteristic roots* of a square matrix $\boldsymbol{B}$ are defined as the roots of the characteristic equation

$$|B - \lambda I| = 0. \tag{33}$$

For example, with $B = \begin{pmatrix} 5 & 2 \\ 2 & 5 \end{pmatrix}$,

$$(34)\quad |B - \lambda I| = \begin{vmatrix} 5-\lambda & 2 \\ 2 & 5-\lambda \end{vmatrix} = 25 - 4 - 10\lambda + \lambda^2 = \lambda^2 - 10\lambda + 21.$$

The degree of this equation is the order of the matrix B and the constant term is $|B|$.

A matrix C is said to be orthogonal if $C'C = I$; it follows that $CC' = I$. Let the vectors $x' = (x_1, \cdots, x_p)$ and $y' = (y_1, \cdots, y_p)$ represent two points in a p-dimensional Euclidean space. The distance squared between them is $D(x, y) = (x - y)'(x - y)$. The transformation $z = Cx$ can be thought of as a change of coordinate axes in the p-dimensional space. If C is orthogonal, the transformation is distance-preserving for

$$(35)\quad \begin{aligned} D(Cx, Cy) &= (Cy - Cx)'(Cy - Cx) \\ &= (y - x)'C'C(y - x) = (y - x)'(y - x) = D(x, y). \end{aligned}$$

Since the angles of a triangle are determined by the lengths of its sides, the transformation $z = Cx$ also preserves angles. It consists of a rotation together with a possible reflection of one or more axes.

THEOREM 2. *Given any symmetric matrix* B, *there exists an orthogonal matrix* C *such that*

$$(36)\quad C'BC = D = \begin{pmatrix} d_1 & 0 & \cdots & 0 \\ 0 & d_2 & \cdots & 0 \\ \cdot & \cdot & & \cdot \\ \cdot & \cdot & & \cdot \\ \cdot & \cdot & & \cdot \\ 0 & 0 & \cdots & d_p \end{pmatrix}.$$

If B *is positive definite, the* d_i *are* > 0.

The proof is given in the discussion of principal components in Section 11.2 for the case of B positive semidefinite. The characteristic equation (33) under transformation by C becomes

$$(37)\quad \begin{aligned} 0 &= |C'||B - \lambda I| \cdot |C| = |C'(B - \lambda I)C| \\ &= |C'BC - \lambda I| = |D - \lambda I| \\ &= \begin{vmatrix} d_1 - \lambda & 0 & \cdots & 0 \\ 0 & d_2 - \lambda & \cdots & 0 \\ \cdot & \cdot & & \cdot \\ \cdot & \cdot & & \cdot \\ \cdot & \cdot & & \cdot \\ 0 & 0 & \cdots & d_p - \lambda \end{vmatrix} = \prod_{i=1}^{p} (d_i - \lambda). \end{aligned}$$

Thus the characteristic roots of $\boldsymbol{B}$ are merely the diagonal elements of the transformed matrix $\boldsymbol{D}$.

Corollary 4. *If $\boldsymbol{B}$ is positive definite, there exists a nonsingular matrix $\boldsymbol{E}$ such that $\boldsymbol{E}'\boldsymbol{B}\boldsymbol{E} = \boldsymbol{I}$.*

Proof. Let

$$\boldsymbol{D}^{\frac{1}{2}} = \begin{pmatrix} \sqrt{d_1} & 0 & \cdots & 0 \\ 0 & \sqrt{d_2} & \cdots & 0 \\ \cdot & \cdot & & \cdot \\ \cdot & \cdot & & \cdot \\ \cdot & \cdot & & \cdot \\ 0 & 0 & \cdots & \sqrt{d_p} \end{pmatrix}. \tag{38}$$

Then it follows from (36) that $\boldsymbol{C}\boldsymbol{D}^{-\frac{1}{2}}$ is such a matrix $\boldsymbol{E}$.

Corollary 5. *If $\boldsymbol{B}$ is positive definite, $|\boldsymbol{B}| > 0$.*

Proof. This follows from Corollary 4 because $|\boldsymbol{B}| = |\boldsymbol{E}'|^{-1} \cdot |\boldsymbol{I}| \cdot |\boldsymbol{E}|^{-1} = |\boldsymbol{E}|^{-2}$.

Corollary 6. *If $\boldsymbol{B}$ is positive definite, every principal minor is positive.*

Proof. A principal minor is the determinant of a matrix formed from $\boldsymbol{B}$ by deleting certain rows and corresponding columns. Corollary 6 follows from Corollaries 3 and 5.

If λ_i is a characteristic root of $\boldsymbol{B}$, then a vector $\boldsymbol{x}_i$ not identically $\boldsymbol{0}$ satisfying

$$(\boldsymbol{B} - \lambda_i \boldsymbol{I})\boldsymbol{x}_i = \boldsymbol{0} \tag{39}$$

is called a characteristic vector of the matrix $\boldsymbol{B}$ corresponding to the characteristic root λ_i. Any scalar multiple of $\boldsymbol{x}_i$ is clearly also a characteristic vector. When $\boldsymbol{B}$ is symmetric $\boldsymbol{x}_i'(\boldsymbol{B} - \lambda_i \boldsymbol{I}) = \boldsymbol{0}$. For the transformed matrix $\boldsymbol{D}$, if $\lambda_1 = d_1$ is the first root, then the corresponding characteristic vector is $\begin{pmatrix} k \\ 0 \\ \cdot \\ \cdot \\ \cdot \\ 0 \end{pmatrix}$ since

$$(\boldsymbol{D} - \lambda_1 \boldsymbol{I}) \begin{pmatrix} k \\ 0 \\ \cdot \\ \cdot \\ \cdot \\ 0 \end{pmatrix} = \begin{pmatrix} 0 & 0 & \cdots & 0 \\ 0 & d_2 - d_1 & \cdots & 0 \\ \cdot & \cdot & & \cdot \\ \cdot & \cdot & & \cdot \\ \cdot & \cdot & & \cdot \\ 0 & 0 & \cdots & d_p - d_1 \end{pmatrix} \begin{pmatrix} k \\ 0 \\ \cdot \\ \cdot \\ \cdot \\ 0 \end{pmatrix} = \boldsymbol{0}. \tag{40}$$

If none of the other d_i are equal to λ_1, then this is the only characteristic vector corresponding to λ_1. Substituting $\boldsymbol{D} = \boldsymbol{C}'\boldsymbol{B}\boldsymbol{C}$, we have

$$(41) \qquad (\boldsymbol{C}'\boldsymbol{B}\boldsymbol{C} - \lambda_1 \boldsymbol{I}) \begin{pmatrix} k \\ 0 \\ \cdot \\ \cdot \\ \cdot \\ 0 \end{pmatrix} = \boldsymbol{0}.$$

Multiplying on the left by $\boldsymbol{C}$ we obtain

$$(42) \qquad \boldsymbol{0} = (\boldsymbol{B}\boldsymbol{C} - \lambda_1 \boldsymbol{I}\boldsymbol{C}) \begin{pmatrix} k \\ 0 \\ \cdot \\ \cdot \\ \cdot \\ 0 \end{pmatrix} = (\boldsymbol{B} - \lambda_1 \boldsymbol{I})\boldsymbol{C} \begin{pmatrix} k \\ 0 \\ \cdot \\ \cdot \\ \cdot \\ 0 \end{pmatrix};$$

so that $\boldsymbol{C} \begin{pmatrix} k \\ 0 \\ \cdot \\ \cdot \\ \cdot \\ 0 \end{pmatrix}$ is the characteristic vector of $\boldsymbol{B}$ corresponding to the root λ_1, namely the first column of $\boldsymbol{C}$. By (40) and (42) we see that a characteristic vector lies in the direction of the principal axis (see Chapter 11). The characteristic roots of $\boldsymbol{B}$ are proportional to the squares of the reciprocals of the lengths of the principal axes of the ellipsoid

$$(43) \qquad \boldsymbol{x}'\boldsymbol{B}\boldsymbol{x} = 1$$

since this becomes under the rotation $\boldsymbol{y} = \boldsymbol{C}\boldsymbol{x}$

$$(44) \qquad 1 = \boldsymbol{y}'\boldsymbol{D}\boldsymbol{y} = \sum_{i=1}^{p} d_i y_i^2.$$

If two or more characteristic roots are equal, then the principal axes are indeterminate.

For a pair of matrices $\boldsymbol{A}$ (nonsingular) and $\boldsymbol{B}$ we shall also consider equations of the form

$$(45) \qquad |\boldsymbol{B} - \lambda \boldsymbol{A}| = 0.$$

The roots of such equations are of interest because of their invariance under certain transformations. In fact, for nonsingular $\boldsymbol{C}$, the roots of

$$(46) \qquad |\boldsymbol{C}'\boldsymbol{B}\boldsymbol{C} - \lambda(\boldsymbol{C}'\boldsymbol{A}\boldsymbol{C})| = 0$$

are the same as those of (45) since

$$|C'BC - \lambda C'AC| = |C'(B - \lambda A)C| = |C'| \cdot |B - \lambda A| \cdot |C| \tag{47}$$

and $|C'| = |C| \neq 0$.

By Corollary 4 we have that if A is positive definite there is a matrix E such that

$$E'AE = I. \tag{48}$$

Let $E'BE = B^*$. From Theorem 2 we deduce that there exists an orthogonal matrix C such that $C'B^*C = D$, where D is diagonal. Defining EC as F, we have the following theorem:

THEOREM 3. *Given* B *positive semidefinite and* A *positive definite, there exists a nonsingular matrix* F *such that*

$$F'BF = \begin{pmatrix} \lambda_1 & 0 & \cdots & 0 \\ 0 & \lambda_2 & \cdots & 0 \\ \cdot & \cdot & & \cdot \\ \cdot & \cdot & & \cdot \\ \cdot & \cdot & & \cdot \\ 0 & 0 & \cdots & \lambda_p \end{pmatrix}, \tag{49}$$

$$F'AF = I, \tag{50}$$

where $\lambda_1 \geq \cdots \geq \lambda_p (\geq 0)$ *are the roots of* (45). *If* B *is positive definite,* $\lambda_i > 0$.

3. PARTITIONED VECTORS AND MATRICES

Consider the matrix A defined by (1). Let

$$\begin{aligned} A_{11} &= (a_{ij}), & i &= 1, \cdots, p;\ j = 1, \cdots, q, \\ A_{12} &= (a_{ij}), & i &= 1, \cdots, p;\ j = q + 1, \cdots, n, \\ A_{21} &= (a_{ij}), & i &= p + 1, \cdots, m;\ j = 1, \cdots, q, \\ A_{22} &= (a_{ij}), & i &= p + 1, \cdots, m;\ j = q + 1, \cdots, n. \end{aligned} \tag{51}$$

Then we can write

$$A = \begin{pmatrix} A_{11} & A_{12} \\ A_{21} & A_{22} \end{pmatrix}. \tag{52}$$

We can say that A has been partitioned into submatrices A_{ij}. Let B ($m \times n$) be partitioned similarly into submatrices $B_{ij} (i, j = 1, 2)$. Then it is easily verified that

$$A + B = \begin{pmatrix} A_{11} + B_{11} & A_{12} + B_{12} \\ A_{21} + B_{21} & A_{22} + B_{22} \end{pmatrix}. \tag{53}$$

Now partition $C\ (n \times r)$ as

$$C = \begin{pmatrix} C_{11} & C_{12} \\ C_{21} & C_{22} \end{pmatrix}, \tag{54}$$

where C_{11} and C_{12} have q rows and C_{11} and C_{21} have s columns. Then

$$\begin{aligned} AC &= \begin{pmatrix} A_{11} & A_{12} \\ A_{21} & A_{22} \end{pmatrix}\begin{pmatrix} C_{11} & C_{12} \\ C_{21} & C_{22} \end{pmatrix} \\ &= \begin{pmatrix} A_{11}C_{11} + A_{12}C_{21} & A_{11}C_{12} + A_{12}C_{22} \\ A_{21}C_{11} + A_{22}C_{21} & A_{21}C_{12} + A_{22}C_{22} \end{pmatrix}. \end{aligned} \tag{55}$$

To verify this consider an element in the first p rows and first s columns of AC. The i, jth element is

$$\sum_{k=1}^{n} a_{ik}c_{kj}, \qquad i \leq p, j \leq s. \tag{56}$$

This sum can be written

$$\sum_{k=1}^{q} a_{ik}c_{kj} + \sum_{k=q+1}^{n} a_{ik}c_{kj}. \tag{57}$$

The first sum is the i, jth element of $A_{11}C_{11}$, the second sum is the i, jth element of $A_{12}C_{21}$, and therefore the entire sum (56) is the i, jth element of $A_{11}C_{11} + A_{12}C_{21}$. In a similar fashion we can verify that the other submatrices of AC can be written as in (55).

We note in passing that if A is partitioned as in (52), then the transpose of A can be written

$$A' = \begin{pmatrix} A'_{11} & A'_{21} \\ A'_{12} & A'_{22} \end{pmatrix}, \tag{58}$$

If $A_{12} = 0$ and $A_{21} = 0$, then for A positive definite and A_{11} square

$$A^{-1} = \begin{pmatrix} A_{11}^{-1} & 0 \\ 0 & A_{22}^{-1} \end{pmatrix}. \tag{59}$$

The matrix on the right exists because A_{11} and A_{22} are nonsingular. That the right-hand matrix is the inverse of A is verified by multiplication

$$\begin{pmatrix} A_{11} & 0 \\ 0 & A_{22} \end{pmatrix}\begin{pmatrix} A_{11}^{-1} & 0 \\ 0 & A_{22}^{-1} \end{pmatrix} = \begin{pmatrix} I & 0 \\ 0 & I \end{pmatrix}, \tag{60}$$

which is a partitioned form of $I\ (p \times p)$.

We also note that

$$\begin{aligned} \begin{vmatrix} A_{11} & 0 \\ 0 & A_{22} \end{vmatrix} &= \begin{vmatrix} A_{11} & 0 \\ 0 & I \end{vmatrix} \cdot \begin{vmatrix} I & 0 \\ 0 & A_{22} \end{vmatrix} \\ &= |A_{11}| \cdot |A_{22}|. \end{aligned} \tag{61}$$

The evaluation of the first determinant in the middle is made by expanding according to minors of the last row; the only nonzero element in the sum is the last which is 1 times a determinant of the same form with I of order one less. The procedure is repeated until $|A_{11}|$ is the minor. Similarly,

$$\begin{vmatrix} A_{11} & A_{12} \\ 0 & A_{22} \end{vmatrix} = \begin{vmatrix} I & 0 \\ 0 & A_{22} \end{vmatrix} \cdot \begin{vmatrix} A_{11} & A_{12} \\ 0 & I \end{vmatrix} = |A_{11}| \cdot |A_{22}|. \tag{62}$$

A useful fact is that if A_1 of q rows and p columns is of rank q, there exists a matrix A_2 of $p-q$ rows and p columns such that

$$A = \begin{pmatrix} A_1 \\ A_2 \end{pmatrix} \tag{63}$$

is nonsingular. This statement is verified by numbering the columns of A so that A_{11} consisting of the first q columns of A_1 is nonsingular (at least one $q \times q$ minor of A_1 is different from zero) and then taking A_2 as $(0 \quad I)$; then

$$|A| = \begin{vmatrix} A_{11} & A_{12} \\ 0 & I \end{vmatrix} = |A_{11}|, \tag{64}$$

which is not equal to zero.

THEOREM 4. *Let the positive definite matrix A be partitioned as in* (52) *so that A_{11} is square, and let*

$$B = \begin{pmatrix} I & -A_{12}A_{22}^{-1} \\ 0 & I \end{pmatrix}. \tag{65}$$

Then

$$BAB' = \begin{pmatrix} A_{11} - A_{12}A_{22}^{-1}A_{21} & 0 \\ 0 & A_{22} \end{pmatrix}. \tag{66}$$

This theorem is proved in Section 2.4. We can use this result to prove that if A is positive definite, there exists a triangular matrix T such that $TAT' = I$. Let B_p be the matrix (65) when $A_{22} = a_{pp}$. Then

$$B_pAB_p' = \begin{pmatrix} A_{11} - A_{12}A_{22}^{-1}A_{21} & 0 \\ 0 & a_{pp} \end{pmatrix} = \begin{pmatrix} A_{11\cdot p} & 0 \\ 0 & a_{pp} \end{pmatrix}. \tag{67}$$

Let B_{p-1} be defined similarly for $A_{11\cdot p}$. Then

$$B_{p-1}A_{11\cdot p}B_{p-1}' = \begin{pmatrix} A_{11\cdot p-1,\, p} & 0 \\ 0 & a_{p-1,\, p-1\cdot p} \end{pmatrix}. \tag{68}$$

Recursively, we define $B_j (j = p - 2, \cdots, 2)$ similarly, so that

$$B_j A_{11\cdot j+1,\cdots,p} B_j' = \begin{pmatrix} A_{11\cdot j,\cdots,p} & 0 \\ 0 & a_{jj\cdot j+1,\cdots,p} \end{pmatrix}. \tag{69}$$

Then let

$$\bar{B}_j = \begin{pmatrix} B_j & 0 \\ 0 & I \end{pmatrix} \tag{70}$$

and $B = \bar{B}_2 \bar{B}_3 \cdots \bar{B}_p$. Then B is triangular (for it is the product of triangular matrices) and

$$BAB' = \begin{pmatrix} a_{11\cdot 2,\cdots,p} & 0 & \cdots & 0 \\ 0 & a_{22\cdot 3,\cdots,p} & \cdots & 0 \\ \cdot & \cdot & & \cdot \\ \cdot & \cdot & & \cdot \\ \cdot & \cdot & & \cdot \\ 0 & 0 & \cdots & a_{pp} \end{pmatrix} = A^*, \tag{71}$$

say. Finally, let $T = (A^*)^{-\frac{1}{2}}B$.

We note that this is essentially the method of pivotal condensation discussed in Section 5 of this appendix.

THEOREM 5. *Let A be partitioned as in (52) so that A_{22} is square, and assume A_{22} is nonsingular. Then*

$$|A| = |A_{11} - A_{12}A_{22}^{-1}A_{21}| \cdot |A_{22}|. \tag{72}$$

Proof. This follows from (66) by taking determinants on both sides. The determinant of the left-hand side is $|A|$ because the determinant of the B matrix is 1; the determinant of the right-hand side of (66) is the right-hand side of (72).

4. SOME MISCELLANEOUS RESULTS

THEOREM 6. *Let C be $p \times p$, positive semidefinite, and of rank r ($\leq p$). Then there is a nonsingular matrix A such that*

$$ACA' = \begin{pmatrix} I & 0 \\ 0 & 0 \end{pmatrix}, \tag{73}$$

where the identity is of order r.

Proof. Since C is of rank r, there is a $(p - r) \times p$ matrix A_2 such that

$$A_2 C = 0. \tag{74}$$

Choose B $(r \times p)$ such that

$$\begin{pmatrix} B \\ A_2 \end{pmatrix} \tag{75}$$

is nonsingular. Then

$$(76)\qquad \begin{pmatrix} \boldsymbol{B} \\ \boldsymbol{A}_2 \end{pmatrix} \boldsymbol{C}(\boldsymbol{B}' \ \boldsymbol{A}_2') = \begin{pmatrix} \boldsymbol{BC} \\ \boldsymbol{0} \end{pmatrix} (\boldsymbol{B}' \ \boldsymbol{A}_2') = \begin{pmatrix} \boldsymbol{BCB}' & \boldsymbol{0} \\ \boldsymbol{0} & \boldsymbol{0} \end{pmatrix}.$$

This matrix is of rank r and therefore $\boldsymbol{BCB}'$ is nonsingular. By Corollary 4 there is a nonsingular matrix $\boldsymbol{D}$ such that $\boldsymbol{D}(\boldsymbol{BCB}')\boldsymbol{D}' = \boldsymbol{I}$. Then

$$(77)\qquad \boldsymbol{A} = \begin{pmatrix} \boldsymbol{DB} \\ \boldsymbol{A}_2 \end{pmatrix} = \begin{pmatrix} \boldsymbol{D} & \boldsymbol{0} \\ \boldsymbol{0} & \boldsymbol{I} \end{pmatrix} \begin{pmatrix} \boldsymbol{B} \\ \boldsymbol{A}_2 \end{pmatrix}$$

is a nonsingular matrix such that (73) holds.

LEMMA 1. *If $\boldsymbol{E}$ is $p \times p$, symmetric, and nonsingular, there is a nonsingular matrix $\boldsymbol{F}$ such that*

$$(78)\qquad \boldsymbol{FEF}' = \begin{pmatrix} \boldsymbol{I} & \boldsymbol{0} \\ \boldsymbol{0} & -\boldsymbol{I} \end{pmatrix},$$

where the order of $\boldsymbol{I}$ is the number of positive characteristic roots of $\boldsymbol{E}$ and the order of $-\boldsymbol{I}$ is the number of negative characteristic roots of $\boldsymbol{E}$.

Proof. From Theorem 2 we know there is an orthogonal matrix $\boldsymbol{G}$ such that

$$(79)\qquad \boldsymbol{GEG}' = \begin{pmatrix} h_1 & 0 & \cdots & 0 \\ 0 & h_2 & \cdots & 0 \\ \vdots & \vdots & & \vdots \\ 0 & 0 & \cdots & h_p \end{pmatrix},$$

where $h_1 \geq \cdots \geq h_q > 0 > h_{q+1} \geq \cdots \geq h_p$ are the characteristic roots of $\boldsymbol{E}$. Let

$$(80)\qquad \mathrm{K} = \begin{pmatrix} 1/\sqrt{h_1} & \cdots & 0 & 0 & \cdots & 0 \\ \vdots & & \vdots & \vdots & & \vdots \\ 0 & \cdots & 1/\sqrt{h_q} & 0 & \cdots & 0 \\ 0 & \cdots & 0 & 1/\sqrt{-h_{q+1}} & \cdots & 0 \\ \vdots & & \vdots & \vdots & & \vdots \\ 0 & \cdots & 0 & 0 & \cdots & 1/\sqrt{-h_p} \end{pmatrix}.$$

Then

$$(81)\qquad \boldsymbol{KGEG'K}' = (\boldsymbol{KG})\boldsymbol{E}(\boldsymbol{KG})' = \begin{pmatrix} \boldsymbol{I} & \boldsymbol{0} \\ \boldsymbol{0} & -\boldsymbol{I} \end{pmatrix}.$$

Corollary 7. *Let $\boldsymbol{C}$ be $p \times p$, symmetric, and of rank r ($\leq p$). Then there is a nonsingular matrix $\boldsymbol{A}$ such that*

$$ACA' = \begin{pmatrix} I & 0 & 0 \\ 0 & -I & 0 \\ 0 & 0 & 0 \end{pmatrix}, \tag{82}$$

where the order of $\boldsymbol{I}$ is the number of positive characteristic roots of $\boldsymbol{C}$ and the order of $-\boldsymbol{I}$ is the number of negative characteristic roots (the sum of the orders being r).

Proof. The proof is the same as that of Theorem 6 except that Lemma 1 is used instead of Corollary 4.

Lemma 2. *Let $\boldsymbol{A}$ be $n \times m$ ($n > m$) such that*

$$A'A = I. \tag{83}$$

There exists an $n \times (n-m)$ matrix $\boldsymbol{B}$ such that $(\boldsymbol{A}\ \boldsymbol{B})$ is orthogonal.

Proof. Since $\boldsymbol{A}$ is of rank m there exists an $n \times (n-m)$ matrix $\boldsymbol{C}$ such that $(\boldsymbol{A}\ \boldsymbol{C})$ is nonsingular. Take $\boldsymbol{D}$ as $\boldsymbol{C} - \boldsymbol{A}\boldsymbol{A}'\boldsymbol{C}$; then $\boldsymbol{D}'\boldsymbol{A} = \boldsymbol{0}$. Let $\boldsymbol{E}$ $(n-m) \times (n-m)$ be such that $\boldsymbol{E}'\boldsymbol{D}'\boldsymbol{D}\boldsymbol{E} = \boldsymbol{I}$. Then $\boldsymbol{B}$ can be taken as $\boldsymbol{D}\boldsymbol{E}$.

Lemma 3. *Let $\boldsymbol{x}$ be a vector of n components. Then there exists an orthogonal matrix $\boldsymbol{O}$ such that*

$$Ox = \begin{pmatrix} c \\ 0 \\ \cdot \\ \cdot \\ \cdot \\ 0 \end{pmatrix}, \tag{84}$$

where $c = \sqrt{\boldsymbol{x}'\boldsymbol{x}}$.

Proof. Let the first row of $\boldsymbol{O}$ be $(1/c)\boldsymbol{x}'$. The other rows may be chosen in any way to make the matrix orthogonal.

Lemma 4. *Let $\boldsymbol{B} = (b_{ij})$ be a $p \times p$ matrix. Then*

$$\frac{\partial |B|}{\partial b_{ij}} = B_{ij}. \tag{85}$$

Proof. The expansion of $|\boldsymbol{B}|$ by elements of the ith row is

$$|B| = \sum_{h=1}^{p} b_{ih} B_{ih}. \tag{86}$$

Since B_{ih} does not contain b_{ij}, the lemma follows.

LEMMA 5. *Let* $b_{ij} = \beta_{ij}(c_1, \cdots, c_n)$ *be the* i, jth *element of a* $p \times p$ *matrix* B. *Then*

$$\frac{\partial |B|}{\partial c_g} = \sum_{i,h=1}^{p} \frac{\partial |B|}{\partial b_{ih}} \cdot \frac{\partial \beta_{ih}(c_1, \cdots, c_n)}{\partial c_g} = \sum_{i,h=1}^{p} B_{ih} \frac{\partial \beta_{ih}(c_1, \cdots, c_n)}{\partial c_g}. \tag{87}$$

THEOREM 7. *If* $A = A'$,

$$\frac{\partial |A|}{\partial a_{ii}} = A_{ii}; \tag{88}$$

$$\frac{\partial |A|}{\partial a_{ij}} = 2A_{ij}, \qquad i \neq j. \tag{89}$$

Proof. (88) follows from the expansion of $|A|$ according to elements of the ith row. To prove (89) let $b_{ij} = b_{ji} = a_{ij}$, $i, j = 1, \cdots, p$; $i \leq j$. Then by Lemma 5,

$$\frac{\partial |B|}{\partial a_{ij}} = B_{ij} + B_{ji}. \tag{90}$$

Since $|A| = |B|$ and $B_{ij} = B_{ji} = A_{ij} = A_{ji}$, (89) follows.

THEOREM 8

$$\frac{\partial}{\partial \boldsymbol{x}} (\boldsymbol{x}' A \boldsymbol{x}) = 2A\boldsymbol{x}, \tag{91}$$

where $\partial/\partial \boldsymbol{x}$ *denotes taking partial derivatives with respect to each component of* $\boldsymbol{x}$ *and arranging the partial derivatives in a column.*

Proof. Let $\boldsymbol{h}$ be a column vector of as many components as $\boldsymbol{x}$. Then

$$\begin{aligned}(\boldsymbol{x} + \boldsymbol{h})' A (\boldsymbol{x} + \boldsymbol{h}) &= \boldsymbol{x}' A \boldsymbol{x} + \boldsymbol{h}' A \boldsymbol{x} + \boldsymbol{x}' A \boldsymbol{h} + \boldsymbol{h}' A \boldsymbol{h} \\ &= \boldsymbol{x}' A \boldsymbol{x} + 2\boldsymbol{h}' A \boldsymbol{x} + \boldsymbol{h}' A \boldsymbol{h}.\end{aligned} \tag{92}$$

The partial derivative vector is clearly the vector multiplying $\boldsymbol{h}'$ in the second term on the right.

DEFINITION 1. *Let* $A = (a_{ij})$ *be a* $p \times p$ *matrix and* $B = (b_{\alpha\beta})$ *be a* $q \times q$ *matrix. The* $pq \times pq$ *matrix with* $a_{ij} b_{\alpha\beta}$ *as the element in the* i, αth *row and the* j, βth *column is called the Kronecker or direct product of* A *and* B *and is denoted by* $A \otimes B$; that is,

$$A \otimes B = \begin{pmatrix} a_{11}B & a_{12}B \cdots a_{1p}B \\ a_{21}B & a_{22}B \cdots a_{2p}B \\ \cdot & \cdot \qquad\quad \cdot \\ \cdot & \cdot \qquad\quad \cdot \\ \cdot & \cdot \qquad\quad \cdot \\ a_{p1}B & a_{p2}B \cdots a_{pp}B \end{pmatrix}. \tag{93}$$

THEOREM 9. *Let the* ith *characteristic root of* $\boldsymbol{A}$ *be* λ_i *and the vector be* $\begin{pmatrix} x_{1i} \\ \cdot \\ \cdot \\ \cdot \\ x_{pi} \end{pmatrix}$ *and let the* αth *root of* $\boldsymbol{B}$ *be* ν_α *and the vector be* $\boldsymbol{y}_\alpha$. *Then the* i, αth *root of* $\boldsymbol{A} \otimes \boldsymbol{B}$ *is* $\lambda_i \nu_\alpha$ *and the vector is* $\begin{pmatrix} x_{1i}\boldsymbol{y}_\alpha \\ \cdot \\ \cdot \\ \cdot \\ x_{pi}\boldsymbol{y}_\alpha \end{pmatrix}$.

Proof.

$$(94) \qquad (\boldsymbol{A} \otimes \boldsymbol{B}) \begin{pmatrix} x_{1i}\boldsymbol{y}_\alpha \\ \cdot \\ \cdot \\ \cdot \\ x_{pi}\boldsymbol{y}_\alpha \end{pmatrix} = \begin{pmatrix} a_{11}\boldsymbol{B} \cdots a_{1p}\boldsymbol{B} \\ \cdot \qquad \cdot \\ \cdot \qquad \cdot \\ \cdot \qquad \cdot \\ a_{p1}\boldsymbol{B} \cdots a_{pp}\boldsymbol{B} \end{pmatrix} \begin{pmatrix} x_{1i}\boldsymbol{y}_\alpha \\ \cdot \\ \cdot \\ \cdot \\ x_{pi}\boldsymbol{y}_\alpha \end{pmatrix}$$

$$= \begin{pmatrix} \sum_j a_{1j}x_{ji}\boldsymbol{B}\boldsymbol{y}_\alpha \\ \cdot \\ \cdot \\ \cdot \\ \sum_j a_{pj}x_{ji}\boldsymbol{B}\boldsymbol{y}_\alpha \end{pmatrix}$$

$$= \begin{pmatrix} \lambda_i x_{1i}\boldsymbol{B}\boldsymbol{y}_\alpha \\ \cdot \\ \cdot \\ \cdot \\ \lambda_i x_{pi}\boldsymbol{B}\boldsymbol{y}_\alpha \end{pmatrix} = \lambda_i \nu_\alpha \begin{pmatrix} x_{1i}\boldsymbol{y}_\alpha \\ \cdot \\ \cdot \\ \cdot \\ x_{pi}\boldsymbol{y}_\alpha \end{pmatrix}.$$

THEOREM 10.

$$(95) \qquad |\boldsymbol{A} \otimes \boldsymbol{B}| = |\boldsymbol{A}|^q |\boldsymbol{B}|^p$$

Proof. The determinant of any matrix is the product of its roots; therefore

$$|\boldsymbol{A} \otimes \boldsymbol{B}| = \prod_{i=1}^{p} \prod_{\alpha=1}^{q} \lambda_i \nu_\alpha = \left(\prod_{i=1}^{p} \lambda_i\right)^q \left(\prod_{\alpha=1}^{q} \nu_\alpha\right)^p.$$

THEOREM 11. *The Jacobian of the transformation* $\boldsymbol{E} = \boldsymbol{Y}^{-1}$ *(from* $\boldsymbol{E}$ *to* $\boldsymbol{Y}$*) is* $|\boldsymbol{Y}|^{-2p}$, *where* p *is the order of* $\boldsymbol{E}$ *and* $\boldsymbol{Y}$.

Proof. From $\boldsymbol{E}\boldsymbol{Y} = \boldsymbol{I}$, we have

$$(96) \qquad \left(\frac{\partial}{\partial \theta}\boldsymbol{E}\right)\boldsymbol{Y} + \boldsymbol{E}\left(\frac{\partial}{\partial \theta}\boldsymbol{Y}\right) = \boldsymbol{0},$$

where

$$(97)\qquad \left(\frac{\partial}{\partial\theta}E\right) = \begin{pmatrix} \frac{\partial e_{11}}{\partial\theta} & \cdots & \frac{\partial e_{1p}}{\partial\theta} \\ \cdot & & \cdot \\ \cdot & & \cdot \\ \cdot & & \cdot \\ \frac{\partial e_{p1}}{\partial\theta} & \cdots & \frac{\partial e_{pp}}{\partial\theta} \end{pmatrix}.$$

Then

$$(98)\qquad \left(\frac{\partial}{\partial\theta}E\right) = -E\left(\frac{\partial}{\partial\theta}Y\right)E = -Y^{-1}\left(\frac{\partial}{\partial\theta}Y\right)Y^{-1}.$$

If $\theta = y_{\alpha\beta}$, then

$$(99)\qquad \left(\frac{\partial}{\partial y_{\alpha\beta}}E\right) = -E\varepsilon_{\alpha\beta}E = -\,e_{\cdot\alpha}e_{\beta\cdot},$$

where $\varepsilon_{\alpha\beta}$ is a $p \times p$ matrix with all elements 0 except the element in the αth row and βth column which is 1 and $e_{\cdot\alpha}$ is the αth column of E and $e_{\beta\cdot}$ is the βth row. Thus $\dfrac{\partial e_{ij}}{\partial y_{\alpha\beta}} = -e_{i\alpha}e_{\beta j}$. Then the Jacobian is the determinant of a $p^2 \times p^2$ matrix

$$(100)\qquad \text{mod}\left|\frac{\partial e_{ij}}{\partial y_{\alpha\beta}}\right| = |e_{i\alpha}e_{\beta j}| = |E \otimes E'| = |E|^p|E'|^p = |E|^{2p} = |Y|^{-2p}.$$

5. ON THE ABBREVIATED DOOLITTLE METHOD AND THE METHOD OF PIVOTAL CONDENSATION FOR SOLVING LINEAR EQUATIONS

In this section we want to prove some of the results stated in Section 8.2.2.

First we would like to show that the method of pivotal condensation and the abbreviated Doolittle method are the same in the sense that the operations are the same but in different orders. We want to verify (35) of Section 8.2, which can be written

$$(101)\qquad \tilde{a}_{gh}^{(j-1)} = a_{gh} - \sum_{i=1}^{j-1} \tilde{a}_{ig}^{(i-1)}\tilde{a}_{ih}^{(i)}, \qquad g, h = j, \cdots, q.$$

To prove (101) by induction we note that this is true for $j = 1$ and 2. Let us assume (101) for $j = k - 1$ and show this implies it for $j = k$. For $j = k$ the right-hand side of (101) is

$$(102)\qquad a_{gh} - \sum_{i=1}^{k-1} \tilde{a}_{ig}^{(i-1)}\tilde{a}_{ih}^{(i)} = a_{gh} - \sum_{i=1}^{(k-1)-1} \tilde{a}_{ig}^{(i-1)}\tilde{a}_{ih}^{(i)} - \tilde{a}_{k-1,g}^{(k-2)}\,\tilde{a}_{k-1,h}^{(k-1)}.$$

By the hypothesis of the induction, this is

$$\tilde{a}_{gh}^{(k-2)} - \tilde{a}_{k-1,g}^{(k-2)}\,\tilde{a}_{k-1,h}^{(k-1)}.$$

Comparison with (31) of Section 8.2 shows that this is $\tilde{a}_{gh}^{(k-1)}$ inasmuch as $\tilde{a}_{gh}^{(j-1)} = \tilde{a}_{hg}^{(j-1)}$, $g, h = j, \cdots, g$. This last symmetry property follows because $a_{gh}^{(0)}$ is symmetric and the operations applied to get $\tilde{a}_{gh}^{(j-1)}$ are symmetric.

The computations of the two procedures can be indicated in matrix form. The operations (31) of Section 8.2, namely

$$\tilde{a}_{gh}^{(j)} = \tilde{a}_{gh}^{(j-1)} - \tilde{a}_{gj}^{(j-1)}\tilde{a}_{jh}^{(j)} = \tilde{a}_{gh}^{(j-1)} - \frac{\tilde{a}_{gj}^{(j-1)}\tilde{a}_{jh}^{(j-1)}}{\tilde{a}_{jj}^{(j-1)}} \tag{103}$$

can be considered to hold for $h = 1, \cdots, j-1$ $(g = j+1, \cdots, q)$ as well as for $h = j, \cdots, q$. Let us see that these elements are identically zero. We observed earlier that $\tilde{a}_{gj}^{(j)} = 0$ for $g = j+1, \cdots, q$; thus in particular $\tilde{a}_{g1}^{(1)} = 0\,(g > 1)$. To prove our desired statement by induction we assume $\tilde{a}_{gh}^{(j)} = 0$ for $h \le j$, $h \le g$, and $j = k-1$. Then

$$\tilde{a}_{gh}^{(k)} = \tilde{a}_{gh}^{(k-1)} - \frac{\tilde{a}_{gk}^{(k-1)}\tilde{a}_{kh}^{(k-1)}}{\tilde{a}_{kk}^{(k-1)}} = 0.$$

Now let us define the method of pivotal condensation in matrix terms. Let $A^{(j)}$ be the matrix with elements of the ith row $(i = 1, \cdots, j)$ being $\tilde{a}_{ih}^{(i-1)}$ and elements of the other rows being $\tilde{a}_{gh}^{(j)}$ $(g > j)$. Then $A^{(j)}$ is the result of the first j steps of the calculation except that at no stage have we performed (and recorded) the operation $\tilde{a}_{ih}^{(i)} = \tilde{a}_{ih}^{(i-1)}/\tilde{a}_{ii}^{(i-1)}$. A recursive definition of $A^{(j)}$ is

$$\begin{aligned} a_{gh}^{(j)} &= a_{gh}^{(j-1)}, & g \le j \\ a_{gh}^{(j)} &= a_{gh}^{(j-1)} - \frac{a_{jg}^{(j-1)}a_{jh}^{(j-1)}}{a_{jj}^{(j-1)}}. \end{aligned} \tag{104}$$

Let $E(i,j)$ be a $q \times q$ matrix with 1 as the i,jth element and 0 elsewhere. Let

$$G(i,j) = I - \frac{a_{ji}^{(j-1)}}{a_{jj}^{(j-1)}}\,E(i,j), \qquad i > j. \tag{105}$$

Each row of $G(i,j)$ is a row of the identity except the ith. Hence, multiplication of any matrix on the left by $G(i,j)$ affects only the ith row of the matrix; in the ith row the resulting element is the original element less the product of $a_{ji}^{(j-1)}/a_{jj}^{(j-1)}$ and the element in the jth row and same column. Thus the resulting element in $G(i,j)\,A^{(j-1)}$ is (104) for $g = i$.

Let

$$G(q,j)G(q-1,j)\cdots G(j+1,j) = F_j. \tag{106}$$

Then

$$A^{(j)} = F_j A^{(j-1)} \tag{107}$$

and

$$\begin{aligned} A^{(q-1)} &= F_{q+1} F_{q-2} \cdots F_2 F_1 A \\ &= FA, \end{aligned} \tag{108}$$

where $F = F_{q-1} \cdots F_1$. Since each F_i is triangular, so is F. We see that $A^{(q-1)} = A^*$ as defined in Section 8.2.2.

The abbreviated Doolittle procedure also amounts to multiplying A by the matrices $G(i, j)$, but in a different order. Let

$$\tilde{A}^{(i)} = H_i \tilde{A}^{(i-1)}, \tag{109}$$

where

$$H_i = G(i + 1, i) G(i + 1, i - 1) \cdots G(i + 1, 1) \tag{110}$$

and $\tilde{A}^{(0)} = A$.

The $(i + 1)$st row of (109) is equivalent to (101) for $g = j = i + 1$. Only the $(i + 1)$st row of $\tilde{A}^{(i-1)}$ is affected by H_i. Then

$$\begin{aligned} \tilde{A}^{(q-1)} &= H_{q-1} \cdots H_1 A \\ &= [G(q, q - 1) \cdots G(q, 1)] \cdots [G(3, 3)G(3, 1)]G(2, 1)A \\ &= G(q, q - 1)[G(q, q - 2)G(q - 1, q - 2)] \cdots \\ &\qquad [G(q, 1) \cdots G(2, 1)]A \\ &= F_{q-1} \cdots F_1 A \\ &= A^{(q-1)}. \end{aligned} \tag{111}$$

This is another demonstration that the abbreviated Doolittle method involves the same operations as in pivotal condensation, but in a different order.

The advantage of the abbreviated Doolittle method is that the products $\sum_{i=1}^{j-1} \tilde{a}_{ig}^{(i-1)} \tilde{a}_{ih}^{(i)}$ can be cumulated on a desk calculator without recording the separate products. All that need be recorded are a_{gh}^*, $g \leq h$, and a_{gh}^{**}, $g < h$.

It was shown in Section 8.2.3 that FAF' was diagonal. This method of reducing A to a diagonal matrix is equivalent to the method given in Section 4 of this appendix.

Since F is triangular with the diagonal elements unity, we have from $FA = A^*$ that $|A| = |A^*| = \prod a_{ii}^*$. Either computational procedure, then, furnishes a way of computing the value of a determinant.

APPENDIX 2

Bibliography

Abruzzi, Adams (1950), *Experimental Procedures and Criteria for Estimating and Evaluating Industrial Productivity*, doctoral dissertation, Columbia University Library.

Adrian, Robert (1808), "Research concerning the probabilities of the errors which happen in making observations, etc.," *The Analyst or Mathematical Museum*, 1.

Aitken, A. C. (1937), "Studies in practical mathematics. II. The evaluation of the latent roots and latent vectors of a matrix," *Proc. Roy. Soc. Edinb.*, 57, 269–304.

——— (1948a), "On a problem in correlated errors," *Proc. Roy. Soc. Edinb., A*, 62, 273–277.

——— (1948b), "On the estimation of many statistical parameters," *Proc. Roy. Soc. Edinb., A*, 62, 369–377.

——— (1949), "On the Wishart distribution in statistics," *Biometrika*, 36, 59–62.

——— (1950), "On the statistical independence of quadratic forms in normal variates," *Biometrika*, 37, 93–96.

Anderson, R. L., and T. A. Bancroft (1952), *Statistical Theory in Research*, New York, McGraw-Hill Book Co.

Anderson, T. W. (1945), *The Non-central Wishart Distribution and Its Application to Problems in Multivariate Statistics*, doctoral dissertation, Princeton University Library.

——— (1946), "The non-central Wishart distribution and certain problems of multivariate statistics," *Ann. Math. Stat.*, 17, 409–431.

——— (1948), "The asymptotic distributions of the roots of certain determinantal equations," *J. Roy. Stat. Soc., B*, 10, 132–139.

——— (1951a), "Classification by multivariate analysis," *Psychometrika*, 16, 31–50.

——— (1951b), "Estimating linear restrictions on regression coefficients for multivariate normal distributions," *Ann. Math. Stat.*, 22, 327–351.

——— (1951c), "The asymptotic distribution of certain characteristic roots and vectors," *Proceedings of the Second Berkeley Symposium on Mathematical Statistics and Probability*, University of California Press, Berkeley and Los Angeles, 103–130.

——— (1955), "Some statistical problems in relating experimental data to predicting performance of a production process," *J. Amer. Stat. Assoc.*, 50, 163–177.

——— (1956), "Asymptotic theory for principal component analysis," unpublished.

——— and M. A. Girshick (1944), "Some extensions of the Wishart distribution," *Ann. Math. Stat.*, 15, 345–357.

Anderson, T. W., and Herman Rubin (1949), "Estimation of the parameters of a single equation in a complete system of stochastic equations," *Ann. Math. Stat.*, 20, 46–63.

——— (1950), "The asymptotic properties of estimates of the parameters of a single equation in a complete system of stochastic equations," *Ann. Math. Stat.*, 21, 570–582.

Anderson, T. W., and Herman Rubin (1956), "Statistical inference in factor analysis," *Proceedings of the Third Berkeley Symposium on Mathematical Statistics and Probability*, Vol. V, University of California Press, Berkeley and Los Angeles, 111–150.

Armitage, P. (1950), "Sequential analysis with more than two alternative hypotheses and its relation to discriminant function analysis," *J. Roy. Stat. Soc., B*, 12, 137–144.

Bacon, H. M. (1948), "A matrix arising in correlation theory," *Ann. Math. Stat.*, 19, 422–424.

Banerjee, D. P. (1952), "On the moments of the multiple correlation coefficient in samples from normal population," *J. Indian Soc. Agric. Stat.*, 4, 88–90.

Barankin, Edward W. (1949), "Extension of the Romanovsky-Bartlett-Scheffé test," *Proceedings of the Berkeley Symposium on Mathematical Statistics and Probability*, University of California Press, Berkeley and Los Angeles, 433–450.

Barnard, M. M. (1935), "The secular variations of skull characters in four series of Egyptian skulls," *Ann. Eugen.*, 6, 352–371.

Barnes, E. W. (1899), "The theory of the Gamma function," *Messeng. Math.*, 29, 64–128.

Bartlett, M. S. (1933), "On the theory of statistical regression," *Proc. Roy. Soc. Edinb.*, 53, 260–283.

——— (1934), "The vector representation of a sample," *Proc. Camb. Phil. Soc.*, 30, 327–340.

——— (1937a), "Properties of sufficiency and statistical tests," *Proc. Roy. Soc., A*, 160, 268–282.

——— (1937b), "The statistical conception of mental factors," *Brit. J. Psych.*, 28, 97–104.

——— (1938), "Further aspects of the theory of multiple regression," *Proc. Camb. Phil. Soc.*, 34, 33–40.

——— (1939a), "A note on tests of significance in multivariate analysis," *Proc. Camb. Phil. Soc.*, 35, 180–185.

——— (1939b), "The standard errors of discriminant function coefficients," *J. Roy. Stat. Soc., Supple.*, 6, 169–173.

——— (1947a), "Multivariate analysis," *J. Roy. Stat. Soc., Supple.*, 9, 176–197.

——— (1947b), "The general canonical correlation distribution," *Ann. Math. Stat.*, 18, 1–17.

——— (1948), "A note on the statistical estimation of demand and supply relations from time series," *Econometrica*, 16, 323–329.

——— (1950), "Tests of significance in factor analysis," *Brit. J. Psych.* (*Stat. Sec.*), 3, 77–85.

——— (1951a), "The goodness of fit of a single hypothetical discriminant function in the case of several groups," *Ann. Eugen.*, 16, 199–214.

——— (1951b), "The effect of standardization on a χ^2 approximation in factor analysis." (With an appendix by W. Ledermann), *Biometrika*, 38, 337–344.

——— (1951c), "A further note on tests of significance in factor analysis," *Brit. J. Psych.* (*Stat. Sec.*), 4, 1–2.

——— (1953), "Factor analysis in psychology as a statistician sees it," *Uppsala Symposium on Psychological Factor Analysis, 17–19 March 1953*, Uppsala, Almqvist and Wiksell, 23–34.

——— (1954), "A note on the multiplying factors for various χ^2-approximations," *J. Roy. Stat. Soc., B*, 16, 296–298.

Beall, Geoffrey (1945), "Approximate methods in calculating discriminant functions," *Psychometrika*, 10, 205–218.

Bennett, B. M. (1951), "Note on a solution of the generalized Behrens-Fisher problem," *Ann. Inst. Stat. Math.*, 2, 87–90.

——— (1955), "On the cumulants of the logarithmic generalized variance and variance ratio," *Skand. Aktuarietidskr.*, 38, 17–21.

Berkson, Joseph (1947), "Cost-utility as a measure of the efficiency of a test," *J. Amer. Stat. Assoc.*, 42, 246–255.

Bhaskara Varma, K. (1951), "On the exact distribution of Wilks' L_{mvc} and L_{vc} criteria," *Bull. Inst. Internat. Stat.*, 33, Part II, 181–214.

Bhattacharyya, A. (1952), "On the uses of the t-distribution in multivariate analysis," *Sankhyā*, 12, 89–104.

Bhattacharyya, D. P., and R. D. Narayan (1939), "Moments of the D^2-statistic for populations with unequal dispersions," *Sankhyā*, 5, 401–412.

Birnbaum, Allan (1955), "Characterizations of complete classes of tests of some multi-parametric hypotheses, with applications to likelihood ratio tests," *Ann. Math. Stat.*, 26, 21–36.

Birnbaum, Z. W. (1950), "Effect of linear truncation on a multinormal population," *Ann. Math. Stat.*, 21, 272–279.

——— and Paul L. Meyer (1953), "On the effect of truncation in some or all co-ordinates of a multinormal population," *J. Indian Soc. Agric. Stat.*, 5, 17–28.

Bishop, D. J. (1939), "On a comprehensive test of the homogeneity of variances and covariances in multivariate problems," *Biometrika*, 31, 31–55.

Blackwell, David, and M. A. Girshick (1954), *Theory of Games and Statistical Decisions*, New York, John Wiley and Sons.

Bose, P. K. (1947a), "Parametric relations in multivariate distributions," *Sankhyā*, 8, 167–171.

——— (1947b), "On recursion formulae, tables and Bessel function populations associated with the distribution of classical D^2-statistic," *Sankhyā*, 8, 235–248.

——— (1951a), "Remarks on computing the incomplete probability integral in multivariate distribution functions," *Bull. Inst. Internat. Stat.*, 33, Part II, 55–64.

——— (1951b), "Corrigenda: On the construction of incomplete probability integral tables of the classical D^2-statistic," *Sankhyā*, 11, 96.

Bose, R. C. (1936a), "On the exact distribution and moment-coefficient of the D^2-statistic," *Sankhyā*, 2, 143–154.

——— (1936b), "A note on the distribution of differences in mean values of two samples drawn from two multivariate normally distributed populations, and the definition of the D^2-statistic," *Sankhyā*, 2, 379–384.

——— and S. N. Roy (1938a), "The distribution of the studentised D^2-statistic," *Sankhyā*, 4, 19–38.

——— (1938b), "The use and distribution of the studentized D^2-statistic when the variances and covariances are based on K samples," *Sankhyā*, 4, 535–542.

Bose, S. N. (1936), "On the complete moment-coefficients of the D^2-statistic," *Sankhyā*, 2, 385–396.

——— (1937), "On the moment-coefficients of the D^2-statistic and certain integral and differential equations connected with the multivariate normal population," *Sankhyā*, 3, 105–124.

Bose, Subhendusekhar (1935), "On the distribution of the ratio of variances of two samples drawn from a given normal bivariate correlated population," *Sankhyā*, 2, 65–72.

Box, G. E. P. (1949), "A general distribution theory for a class of likelihood criteria," *Biometrika*, 36, 317–346.

Bravais, Auguste (1846), "Analyse mathematique sur les probabilités des erreurs de situation d'un point," *Mémoires présentés par divers savants, l'Académie Royale des Sciences de l'Institut de France*, 9, 255–332.

Brown, G. W. (1939), "On the power of the L_1-test for equality of several variances," *Ann. Math. Stat.*, 10, 119–128.

——— (1947), "Discriminant functions," *Ann. Math. Stat.*, 18, 514–528.

——— (1950), "Basic principles for construction and application of discriminators," *J. Clin. Psych.*, 6, 58–61.

Burt, Cyril (1944), "Statistical problems in the evaluation of army tests," *Psychometrika*, 9, 219–236.

Cadwell, J. H. (1951), "The bivariate normal integral," *Biometrika*, 38, 475–479.

Cansado, Enrique (1951), "A study of bivariate distributions," *Trabajos de Estadistica*, 2, 149–178.

Carter, A. H. (1949), "The estimation and comparison of residual regressions where there are two or more related sets of observations," *Biometrika*, 36, 26–46.

Chernoff, Herman (1956), "Large sample theory: parametric case," *Ann. Math. Stat.*, 27, 1–22.

Chown, L. N., and P. A. P. Moran (1951), "Rapid methods for estimating correlation coefficients," *Biometrika*, 38, 464–467.

Cochran, W. G. (1934), "The distribution of quadratic forms in a normal system, with applications to the analysis of variance," *Proc. Camb. Phil. Soc.*, 30, 178–191.

——— (1943), "The comparison of different scales of measurement for experimental results," *Ann. Math. Stat.*, 14, 205–216.

——— and C. I. Bliss (1948), "Discriminant functions with covariance," *Ann. Math. Stat.*, 19, 151–176.

Cook, M. B. (1951a), "Bi-variate *k*-statistics and cumulants of their joint sampling distribution," *Biometrika*, 38, 179–195.

——— (1951b), "Two applications of bivariate *k*-statistics," *Biometrika*, 38, 368–376.

Craig, A. T. (1947), "Bilinear forms in normally correlated variables," *Ann. Math. Stat.*, 18, 565–573.

Cramér, H. (1940), "On the theory of stationary random processes," *Ann. Math., Princeton*, 41, 215–230.

——— (1946), *Mathematical Methods of Statistics*, Princeton, Princeton University Press.

Das, A. C. (1948), "A note on the D^2-statistic when the variances and co-variances are known," *Sankhyā*, 8, 372–374.

Daly, J. F. (1940), "On the unbiased character of likelihood-ratio tests for independence in normal systems," *Ann. Math. Stat.*, 11, 1–32.

David, F. N. (1937), "A note on unbiased limits for the correlation coefficient," *Biometrika*, 29, 157–160.

——— (1938), *Tables of the Ordinates and Probability Integral of the Distribution of the Correlation Coefficient in Small Samples*, London, *Biometrika*.

——— (1953), "A note on the evaluation of the multivariate normal integral," *Biometrika*, 40, 458–459.

Deemer, Walter L., and Ingram Olkin (1951), "The Jacobians of certain matrix transformations useful in multivariate analysis. Based on lectures of P. L. Hsu at the University of North Carolina, 1947," *Biometrika*, 38, 345–367.

DeLury, D. B. (1938), "Note on correlations," *Ann. Math. Stat.*, 9, 149–151.

Des Raj (1953a), "On estimating the parameters of bivariate normal populations from doubly and singly linearly truncated samples," *Sankhyā*, 12, 277–290.

——— (1953b), "On estimating the parameters of binormal populations from linearly truncated samples," *Ganita*, 4, 147–154.

——— (1955), "On optimum selections from multivariate populations," *Sankhyā*, 14, 363–366.

Dunnett, C. W., and M. Sobel (1954), "A bivariate generalization of Student's t-distribution, with tables for certain special cases," *Biometrika*, 41, 153–169.

——— (1955), "Approximations to the probability integral and certain percentage points of a multivariate analogue of Student's t-distribution," *Biometrika*, 42, 258–260.

Durand, David (1941), *Risk Elements in Consumer Installment Financing* (technical ed.), New York, National Bureau of Economic Research, Appendices A, B, C.

Dutton, A. M. (1954), "Application of some multivariate analysis techniques to some data from radiation experiments," *Statistics and Mathematics in Biology*, O. Kempthorne, T. A. Bancroft, J. W. Gowen, and J. L. Lusk (eds.), Ames, Iowa State College Press, 81–92.

Dwyer, P. S. (1949), "Pearsonian correlation coefficients associated with least squares theory," *Ann. Math. Stat.*, 20, 404–416.

——— and M. S. MacPhail (1948), "Symbolic matrix derivatives," *Ann. Math. Stat.*, 19, 517–534.

Elfving, G. (1947), "A simple method of deducing certain distributions connected with multivariate sampling," *Skand. Aktuarietidskr.*, 30, 56–74.

Ezekiel, M. (1941), *Methods of Correlation Analysis*, New York, John Wiley and Sons.

Ferris, C. D., F. E. Grubbs, and C. L. Weaver (1946), "Operating characteristics for the common statistical tests of significance," *Ann. Math. Stat.*, 17, 178–197.

Fieller, E. C., T. Lewis, and E. S. Pearson (1955), *Correlated Random Normal Deviates*, Cambridge, Cambridge University Press.

Finney, D. J. (1938), "The distribution of the ratio of estimates of the two variances in a sample from a normal bi-variate population," *Biometrika*, 30, 190–192.

——— (1946), "The frequency distribution of deviates from means and regression lines in samples from a multivariate normal population," *Ann. Math. Stat.*, 17, 344–349.

——— (1956), "Multivariate analysis and agricultural experiments," *Biometrics*, 12, 67–71.

Fisher, R. A. (1915), "Frequency distribution of the values of the correlation coefficient in samples from an indefinitely large population," *Biometrika*, 10, 507–521.

——— (1921), "On the probable error of a coefficient of correlation deduced from a small sample," *Metron*, 1, Part 4, 3–32.

——— (1924), "The distribution of the partial correlation coefficient," *Metron*, 3, 329–332.

——— (1928), "The general sampling distribution of the multiple correlation coefficient," *Proc. Roy. Soc.*, *A*, 121, 654–673.

——— (1936), "The use of multiple measurements in taxonomic problems," *Ann. Eugen.*, 7, 179–188.

——— (1938), "The statistical utilization of multiple measurements," *Ann. Eugen.*, 8, 376–386.

——— (1939), "The sampling distribution of some statistics obtained from non-linear equations," *Ann. Eugen.*, 9, 238–249.

——— (1940), "The precision of discriminant functions," *Ann. Eugen.*, 10, 422–429.

——— (1947a), *The Design of Experiments* (4th ed.), Edinburgh, Oliver and Boyd.

——— (1947b), "The analysis of covariance method for the relation between a part and the whole," *Biometrics*, 3, 65–68.

——— and F. Yates (1942), *Statistical Tables for Biological, Agricultural and Medical Research* (2nd ed.), Edinburgh, Oliver and Boyd.

Fog, David (1948), "The geometric method in the theory of sampling," *Biometrika*, 39, 46–54.

Fraser, D. A. S. (1951), "Generalized hit probabilities with a Gaussian target," *Ann. Math. Stat.*, 22, 248–255.

Fraser, D. A. S. (1953), "Generalized hit probabilities with a Gaussian target II,' *Ann. Math. Stat.*, 24, 288–294.

Frets, G. P. (1921), "Heredity of head form in man," *Genetica*, 3, 193–384.

Frisch, R. (1929), "Correlation and scatter in statistical variables," *Nordic Stat. J.*, 8, 36–102.

Galton, F. (1889), *Natural Inheritance*, London, Macmillan and Co.

Garrett, Henry E. (1943), "The discriminant function and its use in psychology," *Psychometrika*, 8, 65–79.

Garwood, F. (1933), "The probability integral of the correlation coefficient in samples from a normal bivariate population," *Biometrika*, 25, 71–78.

Gauss, C. F. (1823). *Theory of the Combination of Observations*, Göttingen.

Gayen, A. K. (1951), "The frequency distribution of the product-moment correlation coefficient in random samples of any size drawn from non-normal universes," *Biometrika*, 38, 219–247.

Geary, R. C. (1948), "Studies in relations between economic time series," *J. Roy. Stat. Soc., B*, 10, 140–158.

——— (1949), "Determination of linear relations between economic time series," *Econometrica*, 17, 30–58.

Girshick, M. A. (1936), "Principal components," *J. Amer. Stat. Assoc.*, 31, 519–528.

——— (1939a), "On the sampling theory of roots of determinantal equations," *Ann. Math. Stat.*, 10, 203–224.

——— (1939b), "A scientific preview for standard sizes for children's garments and patterns," *J. Amer. Stat. Assoc.*, 34, 362–364.

——— (1941), "The distribution of the ellipticity statistic L_e when the hypothesis is false," *Terr. Magn. Atmos. Elect.*, 46, 455–457.

——— and T. Haavelmo (1947), "Statistical analysis of the demand for food: examples of simultaneous estimation of structural equations," *Econometrica*, 15, 79–110.

Gnedenko, B. V. (1948), "On a theorem of S. N. Bernstein," *Izvest. Akad. Nauk SSSR Ser. Mat.*, 12, 97–100.

Goldstine, H. H., and J. von Neumann (1951), "Numerical inverting of matrices of high order," *Proc. Amer. Math. Soc.*, 2, 188–202.

Grubbs, Frank E. (1944), "On the distribution of the radial standard deviation," *Ann. Math. Stat.*, 15, 75–81.

Gulliksen, Harold, and S. S. Wilks (1950), "Regression tests for several samples," *Psychometrika*, 15, 91–114.

Gumbel, E. J., and S. B. Littauer (1952), "On the independence of elements of a manufacturing operation," *Columbia Engng. Quart.*, 5, 10–13.

Haavelmo, T. (1943), "The statistical implications of a system of simultaneous equations," *Econometrica*, 11, 1–12.

——— (1944), "The probability approach in econometrics," *Econometrica*, 12, Supple.

Halmos, P. R. (1950), *Measure Theory*, New York, D. van Nostrand.

Hall, P. (1927), "Multiple and partial correlation coefficients in the case of an *m*-fold variate system," *Biometrika*, 19, 100–109.

Hamilton, Max (1950), "The personality of dyspeptics with special reference to gastric and duodenal ulcers," *Brit. J. Med. Psych.*, 23, 182–198.

Harley, B. I. (1954), "A note on the probability integral of the correlation coefficient," *Biometrika*, 41, 278–280.

Hartley, H. O., and E. R. Fitch (1951), "A chart for the incomplete beta-function and the cumulative binomial distribution," *Biometrika*, 38, 423–426.

Herz, C. S. (1955), "Bessel functions of matrix argument," *Ann. Math., Princeton*, 61, 474–523.

Hickman, W. Braddock (1953), *The Volume of Corporate Bond Financing Since* 1900, Princeton, Princeton University Press, 82–90.

Hoel, Paul, G. (1937), "A significance test for component analysis," *Ann. Math. Stat.*, 8, 149–158.

——— and R. P. Peterson (1949), "A solution to the problem of optimum classification," *Ann. Math. Stat.*, 20, 433–438.

Holzinger, K. J., and H. H. Harman (1941), *Factor Analysis. A synthesis of factorial methods.* Chicago, University of Chicago Press.

Hood, W. C., and T. C. Koopmans (eds.) (1953), *Studies in Econometric Method—Cowles Commission Monograph 14*, New York, John Wiley and Sons.

Hooker, R. H. (1907), "The correlation of the weather and crops," *J. Roy. Stat. Soc.*, 70, 1–42.

Horst, Paul, and Stevenson Smith (1950), "The discrimination of two racial samples," *Psychometrika*, 15, 271–290.

Hotelling, Harold (1931), "The generalization of Student's ratio," *Ann. Math. Stat.*, 2, 360–378.

——— (1933), "Analysis of a complex of statistical variables into principal components," *J. Educ. Psych.*, 24, 417–441, 498–520.

——— (1935), "The most predictable criterion," *J. Educ. Psych.*, 26, 139–142.

——— (1936a), "Simplified calculation of principal components," *Psychometrika*, 1, 27–35.

——— (1936b), "Relations between two sets of variates," *Biometrika*, 28, 321–377.

——— (1947), "Multivariate quality control, illustrated by the air testing of sample bombsights," *Techniques of Statistical Analysis*, New York, McGraw-Hill Book Co., 111–184.

——— (1948), "Fitting generalized truncated normal distributions," abstract, *Ann. Math. Stat.*, 19, 596.

——— (1951), "A generalized T test and measure of multivariate dispersion," *Proceedings of the Second Berkeley Symposium on Mathematical Statistics and Probability*, University of California Press, Berkeley and Los Angeles, 23–41.

——— (1953), "New light on the correlation coefficient and its transforms," *J. Roy. Stat. Soc., B*, 15, 193–225.

Hsu, Chung-tsi (1941), "Samples from two bivariate normal populations," *Ann. Math. Stat.*, 12, 279–292.

Hsu, P. L. (1938), "Notes on Hotelling's generalized T," *Ann. Math. Stat.*, 9, 231–243.

——— (1939a), "On the distribution of the roots of certain determinantal equations," *Ann. Eugen.*, 9, 250–258.

——— (1939b), "A new proof of the joint product moment distribution," *Proc. Camb. Phil. Soc.*, 35, 336–338.

——— (1940a), "On generalized analysis of variance (I)," *Biometrika*, 31, 221–237.

——— (1940b), "An algebraic derivation of the distribution of rectangular coordinates," *Proc. Edinb. Math. Soc.*, 6, 185–189.

——— (1941a), "On the limiting distribution of roots of a determinantal equation," *J. London Math. Soc.*, 16, 183–194.

——— (1941b), "On the problem of rank and the limiting distribution of Fisher's test function," *Ann. Eugen.*, 11, 39–41.

——— (1941c), "On the limiting distribution of the canonical correlations," *Biometrika*, 32, 38–45.

——— (1941d), "Canonical reduction of the general regression problem," *Ann. Eugen.*, 11, 42–46.

Hsu, P. L. (1945), "On the power functions for the E^2-test and the T^2-test," *Ann. Math. Stat.*, 16, 278–286.

——— (1949), "The limiting distribution of functions of sample means and applications to testing hypotheses," *Proceedings of the Berkeley Symposium on Mathematical Statistics and Probability*, University of California Press, Berkeley and Los Angeles, 359–402.

Hughes, H. M. (1949), "Estimation of the variance of the bivariate normal distribution," *Univ. Calif. Publ. Stat.*, 1, 37–51.

Ihm, Peter (1955), "Ein Kriterium für zwei Typen zweidimensionaler Normalverteilungen," *Mitt. Math. Stat.*, 7, 46–52.

Immer, F. R., H. D. Hayes, and LeRoy Powers (1934), "Statistical determination of barley varietal adaptation," *J. Amer. Soc. Agron.*, 26, 403–407.

Ingham, A. E. (1933), "An integral that occurs in statistics," *Proc. Camb. Phil. Soc.*, 29, 271–276.

Isserlis, L. (1914), "On the partial correlation ratio. Part I. Theoretical," *Biometrika*, 10, 391–420.

——— (1916), "On the partial correlation ratio. Part II. Numerical," *Biometrika*, 11, 50–66.

——— (1917), "The variation of the multiple correlation coefficient in samples drawn from an infinite population with normal distribution," *Phil. Mag.*, 34 (sixth series), 205–220.

Jambunathan, M. V. (1954), "Some properties of Beta and Gamma distributions," *Ann. Math. Stat.*, 25, 401–405.

James, A. T. (1954), "Normal multivariate analysis and the orthogonal group," *Ann. Math. Stat.*, 25, 40–75.

——— (1955a), "The non-central Wishart distribution," *Proc. Roy. Soc.* (*London*), *A*, 229, 364–366.

——— (1955b), "A generating function for averages over the orthogonal group," *Proc. Roy. Soc.* (*London*), *A*, 229, 367–375.

James, G. S. (1952), "Note on a theorem of Cochran," *Proc. Camb. Phil. Soc.*, 48, 443–446.

——— (1954), "Tests of linear hypotheses in univariate and multivariate analysis when the ratios of the population variances are unknown," *Biometrika*, 41, 19–43.

Johnson, N. L. (1949), "Bivariate distributions based on simple translation systems," *Biometrika*, 36, 297–304.

Kamat, A. R. (1953), "Incomplete and absolute moments of the multivariate normal distribution with some applications," *Biometrika*, 40, 20–34.

Kelley, T. L. (1916), "Tables to facilitate the calculation of partial coefficients of correlation and regression equations," *Bull. Univ. Tex.*, 127.

——— (1928), *Crossroads in the Mind of Man*, Stanford, Stanford University Press.

Kendall, M. G. (1940), "The derivation of multivariate sampling formulae from univariate formulae by symbolic operation," *Ann. Eugen.*, 10, 392–402.

——— (1941), "Proof of relations connected with the tetrachoric series and its generalization," *Biometrika*, 32, 196–198.

——— (1943), *The Advanced Theory of Statistics*, *I*, London, Charles Griffin and Co.

——— (1946), *The Advanced Theory of Statistics*, *II*, London, Charles Griffin and Co.

——— and B. Babington Smith (1950), "Factor analysis," *J. Roy. Stat. Soc.*, *B*, 12, 60–94.

Kolmogorov, A. (1950), *Foundations of the Theory of Probability*, New York, Chelsea Publ. Co.

Kôno, Kasumasa (1952), "On inefficient statistics for measurement of dependency of normal bivariates," *Mem. Fac. Sci. Kyūsyū Univ., A.*, 7, 1–12.

Koopmans, T. C., ed. (1950), *Statistical Inference in Dynamic Economic Models*—Cowles Commission Monograph 10, New York, John Wiley and Sons.

Kossack, Carl F. (1945), "On the mechanics of classification," *Ann. Math. Stat.*, 16, 95–97.

Krull, Wolfgang (1951a), "Zur Korrelationstheorie zweidimensionaler Merkmale," *Mitt. Math. Stat.*, 3, 15–29.

———(1951b), "Korrelationstheorie mehrdimensionaler Merkmale," *Mitt. Math. Stat.*, 3, 185–200.

Kullback, Solomon (1934), "An application of characteristic functions to the distribution problem of statistics," *Ann. Math. Stat.*, 5, 263–307.

——— (1935), "On samples from a multivariate normal population," *Ann. Math. Stat.*, 6, 202–213.

——— (1952), "An application of information theory to multivariate analysis," *Ann. Math. Stat.*, 23, 88–102.

——— (1956), "An application of information theory to multivariate analysis II," *Ann. Math. Stat.*, 27, 122–146.

Laha, R. G. (1954), "On some problems in canonical correlations," *Sankhyā*, 14, 61–66.

——— (1955), "On a characterization of the multivariate normal distribution," *Sankhyā*, 14, 367–368.

Laplace, P. S. (1811), "Mémoire sur les intégrales définies et leur application aux probabilités," *Mémoires de l'Institut Impérial de France, Année 1810*, 279–347.

Lawley, D. N. (1938), "A generalization of Fisher's *z* test," *Biometrika*, 30, 180–187.

—— (1940), "The estimation of factor loadings by the method of maximum likelihood," *Proc. Roy. Soc. Edinb.*, 60, 64–82.

——— (1942), "Further investigations in factor estimation," *Proc. Roy. Soc. Edinb.*, 61, 176–185.

——— (1953), "A modified method of estimation in factor analysis and some large sample results," *Uppsala Symposium on Psychological Factor Analysis, 17–19 March 1953*, Uppsala, Almqvist and Wiksell, 35–42.

——— (1956), "Tests of significance for the latent roots of covariance and correlation matrices," *Biometrika*, 43, 128–136.

Lehmann, E. L. (1959), *Theory of Testing Hypotheses*, New York, John Wiley and Sons.

Lehmer, Emma (1944), "Inverse tables of probabilties of errors of the second kind," *Ann. Math. Stat.*, 15, 388–398.

Lévy, Paul (1948), "The arithmetical character of the Wishart distribution," *Proc. Camb. Phil. Soc.*, 44, 295–297.

Lord, R. D. (1954), "The use of the Hankel transform in statistics," *Biometrika*, 41, 44–55.

Lorge, Irving (1940), "Two group comparisons by multivariate analysis," *Ann. Educ. Research Assoc. Official Report.*

——— and N. Morrison (1938), "The reliability of principal components," *Science*, 87, 491–492.

Lubin, A. (1950), "Linear and non-linear discriminating functions," *Brit. J. Psych.* (*Stat. Sec.*), 3, 90–103.

McFadden, J. A. (1955), "Urn models of correlation and a comparison with the multivariate normal integral," *Ann. Math. Stat.*, 26, 478–489.

Madow, W. G. (1938), "Contributions to the theory of multivariate statistical analysis," *Trans. Amer. Math. Soc.*, 44, 454–495.

Mahalanobis, P. C. (1930), "On tests and measures of group divergence," *J. and Proc. Asiat. Soc. Beng.*, 26, 541–588.

——— (1936), "On the generalized distance in statistics," *Proc. Nat. Inst. Sci. India*, 2, 49–55.

———, R. C. Bose, and S. N. Roy (1937), "Normalisation of statistical variates and the use of rectangular co-ordinates in the theory of sampling distributions," *Sankhyā*, 3, 1–40.

Maritz, J. S. (1953), "Estimation of the correlation coefficient in the case of a bivariate normal population when one of the variables is dichotomized," *Psychometrika*, 18, 97–110.

Marriott, F. H. C. (1952), "Tests of significance in canonical analysis," *Biometrika*, 39, 58–64.

Martin, E. A. (1936), "A study of the Egyptian series of mandibles with special reference to mathematical methods of sexing," *Biometrika*, 28, 149–178.

Masuyama, M. (1939a), "Correlation between tensor quantities," *Proc. Phys.-Math. Soc. Japan*, (3), 21, 638–647.

——— (1939b), "Tensor characteristic of vector set," *Proc. Phys.-Math. Soc. Japan* (3), 21, 648.

——— (1940a), "On the meaning of the symmetric correlation coefficient," *Proc. Phys.-Math. Soc. Japan* (3), 22, 579–585.

——— (1940b), "On the subdependency," *Proc. Phys.-Math. Soc. Japan* (3), 22, 855–858.

——— (1940c), "The variance tensor of a vector set and a nature of the symmetric correlation coefficient," *Proc. Phys.-Math. Soc. Japan* (3), 22, 858–861.

Matusita, Kameo (1949), "Note on the independence of certain statistics," *Ann. Inst. Stat. Math. Tokyo*, 1, 79–82.

Mauchly, J. W. (1940a), "A significance test for ellipticity in the harmonic dial," *Terr. Magn. Atmos. Elect.*, 45, 145–148.

——— (1940b), "Significance test for sphericity of a normal n-variate distribution," *Ann. Math. Stat.*, 11, 204–209.

Mauldon, J. G. (1955), "Pivotal quantities for Wishart's and related distributions and a paradox in fiducial theory," *J. Roy. Stat. Soc., B*, 17, 79–85.

Miner, J. R. (1922), *Tables of* $\sqrt{1 - r^2}$ *and* $1 - r^2$ *for use in Partial Correlations, etc.*, Baltimore, Johns Hopkins Press.

Mises, R. von (1945), "On the classification of observation data into distinct groups," *Ann. Math. Stat.*, 16, 68–73.

Mood, A. M. (1941), "On the joint distribution of the medians in samples from a multivariate population," *Ann. Math. Stat.*, 12, 268–278.

——— (1950), *Introduction to the Theory of Statistics*, New York, McGraw-Hill Book Co.

——— (1951), "On the distribution of the characteristic roots of normal second-moment matrices," *Ann. Math. Stat.*, 22, 266–273.

Moran, P. A. P. (1950), "The distribution of the multiple correlation coefficient," *Proc. Camb. Phil. Soc.*, 46, 521–522.

——— (1956), "The numerical evaluation of a class of integrals," *Proc. Camb. Phil. Soc.*, 52, 230–233.

Morgan, W. A. (1939), "A test for the significance of the difference between the variances in a sample from a normal bivariate population," *Biometrika*, 31, 13–19.

Morrow, Dorothy J. (1948), "On the distribution of the sums of the characteristic roots of a determinantal equation," abstract, *Bull. Amer. Math. Soc.*, 54, 75.

Nabeya, S. (1951), "Absolute moments in 2-dimensional normal distribution," *Ann. Inst. Stat. Math.*, 3, 2–6.

Nair, U. S. (1939), "The application of the moment function in the study of distribution laws in statistics," *Biometrika*, 30, 274–294.

Nanda, D. N. (1948a), "Distribution of a root of a determinantal equation," *Ann. Math. Stat.*, 19, 47–57.

——— (1948b), "Limiting distribution of a root of a determinantal equation," *Ann. Math. Stat.*, 19, 340–350.

——— (1949), "The standard errors of discriminant function coefficients in plant-breeding experiments," *J. Roy. Stat. Soc.*, *B*, 11, 283–290.

——— (1950), "Distribution of the sum of roots of a determinantal equation under a certain condition," *Ann. Math. Stat.*, 21, 432–439.

——— (1951), "Probability distribution tables of the larger root of a determinantal equation with two roots," *J. Indian Soc. Agric. Stat.*, 3, 175–177.

Nandi, H. K. (1946), "On the power function of studentized D^2-statistic," *Bull. Calcutta Math. Soc.*, 38, 79–84.

Narain, R. D. (1948a), "A new approach to sampling distributions of the multivariate normal theory I," *J. Indian Soc. Agric. Stat.*, 1, 59–69.

——— (1948b), "A new approach to sampling distributions of the multivariate normal theory II," *J. Indian Soc. Agric. Stat.*, 1, 137–146.

——— (1949), "Some results on discriminant functions," *J. Indian Soc. Agric. Stat.*, 2, 49–59.

——— (1950), "On the completely unbiased character of tests of independence in multivariate normal systems," *Ann. Math. Stat.*, 21, 293–298.

Neyman, J., and E. S. Pearson (1936), "Contributions to the theory of testing statistical hypotheses," *Stat. Res. Mem.*, 1, 1–37.

Nyquist, H., S. O. Rice, and J. Riordan (1954), "The distribution of random determinants," *Quart. Appl. Math.*, 12, 97–104.

Oberg, E. N. (1947), "Approximate formulas for the radii of circles which include a specified fraction of a normal bivariate distribution," *Ann. Math. Stat.*, 18, 442–447.

Ogasawara, Tōzirō, and Masayuki Takahashi (1951), "Independence of quadratic quantities in a normal system," *J. Sci. Hiroshima Univ.*, *A*, 15, 1–9.

Ogawa, J. (1949), "On the independence of bilinear and quadratic forms of a random sample from a normal population," *Ann. Inst. Stat. Math.*, 1, 83–108.

——— (1950), "On the independence of quadratic forms in a non-central normal system," *Osaka Math. J.*, 2, 151–159.

——— (1953), "On the sampling distributions of classical statistics in multivariate analysis," *Osaka Math. J.*, 5, 13–52.

Olkin, Ingram (1952), "Note on the Jacobians of certain matrix transformations useful in multivariate analysis," *Biometrika*, 40, 43–46.

——— and S. N. Roy (1954), "On multivariate distribution theory," *Ann. Math. Stat.*, 25, 329–339.

Pearson, E. S., and S. S. Wilks (1933), "Methods of statistical analysis appropriate for k samples of two variables," *Biometrika*, 25, 353–378.

Pearson, K. (1896), "Mathematical contributions to the theory of evolution—III. Regression, heredity and panmixia," *Phil. Trans.*, *A*, 187, 253–318.

——— (1900), "On the criterion that a given system of deviations from the probable in the case of a correlated system of variables is such that it can reasonably be supposed to have arisen from random sampling," *Phil. Mag.*, 50, 157–175.

——— (1901), "On lines and planes of closest fit to systems of points in space," *Phil. Mag.*, 2 (sixth series), 559–572.

——— (1926), "On the coefficients of racial likeness," *Biometrika*, 18, 105–117.

Pearson, K. (1928), "Note on standardization of method using the coefficient of racial likeness," *Biometrika*, 20 B, 376–378.

——— (1930), *Tables for Statisticians and Biometricians*, Part I (3rd ed.), London, *Biometrika*.

——— (1931), *Tables for Statisticians and Biometricians*, Part II, London, *Biometrika*.

——— (1934), *Tables of the Incomplete Beta-Function*, London, *Biometrika*.

———, G. B. Jeffry, and E. M. Elderton (1929), "On the distribution of the first product moment coefficient in samples drawn from an indefinitely large normal population," *Biometrika*, 21, 164–201.

Penrose, L. S. (1947), "Some notes on discrimination," *Ann. Eugen.*, 13, 228–237.

Pillai, K. C. S. (1955), "Some new test criteria in multivariate analysis," *Ann. Math. Stat.*, 26, 117–121.

Pitman, E. J. G. (1939), "A note on normal correlation," *Biometrika*, 31, 9–12.

Plackett, R. L. (1947), "An exact test for the equality of variances," *Biometrika*, 34, 311–319.

——— (1954), "A reduction formula for normal multivariate integrals," *Biometrika*, 41, 351–360.

Plana, G. A. A. (1813), "Mémoire sur divers problèmes de probabilité," *Mém. Acad. Impériale de Turin, pour les Années 1811–1812*, 20, 355–498.

Pólya, G. (1949), "Remarks on computing the probability integral in one and two dimensions," *Proceedings of the Berkeley Symposium on Mathematical Statistics and Probability*, University of California Press, Berkeley and Los Angeles, 67–78.

Quenouille, M. H. (1947), "Note on the elimination of insignificant variates in discriminatory analysis," *Ann. Eugen.*, 14, 305–308.

——— (1950), "Multivariate experimentation," *Biometrics*, 6, 303–316.

Rao, C. R. (1944), "Generalized variance of populations," *Proc. Indian Sci. Cong.*

——— (1945), "Studentised tests of linear hypotheses," *Science and Culture*, 11, 202–203.

——— (1946a), "Tests with discriminant functions in multivariate analysis," *Sankhyā*, 7, 407–414.

——— (1946b), "On the problem of *K* samples and *K* multivariate populations with unequal variances and covariances," *Proc. Indian Sci. Cong.*

——— (1947), "A statistical criterion to determine the group to which an individual belongs," *Nature*, 160, 835–836.

——— (1948a), "The utilization of multiple measurements in problems of biological classification," *J. Roy. Stat. Soc.*, *B*, 10, 159–193.

——— (1948b), "Tests of significance in multivariate analysis," *Biometrika*, 35, 58–79.

——— (1949a), "On the distance between two populations," *Sankhyā*, 9, 246–248.

——— (1949b), "On some problems arising out of discrimination with multiple characters," *Sankhyā*, 9, 343–366.

——— (1949c), "Representation of '*p*' dimensional data in lower dimensions," *Sankhyā*, 9, 248–251.

——— (1949d), "On a transformation useful in multivariate computations," *Sankhyā*, 9, 251–253.

——— (1951a), "Statistical inference applied to classificatory problems. II. The problem of selecting individuals for various duties in a specified ratio," *Sankhyā*, 11, 107–116.

Rao, C. R. (1951b), "An asymptotic expansion of the distribution of Wilks' Λ-criterion," *Bull. Inst. Internat. Stat.*, 33, Part II, 177–180.

——— (1952), *Advanced Statistical Methods in Biometric Research*, New York, John Wiley and Sons.

——— (1953), "Discriminant functions for genetic differentiation and selection," (Part IV of "Statistical inference applied to classificatory problems"), *Sankhyā*, 12, 229–246.

——— (1954a), "A general theory of discrimination when the information about alternative population distributions is based on samples," *Ann. Math. Stat.*, 25, 651–670.

——— (1954b), "On the use and interpretation of distance functions in statistics," *Bull. Inst. Internat. Stat.*, 34, Part II, 90–97.

——— (1955a), "Analysis of dispersion for multiply classified data with unequal numbers in cells," *Sankhyā*, 15, 253–280.

——— (1955b), "Estimation and tests of significance in factor analysis," *Psychometrika*, 20, 93–111.

——— and P. Slater (1949), "Multivariate analysis applied to differences between neurotic groups," *Brit. J. Psych. (Stat. Sec.)*, 2, 17–29.

Rao, K. S. (1951), "On the mutual independence of a set of Hotelling's T^2 derivable from a sample of size n from a k-variate normal population," *Bull. Inst. Internat. Stat.*, 33, Part II, 171–176.

——— (1954), "Testing for serial correlation in a stationary multidimensional discrete stochastic process," *Bull. Inst. Internat. Stat.*, 34, Part II, 185–194.

Rasch, G. (1948), "A functional equation for Wishart's distribution," *Ann. Math. Stat.*, 19, 262–266.

——— (1950), "A vectorial t-test in the theory of normal multivariate distributions," *Mat. Tidsskr., B*, 1950, 76–81.

——— (1953), "On simultaneous factor analysis in several populations," *Uppsala Symposium on Psychological Factor Analysis, 17–19 March 1953*, Uppsala, Almqvist and Wicksell, 65–71.

Roy, J. (1951), "The distribution of certain likelihood criteria useful in multivariate analysis," *Bull. Inst. Internat. Stat.*, 33, Part II, 219–230.

Roy, S. N. (1938), "Geometrical note on the use of rectangular coordinates in the theory of sampling distributions connected with a multivariate normal population," *Sankhyā*, 3, 273–384.

——— (1939a), "A note on the distribution of the studentized D^2-statistic," *Sankhyā*, 4, 373–380.

——— (1939b), "p-statistics or some generalisations in analysis of variance appropriate to multivariate problems," *Sankhyā*, 4, 381–396.

——— (1942a), "The sampling distribution of p-statistics and certain allied statistics on the non-null hypothesis," *Sankhyā*, 6, 15–34.

——— (1942b), "Analysis of variance for multivariate normal populations: the sampling distribution of the requisite p-statistics on the null and non-null hypotheses," *Sankhyā*, 6, 35–50.

——— (1945a), "The individual sampling distribution of the maximum, the minimum and any intermediate of the p-statistics on the null-hypothesis," *Sankhyā*, 7, 133–158.

——— (1945*b*), "On a certain class of multiple integrals," *Bull. Calcutta Math. Soc.*, 37, 69–77.

——— (1946a), "Multivariate analysis of variance: the sampling distribution of the numerically largest of the p-statistics on the non-null hypothesis," *Sankhyā*, 8, 15–52.

Roy, S. N. (1946b), "On the individual sampling distribution of p-statistics for testing equality of the dispersion matrices for two multivariate normal populations," *Proc. Indian Sci. Cong.*

——— (1946c), "On the power function of the different p-statistics for multivariate analysis of variance," *Proc. Indian Sci. Cong.*

——— (1946d), "Further studies in multivariate analysis of variance," *Proc. Indian Sci. Cong.*

——— (1946e), "A note on multivariate analysis of variance when the number of variates is greater than the number of linear hypotheses per character," *Sankhyā*, 8, 53–66.

——— (1947), "A note on critical angles between two flats in hyperspace with certain statistical applications," *Sankhyā*, 8, 177–194.

——— (1950a), "Univariate and multivariate analysis as problems in testing of composite hypotheses—I," *Sankhyā*, 10, 29–80.

——— (1952a), "On some aspects of statistical inference," *Proceedings International Congress of Mathematicians, I*, American Mathematical Society, 555–564.

——— (1952b), "Some useful results in Jacobians," *Calcutta Stat. Assoc. Bull.*, 4, 117–122.

——— (1953), "On a heuristic method of test construction and its use in multivariate analysis," *Ann. Math. Stat.*, 24, 220–238.

——— (1954), "Some further results in simultaneous confidence interval estimation," *Ann. Math. Stat.*, 25, 752–761.

——— and P. Bose (1940), "The distribution of the root-mean-square of the second type of the multiple correlation coefficient," *Science and Culture*, 6, 59.

——— and R. C. Bose (1953), "Simultaneous confidence interval estimation," *Ann. Math. Stat.*, 24, 513–536.

Sakamoto, Heihachi (1949), "On the criteria of the independence and degrees of freedom of statistics and their applications to the analysis of variance," *Ann. Inst. Stat. Math.*, 1, 109–122.

Sastry, K. V. Krishna (1948), "On a Bessel function of the second kind and Wilks Z-distribution," *Proc. Indian Acad. Sci., Sect. A*, 28, 532–536.

Sasuly, M. (1930), "Generalized multiple correlation analysis of economic statistical series," *J. Amer. Stat. Assoc.*, 25, 146–152.

Sato, R. (1951), " 'r' distributions and 'r' tests," *Ann. Inst. Stat. Math.*, 2, 91–124.

Scheffé, Henry (1942), "On the ratio of the variances of two normal populations," *Ann. Math. Stat.*, 13, 371–388.

——— (1943), "On solutions of the Behrens-Fisher problem based on the t-distribution," *Ann. Math. Stat.*, 14, 35–44.

Shohat, J. A., and J. D. Tamarkin (1943), *The Problem of Moments*, New York, American Mathematical Society.

Shrivastava, M. P. (1941), "Bivariate correlation surfaces," *Science and Culture*, 6, 615–616.

Simaika, J. B. (1941), "On an optimum property of two important statistical tests," *Biometrika*, 32, 70–80.

Simonsen, W. (1944–45), "On distributions of functions of samples from a normally distributed infinite population," *Skand. Aktuarietidskr.*, 27, 235–261; 28, 20–43.

Sitgreaves, Rosedith (1952), "On the distribution of two random matrices used in classification procedures," *Ann. Math. Stat.*, 23, 263–270.

Skitovič, V. P. (1953), "On a property of a normal distribution," *Doklady Akad. Nauk SSSR (N.S.)*, 89, 217–219.

Smith, C. A. B. (1947), "Some examples of discrimination," *Ann. Eugen.* 13, 272–282.

Smith, H. Fairfield (1936), "A discriminant function for plant selection," *Ann. Eugen.*, 7, 240–250.

Solomon, Herbert (1953), "Distribution of the measure of a random two-dimensional set," *Ann. Math. Stat.*, 24, 650–656.

Soper, H. E., A. W. Young, B. M. Cave, A. Lee, and K. Pearson (1917), "On the distribution of the correlation coefficient in small samples. Appendix II to the papers of 'Student' and R. A. Fisher. A cooperative study," *Biometrika*, 11, 328–413.

Steel, R. G. D. (1951), "Minimum generalized variance for a set of linear functions," *Ann. Math. Stat.*, 22, 456–460.

——— (1955), "Analysis of perennial crop data," *Biometrics*, 11, 201–212.

Steyn, H. S. (1951), "The Wishart distribution derived by solving simultaneous linear differential equations," *Biometrika*, 38, 470–472.

Stone, Richard (1947), "On interdependence of blocks of transactions," *J. Roy. Stat. Soc., Supple.*, 9, 1–32.

Student (W. S. Gosset) (1908), "The probable error of a mean," *Biometrika*, 6, 1–25.

Sverdrup, Erling (1947), "Derivation of the Wishart distribution of the second order sample moments by straightforward integration of a multiple integral," *Skand. Aktuarietidskr.*, 30, 151–166.

Tang, P. C. (1938), "The power function of the analysis of variance tests with tables and illustrations of their use," *Stat. Res. Mem.*, 2, 126–157.

Theil, H. (1954), "Estimation of parameters of econometric models," *Bull. Inst. Internat. Stat.*, 34, Part II, 122–129.

Thomson, G. H. (1951), *The Factorial Analysis of Human Ability* (5th ed.), London, University of London Press.

Thurstone, L. L. (1947), *Multiple-Factor Analysis*, Chicago, University of Chicago Press.

Tintner, G. (1945), "A note on rank, multicollinearity and multiple regression," *Ann. Math. Stat.*, 16, 304–308.

——— (1946), "Some applications of multivariate analysis to economic data," *J. Amer. Stat. Assoc.*, 41, 472–500.

——— (1950a), "Some formal relations in multivariate analysis," *J. Roy. Stat. Soc., B*, 12, 95–101.

——— (1950b), "A test for linear relations between weighted regression coefficients," *J. Roy. Stat. Soc., B*, 12, 273–277.

——— (1952), "Die Anwendung der variate Differenz Methode auf die Probleme der gewogenen Regression und der Multikollinearität," *Mitt. Math. Stat.*, 4, 159–162.

Tukey, J. W. (1949), Dyadic anova, an analysis of variance for vectors," *Hum. Biol.*, 21, 65–110.

——— and S. S. Wilks (1946), "Approximation of the distribution of the product of beta variables by a single beta variable," *Ann. Math. Stat.*, 17, 318–324.

Varma, R. S. (1952), "On the probability function in a normal multivariate distribution," *Quart. J. Mech. Appl. Math.*, 5, 361–362.

Vaswani, Sundri (1950), "Assumptions underlying the use of the tetrachoric correlation coefficient," *Sankhyā*, 10, 269–276.

Vinograde, Bernard (1950), "Canonical positive definite matrices under internal linear transformations," *Proc. Amer. Math. Soc.*, 1, 159–161.

Votaw, D. F., Jr. (1948), "Testing compound symmetry in a normal multivariate distribution," *Ann. Math. Stat.*, 19, 447–473.

———, J. A. Rafferty, and W. L. Deemer (1950), "Estimation of parameters in a truncated trivariate normal distribution," *Psychometrika*, 15, 339–347.

Wald, A. (1943), "Tests of statistical hypotheses concerning several parameters when the number of observations is large," *Trans. Amer. Math. Soc.*, 54, 426–482.

——— (1944), "On a statistical problem arising in the classification of an individual into one of two groups," *Ann. Math. Stat.*, 15, 145–162.

——— (1950), *Statistical Decision Functions*, New York, John Wiley and Sons.

——— and R. J. Brookner (1941), "On the distribution of Wilks' statistic for testing the independence of several groups of variates," *Ann. Math. Stat.*, 12, 137–152.

Wald, A., and J. Wolfowitz (1944), "Statistical tests based on permutations of the observations," *Ann. Math. Stat.*, 13, 358–372.

Walker, Helen M. (1931), *Studies in the History of Statistical Method*, Baltimore, Williams and Wilkins Co.

Wallace, Noel, and R. M. W. Travers (1938), "A psychometric sociological study of a group of specialty salesmen," *Ann. Eugen.*, 8, 266–302.

Waugh, F. V. (1942), "Regressions between sets of variates," *Econometrica*, 10, 290–310.

Weibull, Martin (1953), "The distribution of t- and F-statistics and of correlation and regression coefficients in stratified samples from normal populations with different means," *Skand. Aktuarietidskr.*, 36, *Supple.*, 1–106.

Welch, B. L. (1939), "Note on discriminant functions," *Biometrika*, 31, 218–220.

Wherry, R. J. (1940), "An approximation method for obtaining a maximized multiple criterion," *Psychometrika*, 5, 109–116.

Whittle, P. (1952), "On principal components and least square methods of factor analysis," *Skand. Aktuarietidskr.*, 35, 223–239.

Whittaker, E. T., and G. N. Watson (1943), *A Course of Modern Analysis*, Cambridge University Press, Amer. Ed. (Macmillan Co.).

Wilks, S. S. (1932a), "Certain generalizations in the analysis of variance," *Biometrika*, 24, 471–494.

——— (1932b), "Moments and distributions of estimates of population parameters from fragmentary samples," *Ann. Math. Stat.*, 3, 163–195.

——— (1932c), "On the sampling distribution of the multiple correlation coefficient," *Ann. Math. Stat.*, 3, 196–203.

——— (1932d), "On the distributions of statistics in samples from a normal population of two variables with matched sampling of one variable," *Metron*, 9, Nos. 3–4, 87–126.

——— (1934), "Moment-generating operators for determinants of product moments in samples from a normal system," *Ann. of Math.*, 35, 312–340.

——— (1935a), "On the independence of k sets of normally distributed statistical variables," *Econometrica*, 3, 309–326.

——— (1935b), "Test criteria for statistical hypotheses involving several variables," *J. Amer. Stat. Assoc.*, 30, 549–560.

——— (1936), "The sampling theory of systems of variances, covariances and intraclass covariances," *Amer. J. Math.*, 58, 426–432.

——— (1938), "Weighting systems for linear functions of correlated variables when there is no dependent variable," *Psychometrika*, 3, 23–40.

——— (1943), *Mathematical Statistics*, Princeton, Princeton University Press.

——— (1946), "Sample criteria for testing equality of means, equality of variances, and equality of covariances in a normal multivariate distribution," *Ann. Math. Stat.*, 17, 257–281.

Williams, E. J. (1952), "Some exact tests in multivariate analysis," *Biometrika*, 39, 17–31.

——— (1955), "Significance tests for discriminant functions and linear functional relationships," *Biometrika*, 42, 360–381.

Wishart, John (1928), "The generalized product moment distribution in samples from a normal multivariate population," *Biometrika*, 20 A, 32–52.

——— (1931), "The mean and second moment coefficient of the multiple correlation coefficient in samples from a normal population," *Biometrika*, 22, 353–361.

——— (1948a), "Proofs of the distribution law of the second order moment statistics," *Biometrika*, 35, 55–57.

——— (1948b), "Test of homogeneity of regression coefficients, and its application in the analysis of covariance," presented to the Colloque International de Calcul des Probabilités et de Statistique Mathématique, Lyon.

——— (1955), "Multivariate analysis," *Appl. Stat.*, 4, 103–116.

——— and M. S. Bartlett (1932), "The distribution of second order moment statistics in a normal system," *Proc. Camb. Phil. Soc.*, 28, 455–459.

——— (1933), "The generalised product moment distribution in a normal system," *Proc. Camb. Phil. Soc.*, 29, 260–270.

Wold, H. O. A. (1949), "Statistical estimation of economic relationship," *Econometrica*, 17, Supple., 1–22.

——— (1953), *Demand Analysis, A Study in Econometrics*, (with L. Jureén), New York, John Wiley and Sons.

Woltz, W. G., W. A. Reid, and W. E. Colwell (1948), "Sugar and nicotine in cured bright tobacco as related to mineral element composition," *Proc. Soil Sci. Soc. Amer.*, 13, 385–387.

Yates, F., and W. G. Cochran (1938), "The analysis of groups of experiments," *J. Agric. Sci.*, 28, 556.

Yule, G. U. (1897a), "On the significance of Bravais' Formulae for regression, etc., in the case of skew correlation," *Proc. Roy. Soc.*, 60, 477–489.

——— (1897b), "On the theory of correlation," *J. Roy. Stat. Soc.*, 60, 812–854.

——— (1907), "On the theory of correlation for any number of variables treated by a new system of notation," *Proc. Roy. Soc.*, *A*, 79, 182–193.

Index